# Organic Structures from Spectra

## Fifth Edition

# Organic Structures from Spectra

## Fifth Edition

**L. D. Field**
*University of New South Wales, Australia*

**S. Sternhell**
*University of Sydney, Australia*

**J. R. Kalman**
*University of Technology Sydney, Australia*

A John Wiley & Sons, Ltd., Publication

This edition first published 2013
© 2013 John Wiley & Sons, Ltd.

*Registered Office*

John Wiley & Sons Ltd, The Atrium, Southern Gate, Chichester, West Sussex, PO19 8SQ, United Kingdom

For details of our global editorial offices, for customer services and for information about how to apply for permission to reuse the copyright material in this book please see our website at www.wiley.com.

*Library of Congress Cataloging-in-Publication Data applied for*

HB ISBN: 9781118325452
PB ISBN: 9781118325490

A catalogue record for this book is available from the British Library.

Typeset in 12/18pt Times New Roman by Aptara Inc., New Delhi, India

Printed in Singapore by Markono Print Media Pte Ltd

# CONTENTS

Contents

# PREFACE

The derivation of structural information from spectroscopic data is an integral part of Organic Chemistry courses at all Universities. At the undergraduate level, the principal aim of courses in organic spectroscopy is to teach students to solve simple structural problems efficiently by using combinations of the major techniques (UV, IR, NMR and MS). Over a period more than 30 years, we have evolved courses at the University of Sydney and at the University of New South Wales, which achieve this aim quickly and painlessly. The text is tailored specifically to the needs and philosophy of these courses. As we believe our approach to be successful, we hope that it may be of use in other institutions.

The courses has been taught at the beginning of the third year, at which stage students have completed an elementary course of Organic Chemistry in first year and a mechanistically-oriented intermediate course in second year. Students have also been exposed, in their Physical Chemistry courses, to elementary spectroscopic theory, but are, in general, unable to relate the theory to actually solving spectroscopic problems.

We have delivered courses of about 9 lectures outlining the basic theory, instrumentation and the structure-spectra correlations of the major spectroscopic techniques. The text of this book broadly corresponds to the material presented in the 9 lectures. The treatment is both elementary and condensed and, not surprisingly, the students have great difficulties in solving even the simplest problems at this stage. The lectures are followed by a series of 2-hour problem solving seminars with 5 to 6 problems being presented per seminar. At the conclusion of the course, the great majority of the class is quite proficient and has achieved a satisfactory level of understanding of all methods used. Clearly, the real teaching is done during the hands-on problem seminars, which are organised in a manner modelled on that which we first encountered at the E.T.H. Zurich.

The class (typically 60 - 100 students, attendance is compulsory) is seated in a large lecture theatre in alternate rows and the problems for the day are identified. The students are permitted to work either individually or in groups and may use any written or printed aids they desire. Students solve the problems on their individual copies of this book thereby transforming it into a set of worked examples and most students voluntarily complete many more problems than are set. Staff (generally 4 or 5) wander around giving help and tuition as needed - the empty alternate rows of seats make it possible to speak to each student individually. When an important general point needs to be made, the staff member in charge gives a very brief exposition at the board. There is a 1½ hour examination consisting essentially of 4 problems from the book and the results are in general very satisfactory. Moreover, the students themselves find this a rewarding course since the practical skills acquired are obvious to them. Solving these real puzzles is also addictive - there is a real sense of achievement, understanding and satisfaction, since the challenge in solving the graded problems builds confidence even though the more difficult examples are quite demanding.

Our philosophy can be summarised as follows:

(a)  Theoretical exposition must be kept to a minimum, consistent with gaining of an understanding of the parts of the technique actually used in solving the problems. Our experience indicates that both mathematical detail and description of advanced techniques merely confuse the average student.

(b)  The learning of data must be kept to a minimum. We believe that it is more important to learn to use a restricted range of data well rather than to achieve a nodding acquaintance with more extensive sets of data.

(c)  Emphasis is placed on the concept of identifying "structural elements" and the logic needed to produce a structure out of the structural elements.

We have concluded that the best way to learn how to obtain "structures from spectra" is to practise on simple problems. This book was produced principally to assemble a suitable collection of problems for that purpose.

Problems 1-282 are of the standard "structures from spectra" type and are arranged roughly in order of increasing difficulty. A number of problems deal with related compounds (sets of isomers) which differ mainly in symmetry or the connectivity of the structural elements and are ideally set together. The sets of related examples include: problems 3 and 4; 19 and 20; 31 and 32; 42 and 43; 44, 45 and 46; 47, 48 and 49; 50 and 51; 61, 62 and 63; 64, 65 and 66; 81 and 82; 84 and 85; 99, 100, 101 and 102; 107 and 108; 110, 111, 112 and 113; 114 and 115; 118, 119 and 120; 122 and 123; 127 and 128; 139, 140, 141, 142 and 143; 155, 156, 157, 158, 159 and 160; 179 and 180; 181 and 182; 185 and 186; 215 and 216; 226 and 227; 235, 236 and 237; 276 and 277.

A further group of problems offer practice in the analysis of proton NMR spectra: 19, 20, 29, 37, 58, 75, 79, 90, 92, 93, 94, 99, 101, 123, 137, 146, 159, 163, 164, 183, 187, 192, 195, 205, 208, 236, 237, 238, 239, 248, 250, 251, 252 and 260.

A number of problems (195, 196, 197, 198, 230, 231, 260, 264, 265, 268, 271, 274 and 275) exemplify complexities arising from the presence of chiral centres, or from restricted rotation about peptide bonds (128, 162 and 262), while some problems deal with structures of compounds of biological, environmental, or industrial significance (22, 23, 36, 86, 95, 127, 131, 132, 144, 153, 162, 164, 197, 204, 220, 259, 260, 261, 263, 264, 265, 267, 272, 273, 274 and 275).

Problems 283-288 are again structures from spectra, but with the data presented in a textual form such as might be encountered when reading the experimental section of a paper or report.

Problems 289-296 deal with the use of NMR spectroscopy for quantitative analysis and for the analysis of mixtures of compounds.

Problems 297-323 represent a considerably expanded set of problems dealing with the interpretation of two-dimensional NMR spectra and are a series of graded exercises utilising COSY, NOESY, C-H Correlation, HMBC and TOCSY spectroscopy as aids to spectral analysis and as tools for identifying organic structures from spectra.

Problems 324-346 deal specifically with more detailed analysis of NMR spectra, which tends to be a stumbling block for many students.

In Chapter 9, there are also <u>two worked solutions</u> (to problems 96 and 127) as an illustration of a logical approach to solving problems. However, with the exception that we insist that students perform all routine measurements first, we do not recommend a mechanical attitude to problem solving – intuition has an important place in solving structures from spectra as it has elsewhere in chemistry.

---

*Bona fide* instructors may obtain a list of solutions (at no charge) by writing to the authors or EMAIL: L.Field@unsw.edu.au  or  FAX: (61-2)-9385-8008

---

We wish to thank Dr Alison Magill, and Dr Hsiu Lin Li in the School of Chemistry at the University of New South Wales and Dr Ian Luck at the University of Sydney who helped to assemble the many additional samples and spectra in the 4th and 5th editions of this book. Thanks are also due to the many graduate students and research associates who, over the years, have supplied us with many of the compounds used in the problems.

**L. D. Field**

**S. Sternhell**

**J. R. Kalman** September 2012

# LIST OF TABLES

# LIST OF FIGURES

# 1

# INTRODUCTION

## 1.1    GENERAL PRINCIPLES OF ABSORPTION SPECTROSCOPY

The basic principles of absorption spectroscopy are summarised below.  These are most obviously applicable to UV and IR spectroscopy and are simply extended to cover NMR spectroscopy.  Mass Spectrometry is somewhat different and is not a type of absorption spectroscopy.

*Spectroscopy* is the study of the quantised interaction of energy (typically electromagnetic energy) with matter.  In Organic Chemistry, we typically deal with molecular spectroscopy *i.e.* the spectroscopy of atoms that are bound together in molecules.

A schematic absorption spectrum is given in Figure 1.1.  The absorption spectrum is a plot of absorption of energy (radiation) against its wavelength ($\lambda$) or frequency ($\nu$).

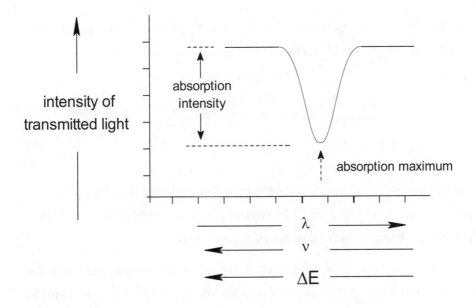

**Figure 1.1    Schematic Absorption Spectrum**

*Organic Structures from Spectra*, Fifth Edition. L. D. Field, S. Sternhell and J. R. Kalman.
© 2013 John Wiley & Sons, Ltd. Published 2013 by John Wiley & Sons, Ltd.

An absorption band can be characterised primarily by two parameters:

*(a)*   the wavelength at which maximum absorption occurs

*(b)*   the intensity of absorption at this wavelength compared to base-line (or background) absorption

A spectroscopic transition takes a molecule from one state to a state of a higher energy.  For any spectroscopic transition between energy states (*e.g.* $E_1$ and $E_2$ in Figure 1.2), the change in energy ($\Delta E$) is given by:

$$\Delta E = h\nu$$

where $h$ is the Planck's constant and $\nu$ is the frequency of the electromagnetic energy absorbed.  Therefore $\nu \propto \Delta E$.

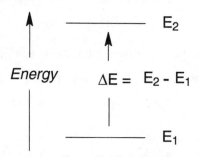

**Figure 1.2    Definition of a Spectroscopic Transition**

It follows that the x-axis in Figure 1.1 is an **energy** scale, since the frequency, wavelength and energy of electromagnetic radiation are interrelated:

$$\nu\lambda = c \text{ (speed of light)}$$

$$\lambda = \frac{c}{\nu}$$

$$\lambda \propto \frac{1}{\Delta E}$$

A spectrum consists of distinct bands or transitions because the absorption (or emission) of energy is quantised.  The energy gap of a transition is a ***molecular property*** and is ***characteristic of molecular structure***.

The y-axis in Figure 1.1 measures the intensity of the absorption band and this depends on the number of molecules observed (the Beer-Lambert Law) and the probability of the transition between the energy levels.  The absorption intensity is also a molecular property and both the frequency and the intensity of a transition can provide structural information.

## 1.2   CHROMOPHORES

In general, any spectral feature, *i.e.* a band or group of bands, is due not to the whole molecule, but to an identifiable part of the molecule, which we loosely call a *chromophore.*

A chromophore may correspond to a functional group (*e.g.* a hydroxyl group or the double bond in a carbonyl group).  However, it may equally well correspond to a single atom within a molecule or to a group of atoms (*e.g.* a methyl group) which is not normally associated with chemical functionality.

The detection of a chromophore permits us to deduce the presence of a *structural fragment* or a *structural element* in the molecule.  The fact that it is the chromophores and not the molecules as a whole that give rise to spectral features is fortunate, otherwise spectroscopy would only permit us to identify known compounds by direct comparison of their spectra with authentic samples.  This "fingerprint" technique is often useful for establishing the identity of known compounds, but the direct determination of molecular structure building up from the molecular fragments is far more powerful.

## 1.3   DEGREE OF UNSATURATION

Traditionally, the molecular formula of a compound was derived from elemental analysis and its molecular weight which was determined independently.  The concept of the **degree of unsaturation** of an organic compound derives simply from the tetravalency of carbon.  For a non-cyclic hydrocarbon (*i.e.* an alkane) the number of hydrogen atoms must be twice the number of carbon atoms plus two, any "deficiency" in the number of hydrogens must be due to the presence of unsaturation, *i.e.* double bonds, triple bonds or rings in the structure.

The degree of unsaturation can be calculated from the molecular formula for all compounds containing C, H, N, O, S or the halogens.  There are 3 basic steps in calculating the degree of unsaturation:

> **Step 1** – take the molecular formula and replace all halogens by hydrogens
>
> **Step 2** – omit all of the sulfur or oxygen atoms
>
> **Step 3** – for each nitrogen, omit the nitrogen and omit one hydrogen

After these 3 steps, the molecular formula is reduced to $C_nH_m$ and the degree of unsaturation is given by:

$$\text{Degree of Unsaturation} = n - \frac{m}{2} + 1$$

The degree of unsaturation indicates the number of $\pi$ bonds or rings that the compound contains. For example, a compound whose molecular formula is $C_4H_9NO_2$ is reduced to $C_4H_8$ which gives a degree of unsaturation of 1 and this indicates that the molecule must have one $\pi$ bond or one ring. Note that any compound that contains an aromatic ring always has a degree of unsaturation greater than or equal to 4, since the aromatic ring contains a ring plus three $\pi$ bonds. Conversely, if a compound has a degree of unsaturation greater than 4, one should suspect the possibility that the structure contains an aromatic ring.

## 1.4   CONNECTIVITY

Even if it were possible to identify sufficient structural elements in a molecule to account for the molecular formula, it may not be possible to deduce the structural formula from a knowledge of the structural elements alone. For example, it could be demonstrated that a substance of molecular formula $C_3H_5OCl$ contains the structural elements:

$$-CH_3$$

$$-Cl$$

$$\diagdown C = O \diagup$$

$$-CH_2-$$

and this leaves two possible structures:

$$\underset{\displaystyle 1}{CH_3-\underset{\displaystyle \underset{O}{\|}}{C}-CH_2-Cl} \qquad \text{and} \qquad \underset{\displaystyle 2}{CH_3-CH_2-\underset{\displaystyle \underset{O}{\|}}{C}-Cl}$$

Not only the presence of various structural elements, but also their juxtaposition, must be determined to establish the structure of a molecule. Fortunately, spectroscopy often gives valuable information concerning the *connectivity* of structural elements and in the above example it would be very easy to determine whether there is a ketonic carbonyl group (as in **1**) or an acid chloride (as in **2**).  In addition, it is possible to determine independently whether the methyl (-CH$_3$) and methylene (-CH$_2$-) groups are separated (as in **1**) or adjacent (as in **2**).

## 1.5   SENSITIVITY

Sensitivity is generally taken to signify the limits of detectability of a chromophore.  Some methods (*e.g.* $^1$H NMR) detect all chromophores accessible to them with equal sensitivity while in other techniques (*e.g.* UV) the range of sensitivity towards different chromophores spans many orders of magnitude.  In terms of overall sensitivity, *i.e.* the amount of sample required, it is generally observed that:

$$MS > UV > IR > {}^1H\ NMR > {}^{13}C\ NMR$$

but considerations of relative sensitivity toward different chromophores may be more important.

## 1.6   PRACTICAL CONSIDERATIONS

The 5 major spectroscopic methods (MS, UV, IR, $^1$H NMR and $^{13}$C NMR) have become established as the principal tools for the determination of the structures of organic compounds, because between them they detect a wide variety of structural elements.

The instrumentation and skills involved in the use of all five major spectroscopic methods are now widely spread, but the ease of obtaining and interpreting the data from each method under real laboratory conditions varies.

In very general terms:

*(a)*    While the ***cost*** of each type of instrumentation differs greatly (NMR instruments cost between $50,000 and several million dollars), as an overall guide, MS and NMR instruments are much more costly than UV and IR spectrometers.  With increasing cost goes increasing difficulty in maintenance and the required operator expertise, thus compounding the total outlay.

*(b)*    In terms of ***ease of usage*** for routine operation, most UV and IR instruments are comparatively straightforward.  NMR Spectrometers are also common as "hands-on" instruments in most chemistry laboratories and the users require routine training and a degree of basic computer literacy.  Similarly some Mass Spectrometers are now designed to be used by researchers as "hands-on" routine instruments.  However, the more advanced NMR Spectrometers and most Mass

Spectrometers are sophisticated instruments that are usually operated and maintained by specialists.

*(c)*   The *scope* of each spectroscopic method can be defined as the amount of useful information it provides.  This is a function of the total amount of information obtainable and also how difficult the data are to interpret.  The scope of each method varies from problem to problem, and each method has its aficionados and specialists, but the overall utility undoubtedly decreases in the order:

$$NMR > MS > IR > UV$$

with the combination of $^1$H and $^{13}$C NMR providing the most useful information.

*(d)*   The theoretical background needed for each method varies with the nature of the experiment, but the minimum overall amount of theory needed decreases in the order:

$$NMR \gg MS > UV \approx IR$$

# 2

# ULTRAVIOLET (UV) SPECTROSCOPY

## 2.1 BASIC INSTRUMENTATION

Basic instrumentation for both UV and IR spectroscopy consists of an energy *source*, a *sample cell,* a *dispersing device* (prism or grating) and a *detector,* arranged as schematically shown in Figure 2.1.

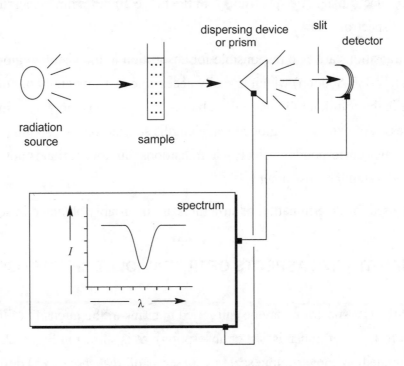

**Figure 2.1   Schematic Representation of an IR or UV Spectrometer**

The drive of the dispersing device is synchronised with the x-axis of the recorder or fed directly to a computer, so that this indicates the wavelength of radiation reaching the detector. The signal from the detector is transmitted to the y-axis of the recorder or to a computer and this indicates how much radiation is absorbed by the sample at any particular wavelength.

*Organic Structures from Spectra*, Fifth Edition. L. D. Field, S. Sternhell and J. R. Kalman.
© 2013 John Wiley & Sons, Ltd. Published 2013 by John Wiley & Sons, Ltd.

In practice, *double-beam* instruments are used where the absorption of a *reference cell*, containing only solvent, is subtracted from the absorption of the sample cell. Double beam instruments also cancel out absorption due to the atmosphere in the optical path as well as the solvent.

The energy source must be appropriate for the wavelengths of radiation being scanned.  The materials from which the dispersing device and the detector are constructed must be as transparent as possible to wavelengths being scanned.  For UV measurements, the cells and optical components are typically made of quartz and ethanol, hexane, water or dioxane are usually chosen as solvents.

## 2.2    THE NATURE OF ULTRAVIOLET SPECTROSCOPY

The term "UV spectroscopy" generally refers to *electronic transitions* occurring in the region of the electromagnetic spectrum ($\lambda$ in the range 200-380 nm) accessible to standard UV spectrometers.

Electronic transitions are also responsible for absorption in the visible region (approximately 380-800 nm) which is easily accessible instrumentally but of less importance in the solution of structural problems, because most organic compounds are colourless.  An extensive region at wavelengths shorter than $\sim$ 200 nm ("vacuum ultraviolet") also corresponds to electronic transitions, but this region is not readily accessible with standard instruments.

UV spectra used for determination of structures are invariably obtained in solution.

## 2.3    QUANTITATIVE ASPECTS OF ULTRAVIOLET SPECTROSCOPY

The y-axis of a UV spectrum may be calibrated in terms of the intensity of transmitted light (*i.e.* percentage of transmission or absorption), as is shown in Figure 2.2, or it may be calibrated on a logarithmic scale *i.e.* in terms of *absorbance* (A) defined in Figure 2.2.

Absorbance is proportional to concentration and path length (the Beer-Lambert Law). The intensity of absorption is usually expressed in terms of *molar absorbance* or the *molar extinction coefficient* ($\varepsilon$) given by:

$$\varepsilon = \frac{M A}{C l}$$

where M is the molecular weight, C the concentration (in grams per litre) and *l* is the path length through the sample in centimetres.

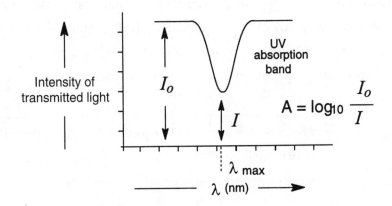

**Figure 2.2    Definition of Absorbance (A)**

UV absorption bands (Figure 2.2) are characterised by the wavelength of the absorption maximum ($\lambda_{max}$) and $\varepsilon$. The values of $\varepsilon$ associated with commonly encountered chromophores vary between 10 and $10^5$. For convenience, extinction coefficients are usually tabulated as $\log_{10}(\varepsilon)$ as this gives numerical values which are easier to manage. The presence of small amounts of strongly absorbing impurities may lead to errors in the interpretation of UV data.

## 2.4    CLASSIFICATION OF UV ABSORPTION BANDS

UV absorption bands have fine structure due to the presence of vibrational sub-levels, but this is rarely observed in solution due to collisional broadening. As the transitions are associated with changes of electron orbitals, they are often described in terms of the orbitals involved, *e.g.*

$$\sigma \rightarrow \sigma^* \qquad <150 \text{ nm}$$
$$\pi \rightarrow \pi^* \qquad 170-210 \text{ nm}$$
$$n \rightarrow \pi^* \qquad 230-290 \text{ nm}$$
$$n \rightarrow \sigma^* \qquad 150-250 \text{ nm}$$

where *n* denotes a non-bonding orbital, the asterisk denotes an antibonding orbital and $\sigma$ and $\pi$ have the usual significance.

Another method of classification uses the symbols:

B    (for benzenoid)

E    (for ethylenic)

R    (for radical-like)

K    (for conjugated - from the German "konjugierte")

A molecule may give rise to more than one band in its UV spectrum, either because it contains more than one chromophore or because more than one transition of a single chromophore is observed.  However, UV spectra typically contain far fewer features (bands) than IR, MS or NMR spectra and therefore have a lower information content. The ultraviolet spectrum of acetophenone in ethanol contains 3 easily observed bands:

| | $\lambda_{max}$ (nm) | $\varepsilon$ | $\log_{10}(\varepsilon)$ | Assignment | |
|---|---|---|---|---|---|
| | 244 | 12,600 | 4.1 | $\pi \rightarrow \pi^*$ | K |
| | 280 | 1,600 | 3.2 | $\pi \rightarrow \pi^*$ | B |
| | 317 | 60 | 1.8 | $n \rightarrow \pi^*$ | R |

acetophenone

## 2.5    SPECIAL TERMS IN UV SPECTROSCOPY

*Auxochromes* (auxiliary chromophores) are groups which have little UV absorption by themselves, but which often have significant effects on the absorption (both $\lambda_{max}$ and $\varepsilon$) of a chromophore to which they are attached.  Generally, auxochromes are atoms with one or more lone pairs *e.g.* -OH, -OR, -NR$_2$, -halogen.

If a structural change, such as the attachment of an auxochrome, leads to the absorption maximum being shifted to a longer wavelength, the phenomenon is termed a *bathochromic shift*.  A shift towards shorter wavelength is called a *hypsochromic shift*.

## 2.6    IMPORTANT UV CHROMOPHORES

Most of the reliable and useful data is due to relatively strongly absorbing chromophores ($\varepsilon > 200$) which are mainly indicative of conjugated or aromatic systems.  Examples listed below encompass most of the commonly encountered effects.

### (1)   Dienes and Polyenes

Extension of conjugation in a carbon chain is always associated with a pronounced shift towards longer wavelength, and usually towards greater intensity (Table 2.1).

**Table 2.1    The Effect of Extended Conjugation on UV Absorption**

| Alkene | $\lambda_{max}$ (nm) | $\varepsilon$ | $\log_{10}(\varepsilon)$ |
|---|---|---|---|
| $CH_2=CH_2$ | 165 | 10,000 | 4.0 |
| $CH_3\text{-}CH_2\text{-}CH=CH\text{-}CH_2\text{-}CH_3$ (trans) | 184 | 10,000 | 4.0 |
| $CH_2=CH\text{-}CH=CH_2$ | 217 | 20,000 | 4.3 |
| $CH_3\text{-}CH=CH\text{-}CH=CH_2$ (trans) | 224 | 23,000 | 4.4 |
| $CH_2=CH\text{-}CH=CH\text{-}CH=CH_2$ (trans) | 263 | 53,000 | 4.7 |
| $CH_3\text{-}(CH=CH)_5\text{-}CH_3$ (trans) | 341 | 126,000 | 5.1 |

When there are more than 8 conjugated double bonds, the absorption maximum of polyenes is such that they absorb light strongly in the visible region of the spectrum.

Empirical rules (Woodward's Rules) of good predictive value are available to estimate the positions of the absorption maxima in conjugated alkenes and conjugated carbonyl compounds.

The stereochemistry and the presence of substituents also influence UV absorption by the diene chromophore.  For example:

$\lambda_{max}$ = 214 nm  

$\varepsilon$ = 16,000  

$\log_{10}(\varepsilon)$ = 4.2  

$\lambda_{max}$ = 253 nm  

$\varepsilon$ = 8,000  

$\log_{10}(\varepsilon)$ = 3.9

## (2) Carbonyl compounds

All carbonyl derivatives exhibit weak ($\varepsilon < 100$) absorption between 250 and 350 nm, and this is only of marginal use in determining structure. However, conjugated carbonyl derivatives always exhibit strong absorption (Table 2.2).

**Table 2.2    UV Absorption Bands in Common Carbonyl Compounds**

| Compound | Structure | $\lambda_{max}$ (nm) | $\varepsilon$ | $\log_{10}(\varepsilon)$ |
|---|---|---|---|---|
| Acetaldehyde | | 293 (hexane solution) | 12 | 1.1 |
| Acetone | | 279 (hexane solution) | 15 | 1.2 |
| Propenal | | 207 / 328 (ethanol solution) | 12,000 / 20 | 4.1 / 1.3 |
| (E)-Pent-3-en-2-one | | 221 / 312 (ethanol solution) | 12,000 / 40 | 4.1 / 1.6 |
| 4-Methylpent-3-en-2-one | | 238 / 316 (ethanol solution) | 12,000 / 60 | 4.1 / 1.8 |
| Cyclohex-2-en-1-one | | 225 | 7,950 | 3.9 |
| Benzoquinone | | 247 / 292 / 363 | 12,600 / 1,000 / 250 | 4.1 / 3.0 / 2.4 |

### (3)   *Benzene derivatives*

Benzene derivatives exhibit medium to strong absorption in the UV region. Bands usually have characteristic fine structure and the intensity of the absorption is strongly influenced by substituents. Examples listed in Table 2.3 include weak auxochromes (-CH$_3$, -Cl, -OCH$_3$), groups which increase conjugation (-CH=CH$_2$, -C(=O)-R, -NO$_2$) and auxochromes whose absorption is pH dependent (-NH$_2$ and -OH).

**Table 2.3     UV Absorption Bands in Common Benzene Derivatives**

| Compound | Structure | $\lambda_{max}$ (nm) | $\varepsilon$ | $\log_{10}(\varepsilon)$ |
|---|---|---|---|---|
| Benzene | | 184 | 60,000 | 4.8 |
| | | 204 | 7,900 | 3.9 |
| | | 256 | 200 | 2.3 |
| Toluene | -CH$_3$ | 208 | 8,000 | 3.9 |
| | | 261 | 300 | 2.5 |
| Chlorobenzene | -Cl | 216 | 8,000 | 3.9 |
| | | 265 | 240 | 2.4 |
| Anisole | -OCH$_3$ | 220 | 8,000 | 3.9 |
| | | 272 | 1,500 | 3.2 |
| Styrene | -CH=CH$_2$ | 244 | 12,000 | 4.1 |
| | | 282 | 450 | 2.7 |
| Acetophenone | -C-CH$_3$ (=O) | 244 | 12,600 | 4.1 |
| | | 280 | 1,600 | 3.2 |
| Nitrobenzene | -NO$_2$ | 251 | 9,000 | 4.0 |
| | | 280 | 1,000 | 3.0 |
| | | 330 | 130 | 2.1 |
| Aniline | -NH$_2$ | 230 | 8,000 | 3.9 |
| | | 281 | 1,500 | 3.2 |
| Anilinium ion | -NH$_3^+$ | 203 | 8,000 | 3.9 |
| | | 254 | 160 | 2.2 |
| Phenol | -OH | 211 | 6,300 | 3.8 |
| | | 270 | 1,500 | 3.2 |
| Phenoxide ion | -O$^-$ | 235 | 9,500 | 4.0 |
| | | 287 | 2,500 | 3.4 |

Aniline and phenoxide ion have strong UV absorptions due to the overlap of the lone pair on the nitrogen (or oxygen) with the $\pi$-system of the benzene ring. This may be expressed in the usual Valence Bond terms:

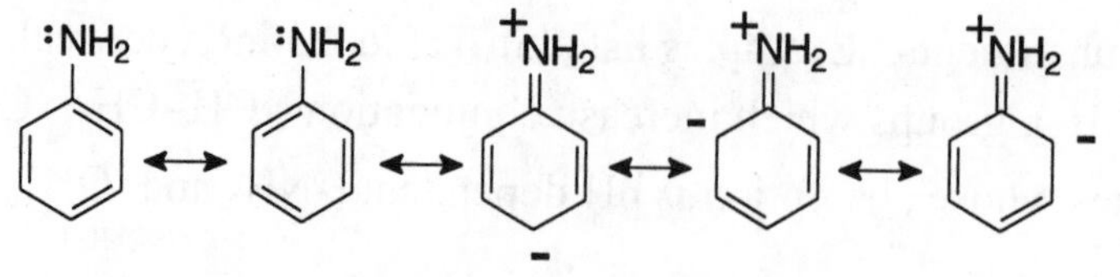

The striking changes in the ultraviolet spectra accompanying protonation of aniline and phenoxide ion are due to loss (or substantial reduction) of the overlap between the lone pairs and the benzene ring.

## 2.7   THE EFFECT OF SOLVENTS

Solvent polarity may affect the absorption characteristics, in particular $\lambda_{max}$, since the polarity of a molecule usually changes when an electron is moved from one orbital to another. Solvent effects of up to 20 nm may be observed with carbonyl compounds. Thus the $n \rightarrow \pi^*$ absorption of acetone occurs at 279 nm in $n$-hexane, 270 nm in ethanol, and at 265 nm in water.

# 3

# INFRARED (IR) SPECTROSCOPY

## 3.1  ABSORPTION RANGE AND THE NATURE OF IR ABSORPTION

Infrared absorption spectra are calibrated in wavelengths expressed in micrometers:

$$1\,\mu m = 10^{-6}\ m$$

or in frequency-related *wave numbers* (cm$^{-1}$) which are reciprocals of wavelengths:

$$\text{wave number } \bar{\nu}\ (cm^{-1}) = \frac{1 \times 10^{4}}{\text{wavelength (in }\mu m)}$$

The range accessible for standard instrumentation is usually:

$$\bar{\nu} = 4000 \text{ to } 666\ cm^{-1}$$

$$\text{or } \quad \lambda = 2.5 \text{ to } 15\ \mu m$$

Infrared absorption intensities are rarely described quantitatively, except for the general classifications of s (strong), m (medium) or w (weak).

The transitions responsible for IR bands are due to *molecular vibrations, i.e.* to periodic motions involving stretching or bending of bonds. Polar bonds are associated with strong IR absorption *while symmetrical bonds may not absorb at all*.

Clearly the vibrational frequency, *i.e.* the position of the IR bands in the spectrum, depends on the nature of the bond. Shorter and stronger bonds have their stretching vibrations at the higher energy end (shorter wavelength) of the IR spectrum than the longer and weaker bonds. Similarly, bonds to lighter atoms (*e.g.* hydrogen), vibrate at higher energy than bonds to heavier atoms.

IR bands often have rotational sub-structure, but this is normally resolved only in spectra taken in the gas phase.

*Organic Structures from Spectra*, Fifth Edition. L. D. Field, S. Sternhell and J. R. Kalman.
© 2013 John Wiley & Sons, Ltd. Published 2013 by John Wiley & Sons, Ltd.

## 3.2    EXPERIMENTAL ASPECTS OF INFRARED SPECTROSCOPY

The basic layout of a simple dispersive IR spectrometer is the same as for an UV spectrometer (Figure 2.1), except that all components must now match the different energy range of electromagnetic radiation.  The more sophisticated Fourier Transform Infrared (FTIR) instruments record an infrared interference pattern generated by a moving mirror and this is transformed by a computer into an infrared spectrum.

Very few substances are transparent over the whole of the IR range: sodium and potassium chloride and sodium and potassium bromide are most common.  The cells used for obtaining IR spectra in solution typically have NaCl windows and liquids can be examined as films on NaCl plates.  Solution spectra are generally obtained in chloroform or carbon tetrachloride but this leads to loss of information at longer wavelengths where there is considerable absorption of energy by the solvent.  Organic solids may also be examined as mulls (fine suspensions) in heavy oils.  The oils absorb infrared radiation but only in well-defined regions of the IR spectrum.  Solids may also be examined as dispersions in compressed KBr or KCl discs.

To a first approximation, the absorption frequencies due to the important IR chromophores are the same in solid and liquid states.

## 3.3    GENERAL FEATURES OF INFRARED SPECTRA

Almost all organic compounds contain C-H bonds and this means that there is invariably an absorption band in the IR spectrum between 2900 and 3100 cm$^{-1}$ at the C-H stretching frequency.

Molecules generally have a large number of bonds and each bond may have several IR-active *vibrational modes*.  IR spectra are complex and have many overlapping absorption bands.  IR spectra are sufficiently complex that the spectrum for each compound is unique and this makes IR spectra very useful for identifying compounds by direct comparison with spectra from authentic samples (*"fingerprinting"*).

The characteristic IR vibrations are influenced strongly by small changes in molecular structure, thus making it difficult to identify structural fragments from IR data alone.  However, there are some groups of atoms that are readily recognised from IR spectra.  IR chromophores are most useful for the determination of structure if:

*(a)*    The chromophore does not absorb in the *most crowded region* of the spectrum (600-1400 cm$^{-1}$) where strong overlapping stretching absorptions from C-X single bonds (X = O, N, S, P and halogens) make assignment difficult.

*(b)*    The chromophores should be *strongly absorbing* to avoid confusion with weak harmonics. However, in otherwise empty regions *e.g.* 1800-2500 cm$^{-1}$, even weak absorptions can be assigned with confidence.

*(c)*    The absorption frequency must be structure dependent in an *interpretable* manner. This is particularly true of the very important bands due to the C=O stretching vibrations, which generally occur between 1630 and 1850 cm$^{-1}$.

## 3.4    IMPORTANT IR CHROMOPHORES

*(1)*    *-O-H Stretch*          Not hydrogen-bonded ("free")          3600 cm$^{-1}$

Hydrogen-bonded          3100 - 3200 cm$^{-1}$

This difference between hydrogen bonded and free OH frequencies is clearly related to the weakening of the O-H bond as a consequence of hydrogen bonding.

*(2)*    *Carbonyl groups* always give rise to **strong** absorption between 1630 and 1850 cm$^{-1}$ due to C=O stretching vibrations. Moreover, carbonyl groups in different functional groups are associated with well-defined regions of IR absorption (Table 3.1).

Even though the ranges for individual types often overlap, it may be possible to make a definite decision from information derived from other regions of the IR spectrum. Esters also exhibit strong C-O stretching absorption between 1200 and 1300 cm$^{-1}$ while carboxylic acids exhibit an additional O-H stretching absorption near 3000 cm$^{-1}$.

The characteristic shift toward lower frequency associated with the introduction of $\alpha$, $\beta$–unsaturation can be rationalised by considering the Valence Bond description of an enone:

The additional structure **C**, which cannot be drawn for an unconjugated carbonyl derivative, implies that the carbonyl band in an enone has more single bond character and is therefore weaker. The involvement of a carbonyl group in hydrogen bonding reduces the frequency of the carbonyl stretching vibration by about 10 cm$^{-1}$. This can be rationalised in a manner analogous to that proposed above for free and H-bonded O-H vibrations.

**Table 3.1  Carbonyl (C=O) IR Absorption Frequencies in Common Functional Groups**

| Carbonyl group | Structure | $\bar{v}$ (cm$^{-1}$) |
|---|---|---|
| Ketones | R–C–R'<br>O | 1700 - 1725 |
| Aldehydes | R–C–H<br>O | 1720 - 1740 |
| Aryl aldehydes or ketones, α, β-unsaturated aldehydes or ketones | Ar–C–R'  R⌁C–R'  R' = alkyl, aryl, or H<br>O      O | 1660 - 1720 |
| Cyclopentanones | (cyclopentanone ring)=O | 1740 - 1750 |
| Cyclobutanones | (cyclobutanone ring)=O | 1760 - 1780 |
| Carboxylic acids [†] | R–C–OH<br>O | 1700 - 1725 |
| α, β-unsaturated and aryl carboxylic acids [†] | Ar–C–OH   R⌁C–OH<br>O        O | 1680 - 1715 |
| Esters [§] | R–C–OR'<br>O | 1735 - 1750 |
| Phenolic Esters [§] | R–C–OAr<br>O | 1760 - 1800 |
| Aryl or α, β–unsaturated Esters [§] | R⌁C–OR'  Ar–C–OR'<br>O        O | 1715 - 1730 |
| δ-Lactones [§] | (δ-lactone ring) O=, O | 1735 - 1750 |
| γ-Lactones [§] | (γ-lactone ring) O=, O | 1760 - 1780 |
| Amides | R–C–NR'R"<br>O | 1630 - 1690 |
| Acid chlorides | R–C–Cl<br>O | 1770 - 1815 |
| Acid anhydrides (two bands) | R–C–O–C–R<br>O    O | 1740 - 1850 |
| Carboxylates | R–C⟨O / O⟩- | 1550 - 1610<br>1300 - 1450 |

[†] Carboxylic acids also exhibit an O-H stretch near 3000 cm$^{-1}$

[§] Esters and lactones also exhibit a strong C-O stretch in the range 1160 – 1250 cm$^{-1}$

*(3)    Other polar functional groups.* Many other functional groups have characteristic IR absorptions (Table 3.2).

**Table  3.2     Characteristic IR Absorption Frequencies for Functional Groups**

| Functional group | Structure | $\bar{\nu}$ (cm$^{-1}$) | Intensity |
|---|---|---|---|
| Amine | >N—H | 3300 - 3500 | |
| Terminal acetylenes | ≡C–H | 3300 | strong |
| Imines | >C=N< | 1480 - 1690 | |
| Enol ethers | >C=C< O—R | 1600 - 1660 | strong |
| Alkenes | R$_1$, R$_3$ C=C R$_2$, R$_4$ | 1640 - 1680 | weak to medium |
| Nitro groups | —N$^+$(O$^-$)=O | 1500 - 1650<br>1250 - 1400 | strong<br>medium |
| Epoxides | (epoxide ring) O | 1250<br>810 - 950 | strong |
| Sulfoxides | >S=O | 1010 - 1070 | strong |
| Sulfones | O=S=O | 1300 - 1350<br>1100 - 1150 | strong<br>strong |
| Sulfonamides and Sulfonate esters | —SO$_2$–N<<br>—SO$_2$–O— | 1140 - 1180<br>1300 - 1370 | strong<br>strong |
| Alcohols | >C–OH | 3000 – 3700<br>1000 - 1260 | strong<br>strong |
| Ethers | —C–OR | 1085 - 1150 | strong |
| Alkyl fluorides | —C–F | 1000 - 1400 | strong |
| Alkyl chlorides | —C–Cl | 580 - 780 | strong |
| Alkyl bromides | —C–Br | 560 - 800 | strong |
| Alkyl iodides | —C—I | 500 - 600 | strong |

Carbon-carbon double bonds in unconjugated alkenes usually exhibit weak to moderate absorptions due to C=C stretching in the range 1660-1640 cm$^{-1}$. Disubstituted, trisubstituted and tetrasubstituted alkenes usually absorb near 1670 cm$^{-1}$.  The more polar carbon-carbon double bonds in enol ethers and enones usually absorb strongly between 1600 and 1700 cm$^{-1}$.  Alkenes conjugated with an aromatic ring absorb strongly near 1625 cm$^{-1}$.

**(4)      *Chromophores absorbing in the region between 1900 and 2600 cm$^{-1}$.*** The absorptions listed in Table 3.3 often yield useful information because, even though some are of only weak or medium intensity, they occur in regions largely devoid of absorption by other commonly occurring chromophores.

**Table  3.3  Common IR Absorption Frequencies in the Region 1900 – 2600 cm$^{-1}$**

| Functional group | Structure | $\bar{\nu}$ (cm$^{-1}$) | Intensity |
|---|---|---|---|
| alkyne | —C≡C— | 2100 - 2300 | weak to medium |
| nitrile | —C≡N | 2215 - 2280 | medium |
| cyanate | —O—C≡N | 2130 - 2270 | strong |
| thiocyanate | —S—C≡N | 2130 - 2175 | medium |
| isocyanate | —N=C=O | 2200 - 2300 | strong broad |
| isothiocyanate | —N=C=S | 2000 - 2200 | strong |
| allene | C=C=C | 1900 - 2000 | strong |

# 4

# MASS SPECTROMETRY

It is possible to determine the masses of individual ions in the gas phase. Strictly speaking, it is only possible to measure their mass/charge ratio ($m/e$), but as multi charged ions are very much less abundant than those with a single electronic charge ($e = 1$), $m/e$ is for all practical purposes equal to the mass of the ion, $m$. The principal experimental problems in mass spectrometry are firstly to volatilise the substrate (which implies high vacuum) and secondly to ionise the neutral molecules to charged species.

## 4.1 IONISATION PROCESSES

The most common method of ionisation involves *Electron Impact* (EI) and there are two general courses of events following a collision of a molecule M with an electron *e*. By far the most probable event involves electron ejection which yields an odd-electron positively charged *cation radical* $[M]^{+\cdot}$ of the same mass as the initial molecule M.

$$M + e \rightarrow \quad [M]^{+\cdot} \quad + \ 2e$$

The cation radical produced is known as the *molecular ion* and its mass gives a direct measure of the molecular weight of a substance. An alternative, far less probable process, also takes place and it involves the capture of an electron to give a negative *anion radical*, $[M]^{-\cdot}$.

$$M + e \rightarrow \quad [M]^{-\cdot}$$

Electron impact mass spectrometers are generally set up to detect only positive ions, but negative-ion mass spectrometry is also possible.

The energy of the electron responsible for the ionisation process can be varied. It must be sufficient to knock out an electron and this threshold, typically about 10-12 eV, is known as the *appearance potential*. In practice much higher energies (~70 eV) are used and this large excess energy (1 eV = 95 kJ mol$^{-1}$) causes further *fragmentation* of the molecular ion.

*Organic Structures from Spectra*, Fifth Edition. L. D. Field, S. Sternhell and J. R. Kalman.
© 2013 John Wiley & Sons, Ltd. Published 2013 by John Wiley & Sons, Ltd.

The two important types of fragmentation are:

$$[M]^{+\cdot} \quad \rightarrow \quad A^+ \text{ (even electron cation) } + \text{ B}^{\cdot} \text{ (radical)}$$

or

$$[M]^{+\cdot} \quad \rightarrow \quad C^{+\cdot} \text{ (cation radical) } + \text{ D (neutral molecule)}$$

As only species bearing a positive charge will be detected, the mass spectrum will show signals due not only to $[M]^{+\cdot}$ but also due to $A^+$, $C^{+\cdot}$ and to fragment ions resulting from subsequent fragmentation of $A^+$ and $C^{+\cdot}$.

As any species may fragment in a variety of ways, the typical mass spectrum consists of many signals.  The mass spectrum consists of a plot of masses of ions against their relative abundance.

There are a number of other methods for ionising the sample in a mass spectrometer. The most important alternative ionisation method to electron impact is *Chemical Ionisation* (CI).  In CI mass spectrometry, an intermediate substance (generally methane or ammonia) is introduced at a higher concentration than that of the substance being investigated.  The carrier gas is ionised by electron impact and the substrate is then ionised by collisions with these ions.  CI is a milder ionisation method than EI and leads to less fragmentation of the molecular ion.

Another common method of ionisation is *Electrospray Ionisation* (ESI).  In this method, the sample is dissolved in a polar, volatile solvent and pumped through a fine metal nozzle, the tip of which is charged with a high voltage.  This produces charged droplets from which the solvent rapidly evaporates to leave naked ions which pass into the mass spectrometer.  ESI is also a relatively mild form of ionisation and is very suitable for biological samples which are usually quite soluble in polar solvents but which are relatively difficult to vaporise in the solid state.  Electrospray ionisation tends to lead to less fragmentation of the molecular ion than EI.

*Matrix Assisted Laser Desorption Ionisation* (MALDI) uses a pulse of laser light to bring about ionisation.  The sample is usually mixed with a highly absorbing compound which acts as a supporting matrix.  The laser pulse ionises and vaporises the matrix and the sample to give ions which pass into the mass spectrometer.  Again MALDI is a relatively mild form of ionisation which tends to give less fragmentation of the molecular ion than EI.

All of the subsequent discussion of mass spectrometry is limited to positive-ion electron-impact mass spectrometry.

## 4.2    INSTRUMENTATION

In a magnetic sector mass spectrometer (Figure 4.1), the positively charged ions of mass, *m,* and charge, *e* (generally *e* = 1) are subjected to an accelerating voltage V and passed through a magnetic field H which causes them to be deflected into a curved path of radius *r.*  The quantities are connected by the relationship:

$$\frac{m}{e} = \frac{H^2 r^2}{2V}$$

The values of H and V are known, *r* is determined experimentally and *e* is assumed to be unity thus permitting us to determine the mass *m.*  In practice the magnetic field is scanned so that streams of ions of different mass pass sequentially to the detecting system (ion collector).  The whole system (Figure 4.1) is under high vacuum (less than $10^{-6}$ Torr) to permit the volatilisation of the sample and so that the passage of ions is not impeded.  The introduction of the sample into the ion chamber at high vacuum requires a complex sample inlet system.

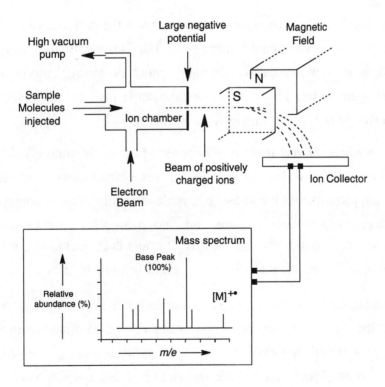

**Figure 4.1    Schematic Diagram of an Electron-Impact Mass Spectrometer**

The magnetic scan is synchronised with the x-axis of a recorder and calibrated to appear as *mass number* (strictly *m/e*).  The amplified current from the ion collector gives the relative abundance of ions on the y-axis.  The signals are usually pre-processed by a computer that assigns a relative abundance of 100% to the strongest peak (*base peak*).

Many modern mass spectrometers do not use a magnet to bend the ion beam to separate ions but rather use the "*time of flight*" (TOF) of an ion over a fixed distance to measure its mass.  In these spectrometers, ions are generated (usually using a very short laser pulse) then accelerated in an electric field.  Lighter ions have a higher velocity as they leave the accelerating field and their time of flight over a fixed distance will vary depending on the speed that they are travelling.  Time of Flight mass spectrometers have the advantage that they do not require large, high-precision magnets to bend and disperse the ion beam so they tend to be much smaller, compact and less complex (desk-top size) instruments.

## 4.3    MASS SPECTRAL DATA

As well as giving the molecular weight of a substance, the molecular ion of a compound may provide additional information.  The **"nitrogen rule"** states that a molecule with an even molecular weight must contain no nitrogen atoms or an even number of nitrogen atoms.  This means that a molecule with an odd molecular weight must contain an odd number of nitrogen atoms.

*(1)      High resolution mass spectra.*  The **mass of an ion** is routinely determined to the nearest unit value.  Thus the mass of $[M]^{+\cdot}$ gives a direct measure of molecular weight.  It is not usually possible to assign a molecular formula to a compound on the basis of the integer *m/e* value of its parent ion.  For example, a parent ion at *m/e* 72 could be due to a compound whose molecular formula is $C_4H_8O$ or one with a molecular formula $C_3H_4O_2$ or one with a molecular formula $C_3H_8N_2$.

However, using a *double-focussing* mass spectrometer or a *time-of-flight* mass spectrometer, the mass of an ion or any fragment can be determined to an accuracy of approximately $\pm 0.00001$ of a mass unit *(a high resolution mass spectrum)*.  Since the masses of the atoms of each element are known to high accuracy, molecules that may have the same mass when measured only to the nearest integer mass unit, can be distinguished when the mass is measured with high precision.  Based on the accurate masses of $^{12}C$, $^{16}O$, $^{14}N$ and $^{1}H$ (Table 4.1) ions with the formulas $C_4H_8O^{+\cdot}$, $C_3H_4O_2^{+\cdot}$ or $C_3H_8N_2^{+\cdot}$ would have accurate masses 72.0573, 72.0210, and 72.0686 so these could easily be distinguished by high resolution mass spectroscopy.  In general, if the mass of any fragment in the mass spectrum can be accurately determined, there is usually only one combination of elements which can give rise to that signal since there are only a limited number of elements and their masses are accurately known. By examining a mass spectrum at sufficiently high resolution, one can obtain the exact composition of *each ion* in a mass spectrum, unambiguously.  Most importantly, determining the accurate mass of [M]$^{+\cdot}$ *gives the molecular formula of the compound.*

**Table  4.1      Accurate Masses of Selected Isotopes**

| Isotope | Natural Abundance (%) | Mass |
|---------|----------------------|------|
| $^1$H | 99.98 | 1.00783 |
| $^2$H | 0.016 | 2.01410 |
| $^{12}$C | 98.9 | 12.0000 |
| $^{13}$C | 1.1 | 13.00336 |
| $^{14}$N | 99.6 | 14.0031 |
| $^{15}$N | 0.37 | 15.0001 |
| $^{16}$O | 99.8 | 15.9949 |
| $^{17}$O | 0.037 | 16.9991 |
| $^{18}$O | 0.20 | 17.9992 |
| $^{19}$F | 100 | 18.99840 |
| $^{28}$Si | 92.28 | 27.9769 |
| $^{29}$Si | 4.7 | 28.9765 |
| $^{30}$Si | 3.02 | 29.9738 |
| $^{31}$P | 100 | 30.97376 |
| $^{32}$S | 95.0 | 31.9721 |
| $^{33}$S | 0.75 | 32.9715 |
| $^{34}$S | 4.2 | 33.9679 |
| $^{35}$Cl | 75.8 | 34.9689 |
| $^{37}$Cl | 24.2 | 36.9659 |
| $^{79}$Br | 50.7 | 78.9183 |
| $^{81}$Br | 49.3 | 80.9163 |
| $^{127}$I | 100 | 126.9045 |

*(2)*     *Molecular Fragmentation.* The **fragmentation pattern** is a molecular fingerprint. In addition to the molecular ion peak, the mass spectrum (see Figure 4.1) consists of a number of peaks at lower mass numbers and these result from fragmentation of the molecular ion. The principles determining the mode of fragmentation are reasonably well understood, and it is possible to derive structural information from the fragmentation pattern in several ways.

*(a)*     The appearance of prominent peaks at certain mass numbers can be correlated empirically with certain structural elements (Table 4.2), *e.g.* a prominent peak at $m/e = 43$ is a strong indication of the presence of a $CH_3$-CO- group in the molecule.

*(b)*     Information can also be obtained from *differences* between the masses of two peaks. Thus a prominent fragment ion that occurs 15 mass numbers below the molecular ion, suggests strongly the loss of a $CH_3$- group and therefore that a methyl group was present in the substance examined.

*(c)*     The knowledge of the principles governing the **mode of fragmentation** of ions makes it possible to confirm the structure assigned to a compound and, quite often, to determine the juxtaposition of structural fragments and to distinguish between isomeric substances. For example, the mass spectrum of benzyl methyl ketone, $Ph$-$CH_2$-CO-$CH_3$ contains a strong peak at $m/e = 91$ due to the stable ion $Ph$-$CH_2^+$, but this ion is absent in the mass spectrum of the isomeric propiophenone $Ph$-CO-$CH_2CH_3$ where the structural elements $Ph$- and -$CH_2$- are separated. Instead, a prominent peak occurs at $m/e = 105$ due to the stable ion $Ph$-$C{\equiv}O^+$.

Electronic databases of the mass spectral fragmentation patterns of known molecules can be rapidly searched by computer. The pattern and intensity of fragments in the mass spectrum is characteristic of an individual compound so comparison of the experimental mass spectrum of a compound with those in a library can be used to positively identify it, if its spectrum has been recorded previously.

**Table 4.2      Common Fragments and their Masses**

| Fragment | Mass | Fragment | Mass | Fragment | Mass |
|---|---|---|---|---|---|
| $CH_3-$ | 15 | $CH_3CH_2-$ | 29 | (H)(O=)C− | 29 |
| NO | 30 | $-CH_2OH$ | 31 | $CH_2=CH-CH_2$ | 41 |
| $CH_3-C(=O)-$ | 43 | $HO-C(=O)-$ | 45 | $-NO_2$ | 46 |
| $C_4H_7$ | 55 | $C_4H_9$ | 57 | $CH_3CH_2-C(=O)-$ | 57 |
| $CH_2=C(OH)(OH)$ | 60 | $C_5H_5$ | 65 | $C_6H_5$ | 77 |
| $C_7H_7$  ($-CH_2-$) | 91 | $C_6H_6N$ ($-CH_2-$) | 92 | $C_7H_5O$ ($-C(=O)-$) | 105 |
| $C_8H_7O$ ($CH_3$... $-C(=O)-$) | 119 | $I-$ | 127 | | |

*(3)*      *Isotope ratios.*  For some elements (most notably bromine and chlorine), there is more than one isotope of high natural abundance *e.g.* bromine has two abundant isotopes - $^{79}Br$ 51 % and $^{81}Br$ 49 %; chlorine also has two abundant isotopes – $^{35}Cl$ 75% and $^{35}Cl$ 25 % (Table 4.1).  The presence of Br or Cl, or other elements that contain significant proportions ($\geq$ 1%) of minor isotopes, is often obvious simply by inspection of ions near the molecular ion.

The relative intensities of the $[M]^{+\cdot}$, $[M+1]^{+\cdot}$ and $[M+2]^{+\cdot}$ ions exhibit a characteristic pattern depending on the specific isotopes that make up the ion.  For any molecular ion (or fragment) which contains one bromine atom, the mass spectrum will contain two peaks separated by two $m/e$ units, one for the ions which contain $^{79}$Br and one for the ions which contain $^{81}$Br.  For bromine-containing fragments, the relative intensities of the two ions will be approximately the same, since the natural abundances of $^{79}$Br and $^{81}$Br are approximately equal.

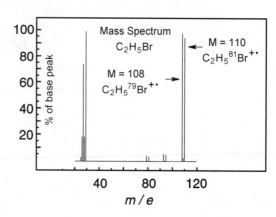

In the mass spectrum of 1-bromoethane, there are two molecular ions of almost equal intensity.  The peak at $m/e$ 108 corresponds to the molecular ions in the sample which contain $^{79}$Br; the peak at $m/e$ 110 corresponds to the molecular ions in the sample which contain $^{81}$Br.

Similarly, for any molecule (or fragment of an molecule) which contains one chlorine atom, the mass spectrum will contain two fragments separated by two $m/e$ units, one for the ions which contain $^{35}$Cl and one for the ions which contain $^{37}$Cl.  For chlorine-containing ions, the relative intensities of the two ions will be approximately 3:1 since this reflects the natural abundances of $^{35}$Cl and $^{37}$Cl.

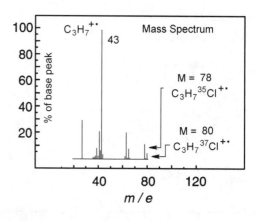

In the mass spectrum of 2-chloropropane, there are two molecular ions at 78 and 80 with intensities approximately in the ratio 3:1.  The peak at $m/e$ 78 corresponds to the molecular ions in the sample which contain $^{35}$Cl; the peak at $m/e$ 80 corresponds to the molecular ions in the sample which contain $^{37}$Cl.  Note that the base peak at $m/e$ 43 is only a single ion so this ion must contain no chlorine.  The pair of ions at $m/e$ 63 and 65 clearly corresponds to a fragment that still contains a chlorine atom.

Any molecular ion (or fragment) which contains 2 bromine atoms will have a pattern of ions M:M+2:M+4 with signals in the ratio 1:2:1 and any molecular ion (or fragment) which contains 2 chlorine atoms will have a pattern of M:M+2:M+4 with signals in the ratio 10:6.5:1 (Figure 4.2).

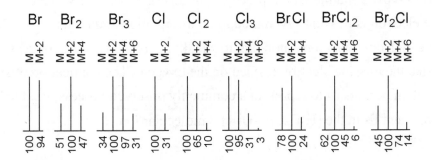

**Figure 4.2    Relative Intensities of the Cluster of Molecular Ions for Molecules Containing Combinations of Bromine and Chlorine Atoms**

*(4)    Chromatography coupled with Mass Spectrometry.*  It is now common to couple an instrument for separating a mixture of organic compounds *e.g.* using gas chromatography (GC) or high performance liquid chromatography (HPLC), directly to the input of a mass spectrometer.  In this way, as each individual compound is separated from the mixture, its mass spectrum can be recorded and compared automatically with the library of known compounds and identified immediately if it is a known compound.

*(5)    Metastable peaks* in a mass spectrum arise if the fragmentation process

$$a^+ \rightarrow b^+ + c \text{ (neutral)}$$

takes place within the ion-accelerating region of the mass spectrometer (Figure 4.1). Ion peaks corresponding to the masses of $a^+$ and to $b^+$ ($m_a$ and $m_b$) may be accompanied by a broader peak at mass $m^*$, such that:

$$m^* = \frac{m_b^2}{m_a}$$

The presence of metastable peaks in a mass spectrum often permits positive identification of a particular fragmentation path.

## 4.4    REPRESENTATION OF FRAGMENTATION PROCESSES

As fragmentation reactions in a mass spectrometer involve the breaking of bonds, they can be represented by the standard "arrow notation" used in organic chemistry.  For some purposes a radical cation (*e.g.* a generalised ion of the molecular ion) can be represented without attempting to localise the missing electron:

$$[M]^{+\cdot} \quad \text{or} \quad [H_3C\text{-}CH_2\text{-}O\text{-}R]^{+\cdot}$$

However, to show a fragmentation process it is generally necessary to indicate "from where the electron is missing" even though no information about this exists.  In the case of the molecular ion corresponding to an alkyl ethyl ether, it can be reasonably inferred that the missing electron resided on the oxygen.  The application of standard arrow notation permits us to represent a commonly observed process, *viz.* the loss of a methyl fragment from the $[H_3C\text{-}CH_2\text{-}O\text{-}R]^{+\cdot}$ molecular ion:

$$CH_3\text{-}CH_2\text{-}\overset{+}{\underset{\cdot\cdot}{O}}\text{-}R \longrightarrow \overset{\cdot}{C}H_3 + H_2C=\overset{+}{\overset{\cdot}{O}}\text{-}R \longleftrightarrow H_2\overset{+}{C}\text{-}O\text{-}R$$

## 4.5   FACTORS GOVERNING FRAGMENTATION PROCESSES

Three factors dominate the fragmentation processes:

*(a)*   **Weak bonds** tend to be broken most easily

*(b)*   **Stable fragments** (not only ions, but also the accompanying radicals and molecules) tend to be formed most readily

*(c)*   Some fragmentation processes depend on the ability of molecules to assume cyclic transition states.

Favourable fragmentation processes naturally occur more often and ions thus formed give rise to strong peaks in the mass spectrum.

## 4.6   EXAMPLES OF COMMON TYPES OF FRAGMENTATION

There are a number of common types of cleavage which are characteristic of various classes of organic compounds.  These result in the loss of well-defined fragments which are characteristic of certain functional groups or structural elements.

*(1)*   *Cleavage at Branch Points.*   Cleavage of aliphatic carbon skeletons at branch points is favoured as it leads to more substituted (and hence more stable) carbocations.  The mass spectrum of 2,2-dimethylpentane shows strong peaks at $m/e = 85$ and $m/e = 57$ where cleavage leads to the formation of stable tertiary carbocations.

$$CH_3-\underset{\underset{CH_3}{|}}{\overset{\overset{CH_3}{|}}{C}}-CH_2-CH_2-CH_3 \longrightarrow CH_3-\underset{\underset{CH_3}{|}}{\overset{\overset{CH_3}{|}}{\overset{+\bullet}{C}}}-CH_2-CH_2-CH_3 \quad m/e = 100$$

$$CH_3-\underset{\underset{CH_3}{|}}{\overset{\overset{CH_3}{|}}{\overset{+\bullet}{C}}}-CH_2-CH_2-CH_3 \longrightarrow \overset{\bullet}{C}H_3 \quad + \quad \underset{\underset{CH_3}{|}}{\overset{\overset{CH_3}{|}}{\overset{+}{C}}}-CH_2-CH_2-CH_3$$

neutral fragment    stable cation    $m/e = 85$

$$CH_3-\underset{\underset{CH_3}{|}}{\overset{\overset{CH_3}{|}}{\overset{+\bullet}{C}}}-CH_2-CH_2-CH_3 \longrightarrow CH_3-\underset{\underset{CH_3}{|}}{\overset{\overset{CH_3}{|}}{\overset{+}{C}}} \quad + \quad \overset{\bullet}{C}H_2-CH_2-CH_3$$

stable cation    $m/e = 57$    neutral fragment

*(2)*   *β - Cleavage.*  Chain cleavage tends to occur β to heteroatoms, double bonds and aromatic rings because relatively stable, delocalised carbocations result in each case.

*(a)*

$$R-\overset{\bullet\bullet}{X}-\overset{|}{\underset{|}{C}}-\overset{|}{\underset{|}{C}}- \xrightarrow{-e^-} R-\overset{+\bullet}{X}-\overset{|}{\underset{|}{C}}-\overset{|}{\underset{|}{C}}-$$

X = O, N, S, halogen

$$R-\overset{+}{X}-C \longleftrightarrow R-\overset{+}{X}=C \qquad \overset{\bullet}{C}-$$

resonance stabilised carbocation    neutral fragment

*(b)*

$$\underset{}{C}=C\overset{}{\underset{}{\diagdown}}\overset{|}{\underset{|}{C}}-\overset{|}{\underset{|}{C}}- \xrightarrow{-e^-} \overset{+}{C}-C\overset{|}{\underset{|}{C}}-\overset{|}{\underset{|}{C}}-$$

$$C=C\overset{+}{\underset{}{C}}- \longleftrightarrow \overset{+}{C}-C\overset{}{=}\underset{}{C}- \qquad \overset{\bullet}{C}-$$

resonance stabilised carbocation    neutral fragment

*(c)*

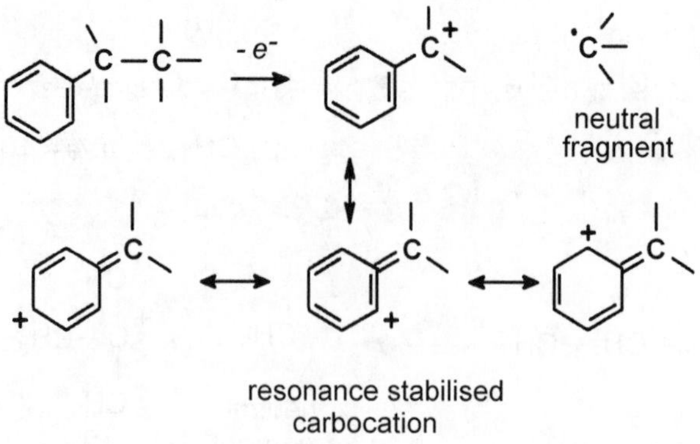

neutral fragment

resonance stabilised carbocation

**(3)  *Cleavage α to carbonyl groups.*** Cleavage tends to occur α to carbonyl groups to give stable acylium cations.  R may be an alkyl, -OH or -OR group.

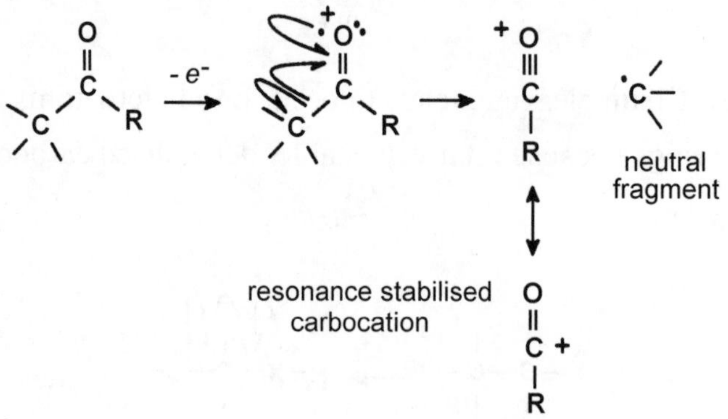

neutral fragment

resonance stabilised carbocation

**(4)  *Cleavage α to heteroatoms.*** Cleavage of chains may also occur α to heteroatoms, *e.g.* in the case of ethers:

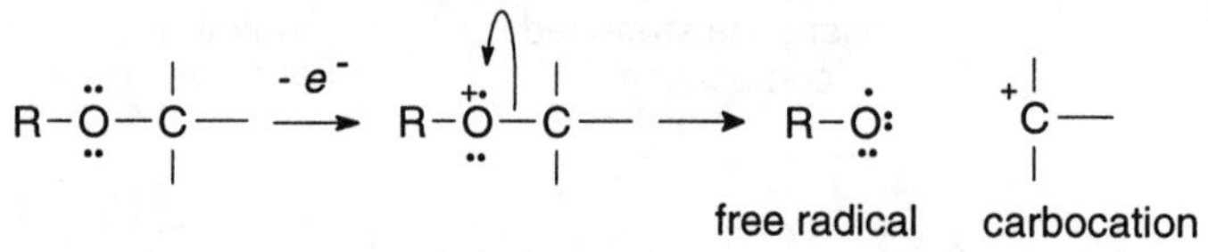

free radical    carbocation

**(5)  *Retro Diels-Alder reaction.*** Cyclohexene derivatives may undergo a retro Diels-Alder reaction:

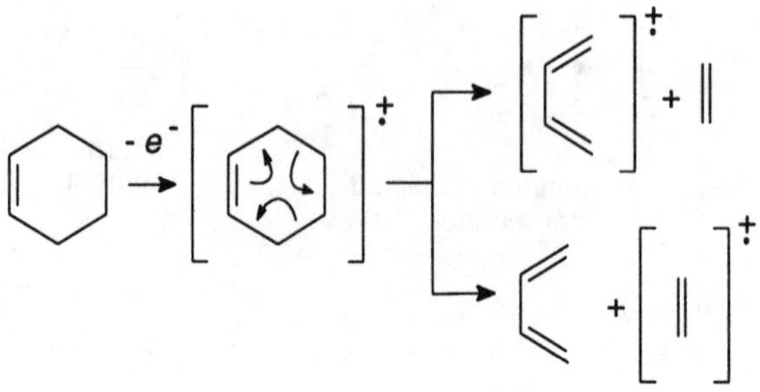

*(6)    The McLafferty rearrangement.*   Compounds where the molecular ion can assume the appropriate 6-membered cyclic transition state usually undergo a cyclic fragmentation, known as the **McLafferty rearrangement.**  This rearrangement involves a transfer of a γ hydrogen atom to an oxygen and is often observed with ketones, acids and esters:

With primary carboxylic acids, R-CH$_2$-COOH, this fragmentation leads to a characteristic peak at *m/e* = 60

$$\left[ H_2C = C - OH \atop OH \right]^{+}_{\bullet}$$

With carboxylic esters, two types of McLafferty rearrangements may be observed and ions resulting from either fragmentation pathway are commonly observed in the mass spectrum:

33

# 5

# NUCLEAR MAGNETIC RESONANCE (NMR) SPECTROSCOPY

## 5.1    THE PHYSICS OF NUCLEAR SPINS AND NMR INSTRUMENTS

### *(1)    The Larmor Equation and Nuclear Magnetic Resonance*

All nuclei have charge because they contain protons and some of them also behave as if they spin.  A spinning charge generates a magnetic dipole and is associated with a small magnetic field **H** (Figure 5.1).  Such nuclear magnetic dipoles are characterised by nuclear magnetic **spin quantum numbers** which are designated by the letter **I** and can take up values equal to 0, $^1/_2$, 1, $^3/_2$ ... *etc.*

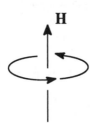

**Figure 5.1     A Spinning Positive Charge Generates a Magnetic Field and Behaves like a Small Magnet**

It is useful to consider three types of nuclei:

*Type 1*:    Nuclei with **I** = 0.  These nuclei do not interact with the applied magnetic field and **are not NMR chromophores**.  Nuclei with **I** = 0 have an even number of protons and even number of neutrons and have no net spin.  This means that nuclear spin is a property characteristic of certain **isotopes** rather than of certain elements.  The most prominent examples of nuclei with **I** = 0 are $^{12}C$ and $^{16}O$, the dominant isotopes of carbon and oxygen.  Both oxygen and carbon also have isotopes that can be observed by NMR spectroscopy.

*Organic Structures from Spectra*, Fifth Edition. L. D. Field, S. Sternhell and J. R. Kalman.
© 2013 John Wiley & Sons, Ltd. Published 2013 by John Wiley & Sons, Ltd.

***Type 2:***   Nuclei with $I = \frac{1}{2}$. These nuclei have a non-zero magnetic moment and are NMR visible and have no nuclear electric quadrupole (Q). The two most important nuclei for NMR spectroscopy belong to this category: $^1H$ (ordinary hydrogen) and $^{13}C$ (a non-radioactive isotope of carbon occurring to the extent of 1.06% at natural abundance). Also, two other commonly observed nuclei $^{19}F$ and $^{31}P$ have $I = \frac{1}{2}$. Together, NMR data for $^1H$ and $^{13}C$ account for well over 90% of all NMR observations in the literature and the discussion and examples in this book mostly refer to these two nuclei. However, the spectra of all nuclei with $I = \frac{1}{2}$ can be understood easily on the basis of common theory.

***Type 3:***   Nuclei with $I > \frac{1}{2}$. These nuclei have both a magnetic moment and an electric quadrupole. This group includes some common isotopes (*e.g.* $^2H$ and $^{14}N$) but they are more difficult to observe and spectra are generally very broad. This group of nuclei will not be discussed further.

The most important consequence of nuclear spin is that in a uniform magnetic field, a nucleus of spin $I$ may assume $2I + 1$ orientations. For nuclei with $I = \frac{1}{2}$, there are just 2 permissible orientations (since $2 \times \frac{1}{2} + 1 = 2$). These two orientations will be of unequal energy (by analogy with the parallel and antiparallel orientations of a bar magnet in a magnetic field) and it is possible to induce a spectroscopic transition (spin-flip) by the absorption of a quantum of electromagnetic energy ($\Delta E$) of the appropriate frequency ($\nu$):

$$\nu = \frac{\Delta E}{h}$$

$(5.1)$

In the case of NMR, the energy required to induce the nuclear spin flip also depends on the strength of the applied field, $H_o$. It is found that:

$$\nu = K H_o \qquad (5.2)$$

where $K$ is a constant characteristic of the nucleus observed. Equation 5.2 is known as the **Larmor equation** and is the fundamental relationship in NMR spectroscopy. Unlike other forms of spectroscopy, in NMR the frequency of the absorbed electromagnetic radiation is not an absolute value for any particular transition, but has a different value depending on the strength of the applied magnetic field. For every value of $H_o$, there is a matching value of $\nu$ corresponding to the condition of *resonance* according to Equation 5.2, and this is the origin of the term *"resonance"* in Nuclear Magnetic Resonance Spectroscopy. Thus for $^1H$ and $^{13}C$, *resonance* *frequencies* corresponding to magnitudes of applied magnetic field ($H_o$) commonly found in commercial instruments are given in Table 5.1.

**Table 5.1     Resonance Frequencies of $^1$H and $^{13}$C Nuclei in Magnetic Fields of Different Strengths**

| ν $^1$H (MHz) | ν $^{13}$C (MHz) | $H_o$ (Tesla) |
|:---:|:---:|:---:|
| 60 | 15.087 | 1.4093 |
| 90 | 22.629 | 2.1139 |
| 100 | 25.144 | 2.3488 |
| 200 | 50.288 | 4.6975 |
| 300 | 75.432 | 7.0462 |
| 400 | 100.577 | 9.3950 |
| 500 | 125.720 | 11.744 |
| 600 | 150.864 | 14.0923 |
| 750 | 188.580 | 17.616 |
| 800 | 201.154 | 18.790 |
| 900 | 226.296 | 21.128 |

In common jargon, NMR spectrometers are commonly known by the frequency they use to observe $^1$H *i.e.* as "60 MHz", "200 MHz" or "400 MHz" instruments, even if the spectrometer is set to observe a nucleus other than $^1$H.

All the frequencies listed in Table 5.1 correspond to the radio frequency region of the electromagnetic spectrum and inserting these values into Equation 5.1 gives the size of the energy gap between the states in an NMR experiment. A resonance frequency of 100 MHz corresponds to an energy gap of approximately $4 \times 10^{-5}$ kJ mol$^{-1}$. This is an extremely small value on the chemical energy scale and this means that NMR spectroscopy is, for all practical purposes, a ground-state phenomenon.

Any absorption signal observed in a spectroscopic experiment must originate from excess of the population in the lower energy state, the so called *Boltzmann excess*, which is equal to $N_\beta$-$N_\alpha$, where $N_\beta$ and $N_\alpha$ are the populations in the lower (β) and upper (α) energy states.

For molar quantities, the general Boltzmann relation (Equation 5.3) shows that:

$$\frac{N_\beta}{N_\alpha} = e^{\frac{\Delta E}{RT}}$$

(5.3)

Clearly, as the energy gap ($\Delta E$) approaches zero, the right hand side of Equation 5.3 approaches 1 and the Boltzmann excess becomes very small.  For the NMR experiment, the population excess in the lower energy state is typically of the order of 1 in $10^5$ which renders NMR spectroscopy an **inherently insensitive** spectroscopic technique.  Equations 5.1 and 5.2 show that the energy gap (and therefore ultimately the Boltzmann excess and sensitivity), increases with increasing applied magnetic field.  This is one of the reasons why it is desirable to use higher and higher magnetic fields in NMR spectrometers.

### (2)   *Nuclear Relaxation*

Even at the highest fields, the NMR experiment would not be practicable if mechanisms did not exist to restore the Boltzmann equilibrium that is perturbed as the result of the absorption of electromagnetic radiation in making an NMR measurement. These mechanisms are known by the general term of **relaxation** and are not confined to NMR spectroscopy.  Because of the small magnitude of the Boltzmann excess in the NMR experiment, relaxation is more critical and more important in NMR than in other forms of spectroscopy.

If relaxation is too efficient (*i.e.* it takes a short time for the nuclear spins to relax after being excited in an NMR experiment) the lines observed in the NMR spectrum are very broad.  If relaxation is too slow (*i.e.* it takes a long time for the nuclear spins to relax after being excited in an NMR experiment) the spins in the sample quickly *saturate* and only a very weak signal can be observed.

The most important relaxation processes in NMR involve interactions with other nuclear spins that are in the state of random thermal motion.  This is called *spin-lattice relaxation* and results in a simple exponential recovery process after the spins are disturbed in an NMR experiment.  The exponential recovery is characterised by a time constant $T_1$ that can be measured for different types of nuclei.  For organic liquids and samples in solution, $T_1$ is typically of the order of several seconds.  In the presence of paramagnetic impurities or in very viscous solvents, relaxation of the spins can be very efficient and NMR spectra obtained become broad.

Nuclei in solid samples typically relax very efficiently and give rise to very broad spectra.  NMR spectra of solid samples can only be acquired using specialised spectroscopic equipment and solid state NMR spectroscopy will not be discussed further.

### (3)   *The Acquisition of an NMR spectrum*

As the NMR phenomenon is not observable in the absence of an applied magnetic field, a magnet is an essential component of any NMR spectrometer.  Magnets for NMR may be permanent magnets (as in many low field routine instruments), electromagnets, or in most modern instruments they are based on superconducting solenoids, cooled by liquid helium.  All magnets used for NMR spectroscopy share the following characteristics:

*(a)*   The magnetic field must be **strong**.  This is partly due to the fact that the sensitivity of the NMR experiment increases as the strength of the magnet increases, but more importantly it ensures adequate **dispersion** of signals and, in the case of $^1$H NMR, also very important **simplification** of the spectrum.

*(b)*   The magnetic field must be extremely **homogeneous** so that all portions of the sample experience exactly the same magnetic field.  Any inhomogeneity of the magnetic field will result in broadening and distortion of spectral bands.  For determining of the structure of organic compounds, the highest attainable degree of magnetic field homogeneity is desirable, because useful information may be lost if the width of the NMR spectral lines exceeds about 0.2 Hz.  Clearly, 0.2 Hz in, say, 100 MHz implies a homogeneity of about 2 parts in $10^9$, and this is a very stringent requirement over the whole volume of an NMR sample.

*(c)*   The magnetic field must be very **stable,** so that it does not drift during the acquisition of the spectrum, which may take from several seconds to several hours.  This also means that there is a requirement for NMR instruments to be relatively isolated from sources of magnetic interference such as the movement of large metal objects (trucks, metal cylinders, heavy machinery, elevators *etc*).

## 5.2   CONTINUOUS WAVE (CW) NMR SPECTROSCOPY

Inspection of the Larmor equation (Equation 5.2) shows that for any nucleus the condition of resonance may be achieved by keeping the field constant and changing (or sweeping) the frequency or, alternatively, by keeping the frequency constant and sweeping the field.  A schematic diagram of a frequency sweep CW NMR spectrometer is given in Figure 5.2.

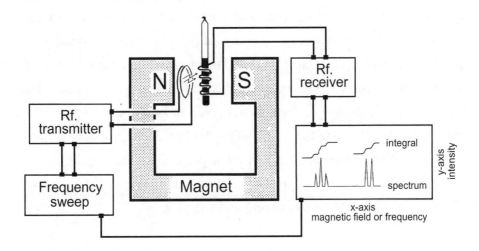

**Figure 5.2     Schematic Representation of a CW NMR Spectrometer**

An NMR spectrum is effectively a graph of the intensity of absorption of Rf radiation (y-axis) against the frequency of the Rf radiation (x-axis).  Since frequency and magnetic field strength are linked by the Larmor equation, the x-axis could also be calibrated in units of magnetic field strength.  In a CW NMR spectrometer, the x-axis of the output device (usually a pen plotter) is coupled to the frequency sweep so that the response of the sample is displayed as the frequency of the Rf transmitter varies.

NMR spectroscopy is a quantitative technique and [1]H NMR spectra are usually recorded with an integral which indicates the relative areas of the absorption peaks in the spectrum.  The area of a peak is proportional to the number of protons which give rise to the signal.  In most NMR spectrometers, the integral is represented as a horizontal line plotted over the spectrum.  Whenever a peak is encountered, the vertical displacement of the integral line is proportional to the area of the peak.  [1]H NMR spectroscopy is an excellent tool for the analysis of mixtures – if a sample contains more than one compound then the areas of the signals belonging to each species in the NMR spectrum will reflect the relative concentrations of the species in the mixture.

## 5.3   FOURIER-TRANSFORM (FT) NMR SPECTROSCOPY

As an alternative to the CW method, an intense short pulse of electromagnetic energy can be used to excite the nuclei in an NMR sample.  The first property of pulsed NMR spectroscopy is that all of the nuclei are excited simultaneously whereas the CW NMR experiment requires a significant period of time (usually several minutes) to sweep or scan through a range of frequencies.  Following the radiofrequency pulse, the magnetism in the sample is sampled as a function of time and, for a single resonance, the detected signal decays exponentially.  The detected signal is called a *free induction decay* or FID (Figure 5.3a) and this type of spectrum (known as a *time-domain* spectrum) is converted into the more usual *frequency-domain* spectrum (Figure 5.3b) by performing a mathematical operation known as *Fourier transformation (FT)*.  Because the signal needs mathematical processing, pulsed NMR spectrometers require a computer and as well as performing the Fourier transformation, the computer also provides a convenient means of storing NMR data and performing secondary data processing and analysis.

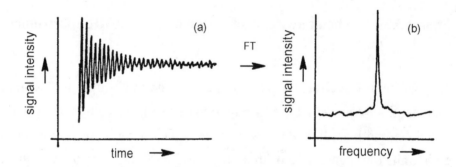

**Figure 5.3      Time Domain and Frequency Domain NMR Spectra**

Most NMR spectra consist of a number of signals and their time-domain spectra appear as a superposition of a number of traces of the type shown in Figure 5.3.  Such spectra are quite uninterpretable by inspection, but Fourier transformation converts them into ordinary frequency-domain spectra.  The time-scale of the FID experiment is of the order of seconds during which the magnetisation may be sampled many thousands of time.  Data sampling is accomplished by a dedicated computer that is also used to perform the Fourier transformation.

The principal advantage of FT NMR spectroscopy is a great *increase in sensitivity per unit time* of the experiment.  A CW scan generally takes of the order of one hundred times as long as the collection of the equivalent FID.  During the time it would have taken to acquire one CW spectrum, a computer can accumulate many FID scans and add them up in its memory.  The sensitivity (signal-to-noise ratio) of the NMR spectrum is proportional to the square root of the number of scans which are added together, so the quality of NMR spectra is vastly improved as more scans are added. It is the increase in sensitivity brought about by the introduction of FT NMR spectroscopy that has permitted the routine observation of $^{13}C$ NMR spectra.

Although it is possible to acquire many spectra in rapid succession using pulsed NMR methods, the speed with which multiple FIDs can be acquired is restricted by the fact that the nuclei in the sample need to relax between acquisitions (Section 5.1).  If successive FIDs are acquired too rapidly, intensity information will be distorted because those nuclei which relax slowly will not be fully relaxed when subsequent scans are acquired and they will contribute less to the resulting signal.  To ensure that the signal intensities are accurate, the repetition rate needs to be such that even any slowly relaxing nuclei in the sample are fully relaxed between scans.

In addition, the FID can be manipulated mathematically to enhance sensitivity (*e.g.* for routine $^{13}C$ NMR) at the expense of resolution, or to enhance resolution (often important for $^{1}H$ NMR) at the expense of sensitivity.  It is also possible to devise **sequences** of Rf pulses to extract specific information from the sample *e.g.* using two-dimensional NMR (see Section 7).

## 5.4    THE NUCLEAR OVERHAUSER EFFECT (NOE)

Irradiation of one nucleus while observing the resonance of another may result in a change in the **amplitude** of the observed resonance *i.e.* an enhancement of the signal intensity.  This is known as the *nuclear Overhauser effect* (NOE).  The NOE is a "through space" effect and its magnitude is inversely proportional to the sixth power of the distance between the interacting nuclei.  Because of the distance dependence of the NOE, it is an important method for establishing which groups are close together in space and because the NOE can be measured quite accurately it is a very powerful means for determining the three dimensional structure (and stereochemistry) of organic compounds.

The intensity of $^{13}C$ resonances may be increased by up to 200% when $^{1}H$ nuclei which are directly bonded to the carbon atom are irradiated.  This effect is very important in increasing the intensity of $^{13}C$ spectra when they are proton-decoupled. The efficiency of the proton/carbon NOE varies from carbon to carbon and this is a factor that contributes to the generally non-quantitative nature of $^{13}C$ NMR.  While the intensity of protonated carbon atoms can be increased significantly by NOE, non-protonated carbons (quaternary carbon atoms) receive little NOE and are usually the weakest signals in a $^{13}C$ NMR spectrum.

## 5.5   CHEMICAL SHIFT IN $^1$H NMR SPECTROSCOPY

It is clear that NMR spectroscopy can be used to readily detect certain nuclei (*e.g.* $^1H$, $^{13}C$, $^{19}F$, $^{31}P$) and, also to estimate them quantitatively.  The real usefulness of NMR spectroscopy in chemistry is based on secondary phenomena, the *chemical shift* and *spin-spin coupling* and, to a lesser extent, on effects related to the *time-scale* of the NMR experiment.  Both the chemical shift and spin-spin coupling reflect the **chemical environment** of the nuclear spins whose spin-flips are observed in the NMR experiment and these can be considered as chemical effects in NMR spectroscopy.

A $^1H$ NMR spectrum is a graph of resonance frequency (chemical shift) vs. the intensity of Rf absorption by the sample.  The spectrum is usually calibrated in dimensionless units called "parts per million" (abbreviated to ppm) although the horizontal scale is a frequency scale, the units are converted to ppm so that the scale has the same numbers **irrespective of the strength of the magnetic field** in which the measurement was made.  The scale in ppm, termed the $\delta$ scale, is usually referenced to the resonance of some standard substance whose frequency is chosen as 0.0 ppm.  The frequency difference between the resonance of a nucleus and the resonance of the reference compound is termed the **chemical shift**.

Tetramethylsilane, $(CH_3)_4Si$, (abbreviated commonly as TMS) is the usual reference compound chosen for both $^1H$ and $^{13}C$ NMR and it is normally added directly to the solution of the substance to be examined.  TMS has the following advantages as a reference compound:

*(a)*   it is a relatively inert low boiling (b.p. 26.5°C) liquid which can be easily removed after use;

*(b)*   it gives a sharp single signal in both $^1H$ and $^{13}C$ because the compound has only one type of hydrogen and one type of carbon;

*(c)*   the chemical environment of both carbon and hydrogen in TMS is unusual due to the presence of silicon and hence the TMS signal occurs outside the normal

range observed for organic compounds so the reference signal is unlikely to overlap a signal from the substance examined;

*(d)*    the chemical shift of TMS is not substantially affected by complexation or solvent effects because the molecule doesn't contain any polar groups.

Chemical shifts can be measured in Hz but are more usually expressed in ppm.

$$\text{chemical shift } (\delta) \text{ in ppm} = \frac{\text{chemical shift from TMS in Hz}}{\text{spectrometer frequency in MHz}}$$

Note that for a spectrometer operating at 200 MHz, 1 ppm corresponds to 200 Hz *i.e.* for a spectrometer operating at *x* MHz, 1.00 ppm corresponds to exactly *x* Hz.

For the majority of organic compounds, the chemical shift range for $^1$H covers approximately the range 0-10 ppm (from TMS) and for $^{13}$C covers approximately the range 0-220 ppm (from TMS). By convention, the $\delta$ scale runs (with increasing values) from right-to-left.

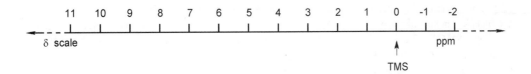

Each $^1$H nucleus is **shielded or screened** by the electrons that surround it. Consequently each nucleus feels the influence of the main magnetic field to a different extent, depending on the efficiency with which it is screened. Each $^1$H nucleus with a different chemical environment has a slightly different shielding and hence a different chemical shift in the $^1$H NMR spectrum. Conversely, the number of different signals in the $^1$H NMR spectrum reflects the number of chemically distinct environments for $^1$H in the molecule. Unless two $^1$H environments are precisely identical (by symmetry) *their chemical shifts must be different.* When two nuclei have identical molecular environments and hence the same chemical shift, they are termed *chemically equivalent* or *isochronous* nuclei. Non-equivalent nuclei that coincidentally have chemical shifts that are so close that their signals are indistinguishable are termed *accidentally equivalent* nuclei.

The chemical shift of a nucleus reflects the molecular structure and it can therefore be used to obtain structural information. Further, as hydrogen and carbon (and therefore $^1$H and $^{13}$C nuclei) are universal constituents of organic compounds the amount of structural information available from $^1$H and $^{13}$C NMR spectroscopy greatly exceeds in value the information available from other forms of molecular spectroscopy.

**Every hydrogen and carbon atom in an organic molecule is "a chromophore"** for NMR spectroscopy.

For $^1$H NMR, the intensity of the signal (which may be measured by electronically measuring the area under individual resonance signals) is directly proportional to the number of nuclei undergoing a spin-flip and **proton NMR spectroscopy is a quantitative method.**

Any effect which alters the density or spatial distribution of electrons around a $^1$H nucleus will alter the degree of shielding and hence its chemical shift. $^1$H chemical shifts are sensitive to both the hybridisation of the atom to which the $^1$H nucleus is attached ($sp^2$, $sp^3$ etc.) and to electronic effects (the presence of neighbouring electronegative/electropositive groups).

Nuclei tend to be deshielded by groups which withdraw electron density. Deshielded nuclei resonate at higher δ values (away from TMS). Conversely shielded nuclei resonate at lower δ values (towards TMS).

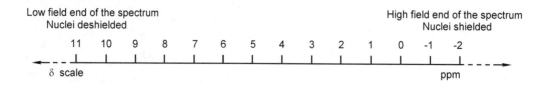

*Electron withdrawing substituents* (-OH, -OCOR, -OR, -NO$_2$, halogen) attached to an aliphatic carbon chain cause a **downfield shift** of 2-4 ppm when present at $C_\alpha$ and have less than half of this effect when present at $C_\beta$.

When $sp^2$ hybridised carbon atoms (carbonyl groups, olefinic fragments, aromatic rings) are present in an aliphatic carbon chain they cause a downfield shift of 1-2 ppm when present at $C_\alpha$. $Sp^2$ hybridised carbons have less than half of this effect when present at $C_\beta$.

Tables 5.2 and 5.3 give characteristic shifts for $^1$H nuclei in some representative organic compounds. Table 5.4 provides approximate ranges for proton chemical shifts in organic compounds. Figure 5.4 gives characteristic chemical shifts for protons in common alkyl derivatives.

Table 5.5 gives characteristic chemical shifts for the olefinic protons in common substituted alkenes. To a first approximation, the shifts induced by substituents attached to an alkene are additive. So, for example, an olefinic proton which is *trans* to a –CN group and has a geminal alkyl group will have a chemical shift of approximately 6.25 ppm [5.25 + 0.55(*trans*–CN) + 0.45(*gem*-alkyl)].

**Table 5.2      Typical $^1$H Chemical Shift Values in Selected Organic Compounds**

| Compound | δ $^1$H (ppm from TMS) |
|---|---|
| $CH_4$ | 0.23 |
| $CH_3Cl$ | 3.05 |
| $CH_2Cl_2$ | 5.33 |
| $CHCl_3$ | 7.27 |
| $CH_3CH_3$ | 0.86 |
| $CH_2=CH_2$ | 5.25 |
| benzene | 7.26 |
| $CH_3CHO$ | 2.20 (-$CH_3$), 9.80 (-CHO) |
| $CH_3CH_2CH_2Cl$ | 1.06 (-$CH_3$), 1.81(-$CH_2$-), 3.47(-$CH_2$-Cl) |

**Table 5.3      Typical $^1$H Chemical Shift Ranges in Organic Compounds**

| Group* | δ $^1$H (ppm from TMS) |
|---|---|
| **Tetramethylsilane** $(CH_3)_4Si$ | 0 |
| **Methyl groups** attached to $sp^3$ hybridised carbon atoms | 0.8 - 1.2 |
| **Methylene groups** attached to $sp^3$ hybridised carbon atoms | 1.0 - 1.5 |
| **Methine groups** attached to $sp^3$ hybridised carbon atoms | 1.2 - 1.8 |
| **Acetylenic protons** | 2 – 3.5 |
| **Olefinic protons** | 5 - 8 |
| **Aromatic and heterocyclic protons** | 6 - 9 |
| **Aldehydic protons** | 9 - 10 |

* –OH protons in alcohols, phenols or carboxylic acids; –SH protons in thiols; –NH protons in amines or amides do not have reliable chemical shift ranges (see page 51).

**Table 5.4**     $^{1}$H Chemical Shifts ($\delta$) for Protons in Common Alkyl Derivatives

| X | CH$_3$—X | CH$_3$CH$_2$—X | | (CH$_3$)$_2$CH—X | |
|---|---|---|---|---|---|
| | —CH$_3$ | —CH$_3$ | —CH$_2$— | —CH$_3$ | CH— |
| —H | 0.23 | 0.86 | 0.86 | 0.91 | 1.33 |
| —CH=CH$_2$ | 1.71 | 1.00 | 2.00 | 1.00 | 1.73 |
| —Ph | 2.35 | 1.21 | 2.63 | 1.25 | 2.89 |
| —Cl | 3.06 | 1.33 | 3.47 | 1.55 | 4.14 |
| —Br | 2.69 | 1.66 | 3.37 | 1.73 | 4.21 |
| —I | 2.16 | 1.88 | 3.16 | 1.89 | 4.24 |
| —OH | 3.39 | 1.18 | 3.59 | 1.16 | 3.94 |
| —OCH$_3$ | 3.24 | 1.15 | 3.37 | 1.08 | 3.55 |
| —O—Ph | 3.73 | 1.38 | 3.98 | 1.31 | 4.51 |
| —OCO—CH$_3$ | 3.67 | 1.21 | 4.05 | 1.22 | 4.94 |
| —OCO—Ph | 3.89 | 1.38 | 4.37 | 1.36 | 5.30 |
| —CO—CH$_3$ | 2.09 | 1.05 | 2.47 | 1.08 | 2.54 |
| —CO—Ph | 2.55 | 1.18 | 2.92 | 1.22 | 3.58 |
| —CO—OCH$_3$ | 2.01 | 1.12 | 2.28 | 1.15 | 2.48 |
| —NH$_2$ | 2.47 | 1.10 | 2.74 | 1.03 | 3.07 |
| —NH—COCH$_3$ | 2.71 | 1.12 | 3.21 | 1.13 | 4.01 |
| —C≡N | 1.98 | 1.31 | 2.35 | 1.35 | 2.67 |
| —NO$_2$ | 4.29 | 1.58 | 4.37 | 1.53 | 4.44 |

**Figure 5.4   Approximate $^1$H Chemical Shift Ranges for Protons in Organic Compounds**

Table 5.5      Approximate $^1$H Chemical Shifts ($\delta$) for Olefinic Protons
C=C-H

$\delta_{C=C-H} = 5.25 + \sigma_{gem} + \sigma_{cis} + \sigma_{trans}$

| X | $\sigma_{gem}$ | $\sigma_{cis}$ | $\sigma_{trans}$ |
|---|---|---|---|
| —H | 0.0 | 0.0 | 0.0 |
| —alkyl | 0.45 | -0.22 | -0.28 |
| —aryl | 1.38 | 0.36 | -0.07 |
| —CH=CH$_2$ | 1.00 | -0.09 | -0.23 |
| —CH=CH–conjugated | 1.24 | 0.02 | -0.05 |
| —C≡C–H | 0.47 | 0.38 | 0.12 |
| —CO–R | 1.10 | 1.12 | 0.87 |
| —CO–OH | 0.80 | 0.98 | 0.32 |
| —CO–OR | 0.78 | 1.01 | 0.46 |
| —C≡N | 0.27 | 0.75 | 0.55 |
| —Cl | 1.08 | 0.18 | 0.13 |
| —Br | 1.07 | 0.45 | 0.55 |
| —OR | 1.22 | -1.07 | -1.21 |
| —NR$_2$ | 0.80 | -1.26 | -1.21 |

Table 5.6 gives characteristic $^1$H chemical shifts for the aromatic protons in benzene derivatives.  To a first approximation, the shifts induced by substituents are additive.  So, for example, an aromatic proton which has a –NO$_2$ group in the *para* position and a –Br group in the *ortho* position will appear at approximately 7.82 ppm [(7.26 + 0.38(*p*-NO$_2$) + 0.18(*o*-Br))].

Table 5.7 gives characteristic chemical shifts for $^1$H nuclei in some polynuclear aromatic compounds and heteroaromatic compounds.

**Table 5.6**   Approximate $^1$H Chemical Shifts ($\delta$) for Aromatic Protons in Benzene Derivatives Ph-X in ppm Relative to Benzene at $\delta$ 7.26 ppm (positive sign denotes a downfield shift)

| X | ortho | meta | para |
|---|-------|------|------|
| —H | 0.0 | 0.0 | 0.0 |
| —CH$_3$ | -0.20 | -0.12 | -0.22 |
| —C(CH$_3$)$_3$ | -0.03 | -0.08 | 0.20 |
| —CH=CH$_2$ | 0.06 | -0.03 | -0.10 |
| —C≡C—H | 0.16 | -0.04 | -0.02 |
| —CO—OR | 0.71 | 0.11 | 0.21 |
| —CO—R | 0.62 | 0.14 | 0.21 |
| —OCO—R | -0.25 | 0.03 | -0.13 |
| —OCH$_3$ | -0.48 | -0.09 | -0.44 |
| —OH | -0.56 | -0.12 | -0.45 |
| —Cl | 0.03 | -0.02 | -0.09 |
| —Br | 0.18 | -0.08 | -0.04 |
| —C≡N | 0.36 | 0.18 | 0.28 |
| —NO$_2$ | 0.95 | 0.26 | 0.38 |
| —NR$_2$ | -0.66 | -0.18 | -0.67 |
| —NH$_2$ | -0.75 | -0.25 | -0.65 |

**Table 5.7**   $^1$H Chemical Shifts ($\delta$) in some Polynuclear Aromatic Compounds and Heteroaromatic Compounds

The chemical shift of a nucleus may also be affected by the presence in its vicinity of a *magnetically anisotropic* group (*e.g.* an aromatic ring or carbonyl group). In an aromatic ring, the "circulation" of electrons effectively forms a current loop which gives rise to an induced magnetic field. This is called the **ring current effect** and the induced field opposes the applied magnetic field of the spectrometer ($B_0$) inside the loop and enhances the field outside the loop. The resonance of a nucleus which is located close to the face of an aromatic ring will be shifted to high field (towards TMS) because it experiences the effect of both the main spectrometer magnetic field but also the magnetic field from the ring current effect of the aromatic ring. Conversely a proton which is in the plane of an aromatic ring is deshielded by the ring current effect.

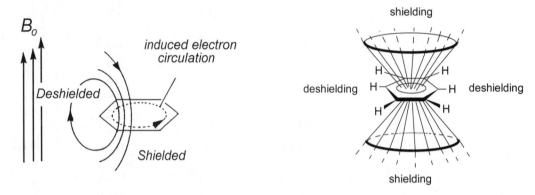

The ring current effect is the main reason that protons attached to aromatic rings typically appear at the low field end of the $^1H$ NMR spectrum since they are in the deshielded zone of the aromatic ring.

There are also a number of common non-aromatic organic functional groups which are magnetically anisotropic and influence the magnetic field experienced by nearby nuclei. The greatest influence comes from multiple bonds and in particular, the C≡C group, the C≡N group, and C=C, N=O and C=O groups have strong magnetic anisotropies. Figure 5.5 depicts the shielding and de-shielding zones around common non-aromatic functional groups

Shielding effects diminish with distance but are useful qualitative indicators of what groups are close by and also their geometric relationship in the three-dimensional structure of the molecule.

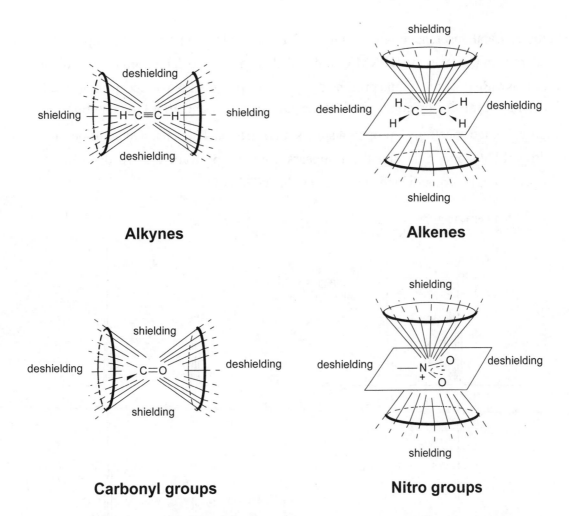

**Figure 5.5    Shielding/deshielding Zones for Common Non-aromatic Functional Groups**

**Labile and Exchangeable protons**.  Protons in groups such as alcohols (R-OH) amines (R-NH-), carboxylic acids (RCOOH), thiols (R-SH) and to a lesser extent amides (R-CO-NH-) are classified as labile or readily exchangeable protons.

Labile protons frequently give rise to broadened resonances in the $^1$H NMR spectrum and their chemical shifts are critically dependent on the solvent, concentration, and on temperature and *they do not have reliable characteristic chemical shift ranges*.

Labile protons exchange rapidly with each other and also with protons in water or with the deuterons in $D_2O$.

$$R-O-H + D_2O \rightleftharpoons R-O-D + H-O-D$$

Labile protons can always be positively identified by *in situ* exchange with $D_2O$.  In practice, a normal $^1$H NMR spectrum is recorded then deuterium exchange of labile protons is achieved by simply adding a drop of deuterated water ($D_2O$) to the NMR sample.  Labile protons in -OH, -COOH, -NH₂ and -SH groups exchange rapidly for deuterons in D₂O and the ¹H NMR is recorded again.  Since deuterium is invisible in the ¹H NMR spectrum, labile protons disappear from the ¹H NMR spectrum and can be readily identified by comparison of the spectra before and after D₂O addition.  In Figure 5.6, the spectrum 1-propanol shows the expected 4 signals.  On addition of 1 drop of D₂O, one of the signals disappears from the spectrum, clearly identifying this as the –OH proton which exchanges with added D₂O.

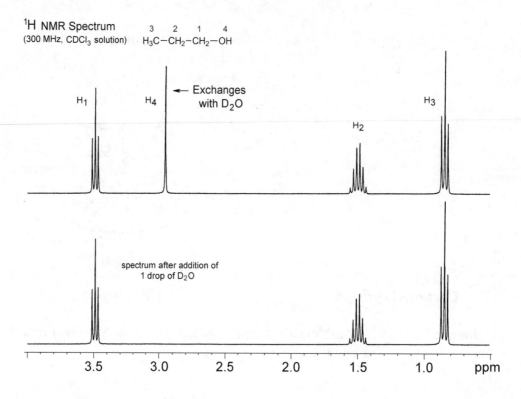

**Figure 5.6     D₂O exchange in the ¹H NMR Spectrum of 1-Propanol**

The N-H protons of primary and secondary amides are slow to exchange and require heating or base catalysis and this is one way an amide functional group can be distinguished from other functional groups.

## 5.6    SPIN-SPIN COUPLING IN $^1$H NMR SPECTROSCOPY

A typical organic molecule contains more than one magnetic nucleus (*e.g.* more than one $^1$H, or $^1$H and $^{31}$P *etc.*).  When one nucleus can sense the presence of other nuclei *through the bonds of the molecule* the signals will exhibit fine structure (*splitting or multiplicity*).  Multiplicity arises because if an observed nucleus can sense the presence of other nuclei with magnetic moments, those nuclei could be in either the $\alpha$ or $\beta$ state.  The observed nucleus is either slightly stabilised or slightly destabilised, depending on which state the remote nuclei are in, and as a consequence nuclei which sense coupled partners with an $\alpha$ state have a slightly different energy to those which sense coupled partners with a $\beta$ state.

The additional fine structure caused by spin-spin coupling is not only the principal cause of difficulty in interpreting $^1$H NMR spectra, but also provides valuable structural information when correctly interpreted.  The **coupling constant** (related to the size of the splittings in the multiplet) is given the symbol $J$ and is measured in Hz.  By convention, a superscript before the symbol '$J$' represents the number of intervening bonds between the coupled nuclei.  Labels identifying the coupled nuclei are usually indicated as subscripts after the symbol '$J$' *e.g.* $^2J_{ab} = 2.7$ Hz would indicate a coupling of 2.7 Hz between nuclei $a$ and $b$ which are separated by two intervening bonds.

Because $J$ depends only on the number, type and spatial arrangement of the bonds separating the two nuclei, it is a property of the molecule and is **independent of the applied magnetic field**.  The magnitude of $J$, or even the mere presence of detectable interaction, constitutes valuable structural information.

Two important observations that relate to $^1$H - $^1$H spin-spin coupling:

*(a)*    No **inter-molecular** spin-spin coupling is observed.  Spin-spin coupling is transmitted through the bonds of a molecule and doesn't occur between nuclei in different molecules.

*(b)*    The effect of coupling falls off as the number of bonds between the coupled nuclei increases.  $^1$H - $^1$H coupling is generally unobservable across more than 3 intervening bonds.  Unexpectedly large couplings across many bonds may occur if there is a particularly favourable bonding pathway *e.g.* extended $\pi$-conjugation or a particularly favourable rigid $\sigma$-bonding skeleton (Table 5.8).

**Table 5.8**     **Typical $^1H - {}^1H$ Coupling Constants**

| Group | $J$(Hz) |
|-------|---------|
| $CH_3CH_2CH_2CH_3$ | $^2J_{HH} \approx -16$ |
| $CH_3CH_2CH_2CH_3$ | $^3J_{HH} = 7.2$ |
| $CH_3CH_2CH_2CH_3$ | $^4J_{HH} = 0.3$ |
| $H_2C=C=C=CH_2$ | $^5J_{HH} = 7$ |
| $H_2C=CH-CH=CH_2$ | $^5J_{HH} = 1.3$ |
| H—⟨structure⟩—H | $^4J_{HH} = 1.5$ |

***Signal Multiplicity - the n+1 rule.***  Spin-spin coupling gives rise to multiplet splittings in $^1H$ NMR spectra.  The NMR signal of a nucleus coupled to ***n*** equivalent hydrogens will be split into a multiplet with (***n***+1) lines.  For simple multiplets, the spacing between the lines (in Hz) is the coupling constant.  The relative intensity of the lines in multiplets will be given by the binomial coefficients of order '***n***' (Table 5.9).

**Table 5.9**     **Relative Line Intensities for Simple Multiplets**

| ***n*** | multiplicity ***n+1*** | relative line intensities | multiplet name |
|---------|------------------------|---------------------------|----------------|
| 0 | 1 | 1 | singlet |
| 1 | 2 | 1 : 1 | doublet |
| 2 | 3 | 1 : 2 : 1 | triplet |
| 3 | 4 | 1 : 3 : 3 : 1 | quartet |
| 4 | 5 | 1 : 4 : 6 : 4 : 1 | quintet |
| 5 | 6 | 1 : 5 : 10 : 10 : 5 : 1 | sextet |
| 6 | 7 | 1 : 6 : 15 : 20 : 15 : 6 : 1 | septet |
| 7 | 8 | 1 : 7 : 21 : 35 : 35 : 21 : 7 : 1 | octet |
| 8 | 9 | 1 : 8 : 28 : 56 : 70 : 56 : 28 : 8 : 1 | nonet |

These simple multiplet patterns give rise to characteristic "fingerprints" for common fragments of organic structures.  A methyl group, -CH$_3$, (isolated from coupling to other protons in the molecule) will always occur as a singlet.  A CH$_3$-CH$_2$- group, (isolated from coupling to other protons in the molecule) will appear as a quartet (-CH$_2$-) and a triplet (CH$_3$-).  Figure 5.7 shows the schematic appearance of the NMR spectra of various common molecular fragments encountered in organic molecules.

**Figure 5.7     Characteristic Multiplet Patterns for Common Organic Fragments**

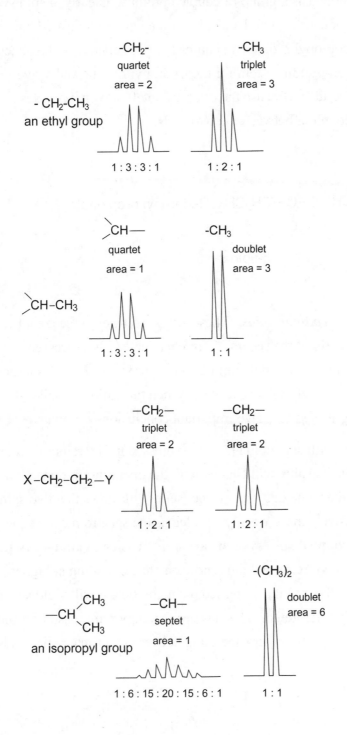

## 5.7   ANALYSIS OF $^1$H NMR SPECTRA

To obtain structurally useful information from NMR spectra, one must solve two separate problems.  Firstly, the spectrum must be **analysed** to obtain the NMR parameters (chemical shifts and coupling constants) for all the protons and, secondly, the values of the coupling constants must be **interpreted** in terms of established relationships between the parameters and structure.

*(1)*      *A spin system* is defined as a group of coupled protons.  Clearly, a spin system cannot extend beyond the bounds of a molecule, but it may not include a whole molecule.  For example, isopropyl propionate comprises **two** separate and isolated proton spin systems, a seven-proton system for the isopropyl residue and a five-proton system for the propionate residue, because the ester group effectively provides a barrier (5 bonds) against coupling between the two parts.

*(2)*      *Strongly and weakly coupled spins.*  These terms refer not to the actual magnitude of *J*, but to the **ratio** of the separation of chemical shifts expressed in Hz ($\Delta v$) to the coupling constant *J* between them.  For most purposes, if $\Delta v/J$ is larger than ~3, the spin system is termed *weakly coupled.*  When this ratio is smaller than ~3, the spins are termed *strongly coupled.*  Two important conclusions follow:

*(a)*   Because the chemical shift separation ($\Delta v$) is expressed in Hz, rather than in the dimensionless $\delta$ units, its value will change with the operating frequency of the spectrometer, while the value of *J* remains constant.  It follows that two spins will become progressively more weakly coupled as the spectrometer frequency increases.  **Weakly coupled spin systems are much easier to analyse than strongly coupled spin systems** and thus spectrometers operating at higher frequencies (and therefore at higher applied magnetic fields) will yield spectra which are more easily interpreted.  This has been an important reason for the development of NMR spectrometers operating at ever higher magnetic fields.

*(b)*   Within a spin system, some pairs of nuclei or groups of nuclei may be strongly coupled and others weakly coupled.  Thus a spin-system may be *partially strongly coupled.*

*(3)   Magnetic equivalence.*  A group of protons is *magnetically equivalent* when they not only have the same chemical shift (chemical equivalence) but also have identical spin-spin coupling to each individual nucleus **outside** the group.

*(4)   Conventions used in naming spin systems.*  Consecutive letters of the alphabet (*e.g.* A, B, C D, .....) are used to describe groups of protons which are strongly coupled.  Subscripts are used to give the number of protons that are magnetically equivalent.  Primes are used to denote protons that are chemically equivalent but not magnetically equivalent.  A break in the alphabet indicates weakly coupled groups.  For example:

> ABC denotes a strongly coupled 3-spin system
>
> AMX denotes a weakly coupled 3-spin system
>
> ABX denotes a *partially* strongly coupled 3-spin system
>
> $A_3$BMXY denotes a spin system in which the three magnetically equivalent A nuclei are strongly coupled to the B nucleus, but weakly coupled to the M, X and Y nuclei.  The nucleus X is strongly coupled to the nucleus Y but weakly coupled to all the other nuclei.  The nucleus M is weakly coupled to all the other 6 nuclei.
>
> AA'XX' is a 4-spin system described by two chemical shift parameters (for the nuclei A and X) but where $J_{AX} \neq J_{AX'}$.  A and A' (as well as X and X') are pairs of nuclei which are chemically equivalent but magnetically non-equivalent.

The process of deriving the NMR parameters ($\delta$ and *J*) from a set of multiplets in a spin system is known as *the analysis of the NMR spectrum*.  In principle, **any** spectrum arising from a spin system, however complicated, can be analysed but some will require calculations or simulations performed by a computer.

Fortunately, in a very large number of cases, multiplets can be correctly analysed by inspection and direct measurements.  These spectra are known as *first order spectra* and **they arise from weakly coupled spin systems.**  At high applied magnetic fields, a large proportion of ¹H NMR spectra are nearly pure first-order and there is a tendency for simple molecules, *e.g.* those exemplified in the problems in this text, to exhibit first-order spectra even at moderate fields.

## 5.8    CHANGING THE MAGNETIC FIELD IN NMR SPECTROSCOPY

Changing the magnetic field of an NMR spectrometer *i.e.* recording the NMR spectrum of a sample on NMR instruments operating at different magnetic fields, highlights several important aspects of NMR spectroscopy.  The chemical shifts of the nuclei (expressed in Hz) are proportional to the strength of the magnetic field (see Section 5.1).  The higher the magnetic field, the higher the resonance frequency of each nucleus in the sample.  Nuclei move further apart (in Hz) from each other as the magnetic field strength increases *i.e.* there is better dispersion in the spectrum.

On the other hand, spin-spin coupling is a molecular property that is independent of the magnetic field of the spectrometer.  So the splittings in an NMR spectrum due to spin-spin coupling, remain constant, irrespective of the magnetic field strength of the spectrometer.

Given below are $^1$H NMR spectra of 2-bromotoluene where spectra on the same sample have been recorded in 3 different NMR spectrometers each operating at a different magnetic field strength.

There are 4 different aromatic proton resonances.  At the lowest field (4.7 Tesla), there is significant overlap of the resonances.  The resonances for H3 and H6 are severely distorted and this is not a 1$^{st}$ order spectrum.  At 14.1 Tesla (600 MHz) dispersion is much better and the spectrum is a 1$^{st}$ order spectrum which could be analysed by 1$^{st}$ order rules.

Aromatic region of the $^1$H NMR spectrum of 2-bromotoluene (acetone-$d_6$ solution).

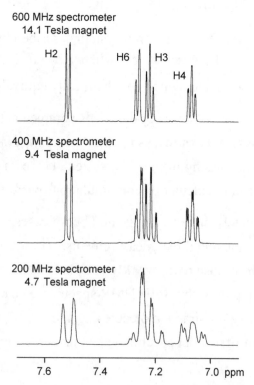

## 5.9    RULES FOR SPECTRAL ANALYSIS OF FIRST ORDER SPECTRA

**Rule 1**    A group of $n$ magnetically equivalent protons will split a resonance of an interacting group of protons into $n+1$ lines.  For example, the resonance due to the A protons in an $A_nX_m$ system will be split into $m+1$ lines, while the resonance due to the X protons will be split into $n+1$ lines.  More generally, splitting by n nuclei of spin quantum number I, results in $2n\mathbf{I}+1$ lines.  This simply reduces to $n+1$ for protons where $\mathbf{I} = \frac{1}{2}$.

**Rule 2**    The spacing (measured in Hz) of the lines in the multiplet will be equal to the coupling constant.  In the above example, all spacings in both parts of the spectrum will be equal to $J_{AX}$.

**Rule 3**    The true chemical shift of each group of interacting protons lies in the centre of the (always symmetrical) multiplet.

**Rule 4**    The relative intensities of the lines within each multiplet will be in the ratio of the binomial coefficients (Table 5.9).  Note that, in the case of higher multiplets, the outside components of multiplets are relatively weak and may be lost in the instrumental noise, *e.g.* a septet may appear as a quintet if the outer lines are not clearly visible.  The intensity relationship is the first to be significantly distorted in non-ideal cases, but this does not lead to serious errors in spectral analysis.

**Rule 5**    When a group of magnetically equivalent protons interacts with more than one group of protons, its resonance will take the form of a *multiplet of multiplets*.  For example, the resonance due to the A protons in a system $A_nM_pX_m$ will have the multiplicity of $(p+1)(m+1)$.  The multiplet patterns are chained *e.g.* a proton coupled to 2 different protons will be split to a doublet by coupling to the first proton then each of the component of the doublet will be split further by coupling to the second proton resulting in a symmetrical multiplet with 4 lines (a doublet of doublets).

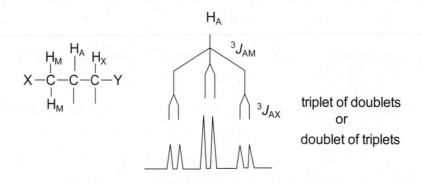

The appropriate coupling constants will control splitting and relative intensities will obey rule 4.

**Rule 6**   Protons that are magnetically equivalent do not split each other. Any system $A_n$ will give rise to a singlet.

**Rule 7**   Spin systems that contain groups of chemically equivalent protons that are not magnetically equivalent **cannot be analysed by first-order methods**.

**Rule 8**   If $\Delta v_{AB}/J_{AB}$ is less than ~3, for **any** pair of nuclei A and B in the spin system, the spectra become distorted from the expected ideal multiplet patterns and the spectra **cannot be analysed by first-order methods**.

## *(1)  Splitting Diagrams*

The knowledge of the rules listed above, permits the development of a simple procedure for the analysis of any spectrum which is suspected of being first order. The first step consists of drawing a *splitting diagram*, from which the line spacings can be measured and identical (hence related) splittings can be identified (Figure 5.8).

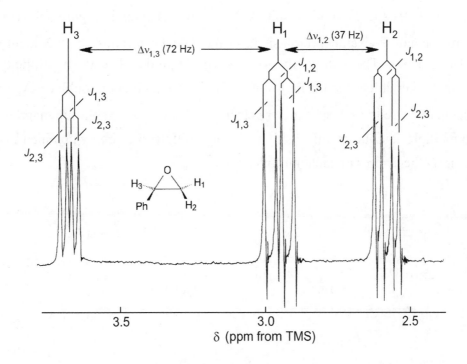

**Figure 5.8    A Portion of the $^1$H NMR Spectrum of Styrene Epoxide (100 MHz as a 5% solution in CCl$_4$)**

The section of the spectrum of styrene epoxide (Figure 5.8) clearly contains the signals from 3 separate protons (identified as H$_1$, H$_2$ and H$_3$) with H$_1$ at δ 2.95, H$_2$ at δ 2.58 and H$_3$ at δ 3.67 ppm.  Each signal appears as a doublet of doublets and the chemical shift of each proton is simply obtained by locating the centre of each multiplet.  The pair of nuclei giving rise to each splitting is clearly indicated by the splitting diagram above each multiplet with $^2J_{H1-H2}$ = 5.9 Hz, $^3J_{H1-H3}$ = 4.0 Hz and $^3J_{H2-H3}$ = 2.5 Hz.

The validity of a first order analysis can be verified by calculating the ratio Δν/$J$ for each pair of nuclei and establishing that it is greater than 3.

61

From Figure 5.8

$$\frac{\Delta v_{12}}{J_{12}} = \frac{37}{5.9} = 6.3 \qquad \frac{\Delta v_{13}}{J_{13}} = \frac{72}{4.0} = 18.0 \qquad \frac{\cdot \Delta v_{23}}{J_{23}} = \frac{109}{2.5} = 43.6$$

Each ratio is greater than 3 so a first order analysis is justified and the 100 MHz spectrum of the aliphatic protons of styrene oxide is indeed a first order spectrum and could be labelled as an AMX spin system.

The 60 MHz $^1$H spectrum of a 4 spin AMX$_2$ system is given in Figure 5.9. This system contains 3 separate proton signals (in the intensity ratios 1 : 1 : 2, identified as $H_A$, $H_M$ and $H_X$). The multiplicity of $H_A$ is a triplet of doublets, the multiplicity of $H_M$ is a triplet of doublets and the multiplicity of $H_X$ is a doublet of doublets. Again, the nuclei giving rise to each splitting are clearly indicated by the splitting diagram above each multiplet and the chemical shifts of each multiplet are simply obtained by measuring the centres of each multiplet.

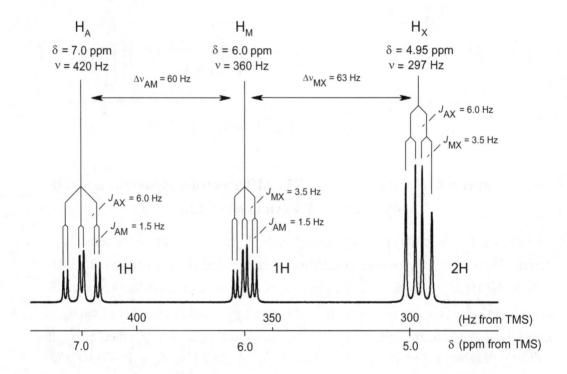

**Figure 5.9    The 60 MHz $^1$H NMR Spectrum of a 4-Spin AMX$_2$ Spin System**

A spin system comprising just two protons (*i.e.* an AX or an AB system) is always exceptionally easy to analyse because, independent of the value of the ratio of $\Delta v/J$, the spectrum always consists of just four lines with each pair of lines separated by the coupling constant $J$.  The only distortion from the first-order pattern consists of the gradual reduction of intensities of the outer lines in favour of the inner lines, a characteristic "sloping" or "tenting" towards the coupling partner.  A series of simulated spectra of two-spin systems are shown in Figure 5.10.

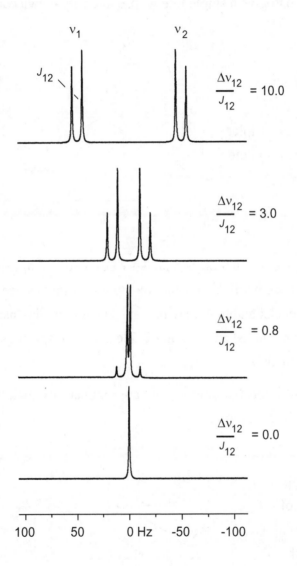

**Figure 5.10    Simulated $^1$H NMR Spectra of a 2-Spin System as the Ratio $\Delta v/J$, is Varied from 10.0 to 0.0**

*Coincidentally equal values of coupling constants.*

**An AMX spin system.**  First order analysis rules predict that the resonance for $H_M$ in an AMX spin system will be a doublet of doublets (4 lines) since $H_M$ will be split by coupling to $H_A$ and to $H_X$.  All lines of the multiplet will have equal intensity and the spacings in the multiplet will be $J_{AM}$ and $J_{AX}$.

However, in the case where $J_{AM}$ and $J_{AX}$ are approximately equal, the central lines of the multiplet overlap to give a single line whose intensity is twice as high as the outer lines.

While the multiplet is technically a doublet of doublets, it appears as a triplet (3 lines) with intensities in the ratio 1:2:1.

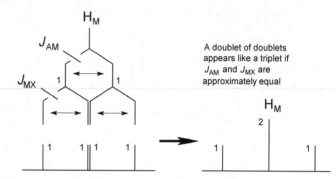

A doublet of doublets appears like a triplet if $J_{AM}$ and $J_{MX}$ are approximately equal

**An $A_2M_2X_2$ spin system.**  First order analysis rules predict that the resonance for $H_M$ in an $A_2M_2X_2$ spin system will be a triplet of triplets (9 lines) since $H_M$ will be split by coupling to 2 x $H_A$ nuclei and to 2 x $H_X$ nuclei.  The relative intensities of the lines in the multiplet can be predicted easily using Table 5.9.  The spacings in the multiplet will be equal to $J_{AM}$ and $J_{AX}$.

However, in the case where $J_{AM}$ and $J_{AX}$ are approximately equal, there is overlap between the lines of the multiplet.

While the multiplet is technically a triplet of triplets, it appears as a quintet (5 lines) with intensities in the ratio 1:4:6:4 1.

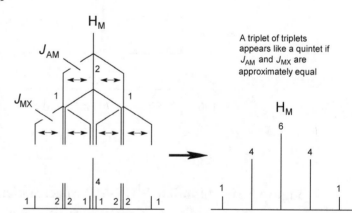

A triplet of triplets appears like a quintet if $J_{AM}$ and $J_{MX}$ are approximately equal

This is not an uncommon situation in flexible alkyl chains ($X\text{-}CH_2\text{-}CH_2\text{-}CH_2\text{-}Y$) since the 3-bond vicinal coupling between protons on adjacent carbons typically falls within a narrow range of about 6-8 Hz.

## *(2) Spin Decoupling*

In the signal of a proton that is a multiplet due to spin-spin coupling, it is possible to remove the splitting effects by irradiating the sample with an additional Rf source at the exact resonance frequency of the proton giving rise to the splitting.  The additional radiofrequency causes rapid flipping of the irradiated nuclei and as a consequence nuclei coupled to them cannot sense them as being in either an α or β state for long enough to cause splitting.  The irradiated nuclei are said to be **decoupled** from other nuclei in the spin system.  Decoupling simplifies the appearance of complex multiplets by removing some of the splittings.  In addition, decoupling is a powerful tool for assigning spectra because the skilled spectroscopist can use a series of decoupling experiments to sequentially identify which nuclei are coupled.

In a 4-spin $AM_2X$ spin system, the signal for proton $H_A$ would appear as a doublet of triplets (with the triplet splitting due to coupling to the 2 M protons and the doublet splitting due to coupling to the X proton).  Irradiation at the frequency of $H_X$ reduces the multiplicity of the A signal to a triplet (with the remaining splitting due to $J_{AM}$) and irradiation at the frequency of $H_M$ reduces the multiplicity of the A signal to a doublet (with the remaining splitting due to $J_{AX}$) (Figure 5.11).

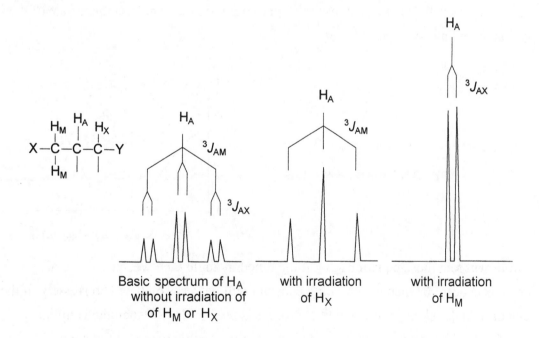

**Figure 5.11    Selective Decoupling in a Simple 4-Spin System**

## 5.10  CORRELATION OF $^1$H – $^1$H COUPLING CONSTANTS WITH STRUCTURE

Interproton spin-spin coupling constants are of obvious value in obtaining structural data about a molecule, in particular information about the connectivity of structural elements and the relative disposition of various protons.

### Non-aromatic Spin Systems.

In saturated systems, the magnitude of the *geminal* coupling constant $^2J_{H-C-H}$ (two protons attached to the same carbon atom) is typically between 10 and 16 Hz but values between 0 and 22 Hz have been recorded in some unusual structures.

$$^2J_{AB} = 10 \text{ -}16 \text{ Hz.}$$

The *vicinal* coupling (protons on adjacent carbon atoms) $^3J_{H-C-C-H}$ can have values 0 - 16 Hz depending mainly on the dihedral angle $\phi$.

The **Karplus relationship** expresses, **approximately**, the angular dependence of the vicinal coupling constant as:

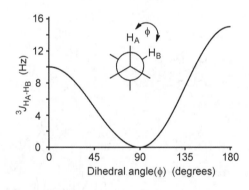

$^3J_{H-C-C-H} = 10 \cos^2 \phi \quad$ for $0 < \phi < 90°$ and

$^3J_{H-C-C-H} = 15 \cos^2 \phi \quad$ for $90 < \phi < 180°$

It follows from these equations that if the dihedral angle $\phi$ between two vicinal protons is near 90° then the coupling constant will be very small and conversely, if the dihedral angle $\phi$ between two vicinal protons is near 0° or 180° then the coupling constant will be relatively large.  The Karplus relationship is of great value in determining the stereochemistry of organic molecules but must be treated with some caution because vicinal coupling constants also depend markedly on the nature of substituents.  In systems that assume an average conformation, such as a flexible hydrocarbon chain, $^3J_{H-H}$ generally lies between 6 and 8 Hz.

In conformationally rigid systems, such as substituted cyclohexanes, there can be pronounced differences in $^3J_{\text{H-H}}$. The axial-axial coupling ($^3J_{\text{H1-H3}}$) between vicinal protons in cyclohexanes, where the dihedral angle is near 180°, is typically large (about 13 Hz). The axial-equatorial and equatorial-equatorial couplings where the dihedral angles are typically closer to 60°, are much smaller, typically in the range of 3-6 Hz.

The coupling constants in unsaturated (olefinic) systems depend on the nature of the substituents attached to the C=C but for the vast majority of substituents, the ranges for $^3J_{\text{H-C=C-H}(cis)}$ and $^3J_{\text{H-C=C-H}(trans)}$ do not overlap. This means that the stereochemistry of the double bond can be determined by measuring the coupling constant between vinylic protons. Where the C=C bond is in a ring, the $^3J_{\text{H-C=C-H}}$ coupling reflects the ring size.

$^3J_{\text{AB}(cis)}$ = 6 - 11 Hz

$^3J_{\text{AC}(trans)}$ = 12 - 19 Hz

$^2J_{\text{BC}(gem)}$ = 0 - 3 Hz

$^3J_{\text{AB}(cis)}$ = 5 - 7 Hz

$^3J_{\text{AB}(cis)}$ = 9 - 11 Hz

In alkyl-substituted alkenes, the long-range allylic couplings, ($^4J_{\text{AB}}$ and $^4J_{\text{AC}}$) are typically in the range 0-3 Hz.

In systems which are stereochemically constrained, the magnitude of the coupling is a function of the dihedral angle between the C-$\text{H}_\text{A}$ bond and the plane of the double bond in a relationship reminiscent of the Karplus relation.

$^4J_{\text{AB}}$, $^4J_{\text{AC}}$ = 0 - 3 Hz

## Aromatic Spin Systems

The coupling constant between protons attached to an aromatic ring is diagnostic of the relative position of the coupled protons *i.e.* whether they are *ortho*, *meta* or *para*.

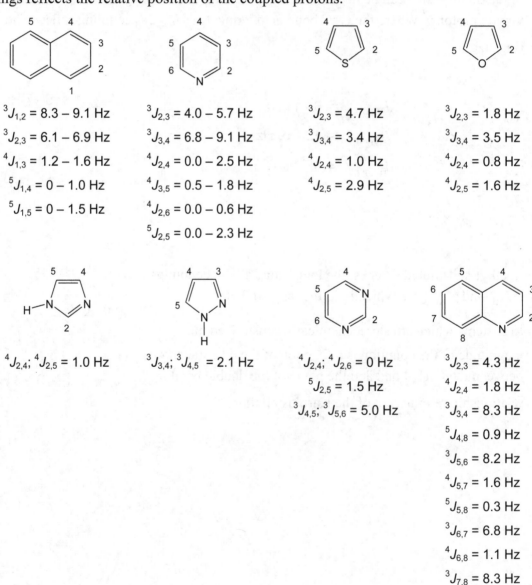

$^3J_{AB(ortho)}$ = 6 - 10 Hz          $^4J_{AB(meta)}$ = 1 - 3 Hz          $^5J_{AB(para)}$ = 0 - 1.5 Hz

Similarly in condensed polynuclear aromatic compounds and heterocyclic compounds, the magnitude of the coupling constants between protons in the aromatic rings reflects the relative position of the coupled protons.

$^3J_{1,2}$ = 8.3 – 9.1 Hz

$^3J_{2,3}$ = 6.1 – 6.9 Hz

$^4J_{1,3}$ = 1.2 – 1.6 Hz

$^5J_{1,4}$ = 0 – 1.0 Hz

$^5J_{1,5}$ = 0 – 1.5 Hz

$^3J_{2,3}$ = 4.0 – 5.7 Hz

$^3J_{3,4}$ = 6.8 – 9.1 Hz

$^4J_{2,4}$ = 0.0 – 2.5 Hz

$^4J_{3,5}$ = 0.5 – 1.8 Hz

$^4J_{2,6}$ = 0.0 – 0.6 Hz

$^5J_{2,5}$ = 0.0 – 2.3 Hz

$^3J_{2,3}$ = 4.7 Hz

$^3J_{3,4}$ = 3.4 Hz

$^4J_{2,4}$ = 1.0 Hz

$^4J_{2,5}$ = 2.9 Hz

$^3J_{2,3}$ = 1.8 Hz

$^3J_{3,4}$ = 3.5 Hz

$^4J_{2,4}$ = 0.8 Hz

$^4J_{2,5}$ = 1.6 Hz

$^4J_{2,4}$; $^4J_{2,5}$ = 1.0 Hz

$^3J_{3,4}$; $^3J_{4,5}$ = 2.1 Hz

$^4J_{2,4}$; $^4J_{2,6}$ = 0 Hz

$^5J_{2,5}$ = 1.5 Hz

$^3J_{4,5}$; $^3J_{5,6}$ = 5.0 Hz

$^3J_{2,3}$ = 4.3 Hz

$^4J_{2,4}$ = 1.8 Hz

$^3J_{3,4}$ = 8.3 Hz

$^5J_{4,8}$ = 0.9 Hz

$^3J_{5,6}$ = 8.2 Hz

$^4J_{5,7}$ = 1.6 Hz

$^5J_{5,8}$ = 0.3 Hz

$^3J_{6,7}$ = 6.8 Hz

$^4J_{6,8}$ = 1.1 Hz

$^3J_{7,8}$ = 8.3 Hz

The splitting patterns of the protons in the aromatic region of the ¹H spectrum are frequently used to establish the substitution pattern of an aromatic ring.  For example, a trisubstituted aromatic ring has 3 remaining protons and they can have relative positions 1,2,3-; 1,2,4-; or 1,3,5- and each has a **characteristic splitting pattern**. Likewise the 1,2-disubstituted and 1,3-disubstituted benzenes also have a characteristic 4-proton splitting pattern (Figure 5.12).

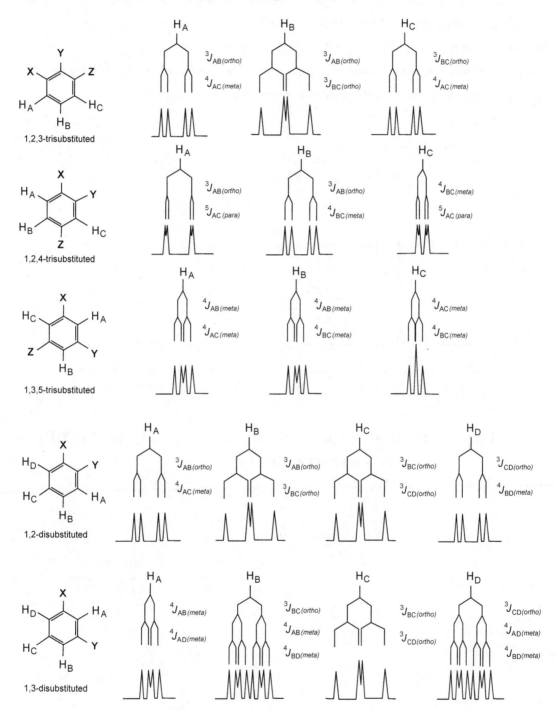

**Figure 5.12    Characteristic Aromatic Splitting Patterns in the ¹H NMR spectra for some Di- and Tri-substituted Benzenes**

### para-Disubstituted benzenes

*para*-Disubstituted benzenes have characteristically "simple" and symmetrical $^1$H NMR spectra in the aromatic region.  Superficially, the spectra of p-disubstituted benzenes always appear as two strong doublets with the line positions symmetrically disposed about a central frequency.  The spectra are in fact far more complex (many lines make up the pattern for the NMR spectrum when it is analysed in detail) but the symmetry of the pattern of lines makes 1,4-disubstituted benzenes very easy to recognise from their $^1$H NMR spectra.  The $^1$H NMR spectrum of *p*-nitrophenylacetylene is given in Figure 5.13.  The expanded section shows the 4 strong prominent signals in the aromatic region, characteristic of 1,4-substitution on a benzene ring.

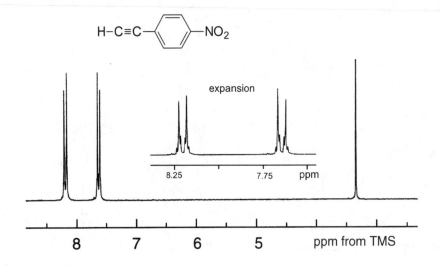

**Figure 5.13**   $^1$**H NMR Spectrum of *p*-Nitrophenylacetylene (200 MHz as a 10% solution in CDCl₃)**

# 6

# $^{13}$C NMR SPECTROSCOPY

The most abundant isotope of carbon ($^{12}$C) cannot be observed by NMR. $^{13}$C is a rare nucleus (1.1% natural abundance) and its low concentration coupled with the fact that $^{13}$C has a relatively low resonance frequency, leads to its relative insensitivity as an NMR-active nucleus (about 1/6000 as sensitive as $^{1}$H). However, with the increasing availability of routine pulsed FT NMR spectrometers, it is now common to acquire many spectra and add them together (Section 5.3), so $^{13}$C NMR spectra of good quality can be obtained readily.

## 6.1    COUPLING AND DECOUPLING IN $^{13}$C NMR SPECTRA

Because the $^{13}$C nucleus is isotopically rare, it is extremely unlikely that any two adjacent carbon atoms in a molecule will *both* be $^{13}$C. As a consequence, **$^{13}$C-$^{13}$C coupling is not observed** in $^{13}$C NMR spectra *i.e.* there is no signal multiplicity or splitting in a $^{13}$C NMR spectrum due to $^{13}$C-$^{13}$C coupling. $^{13}$C couples strongly to any protons that may be attached ($^{1}J_{CH}$ is typically about 125 Hz for saturated carbon atoms in organic molecules). It is the usual practice to irradiate the $^{1}$H nuclei during $^{13}$C acquisition so that all $^{1}$H are fully decoupled from the $^{13}$C nuclei (usually termed **broad band decoupling** or **noise decoupling**). **$^{13}$C NMR spectra usually appear as a series of singlets** (when $^{1}$H is fully decoupled) and *each distinct $^{13}$C environment in the molecule gives rise to a separate signal.*

If $^{1}$H is **not decoupled** from the $^{13}$C nuclei during acquisition, the signals in the $^{13}$C spectrum appear as multiplets where the major splittings are due to the $^{1}J_{C-H}$ couplings (about 125 Hz for *sp$^{3}$* hybridised carbon atoms, about 160 Hz for *sp$^{2}$* hybridised carbon atoms, about 250 Hz for *sp* hybridised carbon atoms). CH$_3$- signals appear as quartets, -CH$_2$- signals appear as triplets, -CH- groups appear as doublets and quaternary C (no attached H) appear as singlets. The **multiplicity information**, taken together with chemical shift data, is useful in identifying and assigning the $^{13}$C resonances.

*Organic Structures from Spectra*, Fifth Edition. L. D. Field, S. Sternhell and J. R. Kalman.
© 2013 John Wiley & Sons, Ltd. Published 2013 by John Wiley & Sons, Ltd.

In $^{13}$C spectra acquired without proton decoupling, there is usually much more *"long range"* coupling information visible in the fine structure of each multiplet. The fine structure arises from coupling between the carbon and protons that are not directly bonded to it (*e.g.* from $^2J_{\text{C-C-H}}$, $^3J_{\text{C-C-C-H}}$). The magnitude of long range C-H coupling is typically < 10 Hz and this is much less than $^1J_{\text{C-H}}$. Sometimes a more detailed analysis of the long-range C-H couplings can be used to provide additional information about the structure of the molecule.

In most $^{13}$C spectra, $^{13}$C nuclei which have directly attached protons receive a significant (but not easily predictable) signal enhancement when the protons are decoupled as a result of the Nuclear Overhauser Effect (see Section 5.4) and as a consequence, peak intensity does not necessarily reflect the number of $^{13}$C nuclei giving rise to the signal.

It is not usually possible to integrate routine $^{13}$C spectra directly unless specific precautions have been taken. However with proper controls, $^{13}$C NMR spectroscopy can be used quantitatively and it is a valuable technique for the analysis of mixtures. To record $^{13}$C NMR spectra where the relative signal intensity can be reliably determined, the spectra must be recorded with techniques to suppress the Nuclear Overhauser Effect and with a long delay between the acquisitions of successive spectra to ensure that all of the carbons in the molecule are completely relaxed between spectral acquisitions.

## 6.2   DETERMINING $^{13}$C SIGNAL MULTIPLICITY USING DEPT

With most modern NMR instrumentation, the DEPT experiment (**D**istortionless **E**nhancement by **P**olarisation **T**ransfer) is the most commonly used method to determine the multiplicity of $^{13}$C signals. The DEPT experiment is a pulsed NMR experiment which requires a series of programmed Rf pulses to both the $^1$H and $^{13}$C nuclei in a sample. The resulting $^{13}$C DEPT spectrum contains only signals arising from protonated carbons (non protonated carbons do not give signals in the $^{13}$C DEPT spectrum). The signals arising from carbons in $CH_3$ and CH groups (*i.e.* those with an odd number of attached protons) appear oppositely phased from those in $CH_2$ groups (*i.e.* those with an even number of attached protons) so signals from $CH_3$ and CH groups point upwards while signals from $CH_2$ groups point downwards (Figure 6.1b).

In more advanced applications, the $^{13}$C DEPT experiment can be used to separate the signals arising from carbons in $CH_3$, $CH_2$ and CH groups. This is termed spectral editing and can be used to produce separate $^{13}$C sub-spectra of just the CH$_3$ carbons, just the CH$_2$ carbons or just the CH carbons.

Figure 6.1 shows various $^{13}$C spectra of methyl cyclopropyl ketone. The $^{13}$C spectrum acquired with full proton decoupling (Figure 6.1a) shows 4 singlet peaks, one for each of the 4 different carbon environments in the molecule.

The DEPT spectrum (Figure 6.1b) shows only the 3 resonances for the protonated carbons. The carbon atoms that have an odd number of attached hydrogens (CH and CH$_3$ groups) point upwards and those with an even number of attached hydrogen atoms (the signals of CH$_2$ groups) point downwards. Note that the carbonyl carbon does not appear in the DEPT spectrum since it has no attached protons.

In the carbon spectrum with no proton decoupling (Figure 6.1c), all of the resonances of protonated carbons appear as multiplets and the multiplet structure is due to coupling to the attached protons. The CH$_3$ (methyl) group appears as a quartet, the CH$_2$ (methylene) groups appear as a triplet and the CH (methine) group appears as a doublet while the carbonyl carbon (with no attached protons) appears as a singlet. In Figure 6.1c, all of the $^1J_{\text{C-H}}$ coupling constants could be measured directly from the spectrum.

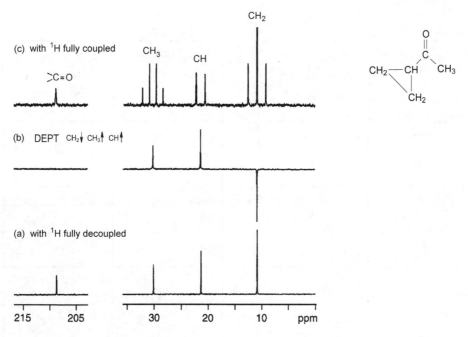

**Figure 6.1**   $^{13}$**C NMR Spectra of Methyl Cyclopropyl Ketone (CDCl$_3$ Solvent, 100 MHz).** *(a)* **with Broad Band Decoupling of $^1$H ;** *(b)* **DEPT Spectrum** *(c)* **with no Decoupling of $^1$H.**

For purposes of assigning a $^{13}$C spectrum, two $^{13}$C spectra are usually obtained. Firstly, a spectrum with complete $^{1}$H decoupling to maximise the intensity of signals and provide sharp singlets to minimise any signal overlap.  This is the best spectrum to **count the number of resonances** and accurately determine their chemical shifts. Secondly, a spectrum which is sensitive to the number of protons attached to each C to permit partial **sorting of the $^{13}$C signals** according to whether they are methyl, methylene, methine or quaternary carbon atoms.  This could be a DEPT spectrum or a $^{13}$C spectrum with no proton decoupling.

The number of resonances visible in a $^{13}$C NMR spectrum immediately indicates **the number of distinct $^{13}$C environments in the molecule** (Table 6.1).  If the number of $^{13}$C environments is less than the number of carbons in the molecule, then the molecule must have some symmetry that dictates that some $^{13}$C nuclei are in identical environments.  This is particularly useful in establishing the **substitution pattern** (position where substituents are attached) in aromatic compounds.

**Table 6.1      The Number of Aromatic $^{13}$C Resonances in Benzenes with Different Substitution Patterns**

| Molecule | Number of aromatic $^{13}$C resonances | Molecule | Number of aromatic $^{13}$C resonances |
|---|---|---|---|
| | 1 | Cl—⟨ ⟩—Cl | 2 |
| ⟨ ⟩—Cl | 4 | Br—⟨ ⟩—Cl | 4 |
| ⟨ ⟩(Cl,Cl) | 3 | ⟨ ⟩(Cl, Br) | 6 |
| ⟨ ⟩(Cl, Cl) | 4 | ⟨ ⟩(Cl, Br) | 6 |

## 6.3   SHIELDING AND CHARACTERISTIC CHEMICAL SHIFTS IN $^{13}$C NMR SPECTRA

The general trends of $^{13}$C chemical shifts somewhat parallel those in $^1$H NMR spectra. However, $^{13}$C nuclei have access to a greater variety of hybridisation states (bonding geometries and electron distributions) than $^1$H nuclei and both hybridisation and changes in electron density have a significantly larger effect on $^{13}$C nuclei than $^1$H nuclei.  As a consequence, the $^{13}$C chemical shift scale spans some 250 ppm, *cf.* the 10 ppm range commonly encountered for $^1$H chemical shifts (Tables 6.2 - 6.4, Figure 6.2).

**Table 6.2      Typical $^{13}$C Chemical Shift Values in Selected Organic Compounds**

| Compound | $\delta$ $^{13}$C (ppm from TMS) |
|:---:|:---:|
| $CH_4$ | -2.1 |
| $CH_3CH_3$ | 7.3 |
| $CH_3OH$ | 50.2 |
| $CH_3Cl$ | 25.6 |
| $CH_2Cl_2$ | 52.9 |
| $CHCl_3$ | 77.3 |
| $CH_3CH_2CH_2Cl$ | 11.5 ($CH_3$) |
| | 26.5 ($-CH_2-$) |
| | 46.7 ($-CH_2-Cl$) |
| $CH_2=CH_2$ | 123.3 |
| $CH_2=C=CH_2$ | 208.5 ($=C=$) |
| | 73.9 ($=CH_2$) |
| $CH_3CHO$ | 31.2 ($-CH_3$) |
| | 200.5 ($-CHO$) |
| $CH_3COOH$ | 20.6 ($-CH_3$), 178.1 ($-COOH$) |
| $CH_3COCH_3$ | 30.7 ($-CH_3$), 206.7 ($-CO-$) |
| | 128.5 |
| | 149.8 (C-2) 123.7 (C-3) 135.9 (C4) |

**Table 6.3        Typical $^{13}$C Chemical Shift Ranges in Organic Compounds**

| Group | $^{13}$C shift (ppm) |
|---|---|
| TMS | 0.0 |
| -CH$_3$ (with only -H or -R at C$_\alpha$ and C$_\beta$) | 0 - 30 |
| -CH$_2$ (with only -H or -R at C$_\alpha$ and C$_\beta$) | 20 - 45 |
| -CH (with only -H or -R at C$_\alpha$ and C$_\beta$) | 30 - 60 |
| C quaternary (with only -H or -R at C$_\alpha$ and C$_\beta$) | 30 - 50 |
| O-CH$_3$ | 50 - 60 |
| N-CH$_3$ | 15 - 45 |
| C≡C | 70 - 95 |
| C=C | 105 - 160 |
| C (aromatic) | 110 - 155 |
| C (heteroaromatic) | 105 - 165 |
| -C≡N | 115 - 125 |
| C=O (acids, acyl halides, esters, amides) | 155 - 185 |
| C=O (aldehydes, ketones) | 185 - 225 |

In $^{13}$C NMR spectroscopy the $^{13}$C signal due to the carbon in CDCl$_3$ appears as a triplet centred at δ 77.0 with peak intensities in the ratio 1:1:1 (due to spin-spin coupling between $^{13}$C and $^2$H).  This resonance serves as a convenient reference for the chemical shifts of $^{13}$C NMR spectra recorded in this solvent.

Table 6.4 gives characteristic $^{13}$C chemical shifts for some $sp^3$-hybridised carbon atoms in common functional groups.  Table 6.5 gives characteristic $^{13}$C chemical shifts for some $sp^2$-hybridised carbon atoms in substituted alkenes and Table 6.6 gives characteristic $^{13}$C chemical shifts for some $sp$-hybridised carbon atoms in alkynes.

## Figure 6.2   Approximate $^{13}$C Chemical Shift Ranges for Carbon Atoms in Organic Compounds

| | | | | | | | | | | | | | |
|---|---|---|---|---|---|---|---|---|---|---|---|---|---|

CR$_3$–I    R = H; alkyl

R–CH$_3$    R = H; alkyl

CH$_3$–(C=O)R    CH$_3$–(C=O)OR    R = alkyl

CR$_3$–Br    R = H; alkyl

R–CH$_2$–R    R = alkyl

R–CH$_2$–NR$_2$    R = –CH$_2$– ; alkyl

CH$_3$–NR$_2$    R = –CH$_2$– ; alkyl

CR$_3$–Cl    R = H; alkyl

$\begin{matrix} R \\ R \end{matrix}$ C $\begin{matrix} R \\ R \end{matrix}$    R = alkyl

$\begin{matrix} R \\ R \end{matrix}$ CH–R    R = alkyl

R–CH$_2$–OH    R–CH$_2$–OR    R–CH$_2$–O(C=O)R    R = alkyl

CR$_3$–NO$_2$    R = H; alkyl

–C≡C–

–C≡N

$\begin{matrix} R \\ R \end{matrix}$ C = C $\begin{matrix} R \\ R \end{matrix}$    R = H; –CH$_2$– ; alkyl

aromatic carbon

–C–X    X = OH; Cl; Br; NH$_2$; NR$_2$; OR    R = alkyl
 ‖
 O

–C–H    aldehydes
 ‖
 O

R–C–R    ketones    R = alkyl
  ‖
  O

δ (ppm)

240   220   200   180   160   140   120   100   80   60   40   20   0   -20

**Table 6.4**     $^{13}$C Chemical Shifts ($\delta$) for $sp^3$ Carbons in Alkyl Derivatives

| X | $CH_3$—X $-CH_3$ | $CH_3CH_2$—X $-CH_3$ | $CH_3CH_2$—X $-CH_2-$ | $(CH_3)_2CH$—X $-CH_3$ | $(CH_3)_2CH$—X $>CH-$ |
|---|---|---|---|---|---|
| —H | -2.3 | 7.3 | 7.3 | 15.4 | 15.9 |
| —CH=CH$_2$ | 18.7 | 13.4 | 27.4 | 22.1 | 32.3 |
| —Ph | 21.4 | 15.8 | 29.1 | 24.0 | 34.3 |
| —Cl | 25.6 | 18.9 | 39.9 | 27.3 | 53.7 |
| —OH | 50.2 | 18.2 | 57.8 | 25.3 | 64.0 |
| —OCH$_3$ | 60.9 | 14.7 | 67.7 | 21.4 | 72.6 |
| —OCO—CH$_3$ | 51.5 | 14.4 | 60.4 | 21.9 | 67.5 |
| —CO—CH$_3$ | 30.7 | 7.0 | 35.2 | 18.2 | 41.6 |
| —CO—OCH$_3$ | 20.6 | 9.2 | 27.2 | 19.1 | 34.1 |
| —NH$_2$ | 28.3 | 19.0 | 36.9 | 26.5 | 43.0 |
| —NH—COCH$_3$ | 26.1 | 14.6 | 34.1 | 22.3 | 40.5 |
| —C≡N | 1.7 | 10.6 | 10.8 | 19.9 | 19.8 |
| —NO$_2$ | 61.2 | 12.3 | 70.8 | 20.8 | 78.8 |

**Table 6.5**     $^{13}$C Chemical Shifts ($\delta$) for $sp^2$ Carbons in Vinyl Derivatives: CH$_2$=CH-X

| X | $CH_2=$ | $=CH-X$ |
|---|---|---|
| —H | 123.3 | 123.3 |
| —CH$_3$ | 115.9 | 136.2 |
| —C(CH$_3$)$_3$ | 108.9 | 149.8 |
| —Ph | 112.3 | 135.8 |
| —CH=CH$_2$ | 116.3 | 136.9 |
| —C≡C—H | 129.2 | 117.3 |
| —CO—CH$_3$ | 128.0 | 137.1 |
| —CO—OCH$_3$ | 130.3 | 129.6 |
| —Cl | 117.2 | 126.1 |
| —OCH$_3$ | 84.4 | 152.7 |
| —OCO-CH$_3$ | 96.6 | 141.7 |
| —C≡N | 137.5 | 108.2 |
| —NO$_2$ | 122.4 | 145.6 |
| —N(CH$_3$)$_2$ | 91.3 | 151.3 |

**Table 6.6**     $^{13}$C Chemical Shifts (δ) for *sp* Carbons in Alkynes:

X-C≡C-Y

| X | Y | X—C≡ | ≡C—Y |
|---|---|---|---|
| H— | —H | 73.2 | 73.2 |
| H— | —CH₃ | 66.9 | 79.2 |
| H— | —C(CH₃)₃ | 67.0 | 92.3 |
| H— | —CH=CH₂ | 80.0 | 82.8 |
| H— | —C≡C—H | 66.3 | 67.3 |
| H— | —Ph | 77.1 | 83.4 |
| H— | —COCH₃ | 81.8 | 78.1 |
| H— | —OCH₂CH₃ | 22.0 | 88.2 |
| CH₃— | —CH₃ | 72.6 | 72.6 |
| CH₃— | —Ph | 79.7 | 85.8 |
| CH₃— | —COCH₃ | 97.4 | 87.0 |
| Ph— | —Ph | 89.4 | 89.4 |
| —COOCH₃ | —COOCH₃ | 74.6 | 74.6 |

Table 6.7 gives characteristic $^{13}$C chemical shifts for the aromatic carbons in benzene derivatives.  To a first approximation, the shifts induced by substituents are additive.  So, for example, an aromatic carbon which has a –NO₂ group in the *para* position and a –Br group in the *ortho* position will appear at approximately 137.9 ppm [(128.5 + 6.1(*p*-NO₂) + 3.3(*o*-Br))].

**Table 6.7** **Approximate $^{13}$C Chemical Shifts (δ) for Aromatic Carbons in Benzene Derivatives Ph-X in ppm relative to Benzene at δ 128.5 ppm (a positive sign denotes a downfield shift)**

| X | ipso | ortho | meta | para |
|---|---|---|---|---|
| —H | 0.0 | 0.0 | 0.0 | 0.0 |
| —NO$_2$ | 19.9 | -4.9 | 0.9 | 6.1 |
| —CO—OCH$_3$ | 2.0 | 1.2 | -0.1 | 4.3 |
| —CO—NH$_2$ | 5.0 | -1.2 | 0.1 | 3.4 |
| —CO—CH$_3$ | 8.9 | 0.1 | -0.1 | 4.4 |
| —C≡N | -16.0 | 3.5 | 0.7 | 4.3 |
| —Br | -5.4 | 3.3 | 2.2 | -1.0 |
| —CH=CH$_2$ | 8.9 | -2.3 | -0.1 | -0.8 |
| —Cl | 5.3 | 0.4 | 1.4 | -1.9 |
| —CH$_3$ | 9.2 | 0.7 | -0.1 | -3.0 |
| —OCO—CH$_3$ | 22.4 | -7.1 | 0.4 | -3.2 |
| —OCH$_3$ | 33.5 | -14.4 | 1.0 | -7.7 |
| —NH$_2$ | 18.2 | -13.4 | 0.8 | -10.0 |

Table 6.8 gives characteristic shifts for $^{13}$C nuclei in some polynuclear aromatic compounds and heteroaromatic compounds.

**Table 6.8** **Characteristic $^{13}$C Chemical Shifts (δ) in some Polynuclear Aromatic Compounds and Heteroaromatic Compounds**

# 7

# 2-Dimensional NMR Spectroscopy

A two-dimensional NMR spectrum has two frequency axes rather than one. A 2D spectrum is acquired using a pulse sequence which contains a delay period '$t_1$' which can be varied systematically as the experiment is repeated. The experiment is repeated many times (typically 512 or 1024), with a different delay '$t_1$' in the pulse sequence for each experiment. One FID is acquired for each experiment giving an array of 'N' individual FID's each of which has been acquired with a slightly different pulse sequence.

Each FID represents the variation of detected signal as a function of time ($t_2$ in the diagram below) and successive FIDs in the array differ as a function of the time variable $t_1$ *within* the preparation period of the pulse sequence.

Fourier transformation of the two-dimensional array of data with respect to $t_2$ affords a series of spectra which vary systematically as a function of $t_1$. A second Fourier transformation, this time with respect to $t_1$, gives a two-dimensional spectral array (which is function of two frequency domains $F_1$ and $F_2$). Two-dimensional spectra are usually represented in terms of a stacked plot or contour plot. The contour plot is a more convenient representation for making measurements or peak assignment.

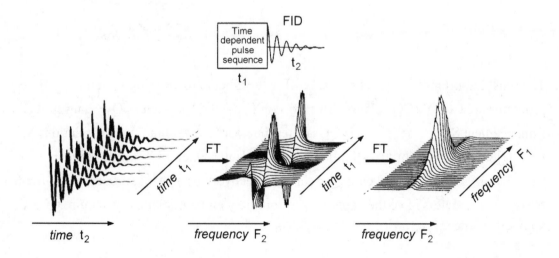

*Organic Structures from Spectra*, Fifth Edition. L. D. Field, S. Sternhell and J. R. Kalman.
© 2013 John Wiley & Sons, Ltd. Published 2013 by John Wiley & Sons, Ltd.

The number of possible two-dimensional experiments is essentially unlimited. Different pulse sequences in the preparation period give rise to different two-dimensional spectra which can be tailored to exhibit various properties of the sample.

The technical detail behind multi-dimensional NMR experiments and the pulse sequences used to generate 2D spectra, is beyond the scope of this book.

Two-dimensional spectra have the appearance of surfaces, generally with two axes corresponding to chemical shift and the third (vertical) axis corresponding to signal intensity.

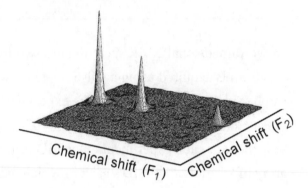

It is usually more useful to plot two-dimensional spectra viewed directly from above (a **contour plot** of the surface) in order to make measurements and assignments.

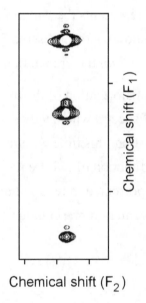

The most important two-dimensional NMR experiments for solving structural problems are COSY (<u>CO</u>rrelation <u>S</u>ectroscop<u>Y</u>), NOESY (<u>N</u>uclear <u>O</u>verhauser <u>E</u>nhancement <u>S</u>pectroscop<u>Y</u>), HSQC (<u>H</u>eteronuclear Single <u>Q</u>uantum <u>C</u>orrelation) or HSC (<u>H</u>eteronuclear <u>S</u>hift <u>C</u>orrelation), HMBC (<u>H</u>eteronuclear <u>M</u>ultiple <u>B</u>ond <u>C</u>orrelation) and TOCSY (<u>TO</u>tal <u>C</u>orrelation <u>S</u>pectroscop<u>Y</u>).  Most modern high-field NMR spectrometers have the capability to routinely and automatically acquire COSY, NOESY, HSQC, HMBC and TOCSY spectra.

## 7.1   COSY (CORRELATION SPECTROSCOPY)

**The COSY spectrum** shows which pairs of protons in a molecule are coupled to each other. The COSY spectrum is a symmetrical spectrum that has the $^1$H NMR spectrum of the substance as both of the chemical shift axes ($F_1$ and $F_2$).

It is usual to plot a normal (one-dimensional) NMR spectrum along each of the $F_1$ and $F_2$ axes to give reference spectra for the peaks that appear in the two-dimensional spectrum. A COSY spectrum of 1-iodobutane ($CH_3CH_2CH_2CH_2I$) is given below (Figure 7.1).

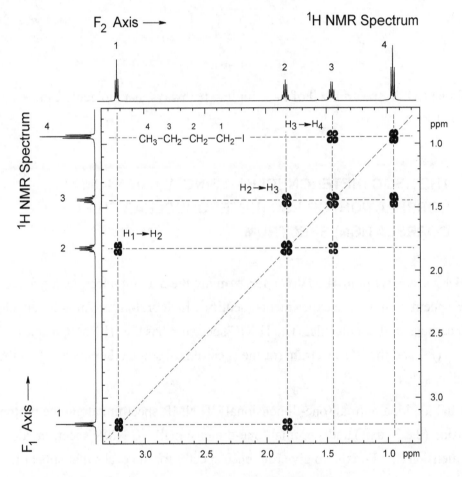

**Figure 7.1**   $^1$**H COSY Spectrum of 1-Iodobutane (CDCl$_3$ solvent, 298K, 600 MHz)**

The COSY spectrum of 1-iodobutane has a set of 4 peaks on the diagonal as well as peaks that are off the diagonal. The COSY spectrum is always symmetrical about the diagonal – the off-diagonal peaks above the diagonal are mirrored on the lower side of the diagonal. The off-diagonal peaks are the important signals, since these occur at positions where there is coupling between a proton on the $F_1$ axis and a proton on the

$F_2$ axis.  The protons which are part of the -$CH_2I$ group ($H_1$) are easy to identify since the halogen substituent characteristically moves these to about δ 3.2 ppm.  There are 3 off-diagonal peaks on each side of the diagonal – one indicates the coupling between $H_1$ and $H_2$; the second indicates coupling between $H_2$ and $H_3$ and the third indicates coupling between $H_3$ and $H_4$.  If you can identify one of the proton signals, then you can identify the protons that are coupled to it and then work sequentially along a coupled network until all the protons which are coupled together are identified.

In a single COSY spectrum all of the coupling pathways in a molecule can be identified.

## 7.2    THE HSQC (HETERONUCLEAR SINGLE QUANTUM CORRELATION) OR HSC (HETERONUCLEAR SHIFT CORRELATION) SPECTRUM

**The HSQC spectrum or the HSC spectrum** are the heteronuclear analogues of the COSY spectrum and these experiments identify which protons are directly attached to which carbons in the molecule.  The HSQC spectrum has the $^1H$ NMR spectrum on one axis ($F_2$) and the $^{13}C$ spectrum (or the spectrum of some other nucleus) on the second axis ($F_1$).

It is usual to plot a normal (one-dimensional) $^1H$ NMR spectrum along the proton dimension (the $F_2$ axis) and a normal (one-dimensional) $^{13}C$ NMR spectrum along the $^{13}C$ dimension (the $F_1$ axis) to give reference spectra for the peaks that appear in the two-dimensional spectrum.  An HSQC spectrum of 1-iodobutane ($CH_3CH_2CH_2CH_2I$) is given in Figure 7.2.

The HSQC spectrum does not have diagonal peaks.  The peaks in an HSQC spectrum occur at positions where a proton in the spectrum on the $F_2$ axis is directly coupled to a carbon in the spectrum on the $F_1$ axis *via a 1-bond C-H coupling.*

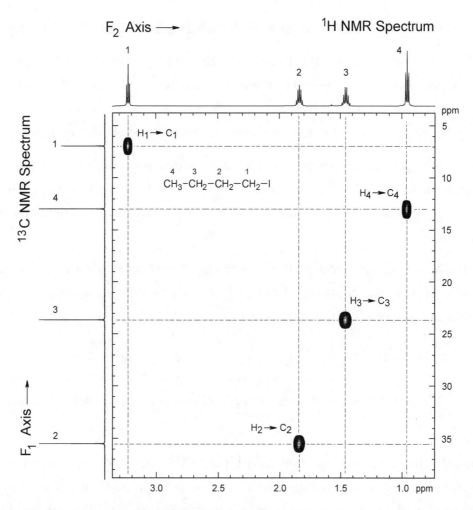

**Figure 7.2      $^1$H – $^{13}$C HSQC Spectrum of 1-Iodobutane (CDCl$_3$ solvent, 298K, $^1$H 600 MHz, $^{13}$C 150 MHz)**

1-Iodobutane has 4 carbon resonances and the HSQC spectrum shows 4 cross peaks.  Having identified all of the resonances in the $^1$H spectrum, then the resonances in the carbon spectrum can simply be identified by the positions of the cross peaks corresponding to each proton resonance,

In the HSQC or HSC experiment, all of the protons which are coupled to carbons can be identified.  It is usually possible to assign all of the resonances in the $^1$H NMR spectrum *i.e.* establish which proton in a molecule gives rise to each signal in the spectrum, using spin-spin coupling information or a COSY experiment, then assign the signals in the $^{13}$C spectrum by correlation to the known proton resonances.

Note that non-protonated carbons (quaternary carbons) do not give rise to signals in the HSQC or HSC spectra.

## 7.3    HMBC (HETERONUCLEAR MULTIPLE BOND CORRELATION)

**The HMBC spectrum** correlates chemical shifts of hydrogen nuclei with carbon nuclei which are separated by two or more chemical bonds.  The HMBC experiment is frequently used to assign quaternary and carbonyl carbons which don't have any directly bound protons so they are "invisible" in the HSQC or HSC experiment.

The HMBC experiment is designed to filter out correlations resulting from large C-H coupling constants (120-160 Hz) resulting from protons directly bound to carbon and to select for smaller couplings (around 10 Hz) which are typical C-H couplings over 2 or 3 bonds.

The HMBC is a very powerful method for making the connection between two parts of a molecule that may be isolated from each other by a carbonyl, an ether, an ester, an amide or by some other functional group.

X = O, S, NH          X = O, S, NH

It is usual to plot a normal (one-dimensional) $^1$H NMR spectrum along the proton dimension (the $F_2$ axis) and a normal (one-dimensional) $^{13}$C NMR spectrum along the $^{13}$C dimension (the $F_1$ axis) to give reference spectra for the peaks that appear in the two-dimensional spectrum.  A HMBC spectrum of 1-iodobutane ($CH_3CH_2CH_2CH_2I$) is given in Figure 7.3.

The HMBC spectrum does not have diagonal peaks.  The peaks in an HMBC spectrum occur at positions where a proton in the spectrum on the $F_2$ axis is coupled to a carbon in the spectrum on the $F_1$ axis *via a 2-bond or 3-bond CH coupling.*

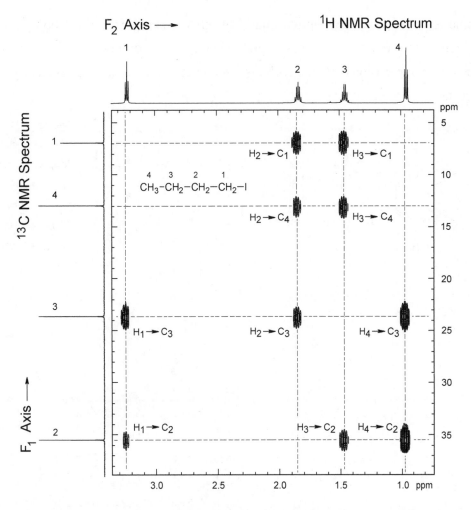

**Figure 7.3**    **$^1$H – $^{13}$C HMBC Spectrum of 1-Iodobutane (CDCl$_3$ solvent, 298K, $^1$H 600 MHz, $^{13}$C 150 MHz)**

1-Iodobutane has 4 carbon resonances and the HMBC spectrum shows 10 peaks. $H_1$ correlates to $C_2$ (2-bond coupling) and to $C_3$ (3-bond coupling). $H_2$ correlates to $C_1$ and to $C_3$ (2-bond couplings) as well as to $C_4$ (3-bond coupling); $H_3$ correlates to $C_2$ and to $C_4$ (2-bond couplings) as well as to $C_1$ (3-bond coupling); $H_4$ correlates to $C_3$ (2-bond coupling) as well as to $C_2$ (3-bond coupling). Because the HMBC provides long-range information it is exceptionally useful in tying together pieces of a molecule as parts are systematically identified.

In aromatic systems, and in heteroaromatic systems, it is typically the 3-bond (*meta*) $J_{CH}$ which is the largest of the long-range C-H couplings, so the HMBC cross peaks between aromatic protons and the carbons which are 3 bonds away are typically the strongest.

$^2J_{H3-C4} = 1$ Hz

$^3J_{H2-C4} = 7$ Hz

$^4J_{H1-C4} = -1$ Hz

$^2J_{H3-C2} = 2$ Hz

$^3J_{H4-C2} = 7$ Hz

$^3J_{H6-C2} = 12$ Hz

$^2J_{H2-C3} = 7$ Hz

$^2J_{H4-C3} = 2$ Hz

$^3J_{H2-C4} = 4$ Hz

$^3J_{H3-C5} = 6$ Hz

$^3J_{H3-C1} = 8$ Hz

$^3J_{H8-C1} = 5$ Hz

$^3J_{H4-C2} = 9$ Hz

$^3J_{H1-C4a} = 8$ Hz

$^3J_{H3-C4a} = 8$ Hz

$^2J_{H3-C2} = 4$ Hz    $^3J_{H6-C8} = 7$ Hz

$^3J_{H4-C2} = 8$ Hz    $^3J_{H3-C4a} = 7$ Hz

$^2J_{H2-C3} = 10$ Hz   $^3J_{H6-C4a} = 7$ Hz

$^3J_{H2-C4} = 5$ Hz    $^3J_{H8-C4a} = 5$ Hz

$^3J_{H5-C4} = 5$ Hz    $^3J_{H2-C8a} = 13$ Hz

$^3J_{H4-C5} = 5$ Hz    $^3J_{H4-C8a} = 6$ Hz

$^3J_{H7-C5} = 7$ Hz    $^3J_{H5-C8a} = 5$ Hz

$^3J_{H8-C6} = 9$ Hz    $^3J_{H7-C8a} = 7$ Hz

$^3J_{H5-C7} = 9$ Hz

The HMBC is a powerful method for assigning the $^{13}C$ chemical shifts of the substituted (*i.e.* non-protonated) carbons in an aromatic framework.

The HMBC spectrum showing the aromatic protons and carbons of 2-bromophenol is given in Figure 7.4.

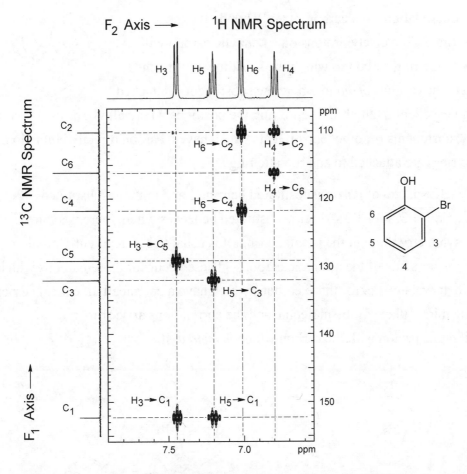

**Figure 7.4**      **$^1$H – $^{13}$C HMBC Spectrum of 2-Bromophenol (CDCl$_3$ solvent, 298K, $^1$H 400 MHz, $^{13}$C 100 MHz)**

2-Bromophenol is a disubstituted aromatic compound.  It has 4 aromatic protons and 6 aromatic carbons.  There are 4 protonated carbons and 2 substituted carbons.  The proton assignments are given in Figure 7.4.

The HMBC spectrum of 2-bromophenol shows 8 strong correlations and all of the strong peaks correspond to 3-bond *meta* couplings.  H$_3$ correlates to C$_1$ and to C$_5$;  H$_4$ correlates to C$_2$ and to C$_6$;  H$_5$ correlates to C$_3$ and to C$_1$;  H$_6$ correlates to C$_2$ and to C$_4$.

In substituted aromatic systems, the protons *ortho* to a substituent will show correlations to a carbon directly bonded to the aromatic ring.  Conversely protons on a benzylic carbon will correlate to the ring carbon to which the substituent is attached and also to the carbons *ortho* to where the substituent is attached. This is an excellent method for identifying the positions where alkyl-, acyl-, vinyl- or alkynyl- substituents (or other substituents which have a carbon directly bound to the aromatic ring) are attached to an aromatic ring.

In the HMBC spectra of *para*-disubstituted benzenes and in monosubstituted benzenes, or in 1,3,5- or 1,3,4,5-tetrasubstituted benzenes where there is a mirror plane of symmetry through the aromatic ring, it is usual to see apparent 1-bond correlations for some of the aromatic carbons.  These occur for those carbons which are symmetry related i.e. for those carbons which have a chemically identical carbon across the mirror plane of the molecule and the correlations arise from the $^3J_{H-C}$ interaction of a proton with the carbon which is *meta* to it.

It is also common to observe apparent 1-bond C-H correlations in the HMBC spectra of the methyl groups in *t*-butyl groups, isopropyl groups or in compounds with *gem*-dimethyl groups.  Again while the correlation appears to be a 1-bond correlation, the HMBC correlation arises from the $^3J_{H-C}$ interaction of the protons of one of the methyl groups with the chemically equivalent carbon which is 3 bonds away.

## 7.4   NOESY (NUCLEAR OVERHAUSER EFFECT SPECTROSCOPY)

**The NOESY spectrum** relies on the Nuclear Overhauser Effect (Section 5.4) and shows which pairs of nuclei in a molecule are close together in space.

The NOESY spectrum is very similar in appearance to a COSY spectrum. The NOESY spectrum is symmetrical about the diagonal and has the $^1$H NMR spectrum as both of the chemical shift axes ($F_1$ and $F_2$). Again, it is usual to plot a normal (one-dimensional) NMR spectrum along each of the axes to give reference spectra for the peaks that appear in the two-dimensional spectrum.

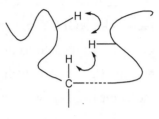

From the analysis of a NOESY spectrum, it is possible to determine the three dimensional structure of a molecule or parts of a molecule. The NOESY spectrum is particularly useful for establishing the stereochemistry (*e.g.* the *cis/trans* configuration of a double bond or a ring junction) of a molecule where more than one possible stereoisomer exists. A NOESY spectrum of β-butyrolactone is given in Figure 7.5.

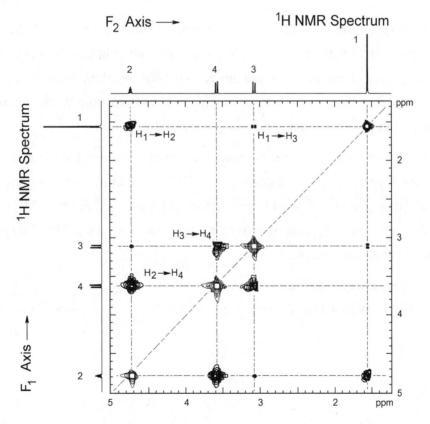

**Figure 7.5**    $^1$**H NOESY Spectrum of β-Butyrolactone (CDCl$_3$ solvent, 298K, $^1$H 600 MHz)**

β-Butyrolactone is a cyclic compound (4-membered ring) with a methyl group and 3 protons.  The ring structure dictates that the methyl group ($H_1$) and one of the protons ($H_3$) are on one face of the molecule and two protons ($H_2$ and $H_4$) are on the other face.  The NOESY spectrum contains strong cross peaks between $H_1$ and $H_2$ (geminal groups *i.e.* attached to the same carbon); strong cross peaks between $H_3$ and $H_4$ (also germinal groups) as well as strong cross peaks between $H_2$ and $H_4$ (same face of the molecule) and cross peaks between $H_1$ and $H_3$ (same face of the molecule).

From the NOESY spectrum it is very clear which proton signal derives from $H_3$ and which from $H_4$.  The spectrum also shows weak cross peaks between $H_2$ and $H_3$ indicating that these protons are still sufficiently close in space to have a weak NOE interaction.

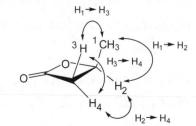

## 7.5    TOCSY (TOTAL CORRELATION SPECTROSCOPY)

**The TOCSY spectrum** is useful in identifying all of the protons which belong to the same isolated spin system.  The experiment relies on spin-spin coupling but rather than showing pairs of nuclei which are directly coupled together, the TOCSY shows a cross peak (off-diagonal peak) for every nucleus which is part of the same coupled spin system - not just those that are directly coupled to each other.

Like the COSY spectrum, the TOCSY has peaks along a diagonal at the frequencies of all of the resonances in the spectrum.  The TOCSY spectrum is also symmetrical around the diagonal and it is usual to plot a normal (one-dimensional) NMR spectrum along each of the $F_1$ and $F_2$ axes to give reference spectra for the peaks that appear in the two-dimensional spectrum.

A TOCSY spectrum of butyl ethyl ether ($CH_3CH_2CH_2CH_2OCH_2CH_3$) is given in Figure 7.6.

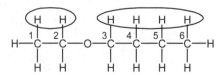

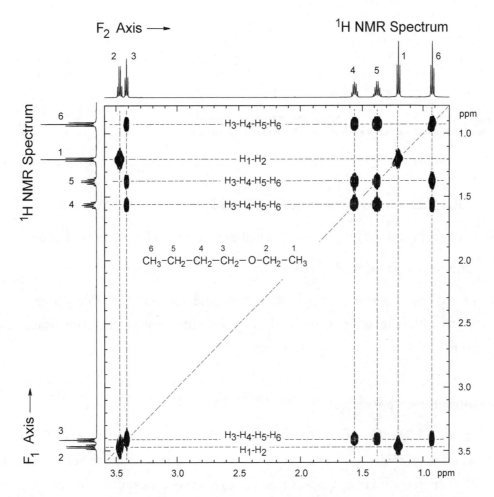

**Figure 7.6**     **¹H TOCSY Spectrum of Butyl Ethyl Ether (CDCl₃ solvent, 298K, ¹H 600 MHz)**

Butyl ethyl ether has two separate spin systems – a butyl fragment and an ethyl fragment which are isolated from each other by the ether oxygen bridge.  Butyl ethyl ether has 6 proton resonances and the TOCSY shows 6 peaks on the diagonal.  There are 7 off-diagonal peaks above the diagonal and a corresponding set of mirrored peaks below the diagonal.

At the frequency corresponding to each resonance in F1, all of the nuclei in the same coupled spin system are identified by a series of peaks.  At the frequency corresponding to $H_6$ in the $F_1$ dimension, a horizontal line will show peaks in the $F_2$ dimension for $H_3$, $H_4$, $H_5$ and $H_6$ which are all the signals for the butyl fragment of butyl ethyl ether.  At the frequency corresponding to $H_1$ in the $F_1$ dimension, a horizontal line will show peaks in the F2 dimension for H1 and H2 which are all the signals for the ethyl fragment of butyl ethyl ether.  There is a significant amount of redundant information in a TOCSY spectrum.

The TOCSY spectrum is useful where there are a number of isolated spin systems in the sample, particularly if the spectra are heavily overlapped.

# 8

# MISCELLANEOUS TOPICS

## 8.1   SOLVENTS FOR NMR SPECTROSCOPY

NMR spectra are almost invariably obtained in solution.  The solvents of choice:

*(a)*    should have adequate dissolving power.

*(b)*    should not associate strongly with solute molecules as this is likely to produce appreciable effects on chemical shifts.  This requirement must sometimes be sacrificed to achieve adequate solubility.

*(c)*    should be essentially free of interfering signals.  Thus for $^1$H NMR, the best solvents are proton-free.

*(d)*    should preferably contain deuterium, $^2$H.  Deuterium is an isotope of hydrogen which is relatively easy to obtain and incorporate into common solvents in place of hydrogen with insignificant changes to the properties of the solvent.  Almost all NMR instruments use deuterium as a convenient "locking" signal to stabilise the magnetic field of the NMR magnet and to provide a strong signal to tune the homogeneity of the NMR magnet to produce the optimum spectrum.

The most commonly used organic solvent is **deuterochloroform**, $CDCl_3$, which is an excellent solvent and is only weakly associated with most organic substrates.  $CDCl_3$ contains no protons and has a deuterium atom.  For ionic compounds or hydrophilic compounds, the most common solvent is deuterated water, $D_2O$.

Almost all deuterated solvents are not 100% deuterated and they contain a residual protonated impurity.  With the sensitivity of modern NMR instruments, the signal from residual protons in the deuterated solvent is usually visible in the $^1$H NMR spectrum.  Table 8.1 provides the chemical shifts of the residual signals from solvents which are commonly used in NMR spectroscopy.

For many spectra, the signal from residual protons or the $^{13}$C signal from the solvent can be used as a reference signal (instead of adding TMS) since the chemical shifts of most common solvents are known accurately.  In $CDCl_3$, the residual $CHCl_3$ has a shift of 7.27 ppm in the $^1$H NMR spectrum and the $^{13}$C shift of $CDCl_3$ is 77.0 ppm.

Solvents that are miscible with water (and are difficult to "dry" completely) e.g. $CD_3(CO)CD_3$, $CD_3(SO)CD_3$, and $D_2O$, also commonly contain a small amount of residual water. The residual water typically appears as a broad resonance in the region 2.5 – 5 ppm in the $^1H$ NMR spectrum.

**Table 8.1      $^1H$ and $^{13}C$ Chemical Shifts for Common NMR solvents***

| Solvent | Formula | Residual $^1H$ signal(s)* (multiplicity) | $^{13}C$ signal(s)* (multiplicity) |
|---|---|---|---|
| acetone-$d_6$ | $CD_3(C=O)CD_3$ | 2.05 (5) | 29.9 (7); 206.7 (1) |
| acetonitrile-$d_3$ | $CD_3C\equiv N$ | 1.94 (5) | 1.4 (7); 118.7 (1) |
| benzene-$d_6$ | $C_6D_6$ | 7.16 (1) | 128.4 (3) |
| chloroform-$d_1$ | $CDCl_3$ | 7.27 (1) | 77.0 (3) |
| cyclohexane-$d_{12}$ | $C_6D_{12}$ | 1.38 (3) | 26.4 (5) |
| deuterium oxide-$d_2$ | $D_2O$ | 4.81 (1) | - |
| dichloromethane-$d_2$ | $CD_2Cl_2$ | 5.32 (3) | 53.8 (5) |
| dimethylsulfoxide-$d_6$ | $CD_3(S=O)CD_3$ | 2.50 (5) | 39.5 (7) |
| 1,4-dioxane-$d_6$ | $C_4D_8O_2$ | 3.53 (3) | 66.7 (5) |
| methanol-$d_4$ | $CD_3OD$ | 4.87 (1); 3.31 (5) | 49.2 (7) |
| pyridine-$d_5$ | $C_6D_5N$ | 8.74 (1); 7.58 (1); 7.22 (1) | 150.4 (3); 135.9 (3); 123.9 (3) |
| toluene-$d_8$ | $C_6D_5CD_3$ | 7.09 (1); 7.00 (1); 6.98 (1); 2.09 (5) | 137.9 (1); 129.4 (3); 128.3 (3); 125.5(3); 20.4 (7) |

*   ppm from TMS

## 8.2    SOLVENT INDUCED SHIFTS

Generally solvents chosen for NMR spectroscopy do not associate with the solute. However, solvents which are capable of both association and inducing differential chemical shifts in the solute are sometimes deliberately used to remove accidental chemical equivalence. The most useful solvents for the purpose of inducing *solvent-shifts* are aromatic solvents, in particular hexadeuterobenzene ($C_6D_6$), and the effect is called *aromatic solvent induced shift* (ASIS). The numerical values of ASIS are usually of the order of 0.1 - 0.5 ppm and they vary with the molecule studied depending mainly on the geometry of the solvent/solute complexation.

## 8.3    DYNAMIC NMR SPECTROSCOPY: THE NMR TIME-SCALE

Two magnetic nuclei situated in different molecular environments must give rise to separate signals in the NMR spectrum, say $\Delta v$ Hz apart (Figure 8.1a). However, if some process interchanges the environments of the two nuclei at a rate ($k$) much faster than $\Delta v$ times per second, the two nuclei will be observed as a single signal at an intermediate frequency (Figures 8.1d and 8.1e). When the rates ($k$) of the exchange process are comparable to $\Delta v$, *exchange broadened* spectra (Figure 8.1b) are observed. From the exchange broadened spectra, the rate constants for the exchange process (and hence the activation parameters $\Delta G^{\neq}$, $\Delta H^{\neq}$, $\Delta S^{\neq}$) can be derived. Where signals coalesce (Figure 8.1c) from being two separate signals to a single averaged signal, the rate constant for the exchange can be approximated as $k = \pi.\Delta v / \sqrt{2}$.

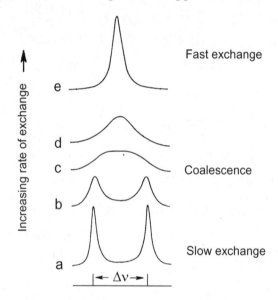

**Figure 8.1    Schematic NMR Spectra of Two Exchanging Nuclei**

In practice, a compound where an exchange process operates can give rise to a series of spectra of the type shown in Figure 8.1, if the NMR spectra are recorded at different temperatures. Changing the sample temperature alters the rate constant for the exchange (increasing the temperature increases the rate of an exchange process) and the spectra will have a different appearance depending on whether the rate constant, $k$ (expressed in sec$^{-1}$) is large or small compared to the chemical shift differences between exchanging nuclei ($\Delta v$ expressed in Hz). *Molecules where there are exchange processes taking place may also give rise to different NMR spectra in different NMR spectrometers* because $\Delta v$ depends on the strength of the magnetic field. An NMR spectrum which shows exchange broadening will tend to give a slow exchange spectrum if the spectrum is re-run in a spectrometer with a stronger magnetic field.

The averaging effects of exchange apply to any dynamic process that takes place in a molecule (or between molecules).  However, many processes occur at rates that are too fast or too slow to give rise to visible broadening of NMR spectra.  The **NMR time-scale** happens to coincide with the rates of a number of common chemical processes that give rise to variation of the appearance of NMR spectra with temperature and these include:

*(1)     Conformational exchange processes.*  Conformational processes can give rise to exchange broadening in NMR spectra when a molecule exchanges between two or more conformations.  Fortunately most conformational processes are so fast on the NMR time-scale that normally only averaged spectra are observed.  In particular, in molecules which are not unusually sterically bulky, the rotation about C-C single bonds is normally fast on the NMR time scale so, for example, the 3 hydrogen atoms of a methyl group appear as a singlet as a result of averaging of the various rotational conformers.

In molecules where there are very bulky groups, steric hindrance can slow the rotation about single bonds and give rise to broadening in NMR spectra.  In molecules containing rings, the exchange between various ring conformations (*e.g.* chair-boat-chair) can exchange nuclei.

For example, cyclohexane gives a single averaged resonance in the $^1$H NMR at room temperature, but separate signals are seen for the axial and equatorial hydrogens when spectra are acquired at very low temperature.

*(2) Intermolecular interchange of labile (slightly acidic) protons.*  Functional groups such as -OH, -COOH, -NH$_2$ and -SH have labile protons which exchange with each other in solution.  The -OH protons of a mixture of two different alcohols may give rise to either an averaged signal or to separate signals depending on the rate of exchange and this depends on many factors including temperature, the polarity of the solvent, the concentrations of the solutes and the presence of acidic or basic catalysts.

$$R\text{-}OH + R'\text{-}OH' \rightleftharpoons R\text{-}OH' + R'\text{-}OH$$

### (3)  *Rotation about partial double bonds*.

Exchange broadening is frequently observed in amides due to restricted rotation about the N-C bond of the amide group.

The restricted rotation about amide bonds often occurs at a rate that gives rise to observable broadening in NMR spectra.

The restricted rotation in amide bonds results from the partial double bond character of the C-N bond.

## 8.4   THE EFFECT OF CHIRALITY

In an achiral solvent, enantiomers will give identical NMR spectra.  However in a chiral solvent or in the presence of a chiral additive to the NMR solvent, enantiomers will have different spectra and this is frequently used to establish the enantiomeric purity of compounds.  The resonances of one enantiomer can be integrated against the resonances of the other to quantify the enantiomeric purity of a compound.

In molecules that contain a stereogenic centre, the NMR spectra can sometimes be more complex than would otherwise be expected.  Groups such as $-CH_2-$ groups (or any $-CX_2-$ group such as $-C(Me)_2-$ or $-CR_2-$) require particular attention in molecules which contain a stereogenic centre.  The carbon atom of a $-CX_2-$ group is termed a **prochiral carbon** if there is a stereogenic centre (a chiral centre) elsewhere in the molecule.  A prochiral carbon atom is a carbon in a molecule that would be chiral if one of its substituents was replaced by a different substituent.  From an NMR perspective, the important fact is that the presence of a stereogenic centre makes the substituents on a prochiral carbon atom **chemically non-equivalent**.  So whereas the protons of a $-CH_2-$ group in an acyclic aliphatic compound would normally be expected to be equivalent and resonate at the same frequency in the $^1H$ NMR spectrum, if there is a stereogenic centre in the molecule, each of the protons of the $-CH_2-$ group will appear at different chemical shifts.  Also, since they are non-equivalent, the protons will couple to each other typically with a large coupling of about 15 Hz.

The effect of chirality is particularly important in the spectra of natural products including amino acids, proteins or peptides. Many molecules derived from natural sources contain a stereogenic centre and they are typically obtained as a single pure enantiomer. In these molecules, the resonances for all of the methylene groups (*i.e.* -CH$_2$- groups) in the molecule will be complicated by the fact that the two protons of the methylene groups will be non-equivalent. Figure 8.2 shows the aliphatic protons in the $^1$H NMR spectrum of the amino acid cysteine (HSCH$_2$CHNH$_3^+$COO$^-$). Cysteine has a stereogenic centre and the signals of the methylene group appear as separate signals at δ 3.18 and δ 2.92 ppm. Each of the methylene protons is split into a doublet of doublets due to coupling firstly to the other methylene proton and secondly to the proton on the α-carbon (H$_c$).

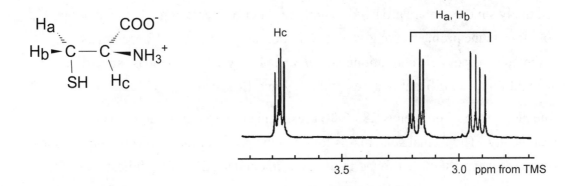

**Figure 8.2     $^1$H NMR Spectrum of the Aliphatic Region of Cysteine**

## 8.5    THE NMR SPECTRA OF "OTHER NUCLEI"

$^1$H and $^{13}$C NMR spectroscopy accounts for the overwhelming proportion of all NMR observations. However, there are many other isotopes which are NMR observable and they include the common isotopes $^{19}$F, $^{31}$P and $^2$H. The NMR spectroscopy of these "other nuclei" has had surprisingly little impact on the solution of structural problems in organic chemistry and will not be discussed here. It is however important to be alert for the presence of other magnetic nuclei in the molecule, because they often cause additional multiplicity in $^1$H and $^{13}$C NMR spectra due to spin-spin coupling.

# 9

# DETERMINING THE STRUCTURE OF ORGANIC COMPOUNDS FROM SPECTRA

The main purpose of this book is to present a collection of suitable problems to teach and train researchers in the general important methods of spectroscopy.

Problems 1 - 282 are all of the basic "structures from spectra" type, are generally relatively simple and are arranged roughly in order of increasing complexity. No solutions to the problems are given. It is important to assign NMR spectra as completely as possible and rationalise *all numbered peaks* in the mass spectrum and account for all significant features of the UV and IR spectra.

The next group of problems (283-288) present data in text form rather than graphically. The formal style that is found in the presentation of spectral data in these problems is typical of that found in the experimental section of a publication or thesis. This is a completely different type of data presentation, and one that students will encounter frequently. Problems 289 - 296 involve the quantitative analysis of mixtures using $^1$H and $^{13}$C NMR. These problems demonstrate the power of NMR in analysing samples that are not pure compounds and also develop skills in using spectral integration.

Problems 297 - 323 are a graded series of exercises in two-dimensional NMR (COSY, NOESY, HSQC, HMBC and TOCSY spectra) ranging from very simple examples to demonstrate each of the techniques, to complex examples where a combination of 1D and 2D methods is used to establish structure and distinguish between stereoisomers.

Problem 324 deals with molecular symmetry and is a useful exercise to establish how symmetry in a molecule can be established from the number of resonances in $^1$H and $^{13}$C NMR spectra. The last group of problems (325-346) are of a different type and deal with interpretation of simple $^1$H NMR spin-spin multiplets. To the best of our knowledge, problems of this type are not available in other collections and they are included here because we have found that the interpretation of multiplicity in $^1$H NMR spectra is probably the greatest single cause of confusion in the minds of students.

*Organic Structures from Spectra*, Fifth Edition. L. D. Field, S. Sternhell and J. R. Kalman.
© 2013 John Wiley & Sons, Ltd. Published 2013 by John Wiley & Sons, Ltd.

**Collection of Spectra.**

The spectra presented in the problems were obtained under conditions stated on the individual problem sheets. Mass spectra were obtained on an AEI MS-9 spectrometer or a Hewlett Packard MS-Engine mass spectrometer. 100 MHz $^1$H NMR spectra were obtained on a Varian XL-100 spectrometer, 200 MHz $^1$H NMR spectra and 50 MHz $^{13}$C NMR spectra were obtained on a Bruker AC-200 spectrometer, 300 MHz $^1$H NMR spectra and 75 MHz $^{13}$C NMR spectra were obtained on a Bruker DRX-300 spectrometer, 400 MHz $^1$H NMR spectra and 100 MHz $^{13}$C NMR spectra were obtained on Bruker AMX-400 or DRX-400 spectrometers, and 500 and 600 MHz $^1$H NMR spectra were obtained on Bruker DRX-500 or AMX-600 or DRX-600 spectrometers.

Ultraviolet spectra were recorded on a Perkin-Elmer 402 UV spectrophotometer or Hitachi 150-20 UV spectrophotometer and Infrared spectra on a Perkin-Elmer 710B or a Perkin-Elmer 1600 series FTIR spectrometer.

The following collections are useful sources of spectroscopic data on organic compounds and some of the data for literature compounds have been derived from these collections:

*(a)* http://riodb01.ibase.aist.go.jp/sdbs/cgi-bin/cre_index.cgi?lang=eng website maintained by the National Institute of Advanced Industrial Science and Technology, Tsukuba, Ibaraki, Japan;

*(b)* http://webbook.nist.gov/chemistry/ website which is the NIST Chemistry WebBook, NIST Standard Reference Database Number 69, June 2005, Eds. P.J. Linstrom and W.G. Mallard.

*(c)* E Pretch, P Bühlmann and C Affolter, "Structure Determination of Organic Compounds, Tables of Spectral Data", 3$^{rd}$ edition, Springer, Berlin 2000.

## 9.1    SOLVING PROBLEMS

While there is no doubt in our minds that the only way to acquire expertise in obtaining "organic structures from spectra" is to practise, some students have found the following **general approach to solving structural problems by a combination of spectroscopic methods** helpful:

*(1)*    Perform all **routine operations:**

   *(a)*    Determine the molecular weight from the Mass Spectrum.

   *(b)*    Determine relative numbers of protons in different environments from the $^1$H NMR spectrum.

   *(c)*    Determine the number of carbons in different environments and the number of quaternary carbons, methine carbons, methylene carbons and methyl carbons from the $^{13}$C NMR spectrum.

   *(d)*    Examine the problem for any additional data concerning composition and determine the molecular formula if possible.  From the molecular formula, determine the degree of unsaturation.

   *(e)*    Determine the molar absorbance in the UV spectrum, if applicable.

*(2)*    Examine each spectrum (IR, mass spectrum, UV, $^{13}$C NMR, $^1$H NMR) in turn for obvious **structural elements:**

   *(a)*    Examine the IR spectrum for the presence or absence of groups with diagnostic absorption bands *e.g.* carbonyl groups, hydroxyl groups, NH groups, C≡C or C≡N, *etc.*

   *(b)*    Examine the mass spectrum for typical fragments *e.g.* $PhCH_2$-, $CH_3CO$-, $CH_3$-, *etc.*

   *(c)*    Examine the UV spectrum for evidence of conjugation, aromatic rings *etc.*

   *(d)*    Examine the $^1$H NMR spectrum for $CH_3$- groups, $CH_3CH_2$- groups, aromatic protons, -$CH_nX$, exchangeable protons *etc.*

*(3)*    Write down all structural elements you have determined.  Note that some are monofunctional (*i.e.* must be end-groups, such as -$CH_3$, -C≡N, -$NO_2$) whereas some are bifunctional (*e.g.* -CO-, -$CH_2$-, -COO-), or trifunctional (*e.g.* CH, N).

   Add up the atoms of each structural element and compare the total with the molecular formula of the unknown.  The difference (if any) may give a clue to the nature of the undetermined structural elements (*e.g.* an ether oxygen). At this stage, elements of symmetry may become apparent.

*(4)* Try to assemble the structural elements. Note that **there may be more than one way of fitting them together.** Spin-spin coupling data or information about conjugation may enable you to make a definite choice between possibilities.

*(5)* Return to each spectrum (IR, UV, mass spectrum, $^{13}C$ NMR, $^1H$ NMR) in turn and *rationalise all major features* (especially all major fragments in the mass spectrum and all features of the NMR spectra) in terms of your proposed structure. Ensure that no spectral features are inconsistent with your proposed structure.

**Note on the use of data tables.** Tabulated data typically give characteristic absorptions or chemical shifts for representative compounds and these may not correlate *exactly* with those from an unknown compound. The data contained in data tables should always be used indicatively (not mechanically).

## 9.2   WORKED EXAMPLES

This section works through two problems from the text to indicate a reasonable process for obtaining the structure of the unknown compound from the spectra provided. It should be emphasised that the logic used here is by no means the only way to arrive at the correct solution but it does provide a systematic approach to obtaining structures by assembling structural fragments identified by each type of spectroscopy.

## PROBLEM 96

*(1)*   *Perform all Routine Operations*

   *(a)*   From the molecular ion, the molecular weight is 198/200. The molecular ion has two peaks of approximately equal intensity separated by two mass units. This is the characteristic pattern for a compound containing one bromine atom.

   *(b)*   The molecular formula is $C_9H_{11}Br$ so one can determine the degree of unsaturation (see Section 1.3). Replace the Br by H to give an effective molecular formula of $C_9H_{12}$ ($C_nH_m$) which gives the degree of unsaturation as $(n - m/2 +1) = 9 - 6 + 1 = 4$. The compound must contain the equivalent or 4 $\pi$ bonds and/or rings. This degree of unsaturation would be consistent with one aromatic ring (with no other elements of unsaturation).

   *(c)*   The total integral <u>across all peaks</u> in the $^1H$ spectrum is 43 mm. From the molecular formula, there are 11 protons in the structure so this corresponds

to 3.9 mm per proton. The relative numbers of protons in different environments:

| $\delta$ $^1$H (ppm) | Integral (mm) | Relative No. of hydrogens (rounded) |
|---|---|---|
| ~ 7.2 | 19 | 4.9 (5H) |
| ~ 3.3 | 8 | 2  (2H) |
| ~ 2.8 | 8 | 2  (2H) |
| ~ 2.2 | 8 | 2  (2H) |

Note that this analysis gives a total of 5+2+2+2 = 11 protons which is consistent with the molecular formula provided.

*(d)* From the $^{13}$C spectrum there are 7 carbon environments: 4 carbons are in the typical aromatic/olefinic chemical shift range and 3 carbons in the aliphatic chemical shift range. The molecular formula is $C_9H_{11}Br$ so there must be an element (or elements) of symmetry to account for the 2 carbons not apparent in the $^{13}$C spectrum.

*(e)* From the $^{13}$C DEPT spectrum there are 3 × CH resonances in the aromatic/olefinic chemical shift range and 3 × -$CH_2$- carbons in the aliphatic chemical shift range

*(f)* Calculate the extinction coefficient from the UV spectrum:

$$\varepsilon_{255} = \frac{199 \times 0.95}{0.53 \times 1.0} = 357$$

## *(2)*    *Identify any Structural Elements*

*(a)* There is no useful additional information from the infrared spectrum.

*(b)* In the mass spectrum there is a strong fragment at $m/e = 91$ and this indicates a possible Ph-$CH_2$- group.

*(c)* The ultraviolet spectrum shows a typical benzenoid absorption without further conjugation or auxochromes. This would also be consistent with the Ph-$CH_2$- group.

*(d)* From the $^{13}$C NMR spectrum, there is one resonance in the $^{13}$C {$^1$H} spectrum which does not appear in the $^{13}$C DEPT spectrum. This indicates one quaternary (non-protonated) carbon. There are 4 × $^{13}$C resonances in the aromatic region, 3 × CH and 1 × quaternary carbon, which is typical of a monosubstituted benzene ring.

*(e)* From the $^1$H NMR, there are 5 protons near $\delta$ ~7.2 which strongly suggests a monosubstituted benzene ring, consistent with *(b), (c)* and *(d)*. The Ph-$CH_2$- group is confirmed.

The triplet at approximately δ 3.3 of intensity 2H suggests a –$CH_2$- group. The downfield chemical shift suggests a –$CH_2$- X group with X being an electron withdrawing group (probably bromine). The triplet splitting indicates that there must be another –$CH_2$- as a neighbouring group. In the expanded proton spectrum 1 ppm = 42 mm and since this is a 200 MHz NMR spectrum, therefore 200 Hz = 42 mm. The triplet spacing is measured to be 1.5 mm *i.e.* 7.1 Hz and this is typical of vicinal coupling ($^3J_{HH}$).

The triplet at approximately δ 2.8 ppm of intensity 2H in the $^1$H NMR spectrum suggests a –$CH_2$- with one –$CH_2$- as a neighbour. The spacing of this triplet is almost identical with that observed for the triplet near δ 3.3.

The quintet at approximately δ 2.2 of intensity 2H has the same spacings as observed in the triplets near δ 2.8 and δ 3.3. This signal is consistent with a –$CH_2$- group coupled to *two* flanking –$CH_2$- groups. A sequence -$CH_2$-$CH_2$-$CH_2$- emerges in agreement with the $^{13}$C data.

Thus the structural elements are:

1.  Ph-$CH_2$-

2.  -$CH_2$-$CH_2$-$CH_2$-

3.  -Br


### *(3)   Assemble the Structural Elements*

Clearly there must be some common segments in these structural elements since the total number of C and H atoms adds to more than is indicated in the molecular formula. One of the –$CH_2$- groups in structural element (2) must be the benzylic -$CH_2$- group of structural element (1).

The structural elements can be assembled in only one way and this identifies the compound as 1-bromo-3-phenylpropane.

$$\text{Ph}-CH_2-CH_2-CH_2-\text{Br}$$


### *(4)   Check that the answer is consistent with all spectra.*

There are no additional strong fragments in the mass spectrum.

In the infrared spectrum there are two strong absorptions between 600 and 800 cm$^{-1}$ which are consistent with the C-Br stretch of alkyl bromide.

# PROBLEM 127

*(1)*    **Perform all Routine operations**

*(a)*    The molecular formula is given as $C_9H_{11}NO_2$. The molecular ion in the mass spectrum gives the molecular weight as 165.

*(b)*    From the molecular formula, $C_9H_{11}NO_2$, determine the degree of unsaturation (see Section 1.3). Ignore the O atoms and ignore the N and remove one H to give an effective molecular formula of $C_9H_{10}$ ($C_nH_m$) which gives the degree of unsaturation as $(n – m/2 +1) = 9 – 5 + 1 = 5$. The compound must contain the equivalent or 5 π bonds and/or rings. This degree of unsaturation would be consistent with one aromatic ring with one other ring or double bond.

*(c)*    The total integral across all peaks in the $^1H$ spectrum is 97.5 mm. From the molecular formula, there are 11 protons in the structure so this corresponds to 8.9 mm per proton. The relative numbers of protons in different environments:

| δ $^1H$ (ppm) | Integral (mm) | Relative No. of hydrogens (rounded) |
|---|---|---|
| ~ 7.9 | 17.5 | 2.0 (2H) |
| ~ 6.6 | 17.5 | 2.0 (2H) |
| ~ 4.3 | 18 | 2.0 (2H) |
| ~ 4.0 | 17.5 | 2.0 (2H) |
| ~ 1.4 | 27 | 3.0 (3H) |

Note that this analysis gives a total of $2+2+2+2 + 3 = 11$ protons which is consistent with the molecular formula provided.

*(d)*    From the $^{13}C$ spectrum there are 7 carbon environments: 4 carbons are in the typical aromatic/olefinic chemical shift range, 2 carbons in the aliphatic chemical shift range and 1 carbon at low field (167 ppm) characteristic of a carbonyl carbon. The molecular formula is given as $C_9H_{11}NO_2$ so there must be an element (or elements) of symmetry to account for the 2 carbons not apparent in the $^{13}C$ spectrum.

*(e)*    From the $^{13}C$ DEPT spectrum there are 2 × CH resonances in the aromatic/olefinic chemical shift range, 1 × -CH$_2$- and 1 × –CH$_3$ carbon in the aliphatic chemical shift range.

*(f)*    Calculate the extinction coefficient from the UV spectrum:

$$\varepsilon_{292} = \frac{165 \times 0.90}{0.0172 \times 0.5} = 17,267$$

*(2)*    *Identify any __Structural Elements__*

*(a)*    From the infrared spectrum, there is a strong absorption at 1680 cm$^{-1}$ and this is probably a C=O stretch at an unusually low frequency (such as an amide or strongly conjugated ketone).

*(b)*    In the mass spectrum there are no obvious fragment peaks, but the difference between 165 (M) and 137 = 28 suggests loss of ethylene (CH$_2$=CH$_2$) or CO.

*(c)*    In the UV spectrum, the presence of extensive conjugation is apparent from the large extinction coefficient ($\varepsilon \approx 17,000$).

*(d)*    In the $^1$H NMR spectrum:

The appearance of a 4 proton symmetrical pattern in the aromatic region near $\delta$ 7.9 and 6.6 is strongly indicative of a *para* disubstituted benzene ring. This is confirmed by the presence of two quaternary $^{13}$C resonances at $\delta$ 152 and 119 in the $^{13}$C spectrum and two CH $^{13}$C resonances at $\delta$ 131 and 113.

Note that the presence of a *para* disubstituted benzene ring also accounts for the element of symmetry identified above. The triplet of 3H intensity at approximately $\delta \sim 1.4$ and the quartet of 2H intensity at approximately $\delta \sim 4.3$ have the same spacings. On this 400 MHz NMR spectrum, 400 Hz (1 ppm) corresponds to 90 mm so the measured splitting of 1.5 mm corresponds to a coupling of about 6.7 Hz that is typical of a vicinal coupling constant. The triplet and quartet clearly correspond to an ethyl group and the downfield shift of the CH$_2$ resonance ($\delta \sim 4.3$) indicates that it must be attached to a heteroatom so this is possibly an –O-CH$_2$-CH$_3$ group.

*(e)*    In the $^{13}$C NMR spectrum:

The signals at $\delta$ 14 (-CH$_3$) and $\delta$ 60 (-CH$_2$-) in the $^{13}$C NMR spectrum confirm the presence of the ethoxy group and the 4 resonances in the aromatic region (2 × CH and 2 × quaternary carbons) confirm the presence of a *p*-disubstituted benzene ring.

The quaternary carbon signal at $\delta$ 167 in the $^{13}$C NMR spectrum indicates an ester or an amide carbonyl group.

The following structural elements have been identified so far:

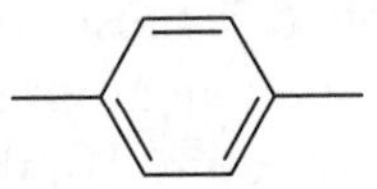

C$_6$H$_4$

ethoxy group            C$_2$H$_5$O

carbonyl group          CO

In total this accounts for C$_9$H$_9$O$_2$ and this differs from the given molecular formula only by NH$_2$. The presence of an -NH$_2$ group is confirmed by the exchangeable signal at $\delta \sim 4.1$ in the $^1$H NMR spectrum and the characteristic N-H stretching vibrations at 3200 - 3350 cm$^{-1}$ in the IR spectrum.

*(f)* The presence of one aromatic ring plus the double bond in the carbonyl group is consistent with the calculated degree of unsaturation – there can be no other rings or multiple bonds in the structure.

## *(3)* *Assemble the Structural Elements*

The structural elements:

can be assembled as either as:

or

| (A) | (B) |
|-----|-----|

These possibilities can be distinguished because:

*(a)* The **amine** –NH$_2$ group in (A) is "exchangeable with D$_2$O" as stated in the data but the **amide** –NH$_2$ group in (B) would require heating or base catalysis.

*(b)* From Table 5.4, the $^1$H chemical shift of the –O-CH$_2$- group fits better to the ester structure in (A) than the phenoxy ether structure in (B) given the models:

$$\delta\ 1.38 \quad 4.37$$

$$CH_3-CH_2-O-\underset{\underset{O}{\|}}{C}-C_6H_5$$

$$\delta\ 1.38 \quad 3.98$$

$$CH_3-CH_2-O-C_6H_5$$

*(c)* The $^{13}$C chemical shifts of the quaternary carbons in the aromatic ring are at approximately 152 and 119 ppm. From Table 6.7, these shifts would be consistent with an –NH$_2$ and an ester substituent on an aromatic ring (structure A) but for an -OEt substituent (as in structure B), the *ipso* carbon would be expected at much lower field (between 160 and 170 ppm). The $^{13}$C chemical shifts are consistent with structure (A).

*(d)*    The fragmentation pattern in the mass spectrum shown below fits (A) but not (B).  The key fragments at *m/e* 137, 120 and 92 can be rationalised only from (A).  This is decisive and ethyl 4-aminobenzoate (A) must be the correct answer.

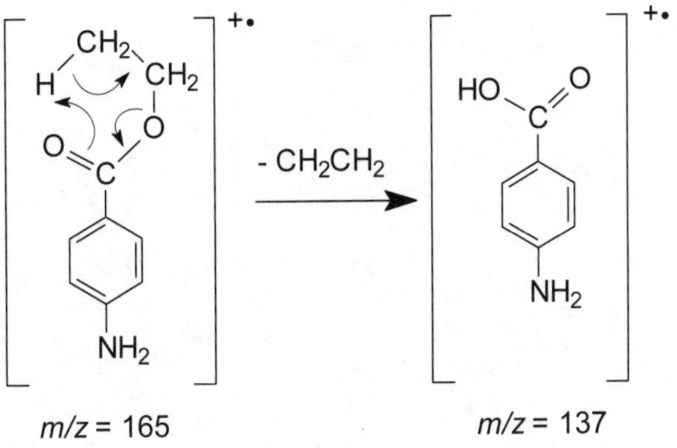

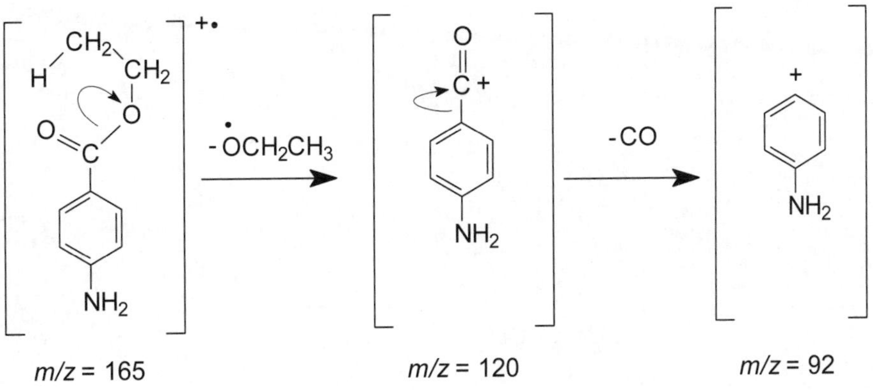

*m/z* = 165                    *m/z* = 137

*m/z* = 165                    *m/z* = 120                    *m/z* = 92

# 10

# PROBLEMS

# 10.1

# SPECTROSCOPIC IDENTIFICATION OF ORGANIC COMPOUNDS

*Organic Structures from Spectra*, Fifth Edition. L. D. Field, S. Sternhell and J. R. Kalman.
© 2013 John Wiley & Sons, Ltd. Published 2013 by John Wiley & Sons, Ltd.

# Problem 1

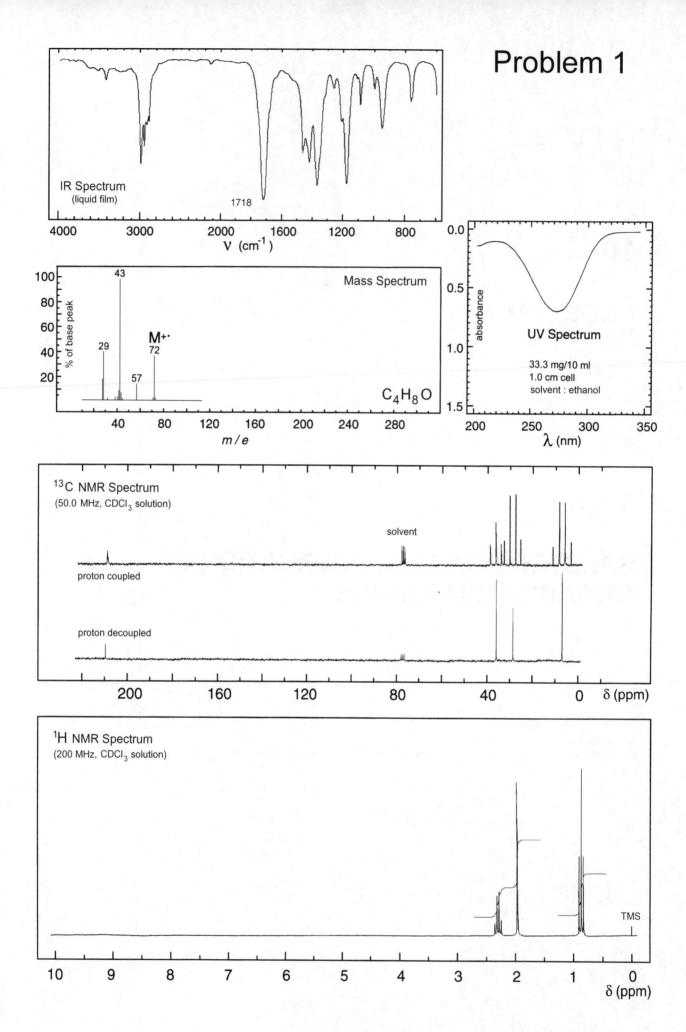

IR Spectrum
(liquid film)

1718

ν (cm⁻¹)

Mass Spectrum

C₄H₈O

% of base peak

43

29

57

M⁺·
72

m/e

UV Spectrum

33.3 mg/10 ml
1.0 cm cell
solvent : ethanol

λ (nm)

absorbance

¹³C NMR Spectrum
(50.0 MHz, CDCl₃ solution)

solvent

proton coupled

proton decoupled

δ (ppm)

¹H NMR Spectrum
(200 MHz, CDCl₃ solution)

TMS

δ (ppm)

## Problem 2

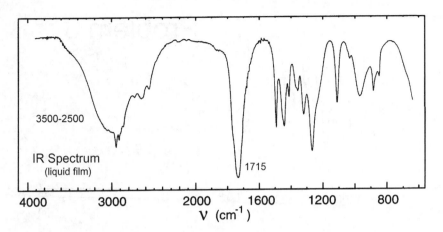

IR Spectrum
(liquid film)

3500-2500

1715

ν (cm⁻¹)

4000   3000   2000   1600   1200   800

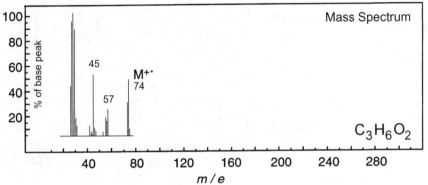

Mass Spectrum

% of base peak

45

57

M⁺·
74

C₃H₆O₂

No significant UV
absorption above 220 nm

100
80
60
40
20

40   80   120   160   200   240   280

m/e

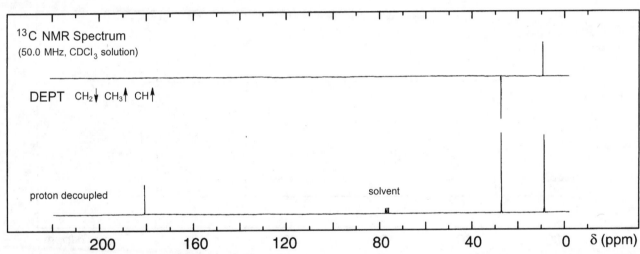

¹³C NMR Spectrum
(50.0 MHz, CDCl₃ solution)

DEPT   CH₂↓ CH₃↑ CH↑

proton decoupled

solvent

200   160   120   80   40   0   δ (ppm)

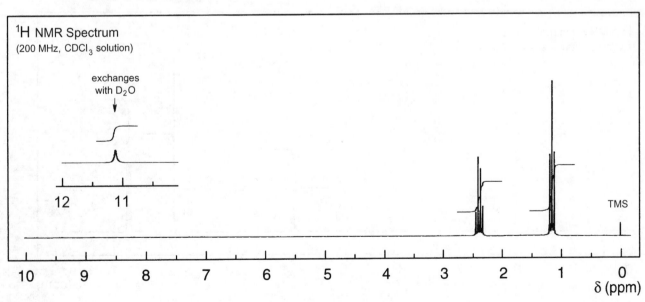

¹H NMR Spectrum
(200 MHz, CDCl₃ solution)

exchanges
with D₂O

12   11

TMS

10   9   8   7   6   5   4   3   2   1   0
δ (ppm)

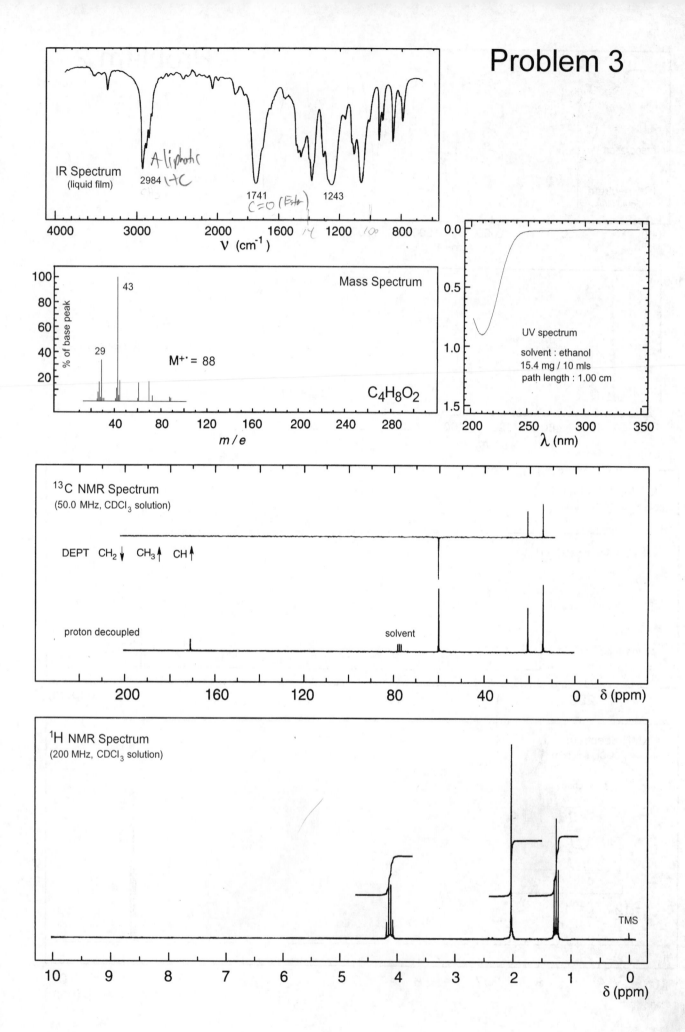

# Problem 3

**IR Spectrum** (liquid film)

2984 Aliphatic HC

1741 C=O (Ester)   1243

**Mass Spectrum**

43

29

M+· = 88

$C_4H_8O_2$

**UV spectrum**

solvent : ethanol
15.4 mg / 10 mls
path length : 1.00 cm

**$^{13}$C NMR Spectrum** (50.0 MHz, CDCl$_3$ solution)

DEPT   CH$_2$ ↓   CH$_3$ ↑   CH ↑

proton decoupled

solvent

**$^1$H NMR Spectrum** (200 MHz, CDCl$_3$ solution)

TMS

# Problem 4

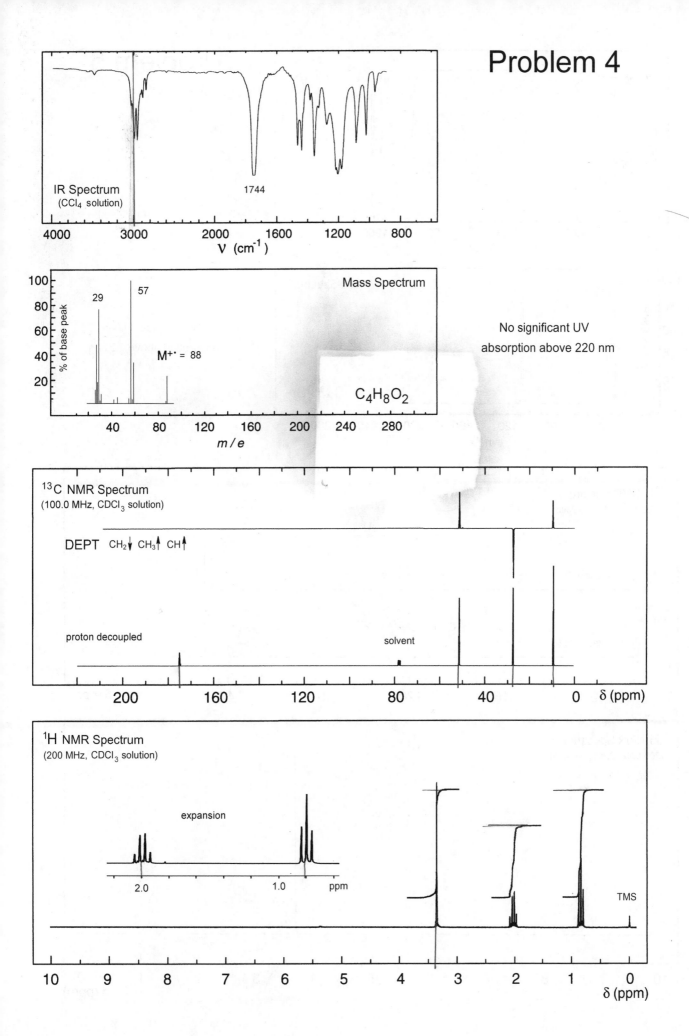

IR Spectrum
(CCl₄ solution)

1744

ν (cm⁻¹)

Mass Spectrum

29

57

M⁺· = 88

% of base peak

m/e

No significant UV
absorption above 220 nm

$C_4H_8O_2$

¹³C NMR Spectrum
(100.0 MHz, CDCl₃ solution)

DEPT    CH₂↓ CH₃↑ CH↑

proton decoupled

solvent

δ (ppm)

¹H NMR Spectrum
(200 MHz, CDCl₃ solution)

expansion

2.0

1.0

ppm

TMS

δ (ppm)

115

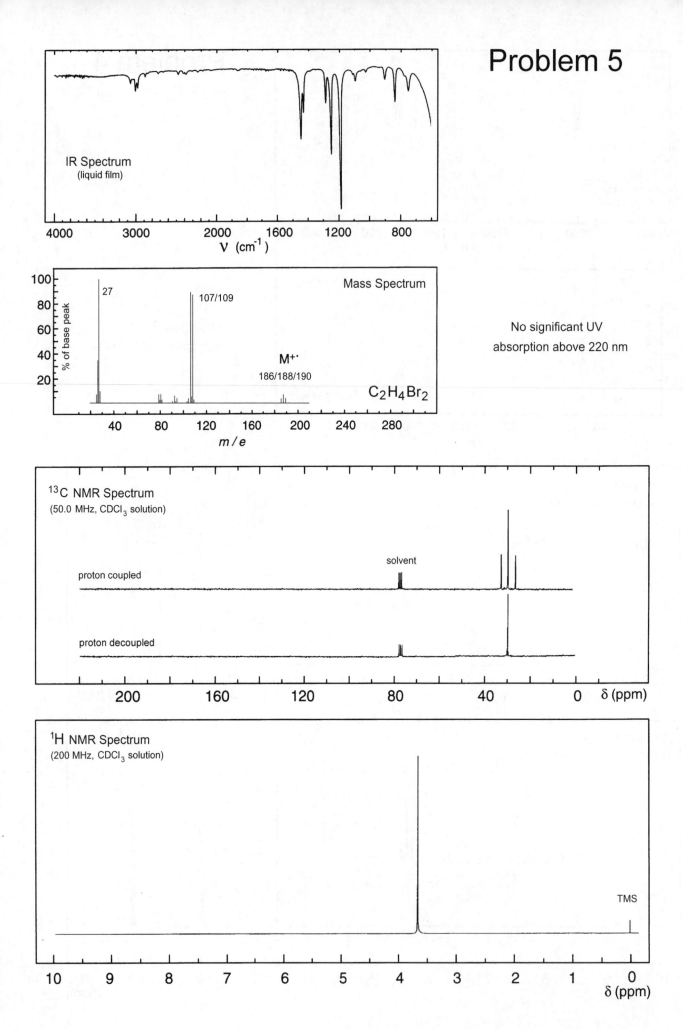

IR Spectrum
(liquid film)

4000   3000   2000   1600   1200   800

$\nu$ (cm$^{-1}$)

Mass Spectrum

100
80
60
40
20

% of base peak

27

107/109

M$^{+\cdot}$
186/188/190

C$_2$H$_4$Br$_2$

40   80   120   160   200   240   280

$m/e$

No significant UV
absorption above 220 nm

$^{13}$C NMR Spectrum
(50.0 MHz, CDCl$_3$ solution)

proton coupled

solvent

proton decoupled

200   160   120   80   40   0   $\delta$ (ppm)

$^1$H NMR Spectrum
(200 MHz, CDCl$_3$ solution)

TMS

10   9   8   7   6   5   4   3   2   1   0

$\delta$ (ppm)

116

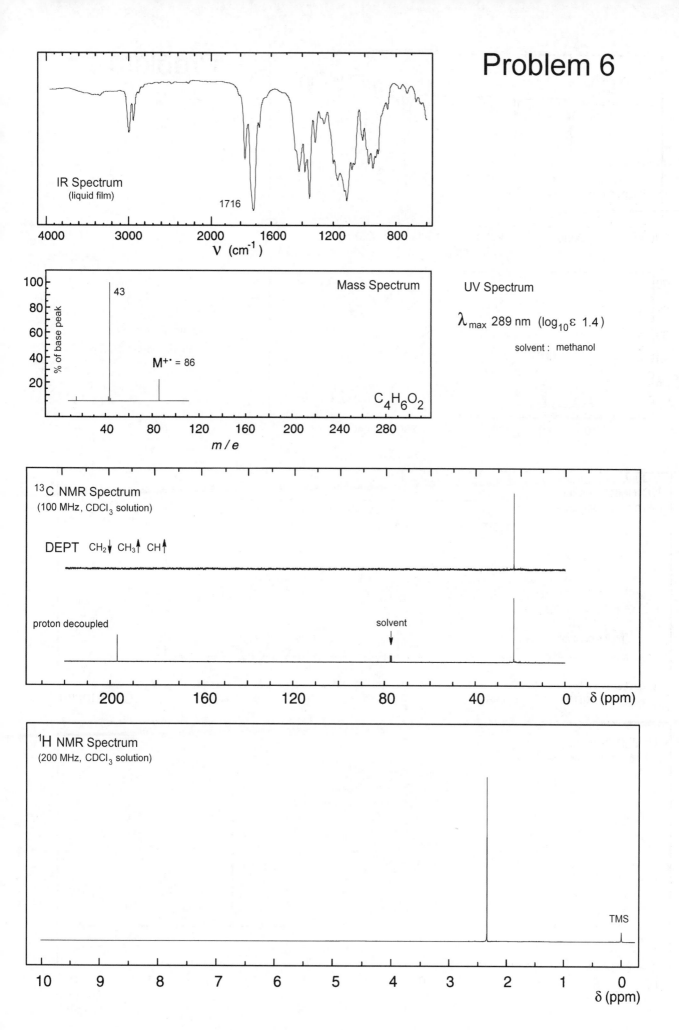

# Problem 6

IR Spectrum
(liquid film)

1716

ν (cm⁻¹)

Mass Spectrum

43

M⁺· = 86

$C_4H_6O_2$

m/e

% of base peak

UV Spectrum

$\lambda_{max}$ 289 nm  ($\log_{10}\varepsilon$  1.4)

solvent : methanol

¹³C NMR Spectrum
(100 MHz, CDCl₃ solution)

DEPT   CH₂↓ CH₃↑ CH↑

proton decoupled

solvent

δ (ppm)

¹H NMR Spectrum
(200 MHz, CDCl₃ solution)

TMS

δ (ppm)

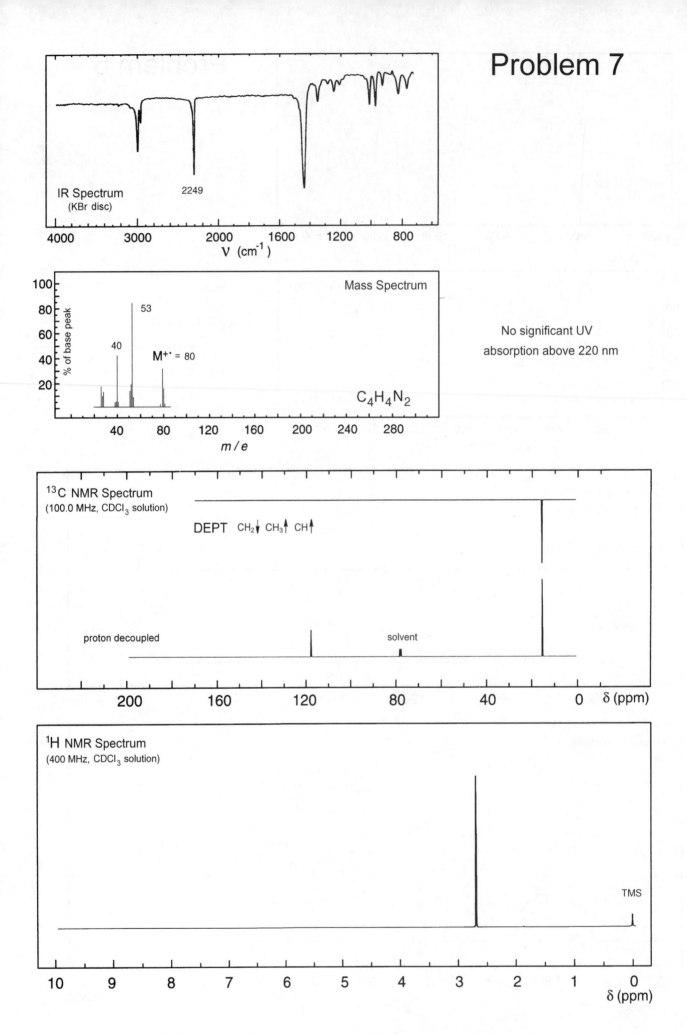

## Problem 7

IR Spectrum
(KBr disc)

2249

ν (cm⁻¹)

Mass Spectrum

53

40

M⁺˙ = 80

% of base peak

m/e

No significant UV
absorption above 220 nm

C₄H₄N₂

¹³C NMR Spectrum
(100.0 MHz, CDCl₃ solution)

DEPT   CH₂↓ CH₃↑ CH↑

proton decoupled

solvent

δ (ppm)

¹H NMR Spectrum
(400 MHz, CDCl₃ solution)

TMS

δ (ppm)

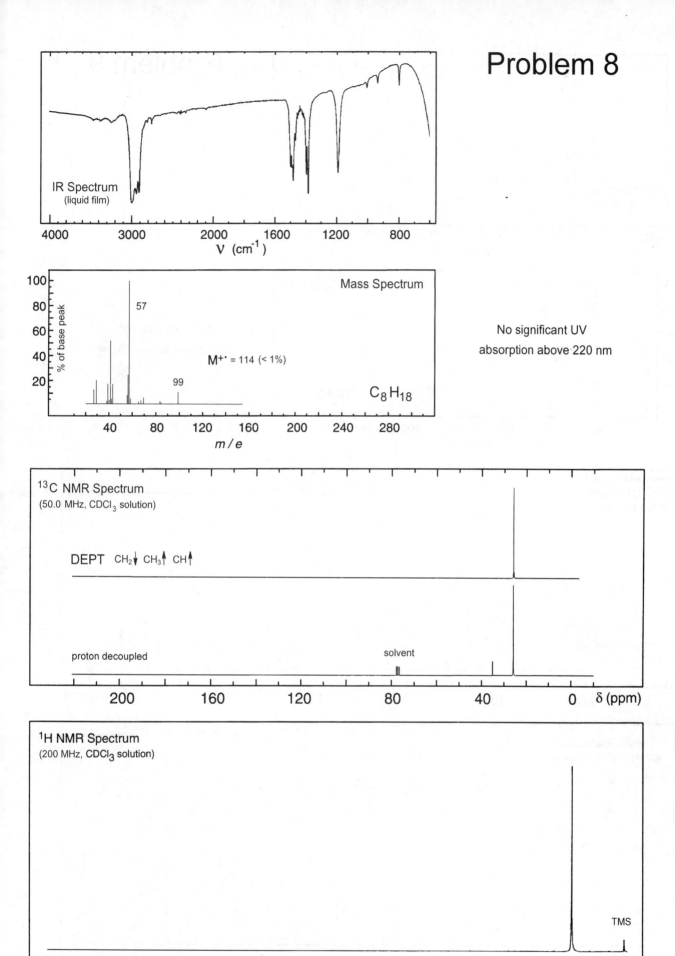

# Problem 8

**IR Spectrum**
(liquid film)

4000  3000  2000  1600  1200  800

$\nu$ (cm$^{-1}$)

**Mass Spectrum**

57

$M^{+\cdot} = 114$ (< 1%)

99

$C_8H_{18}$

% of base peak

40  80  120  160  200  240  280

*m/e*

No significant UV
absorption above 220 nm

**$^{13}$C NMR Spectrum**
(50.0 MHz, CDCl$_3$ solution)

DEPT  CH$_2\downarrow$ CH$_3\uparrow$ CH$\uparrow$

proton decoupled

solvent

200  160  120  80  40  0  $\delta$ (ppm)

**$^1$H NMR Spectrum**
(200 MHz, CDCl$_3$ solution)

TMS

10  9  8  7  6  5  4  3  2  1  0

$\delta$ (ppm)

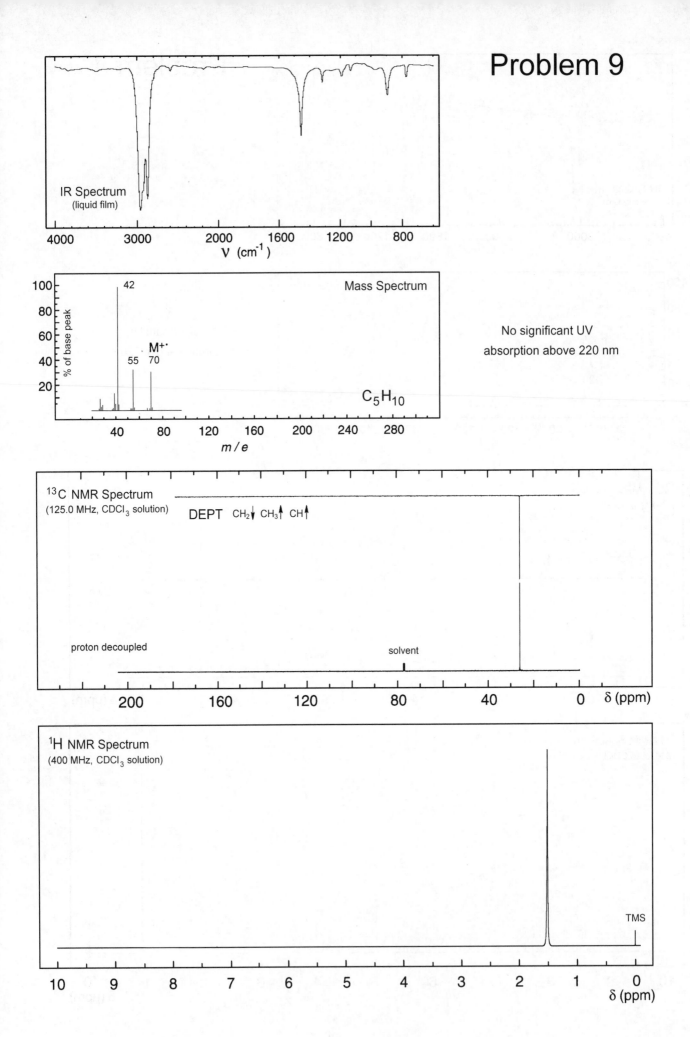

# Problem 9

**IR Spectrum**
(liquid film)

$\nu$ (cm$^{-1}$)

**Mass Spectrum**

% of base peak

42

55

**M$^{+\cdot}$**

70

$m/e$

No significant UV
absorption above 220 nm

$C_5H_{10}$

**$^{13}$C NMR Spectrum**
(125.0 MHz, CDCl$_3$ solution)

DEPT  CH$_2$↓ CH$_3$↑ CH↑

proton decoupled

solvent

$\delta$ (ppm)

**$^1$H NMR Spectrum**
(400 MHz, CDCl$_3$ solution)

TMS

$\delta$ (ppm)

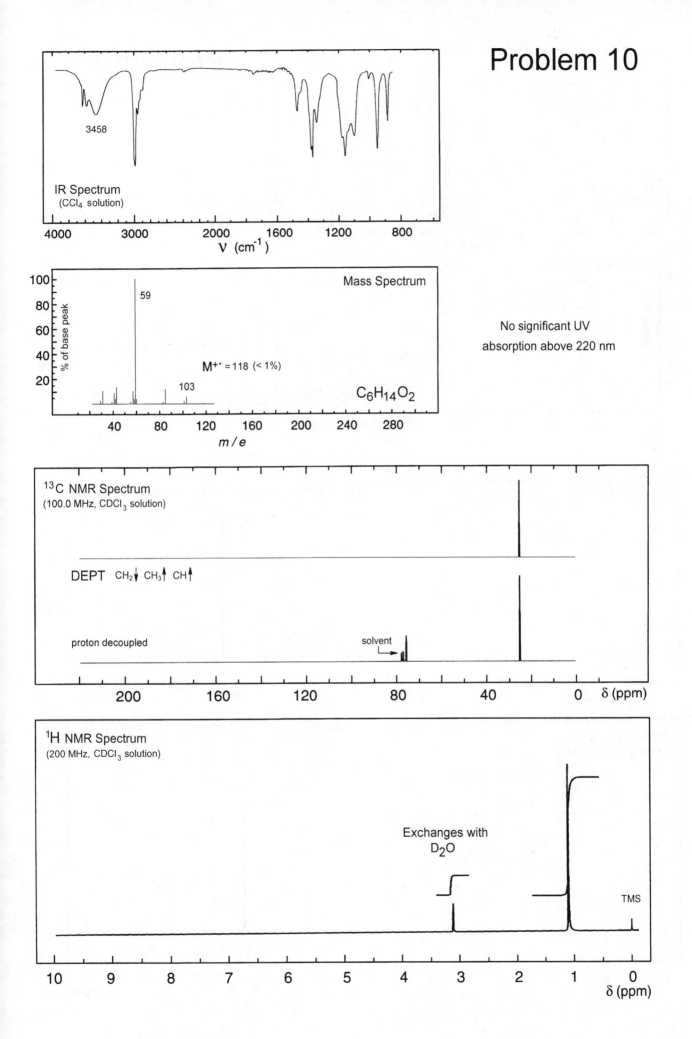

## Problem 10

IR Spectrum
(CCl₄ solution)

3458

ν (cm⁻¹)

Mass Spectrum

No significant UV
absorption above 220 nm

59

M⁺• = 118 (< 1%)

103

$C_6H_{14}O_2$

% of base peak

m / e

¹³C NMR Spectrum
(100.0 MHz, CDCl₃ solution)

DEPT   CH₂↓ CH₃↑ CH↑

proton decoupled

solvent

δ (ppm)

¹H NMR Spectrum
(200 MHz, CDCl₃ solution)

Exchanges with
D₂O

TMS

δ (ppm)

# Problem 11

IR Spectrum
(KBr disc)

1693

ν (cm⁻¹)

4000    3000    2000    1600    1200    800

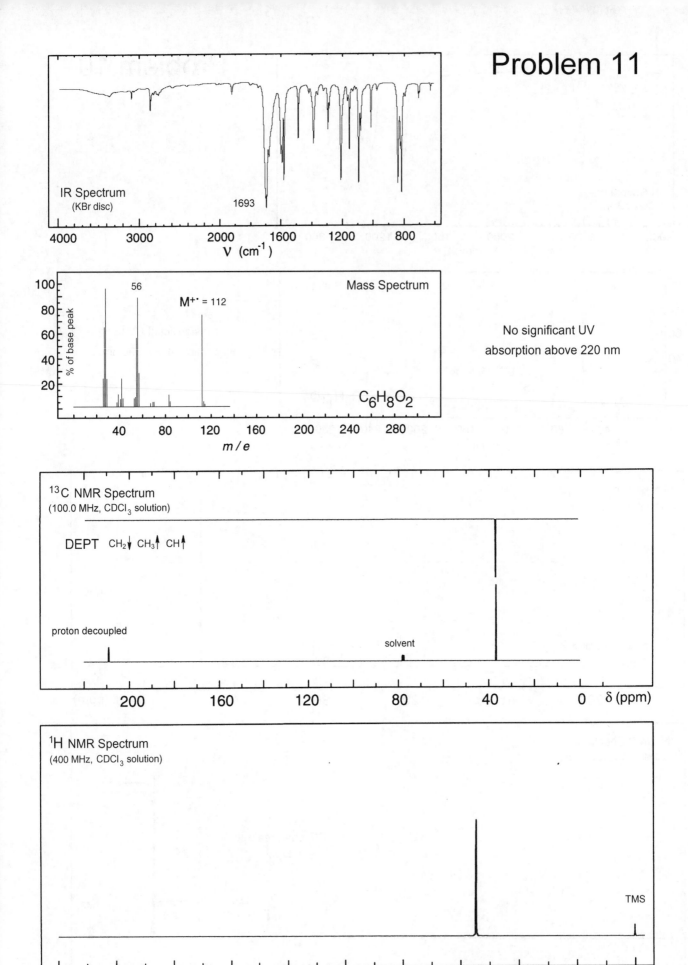

Mass Spectrum

56

M⁺˙ = 112

% of base peak

C₆H₈O₂

40    80    120    160    200    240    280

m / e

No significant UV
absorption above 220 nm

¹³C NMR Spectrum
(100.0 MHz, CDCl₃ solution)

DEPT  CH₂↓ CH₃↑ CH↑

proton decoupled

solvent

200    160    120    80    40    0    δ (ppm)

¹H NMR Spectrum
(400 MHz, CDCl₃ solution)

TMS

10    9    8    7    6    5    4    3    2    1    0    δ (ppm)

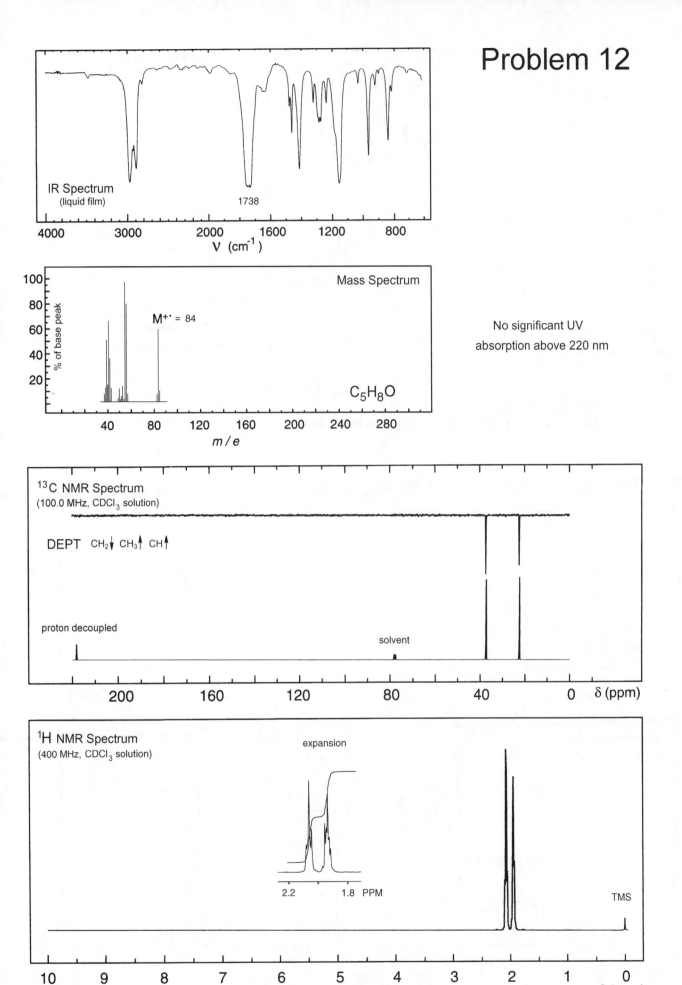

# Problem 12

IR Spectrum
(liquid film)

1738

ν (cm⁻¹)

Mass Spectrum

M⁺· = 84

C₅H₈O

% of base peak

m/e

No significant UV
absorption above 220 nm

¹³C NMR Spectrum
(100.0 MHz, CDCl₃ solution)

DEPT   CH₂↓ CH₃↑ CH↑

proton decoupled

solvent

δ (ppm)

¹H NMR Spectrum
(400 MHz, CDCl₃ solution)

expansion

2.2      1.8  PPM

TMS

δ (ppm)

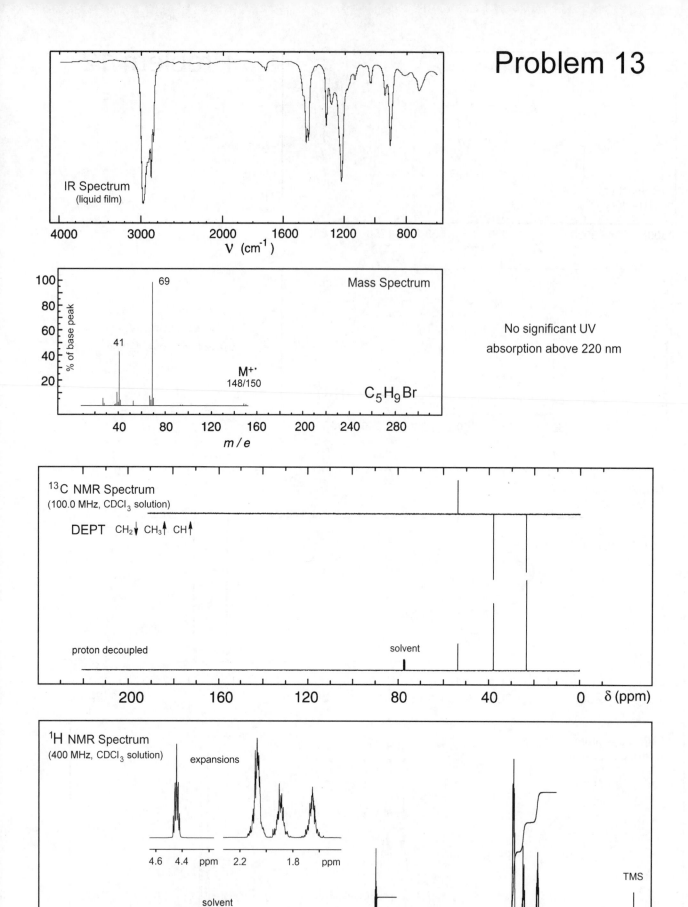

# Problem 13

IR Spectrum
(liquid film)

ν (cm⁻¹)

Mass Spectrum

No significant UV
absorption above 220 nm

M⁺·
148/150

C₅H₉Br

m/e

¹³C NMR Spectrum
(100.0 MHz, CDCl₃ solution)

DEPT  CH₂↓ CH₃↑ CH↑

proton decoupled

solvent

δ (ppm)

¹H NMR Spectrum
(400 MHz, CDCl₃ solution)

expansions

4.6  4.4  ppm    2.2    1.8    ppm

solvent
residual

TMS

δ (ppm)

124

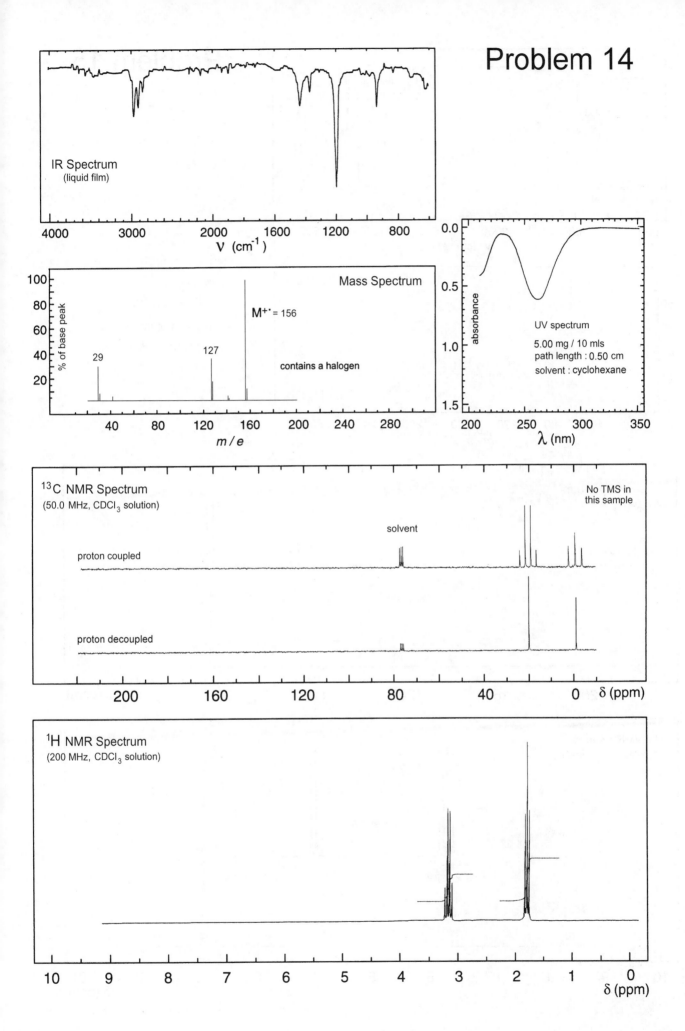

# Problem 14

**IR Spectrum** (liquid film)

$\nu$ (cm$^{-1}$)

**Mass Spectrum**

$M^{+\cdot} = 156$

contains a halogen

% of base peak

29

127

$m/e$

**UV spectrum**

5.00 mg / 10 mls
path length : 0.50 cm
solvent : cyclohexane

absorbance

$\lambda$ (nm)

**$^{13}$C NMR Spectrum**
(50.0 MHz, CDCl$_3$ solution)

No TMS in this sample

solvent

proton coupled

proton decoupled

$\delta$ (ppm)

**$^1$H NMR Spectrum**
(200 MHz, CDCl$_3$ solution)

$\delta$ (ppm)

# Problem 15

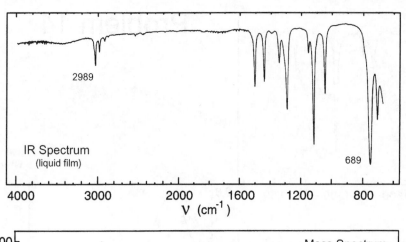

IR Spectrum
(liquid film)

2989

689

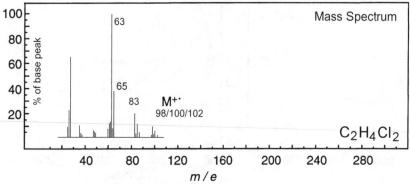

Mass Spectrum

63

65

83

M⁺·
98/100/102

$C_2H_4Cl_2$

No significant UV
absorption above 220 nm

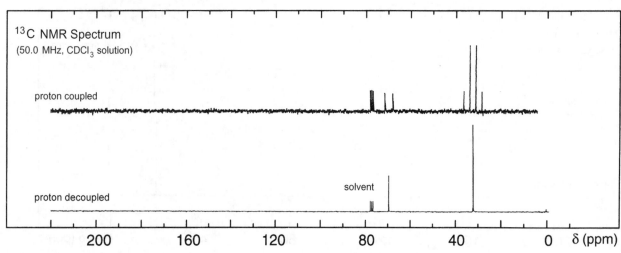

$^{13}C$ NMR Spectrum
(50.0 MHz, CDCl₃ solution)

proton coupled

proton decoupled

solvent

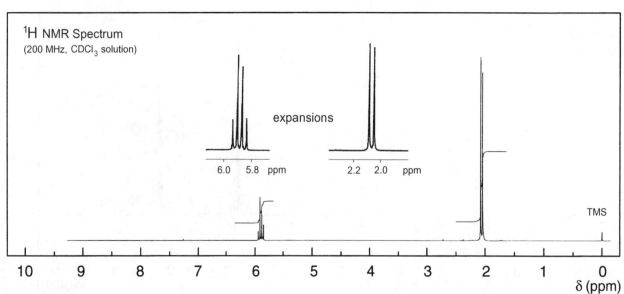

$^1H$ NMR Spectrum
(200 MHz, CDCl₃ solution)

expansions

6.0   5.8  ppm

2.2   2.0  ppm

TMS

# Problem 16

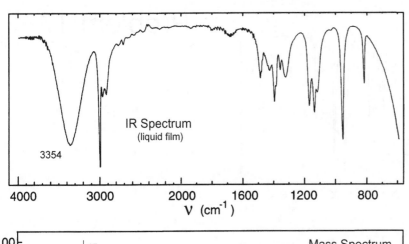

IR Spectrum
(liquid film)

3354

$\nu$ (cm$^{-1}$)

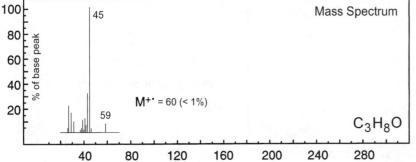

Mass Spectrum

% of base peak

45

59

M$^{+\cdot}$ = 60 (< 1%)

$C_3H_8O$

No significant UV
absorption above 220 nm

$m/e$

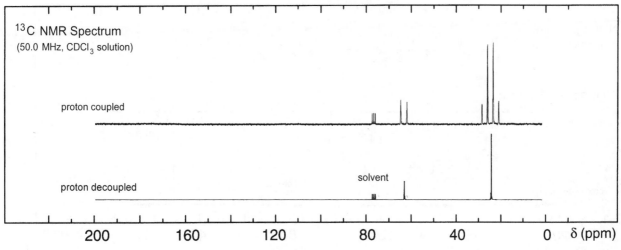

$^{13}$C NMR Spectrum
(50.0 MHz, CDCl$_3$ solution)

proton coupled

proton decoupled

solvent

$\delta$ (ppm)

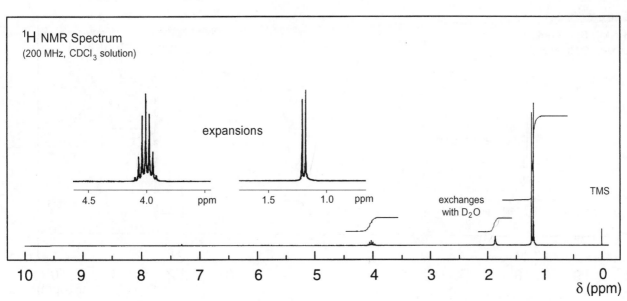

$^1$H NMR Spectrum
(200 MHz, CDCl$_3$ solution)

expansions

ppm

ppm

exchanges
with D$_2$O

TMS

$\delta$ (ppm)

# Problem 17

## IR Spectrum
(liquid film)

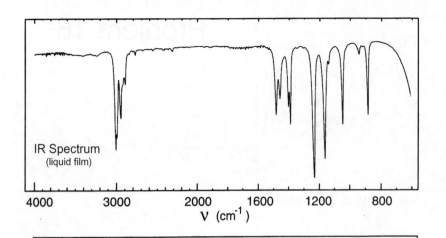

$\nu$ (cm$^{-1}$)

## Mass Spectrum

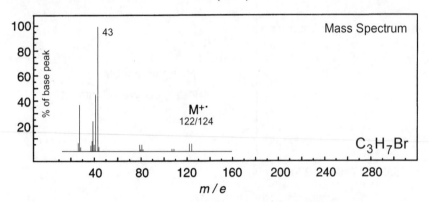

% of base peak

43

M$^{+\cdot}$
122/124

$C_3H_7Br$

m/e

No significant UV
absorption above 220 nm

## $^{13}C$ NMR Spectrum
(50.0 MHz, CDCl$_3$ solution)

DEPT  CH$_2$↓ CH$_3$↑ CH↑

proton decoupled

solvent

$\delta$ (ppm)

## $^1H$ NMR Spectrum
(200 MHz, CDCl$_3$ solution)

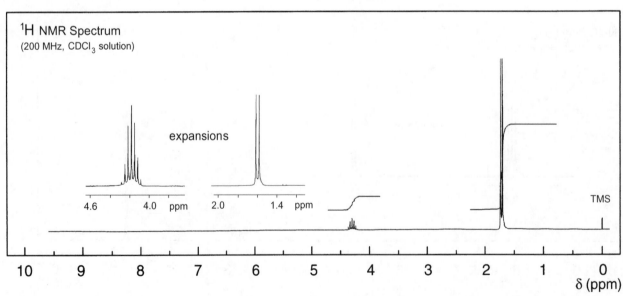

expansions

4.6    4.0  ppm

2.0    1.4  ppm

TMS

$\delta$ (ppm)

# Problem 18

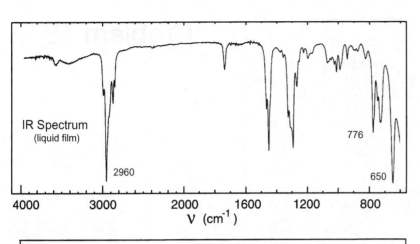

IR Spectrum
(liquid film)

2960
776
650

ν (cm⁻¹)

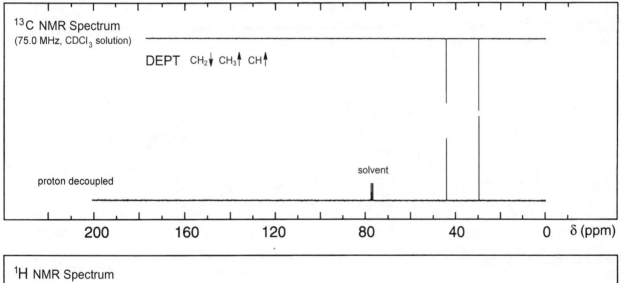

Mass Spectrum

% of base peak

55

41

90

92

M⁺· 126/128/130 < 1%

$C_4H_8Cl_2$

m/e

No significant UV
absorption above 220 nm

## ¹³C NMR Spectrum
(75.0 MHz, CDCl₃ solution)

DEPT   CH₂↓  CH₃↑  CH↑

proton decoupled

solvent

δ (ppm)

## ¹H NMR Spectrum
(300 MHz, CDCl₃ solution)

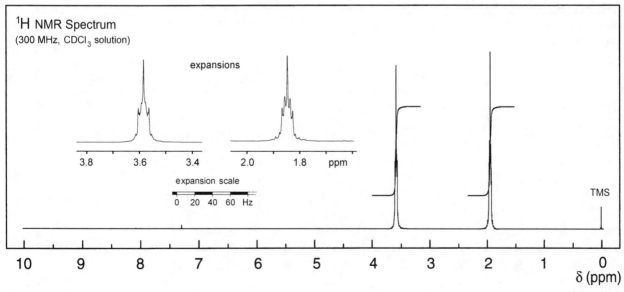

expansions

3.8   3.6   3.4

2.0   1.8   ppm

expansion scale

0   20   40   60  Hz

TMS

δ (ppm)

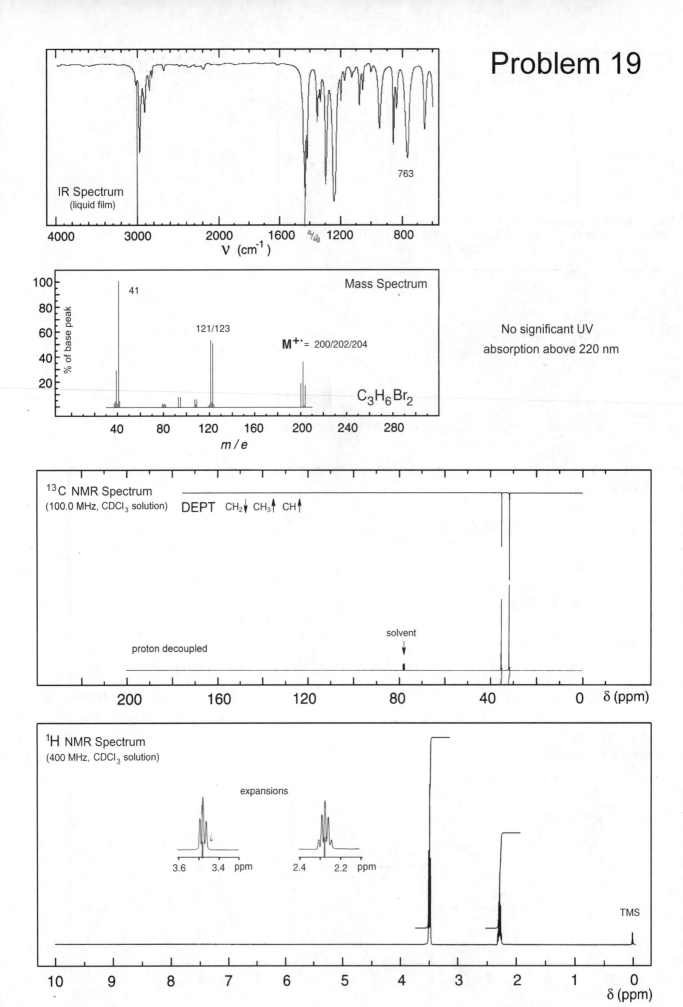

# Problem 19

IR Spectrum
(liquid film)

763

$\nu$ (cm$^{-1}$)

Mass Spectrum

41

121/123

$M^{+\cdot}$= 200/202/204

% of base peak

$C_3H_6Br_2$

m/e

No significant UV
absorption above 220 nm

$^{13}$C NMR Spectrum
(100.0 MHz, CDCl$_3$ solution)

DEPT   CH$_2$↓ CH$_3$↑ CH↑

proton decoupled

solvent

$\delta$ (ppm)

$^1$H NMR Spectrum
(400 MHz, CDCl$_3$ solution)

expansions

3.6    3.4  ppm

2.4    2.2  ppm

TMS

$\delta$ (ppm)

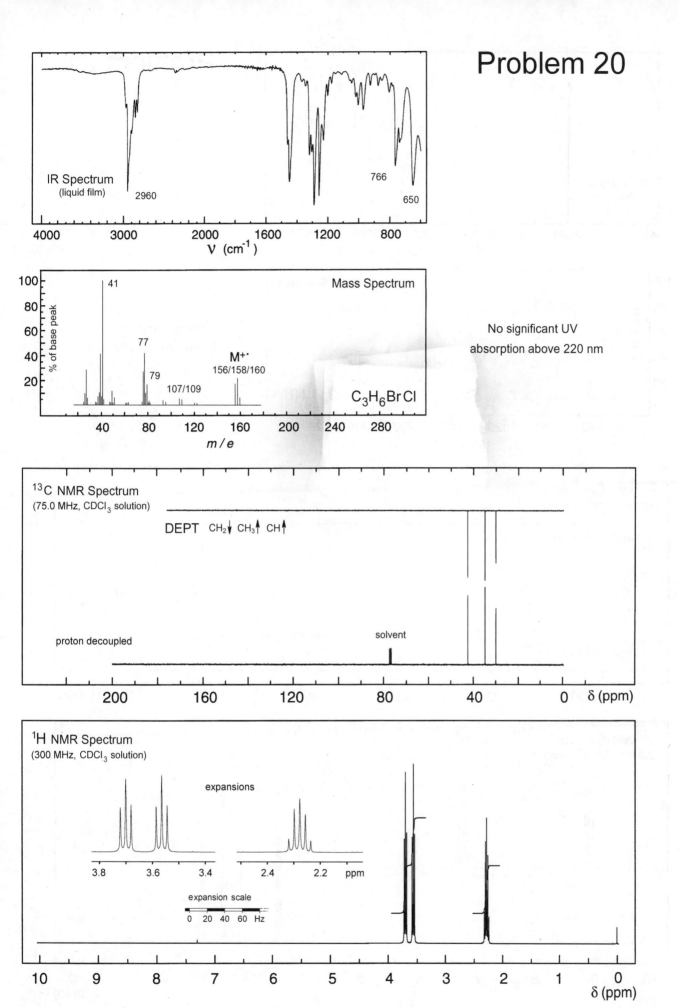

# Problem 20

**IR Spectrum**
(liquid film)
2960
766
650

**Mass Spectrum**
41
77
79
107/109
M⁺·
156/158/160

$C_3H_6BrCl$

No significant UV
absorption above 220 nm

**¹³C NMR Spectrum**
(75.0 MHz, CDCl₃ solution)

DEPT  CH₂↓ CH₃↑ CH↑

proton decoupled

solvent

δ (ppm)

**¹H NMR Spectrum**
(300 MHz, CDCl₃ solution)

expansions

3.8  3.6  3.4

2.4  2.2  ppm

expansion scale

0  20  40  60  Hz

δ (ppm)

# Problem 21

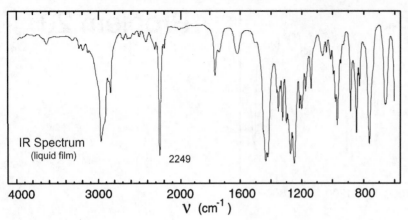

IR Spectrum
(liquid film)

2249

ν (cm⁻¹)

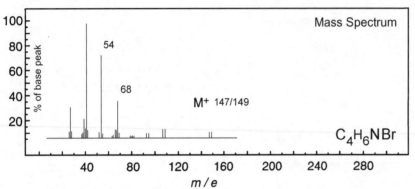

Mass Spectrum

% of base peak

54

68

M⁺ 147/149

C₄H₆NBr

m/e

No significant UV
absorption above 220 nm

¹³C NMR Spectrum
(100 MHz, CDCl₃ solution)

DEPT  CH₂↓  CH₃↑  CH↑

solvent

proton decoupled

δ (ppm)

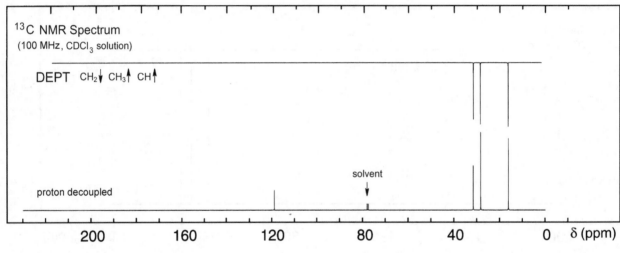

¹H NMR Spectrum
(200 MHz, CDCl₃ solution)

expansion

3.4    3.0    2.6    2.2  ppm

TMS

δ (ppm)

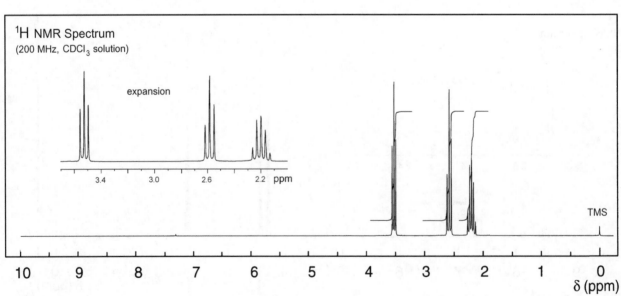

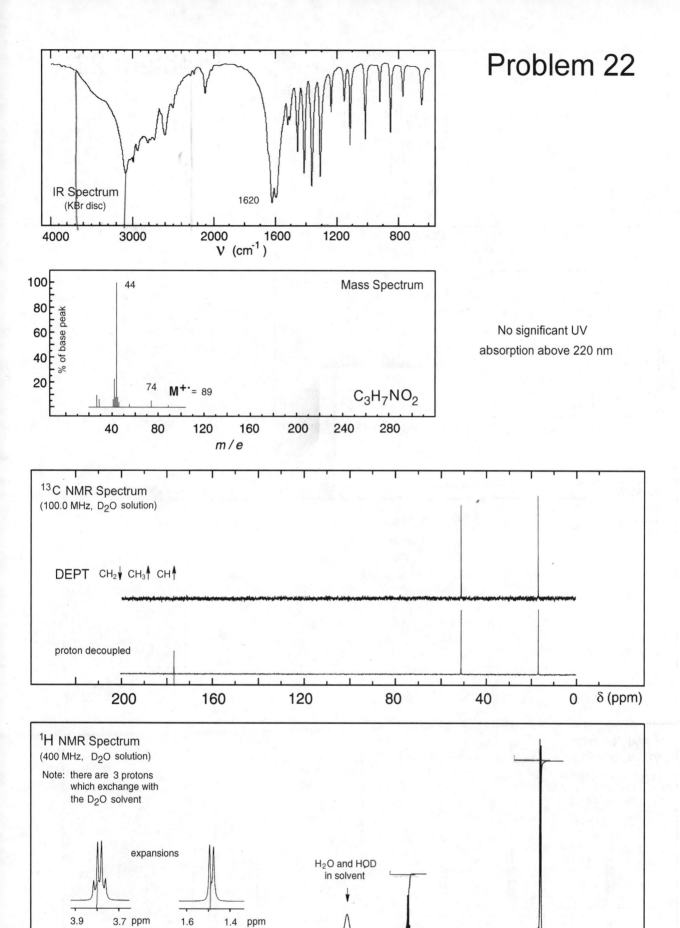

# Problem 22

IR Spectrum
(KBr disc)

1620

ν (cm⁻¹)

Mass Spectrum

44

% of base peak

74   M⁺· = 89

C₃H₇NO₂

m/e

No significant UV
absorption above 220 nm

¹³C NMR Spectrum
(100.0 MHz, D₂O solution)

DEPT   CH₂↓ CH₃↑ CH↑

proton decoupled

δ (ppm)

¹H NMR Spectrum
(400 MHz,  D₂O solution)

Note:  there are  3 protons
which exchange with
the D₂O solvent

expansions

3.9        3.7 ppm        1.6      1.4  ppm

H₂O and HOD
in solvent

δ (ppm)

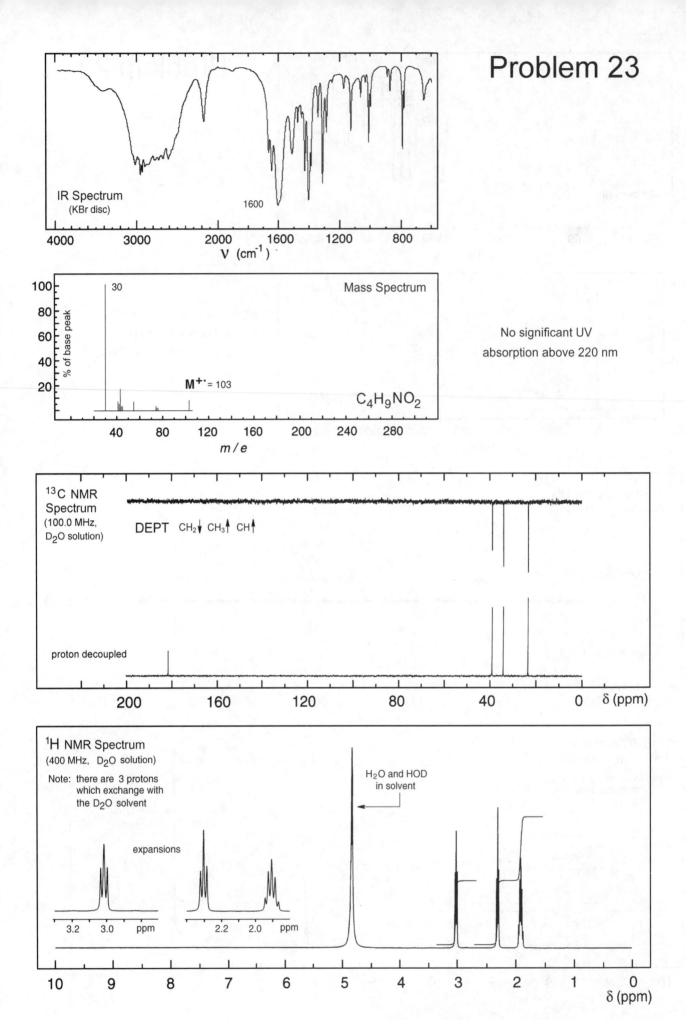

IR Spectrum
(KBr disc)

1600

4000    3000    2000    1600    1200    800

ν (cm⁻¹)

Mass Spectrum

No significant UV
absorption above 220 nm

30

% of base peak

M⁺˙ = 103

$C_4H_9NO_2$

40    80    120    160    200    240    280

m / e

¹³C NMR
Spectrum
(100.0 MHz,
D₂O solution)

DEPT   CH₂↓ CH₃↑ CH↑

proton decoupled

200    160    120    80    40    0    δ (ppm)

¹H NMR Spectrum
(400 MHz,  D₂O solution)

Note:  there are 3 protons
which exchange with
the D₂O solvent

H₂O and HOD
in solvent

expansions

3.2    3.0    ppm        2.2    2.0    ppm

10    9    8    7    6    5    4    3    2    1    0

δ (ppm)

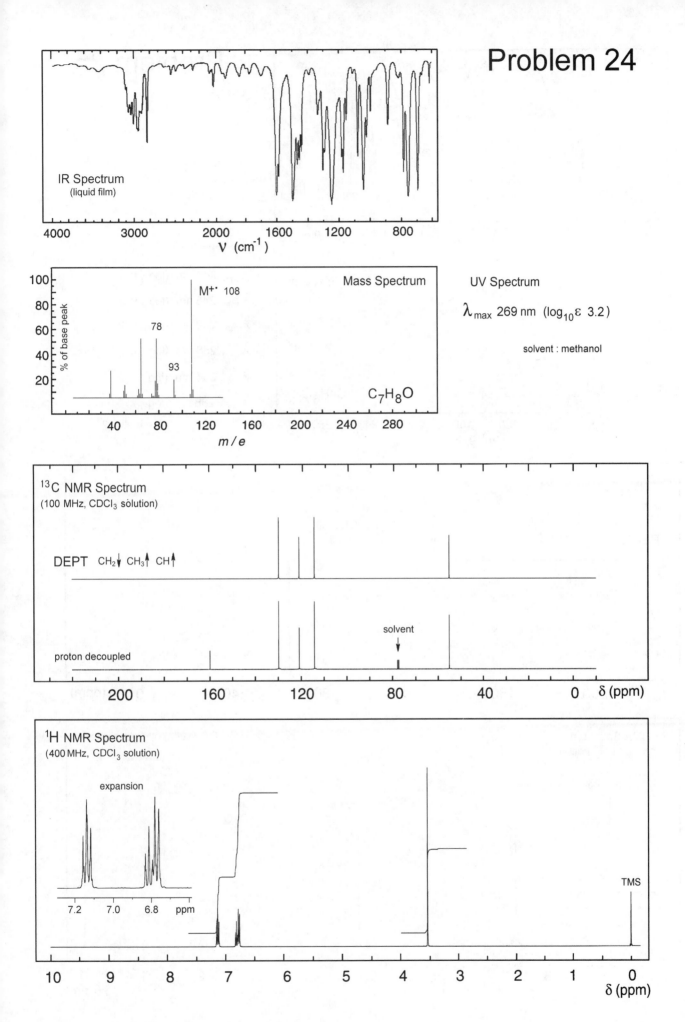

## Problem 24

**IR Spectrum**
(liquid film)

ν (cm⁻¹)

$\nu$ (cm$^{-1}$)

Mass Spectrum

M⁺· 108

78

93

$C_7H_8O$

UV Spectrum

$\lambda_{max}$ 269 nm (log₁₀ε 3.2)

$\lambda_{max}$ 269 nm ($\log_{10}\varepsilon$ 3.2)

solvent : methanol

% of base peak

m/e

**¹³C NMR Spectrum**
(100 MHz, CDCl₃ solution)

DEPT  CH₂↓ CH₃↑ CH↑

solvent

proton decoupled

δ (ppm)

**¹H NMR Spectrum**
(400 MHz, CDCl₃ solution)

expansion

7.2  7.0  6.8  ppm

TMS

δ (ppm)

# Problem 25

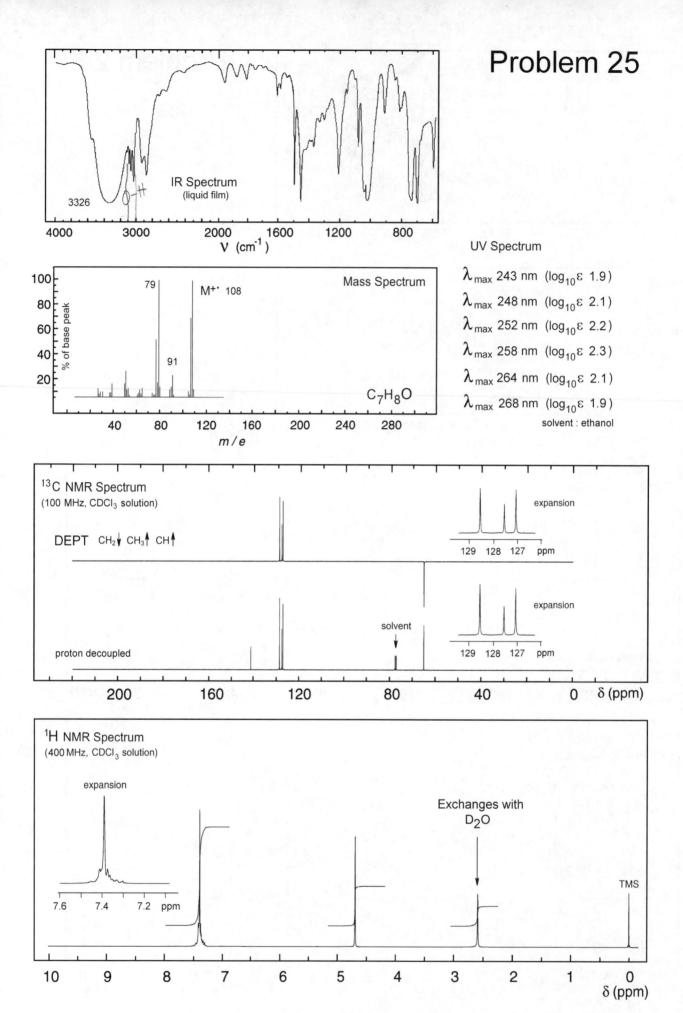

IR Spectrum
(liquid film)

3326

ν (cm⁻¹)

Mass Spectrum

79

91

M⁺· 108

C₇H₈O

## UV Spectrum

$\lambda_{max}$ 243 nm ($\log_{10}\varepsilon$ 1.9)

$\lambda_{max}$ 248 nm ($\log_{10}\varepsilon$ 2.1)

$\lambda_{max}$ 252 nm ($\log_{10}\varepsilon$ 2.2)

$\lambda_{max}$ 258 nm ($\log_{10}\varepsilon$ 2.3)

$\lambda_{max}$ 264 nm ($\log_{10}\varepsilon$ 2.1)

$\lambda_{max}$ 268 nm ($\log_{10}\varepsilon$ 1.9)

solvent : ethanol

## ¹³C NMR Spectrum
(100 MHz, CDCl₃ solution)

DEPT   CH₂↓ CH₃↑ CH↑

expansion

solvent

expansion

proton decoupled

δ (ppm)

## ¹H NMR Spectrum
(400 MHz, CDCl₃ solution)

expansion

Exchanges with
D₂O

TMS

δ (ppm)

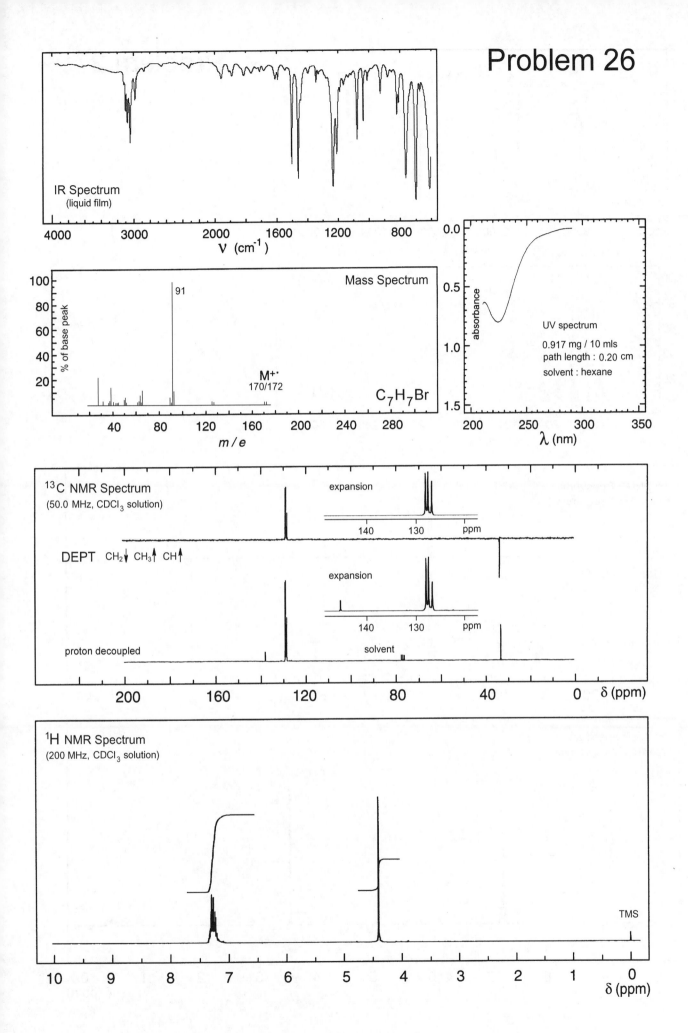

# Problem 26

**IR Spectrum**
(liquid film)

ν (cm⁻¹): $\nu$ (cm$^{-1}$)

Mass Spectrum

91

M⁺˙
170/172

$C_7H_7Br$

% of base peak

m/e: $m/e$

UV spectrum

0.917 mg / 10 mls
path length : 0.20 cm
solvent : hexane

λ (nm): $\lambda$ (nm)

absorbance

**¹³C NMR Spectrum**
(50.0 MHz, CDCl₃ solution)

expansion

140    130    ppm

DEPT    CH₂↓ CH₃↑ CH↑

expansion

140    130    ppm

proton decoupled                    solvent

δ (ppm): $\delta$ (ppm)

**¹H NMR Spectrum**
(200 MHz, CDCl₃ solution)

TMS

δ (ppm): $\delta$ (ppm)

# Problem 27

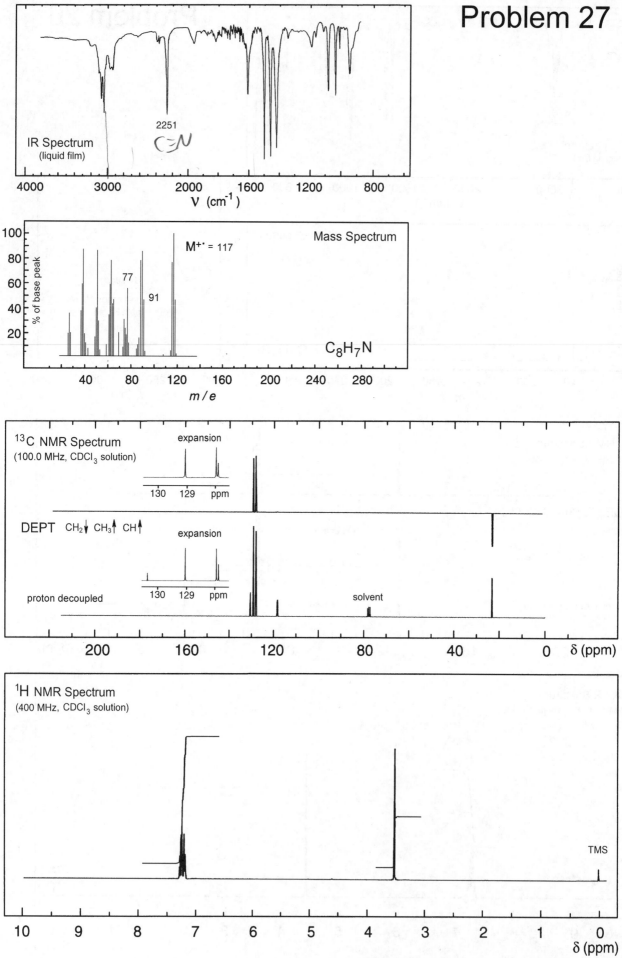

IR Spectrum
(liquid film)

2251
C≡N

Mass Spectrum

M⁺· = 117

77

91

C₈H₇N

¹³C NMR Spectrum
(100.0 MHz, CDCl₃ solution)

expansion

130    129    ppm

DEPT    CH₂↓ CH₃↑ CH↑

expansion

proton decoupled

130    129    ppm

solvent

¹H NMR Spectrum
(400 MHz, CDCl₃ solution)

TMS

# Problem 28

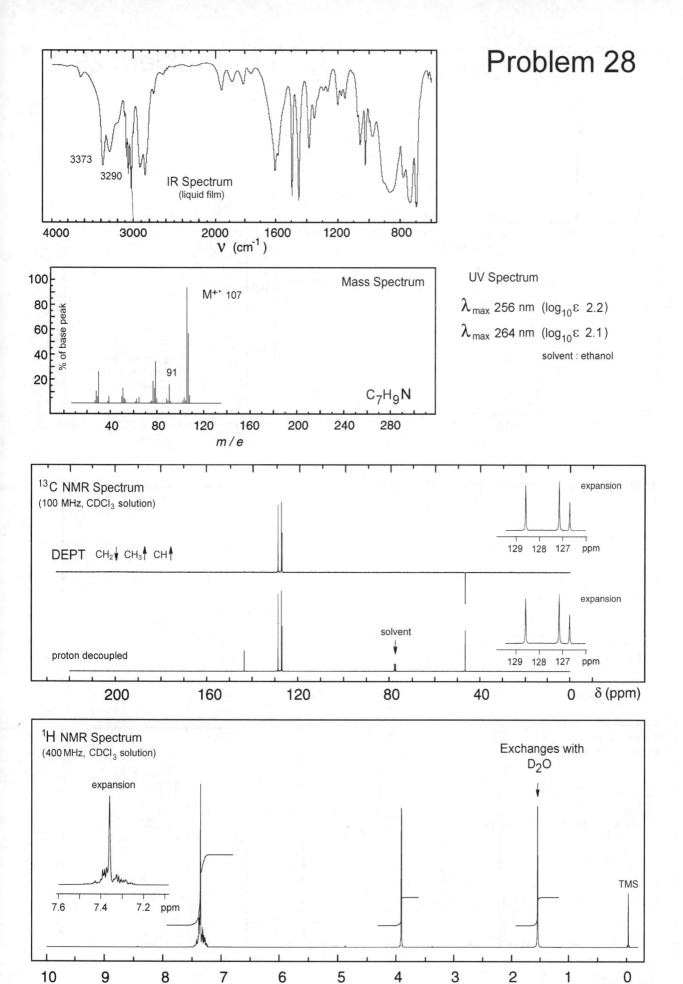

IR Spectrum
(liquid film)

3373
3290

ν (cm⁻¹)

Mass Spectrum

M⁺· 107

91

$C_7H_9N$

UV Spectrum

$\lambda_{max}$ 256 nm (log₁₀ε 2.2)

$\lambda_{max}$ 264 nm (log₁₀ε 2.1)

solvent : ethanol

¹³C NMR Spectrum
(100 MHz, CDCl₃ solution)

expansion

DEPT  CH₂↓ CH₃↑ CH↑

proton decoupled

solvent

expansion

δ (ppm)

¹H NMR Spectrum
(400 MHz, CDCl₃ solution)

Exchanges with
D₂O

expansion

TMS

δ (ppm)

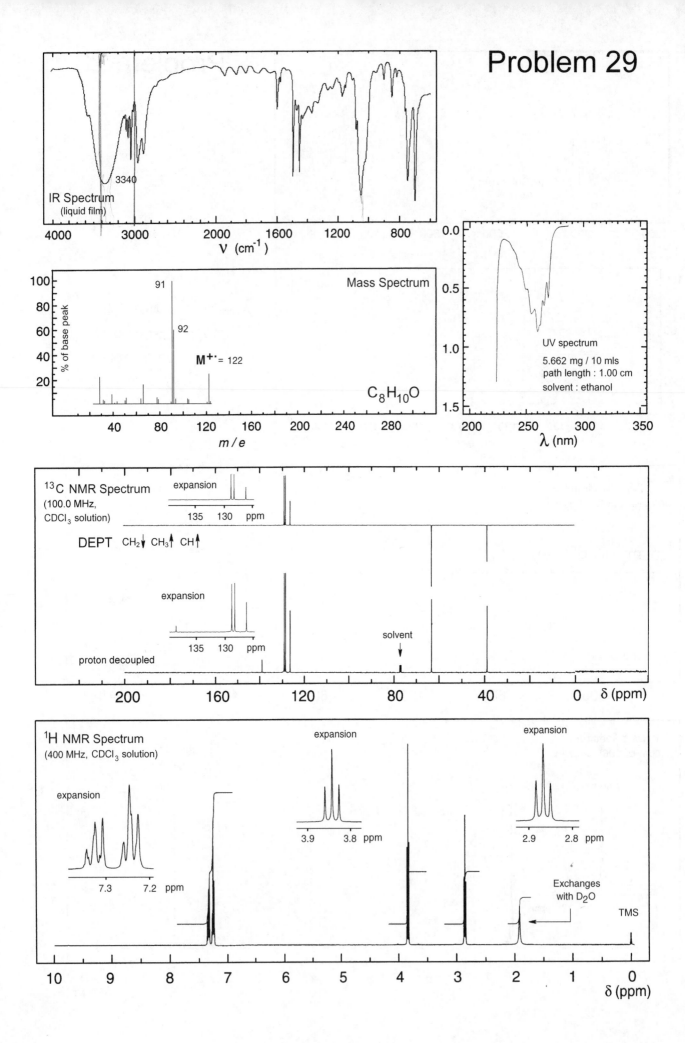

# Problem 29

IR Spectrum
(liquid film)

3340

ν (cm⁻¹)

Mass Spectrum

% of base peak

91

92

M⁺˙ = 122

C₈H₁₀O

m/e

UV spectrum

5.662 mg / 10 mls
path length : 1.00 cm
solvent : ethanol

λ (nm)

¹³C NMR Spectrum
(100.0 MHz,
CDCl₃ solution)

expansion

135   130   ppm

DEPT   CH₂↓ CH₃↑ CH↑

expansion

135   130   ppm

solvent

proton decoupled

δ (ppm)

¹H NMR Spectrum
(400 MHz, CDCl₃ solution)

expansion

expansion

3.9   3.8 ppm

expansion

2.9   2.8 ppm

expansion

7.3   7.2 ppm

Exchanges
with D₂O

TMS

δ (ppm)

# Problem 30

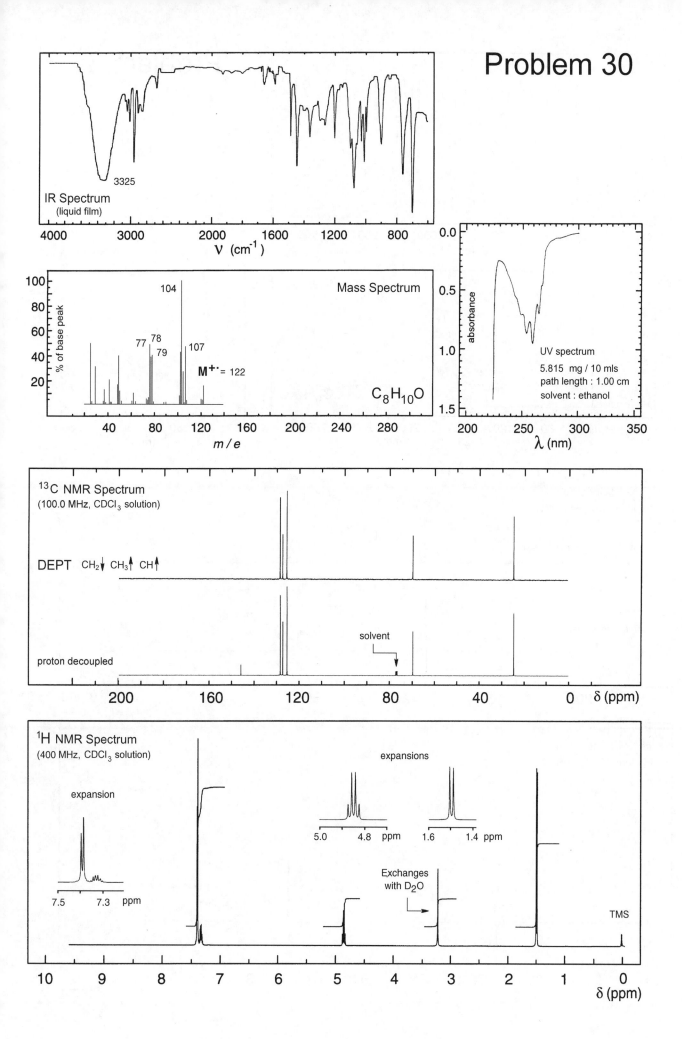

IR Spectrum
(liquid film)
3325

ν (cm⁻¹)

Mass Spectrum

% of base peak

104
77 78
79
107
M⁺· = 122

C₈H₁₀O

m/e

UV spectrum
5.815 mg / 10 mls
path length : 1.00 cm
solvent : ethanol

absorbance

λ (nm)

¹³C NMR Spectrum
(100.0 MHz, CDCl₃ solution)

DEPT   CH₂↓ CH₃↑ CH↑

proton decoupled

solvent

δ (ppm)

¹H NMR Spectrum
(400 MHz, CDCl₃ solution)

expansion

expansions

7.5    7.3  ppm

5.0    4.8  ppm

1.6    1.4  ppm

Exchanges
with D₂O

TMS

δ (ppm)

# Problem 31

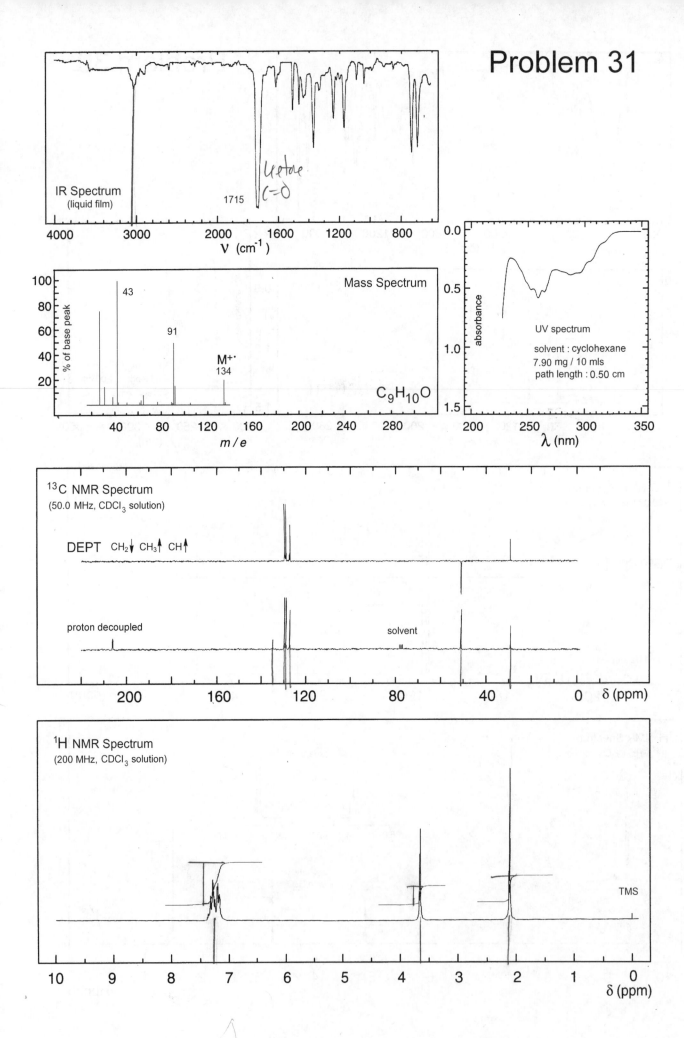

IR Spectrum
(liquid film)

1715  Uetore C=O

ν (cm⁻¹)

Mass Spectrum

43

91

M⁺·
134

$C_9H_{10}O$

% of base peak

m/e

UV spectrum

solvent : cyclohexane
7.90 mg / 10 mls
path length : 0.50 cm

absorbance

λ (nm)

¹³C NMR Spectrum
(50.0 MHz, CDCl₃ solution)

DEPT   CH₂↓ CH₃↑ CH↑

proton decoupled

solvent

δ (ppm)

¹H NMR Spectrum
(200 MHz, CDCl₃ solution)

TMS

δ (ppm)

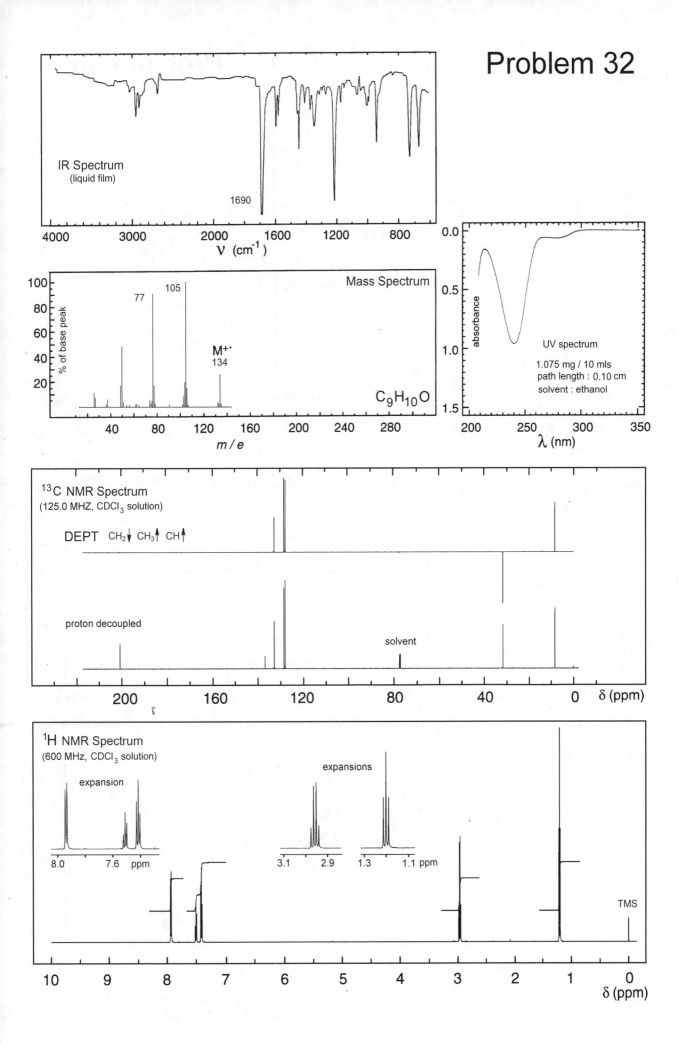

Problem 32

IR Spectrum
(liquid film)

1690

ν (cm⁻¹)

Mass Spectrum

% of base peak

77

105

M⁺˙
134

C₉H₁₀O

m/e

UV spectrum

1.075 mg / 10 mls
path length : 0.10 cm
solvent : ethanol

λ (nm)

¹³C NMR Spectrum
(125.0 MHZ, CDCl₃ solution)

DEPT   CH₂↓ CH₃↑ CH↑

proton decoupled

solvent

δ (ppm)

¹H NMR Spectrum
(600 MHz, CDCl₃ solution)

expansion

expansions

TMS

δ (ppm)

# Problem 33

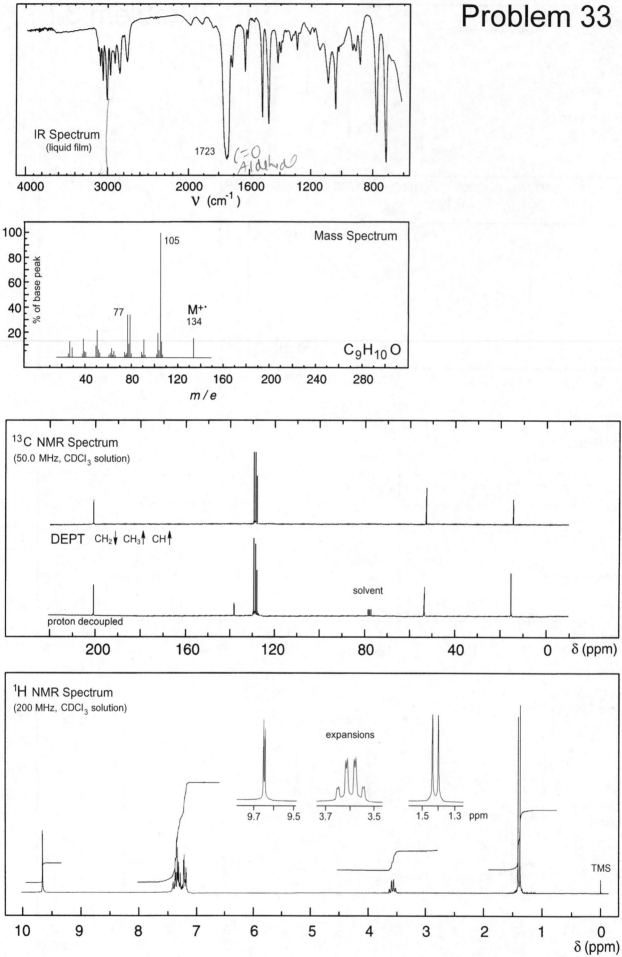

IR Spectrum
(liquid film)

1723 C=O
Aldehyd

ν (cm⁻¹)

Mass Spectrum

100
80
60
40
20

% of base peak

105

77

M⁺·
134

C₉H₁₀O

m/e

¹³C NMR Spectrum
(50.0 MHz, CDCl₃ solution)

DEPT   CH₂↓ CH₃↑ CH↑

solvent

proton decoupled

δ (ppm)

¹H NMR Spectrum
(200 MHz, CDCl₃ solution)

expansions

9.7  9.5   3.7  3.5   1.5  1.3 ppm

TMS

δ (ppm)

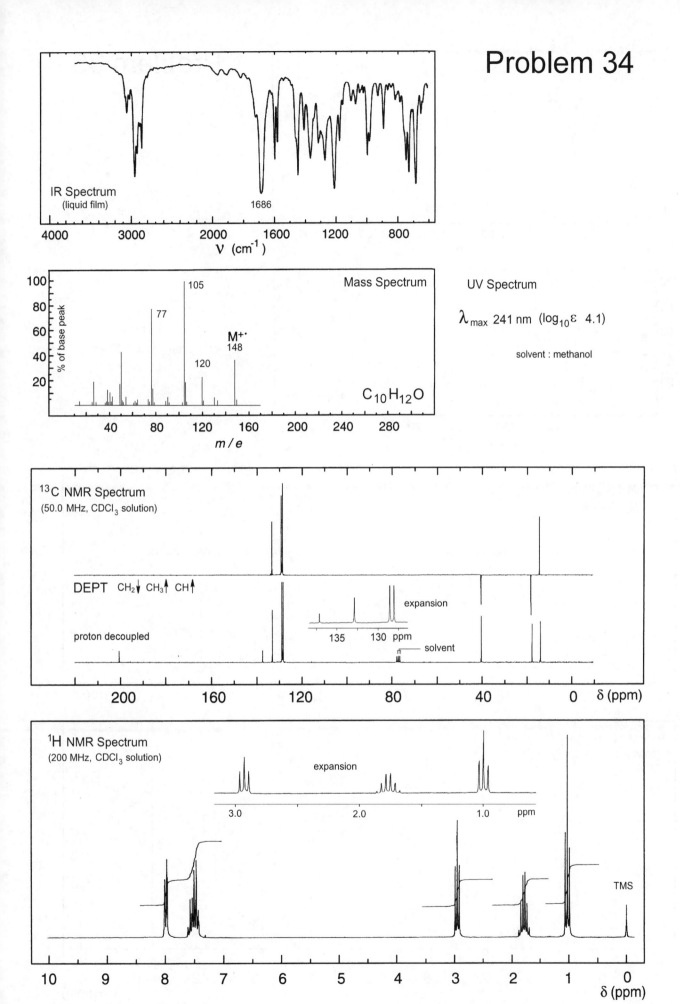

# Problem 34

**IR Spectrum**
(liquid film)

1686

ν (cm⁻¹)

**Mass Spectrum**

105

77

M⁺˙
148

120

% of base peak

m/e

$C_{10}H_{12}O$

**UV Spectrum**

$\lambda_{max}$ 241 nm  ($\log_{10}\varepsilon$  4.1)

solvent : methanol

**¹³C NMR Spectrum**
(50.0 MHz, CDCl₃ solution)

DEPT  CH₂↓ CH₃↑ CH↑

expansion

135    130  ppm

proton decoupled

solvent

δ (ppm)

**¹H NMR Spectrum**
(200 MHz, CDCl₃ solution)

expansion

3.0        2.0        1.0    ppm

TMS

δ (ppm)

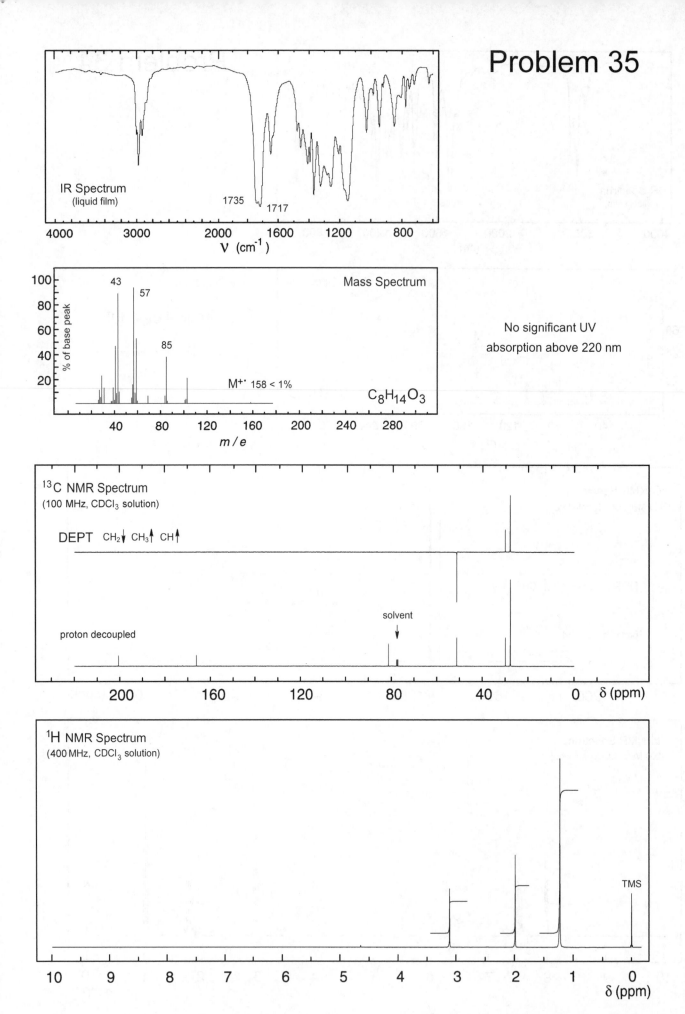

IR Spectrum
(liquid film)
1735
1717

ν (cm$^{-1}$)

Mass Spectrum

43
57
85

M$^{+\cdot}$ 158 < 1%

C$_8$H$_{14}$O$_3$

% of base peak

m/e

No significant UV
absorption above 220 nm

$^{13}$C NMR Spectrum
(100 MHz, CDCl$_3$ solution)

DEPT  CH$_2$↓ CH$_3$↑ CH↑

proton decoupled

solvent

δ (ppm)

$^1$H NMR Spectrum
(400 MHz, CDCl$_3$ solution)

TMS

δ (ppm)

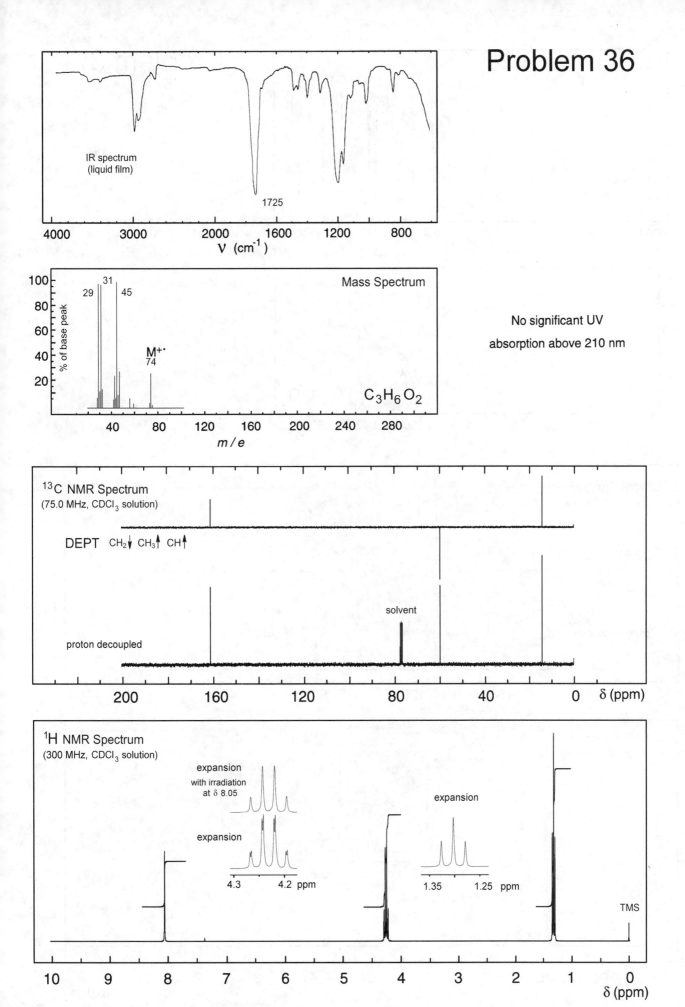

IR spectrum
(liquid film)

1725

$\nu$ (cm$^{-1}$)

Mass Spectrum

29
31
45

M$^{+ \cdot}$
74

% of base peak

m/e

No significant UV
absorption above 210 nm

$C_3H_6O_2$

$^{13}$C NMR Spectrum
(75.0 MHz, CDCl$_3$ solution)

DEPT    CH$_2\downarrow$  CH$_3\uparrow$  CH$\uparrow$

solvent

proton decoupled

$\delta$ (ppm)

$^1$H NMR Spectrum
(300 MHz, CDCl$_3$ solution)

expansion
with irradiation
at $\delta$ 8.05

expansion

expansion

4.3        4.2 ppm

1.35    1.25  ppm

TMS

$\delta$ (ppm)

# Problem 37

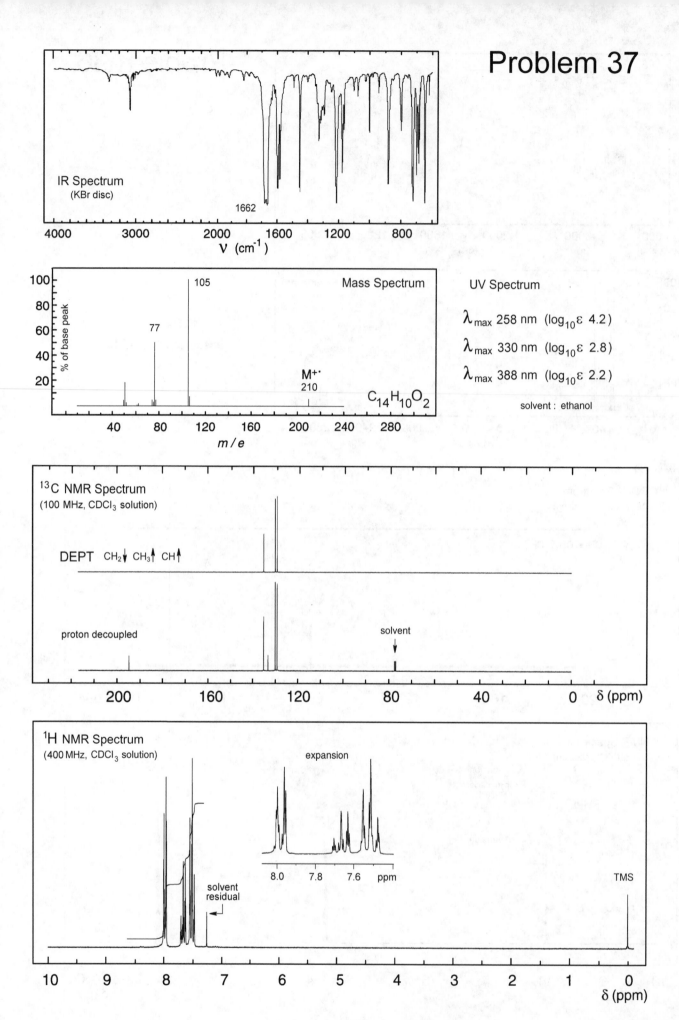

IR Spectrum
(KBr disc)

1662

ν (cm⁻¹)

Mass Spectrum

105

77

M⁺·
210

$C_{14}H_{10}O_2$

% of base peak

m/e

UV Spectrum

$\lambda_{max}$ 258 nm  (log₁₀ε  4.2)

$\lambda_{max}$ 330 nm  (log₁₀ε  2.8)

$\lambda_{max}$ 388 nm  (log₁₀ε  2.2)

solvent : ethanol

¹³C NMR Spectrum
(100 MHz, CDCl₃ solution)

DEPT   CH₂↓ CH₃↑ CH↑

proton decoupled

solvent

δ (ppm)

¹H NMR Spectrum
(400 MHz, CDCl₃ solution)

expansion

solvent
residual

TMS

δ (ppm)

δ (ppm)

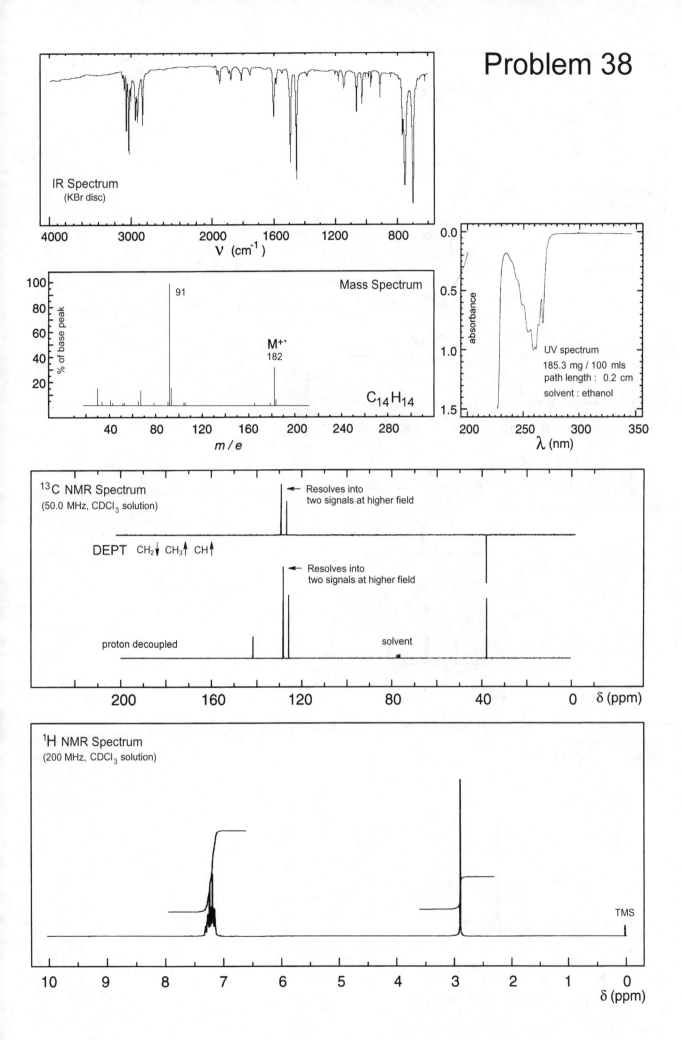

# Problem 38

**IR Spectrum**
(KBr disc)

ν (cm⁻¹)

**Mass Spectrum**

91

M⁺·
182

C₁₄H₁₄

% of base peak

m/e

UV spectrum
185.3 mg / 100 mls
path length : 0.2 cm
solvent : ethanol

absorbance

λ (nm)

¹³C NMR Spectrum
(50.0 MHz, CDCl₃ solution)

← Resolves into
two signals at higher field

DEPT  CH₂↓ CH₃↑ CH↑

← Resolves into
two signals at higher field

proton decoupled

solvent

δ (ppm)

¹H NMR Spectrum
(200 MHz, CDCl₃ solution)

TMS

δ (ppm)

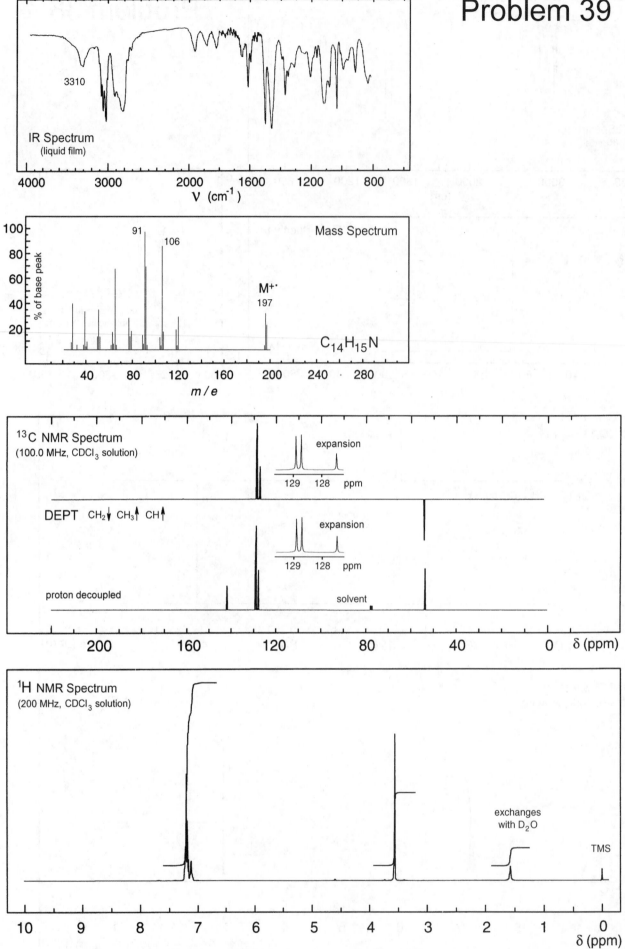

# Problem 39

**IR Spectrum**
(liquid film)

3310

V (cm⁻¹) — 4000 3000 2000 1600 1200 800

**Mass Spectrum**

% of base peak

91
106
M⁺·
197

C₁₄H₁₅N

m/e — 40 80 120 160 200 240 280

**¹³C NMR Spectrum**
(100.0 MHz, CDCl₃ solution)

expansion

129 128 ppm

**DEPT** CH₂↓ CH₃↑ CH↑

expansion

129 128 ppm

proton decoupled          solvent

δ (ppm) — 200 160 120 80 40 0

**¹H NMR Spectrum**
(200 MHz, CDCl₃ solution)

exchanges
with D₂O

TMS

δ (ppm) — 10 9 8 7 6 5 4 3 2 1 0

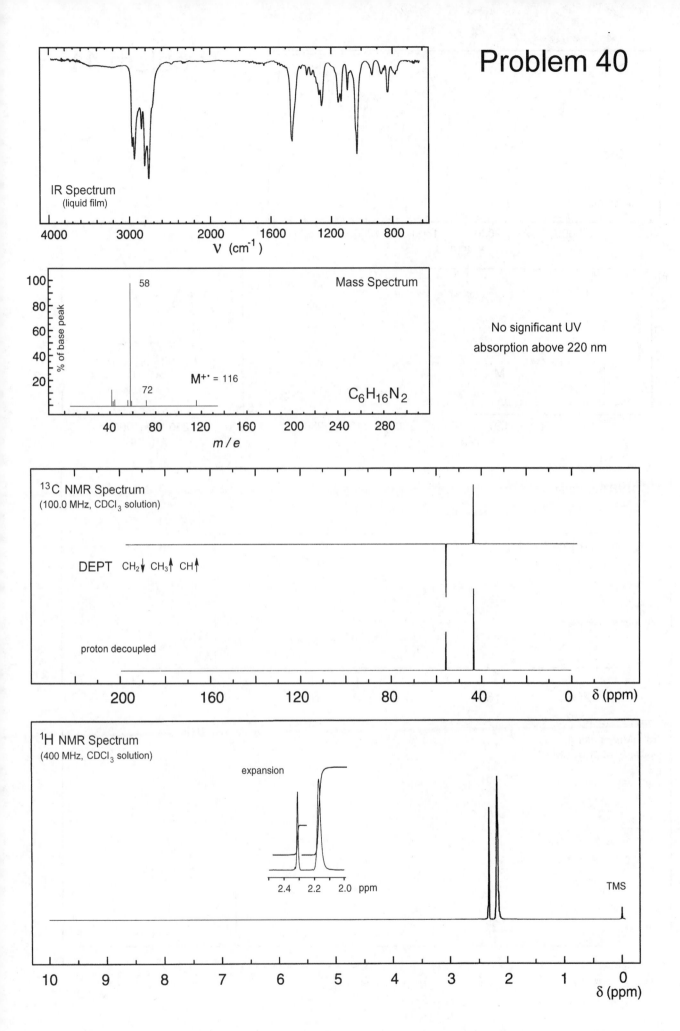

# Problem 40

IR Spectrum
(liquid film)

ν (cm$^{-1}$)

Mass Spectrum

No significant UV
absorption above 220 nm

$M^{+\cdot} = 116$

$C_6H_{16}N_2$

% of base peak

m/e

$^{13}$C NMR Spectrum
(100.0 MHz, CDCl$_3$ solution)

DEPT  CH$_2\downarrow$ CH$_3\uparrow$ CH$\uparrow$

proton decoupled

δ (ppm)

$^1$H NMR Spectrum
(400 MHz, CDCl$_3$ solution)

expansion

ppm

TMS

δ (ppm)

# Problem 41

IR Spectrum
(liquid film)

1713

ν (cm⁻¹)

Mass Spectrum

43

99

M⁺·
114

$C_6H_{10}O_2$

% of base peak

m/e

UV spectrum

104.1 mg / 10 mls
path length : 0.2 cm
solvent : ethanol

absorbance

λ (nm)

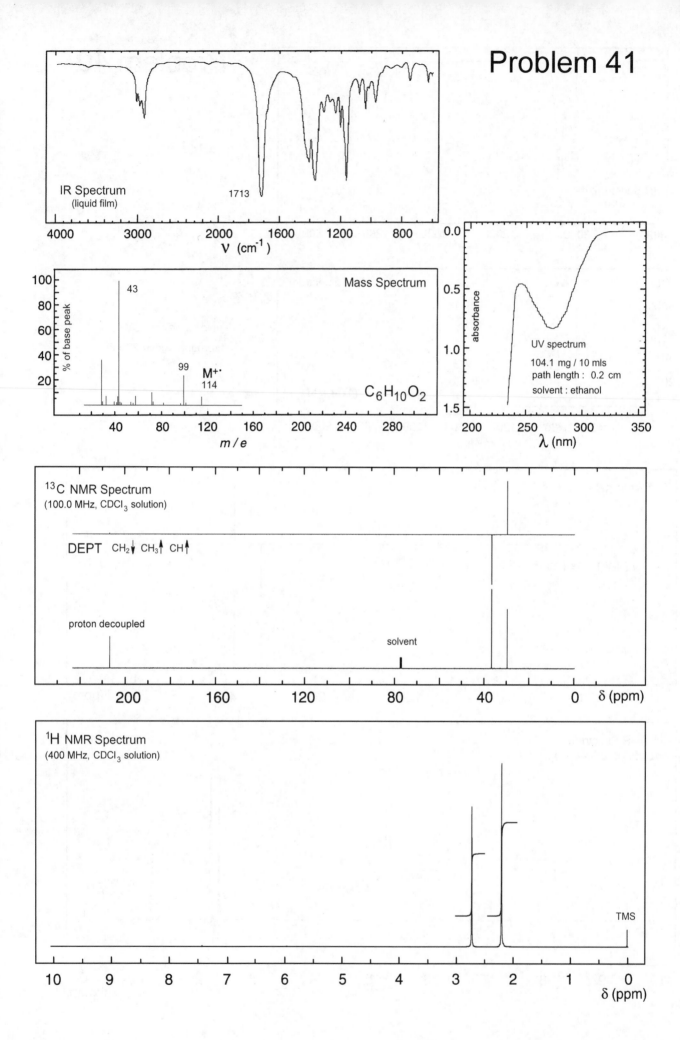

¹³C NMR Spectrum
(100.0 MHz, CDCl₃ solution)

DEPT   CH₂↓ CH₃↑ CH↑

proton decoupled

solvent

δ (ppm)

¹H NMR Spectrum
(400 MHz, CDCl₃ solution)

TMS

δ (ppm)

# Problem 42

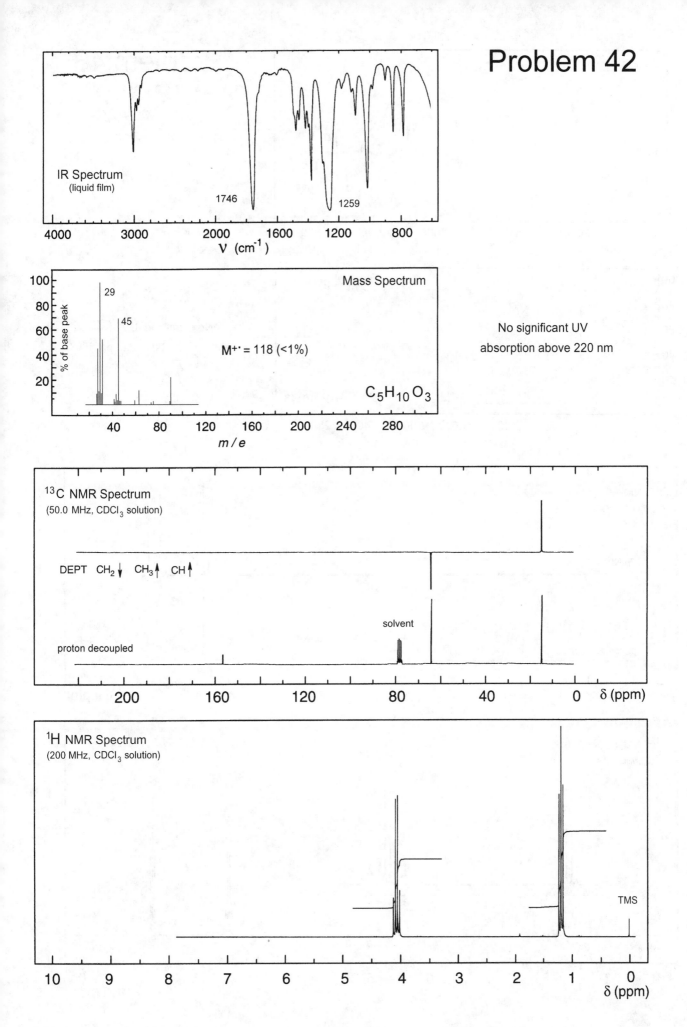

IR Spectrum
(liquid film)

1746    1259

Mass Spectrum

29

45

$M^{+\cdot} = 118$ (<1%)

$C_5H_{10}O_3$

% of base peak

$m/e$

No significant UV
absorption above 220 nm

$^{13}C$ NMR Spectrum
(50.0 MHz, CDCl$_3$ solution)

DEPT   CH$_2$↓   CH$_3$↑   CH↑

solvent

proton decoupled

$\delta$ (ppm)

$^1$H NMR Spectrum
(200 MHz, CDCl$_3$ solution)

TMS

$\delta$ (ppm)

# Problem 43

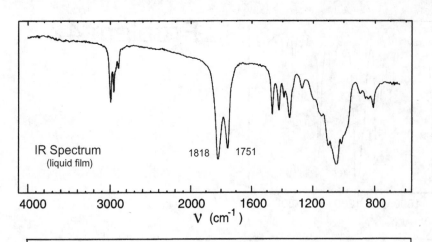

IR Spectrum
(liquid film)

1818   1751

Mass Spectrum

No significant UV
absorption above 220 nm

% of base peak

100
80
60
40
20

57

29

74

M⁺· = 130 (< 1%)     $C_6H_{10}O_3$

m / e

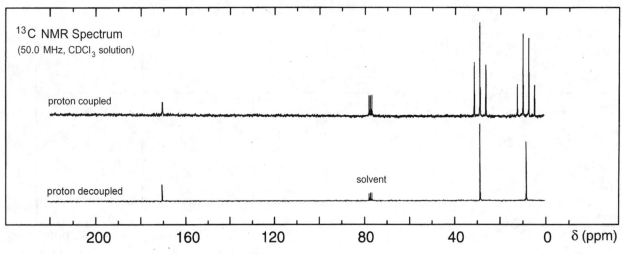

¹³C NMR Spectrum
(50.0 MHz, CDCl₃ solution)

proton coupled

proton decoupled

solvent

δ (ppm)

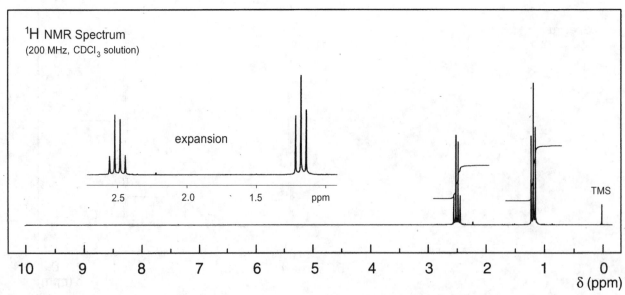

¹H NMR Spectrum
(200 MHz, CDCl₃ solution)

expansion

2.5     2.0     1.5     ppm

TMS

δ (ppm)

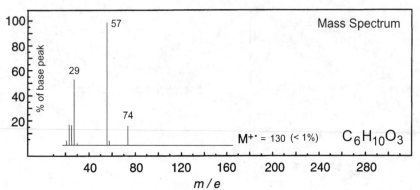

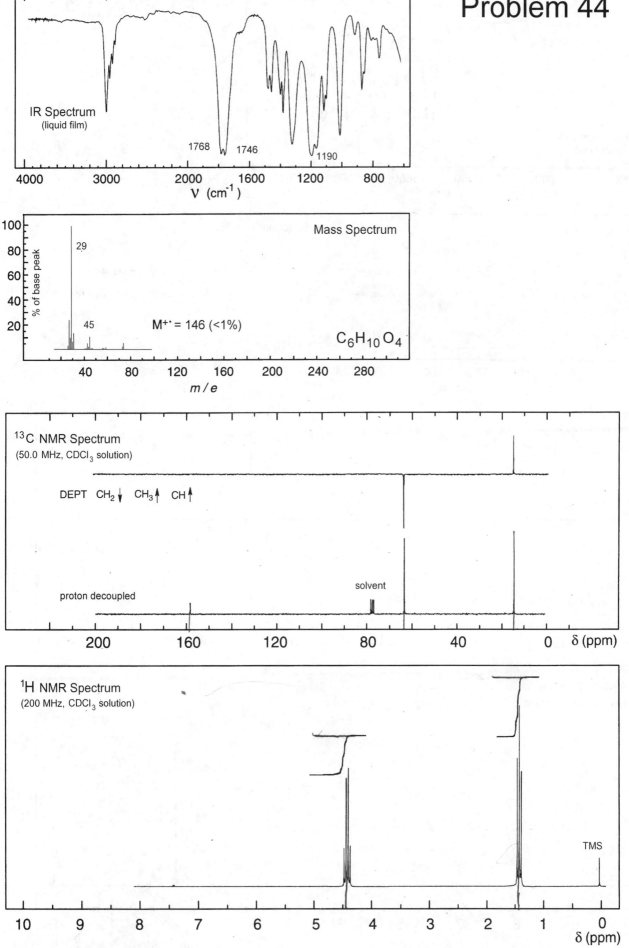

# Problem 44

**IR Spectrum**
(liquid film)

1768    1746    1190

$\nu$ (cm$^{-1}$)

4000    3000    2000    1600    1200    800

**Mass Spectrum**

% of base peak

29

45

M$^{+\cdot}$ = 146 (<1%)

$C_6H_{10}O_4$

40    80    120    160    200    240    280

$m/e$

**$^{13}$C NMR Spectrum**
(50.0 MHz, CDCl$_3$ solution)

DEPT   CH$_2\downarrow$   CH$_3\uparrow$   CH$\uparrow$

proton decoupled

solvent

200    160    120    80    40    0    $\delta$ (ppm)

**$^1$H NMR Spectrum**
(200 MHz, CDCl$_3$ solution)

TMS

10    9    8    7    6    5    4    3    2    1    0    $\delta$ (ppm)

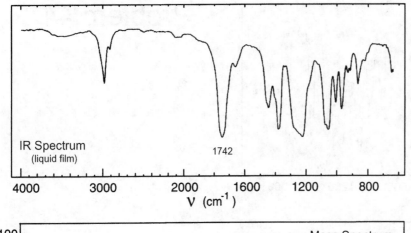

IR Spectrum
(liquid film)

1742

ν (cm⁻¹)

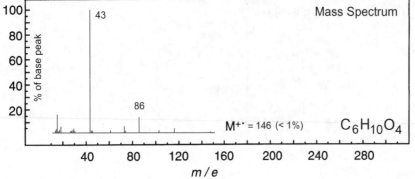

Mass Spectrum

% of base peak

43

86

M⁺˙ = 146 (< 1%)    $C_6H_{10}O_4$

m/e

No significant UV
absorption above 220 nm

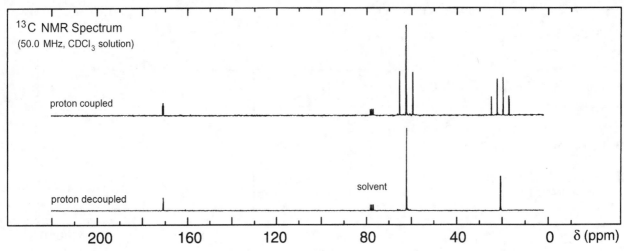

¹³C NMR Spectrum
(50.0 MHz, CDCl₃ solution)

proton coupled

solvent

proton decoupled

δ (ppm)

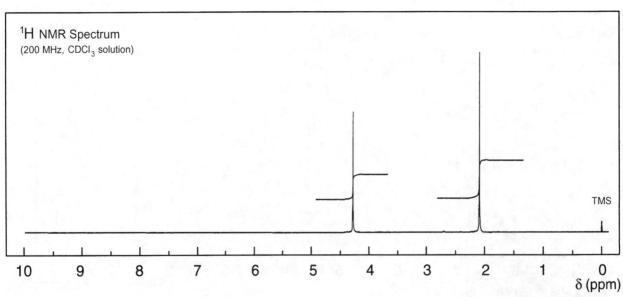

¹H NMR Spectrum
(200 MHz, CDCl₃ solution)

TMS

δ (ppm)

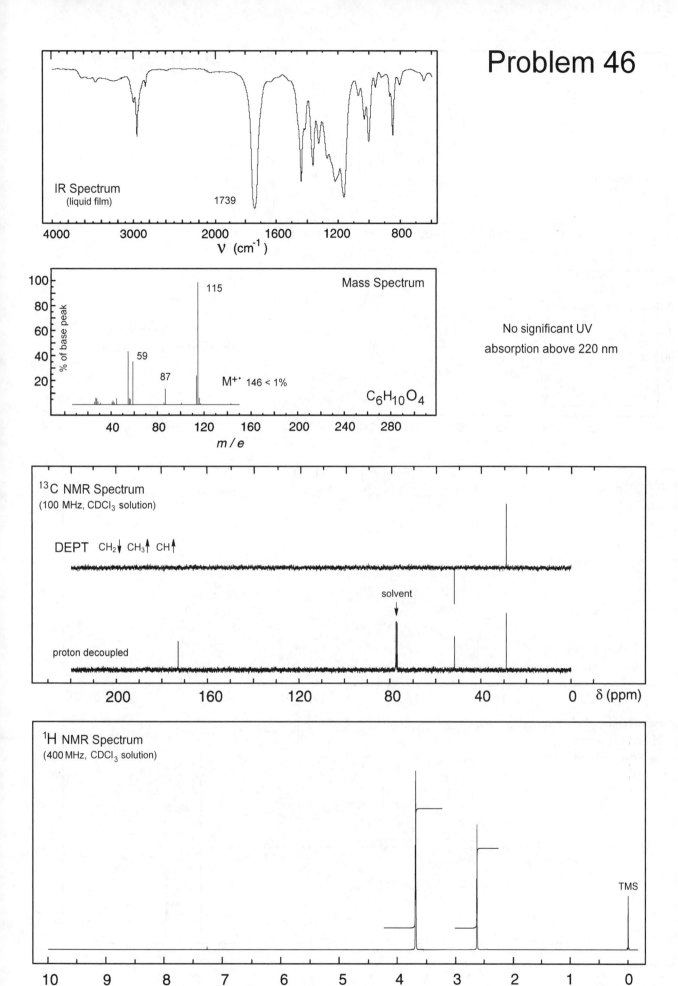

## Problem 46

### IR Spectrum
(liquid film)

1739

ν (cm⁻¹)

$\nu$ (cm$^{-1}$)

### Mass Spectrum

% of base peak

115

59

87

M⁺· 146 < 1%

$C_6H_{10}O_4$

m/e

$m/e$

No significant UV
absorption above 220 nm

### ¹³C NMR Spectrum
(100 MHz, CDCl₃ solution)

DEPT  CH₂↓ CH₃↑ CH↑

solvent

proton decoupled

δ (ppm)

### ¹H NMR Spectrum
(400 MHz, CDCl₃ solution)

TMS

δ (ppm)

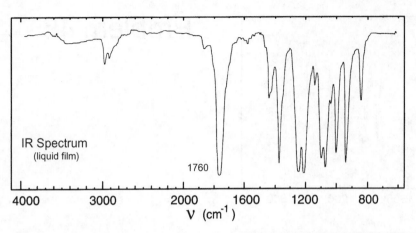

IR Spectrum
(liquid film)

1760

ν (cm⁻¹)

4000    3000    2000    1600    1200    800

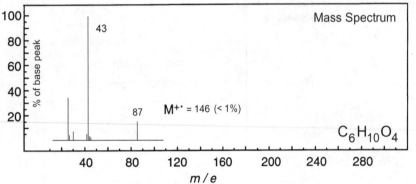

Mass Spectrum

43

% of base peak

100
80
60
40
20

87

$M^{+\cdot}$ = 146 (< 1%)

$C_6H_{10}O_4$

No significant UV
absorption above 220 nm

40    80    120    160    200    240    280

m / e

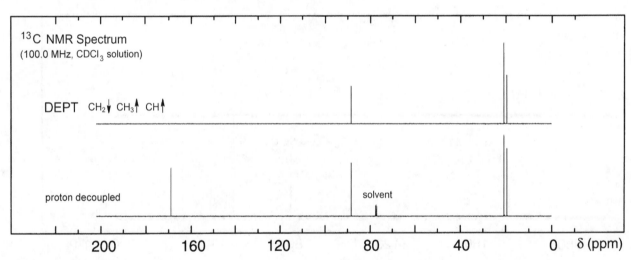

¹³C NMR Spectrum
(100.0 MHz, CDCl₃ solution)

DEPT   CH₂↓ CH₃↑ CH↑

proton decoupled

solvent

200    160    120    80    40    0   δ (ppm)

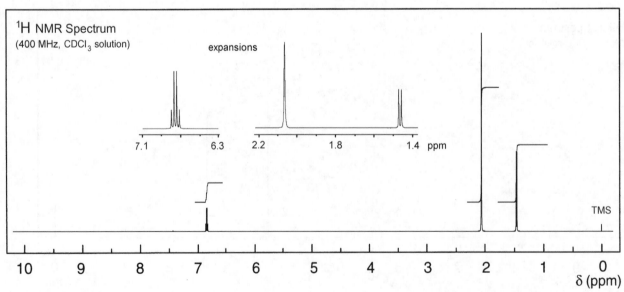

¹H NMR Spectrum
(400 MHz, CDCl₃ solution)

expansions

7.1        6.3      2.2        1.8        1.4  ppm

TMS

10    9    8    7    6    5    4    3    2    1    0
                                                δ (ppm)

# Problem 48

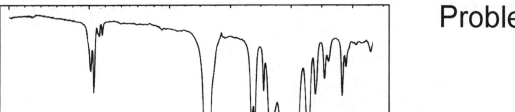

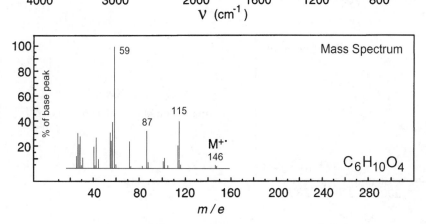

IR Spectrum
(liquid film)

1737

ν (cm⁻¹)

Mass Spectrum

100
80
60
40
20

% of base peak

59

87

115

M⁺·
146

$C_6H_{10}O_4$

m / e

No significant UV
absorption above 220 nm

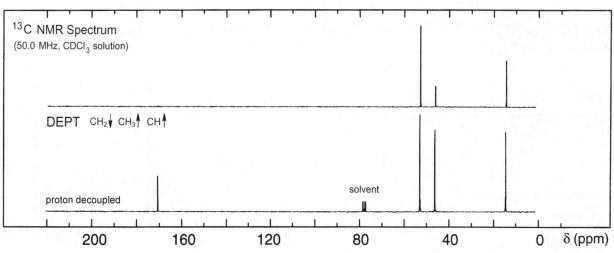

$^{13}$C NMR Spectrum
(50.0 MHz, CDCl₃ solution)

DEPT  CH₂↓ CH₃↑ CH↑

solvent

proton decoupled

δ (ppm)

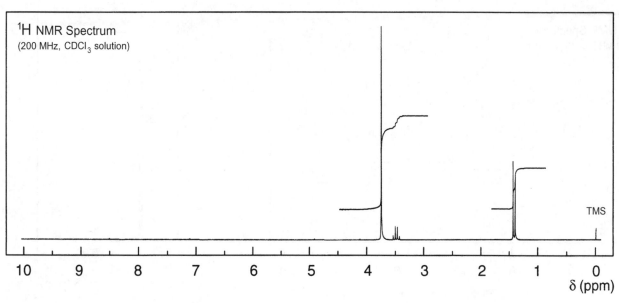

$^1$H NMR Spectrum
(200 MHz, CDCl₃ solution)

TMS

δ (ppm)

# Problem 49

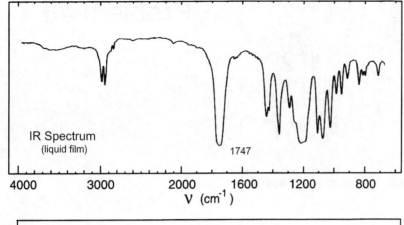

IR Spectrum
(liquid film)

1747

ν (cm⁻¹)

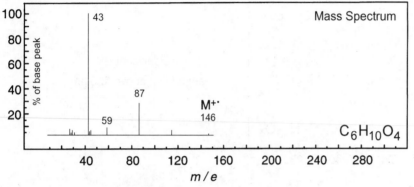

Mass Spectrum

% of base peak

43

59

87

M⁺˙
146

$C_6H_{10}O_4$

m/e

No significant UV
absorption above 220 nm

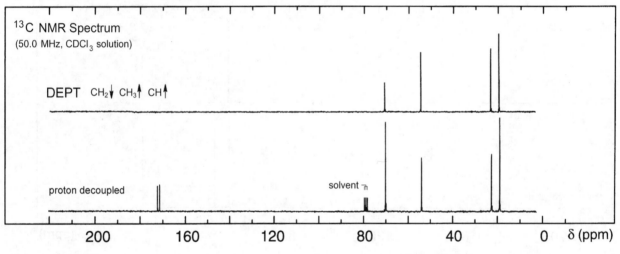

¹³C NMR Spectrum
(50.0 MHz, CDCl₃ solution)

DEPT   CH₂↓ CH₃↑ CH↑

proton decoupled

solvent

δ (ppm)

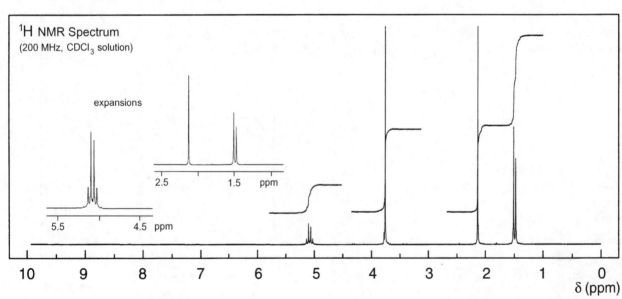

¹H NMR Spectrum
(200 MHz, CDCl₃ solution)

expansions

2.5        1.5    ppm

5.5        4.5  ppm

δ (ppm)

# Problem 50

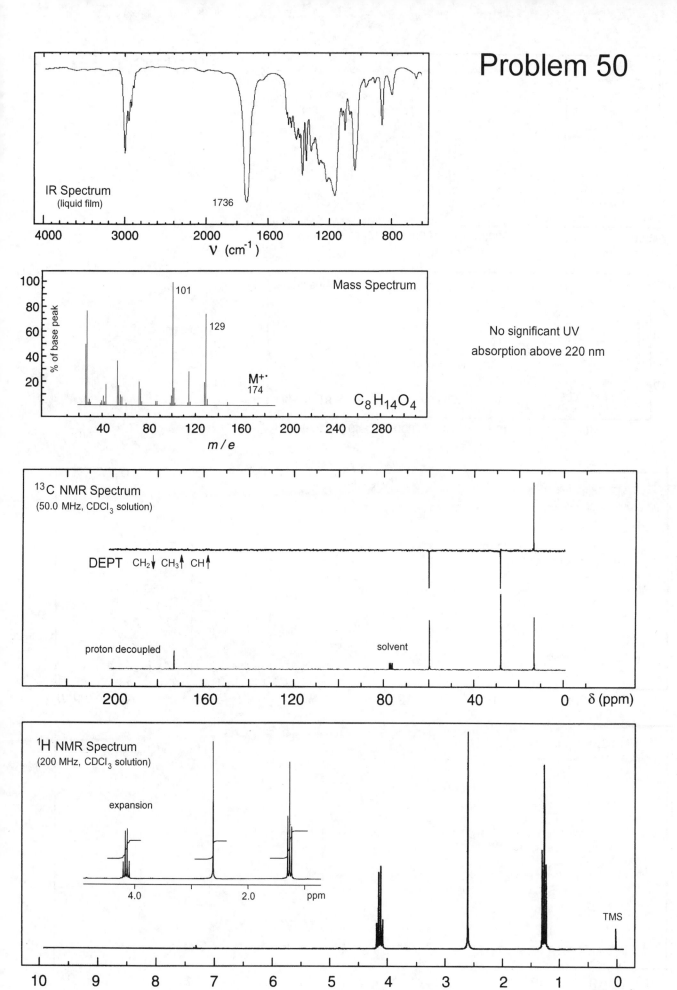

IR Spectrum
(liquid film)

1736

ν (cm⁻¹)

4000  3000  2000  1600  1200  800

Mass Spectrum

101

129

M⁺·
174

C₈H₁₄O₄

% of base peak

40  80  120  160  200  240  280

m/e

No significant UV
absorption above 220 nm

¹³C NMR Spectrum
(50.0 MHz, CDCl₃ solution)

DEPT   CH₂↓ CH₃↑ CH↑

proton decoupled                    solvent

200  160  120  80  40  0   δ (ppm)

¹H NMR Spectrum
(200 MHz, CDCl₃ solution)

expansion

4.0        2.0      ppm

TMS

10  9  8  7  6  5  4  3  2  1  0
δ (ppm)

# Problem 51

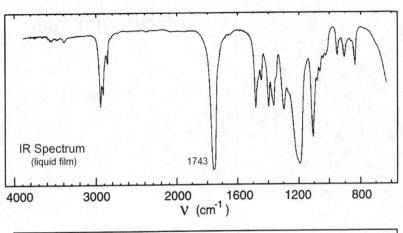

IR Spectrum
(liquid film)

1743

$\nu$ (cm$^{-1}$)

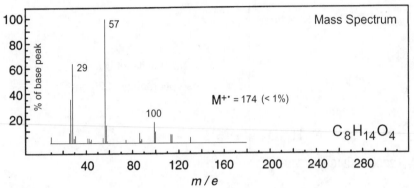

Mass Spectrum

57

29

100

M$^{+\cdot}$ = 174 (< 1%)

$C_8H_{14}O_4$

% of base peak

m / e

No significant UV
absorption above 220 nm

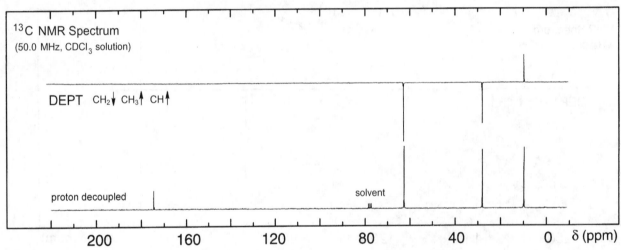

$^{13}$C NMR Spectrum
(50.0 MHz, CDCl$_3$ solution)

DEPT   CH$_2$↓ CH$_3$↑ CH↑

proton decoupled

solvent

$\delta$ (ppm)

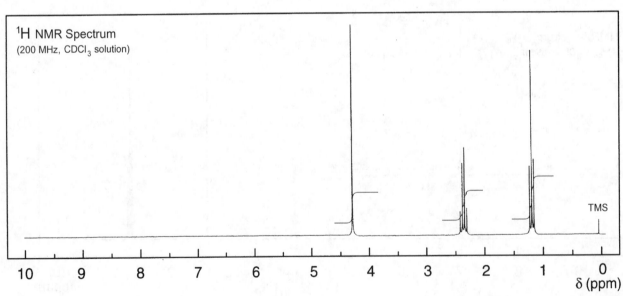

$^1$H NMR Spectrum
(200 MHz, CDCl$_3$ solution)

TMS

$\delta$ (ppm)

# Problem 52

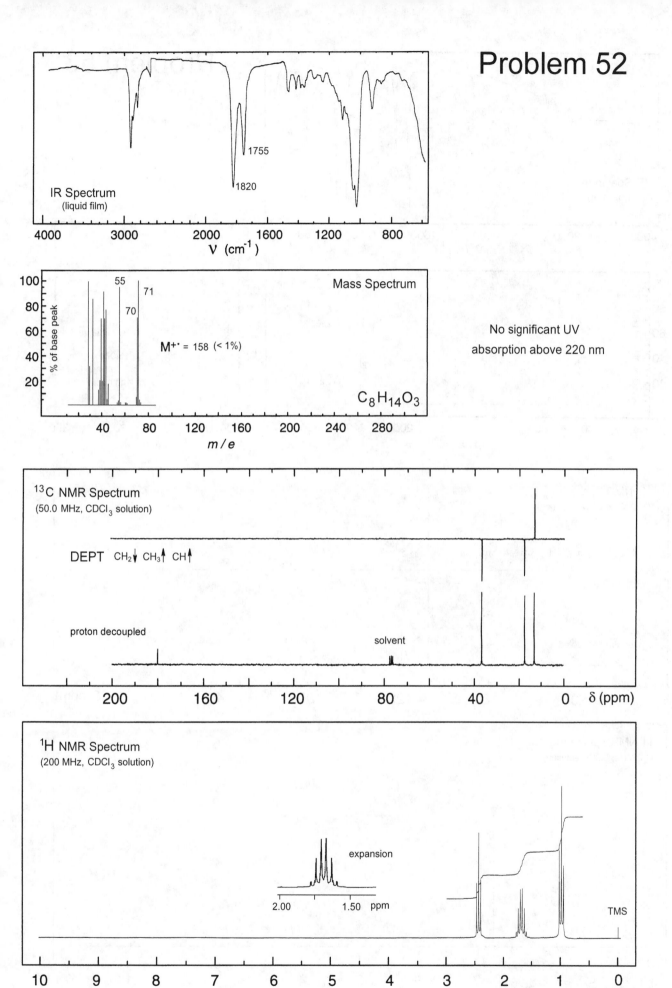

IR Spectrum
(liquid film)

1820
1755

ν (cm⁻¹)

$\nu$ (cm$^{-1}$)

Mass Spectrum

55
71
70

M⁺• = 158 (< 1%)

$M^{+\cdot} = 158$ (< 1%)

$C_8H_{14}O_3$

% of base peak

m/e

No significant UV
absorption above 220 nm

¹³C NMR Spectrum
(50.0 MHz, CDCl₃ solution)

$^{13}C$ NMR Spectrum
(50.0 MHz, CDCl$_3$ solution)

DEPT  CH₂↓ CH₃↑ CH↑

proton decoupled

solvent

δ (ppm)

¹H NMR Spectrum
(200 MHz, CDCl₃ solution)

$^1H$ NMR Spectrum
(200 MHz, CDCl$_3$ solution)

expansion

2.00    1.50  ppm

TMS

δ (ppm)

163

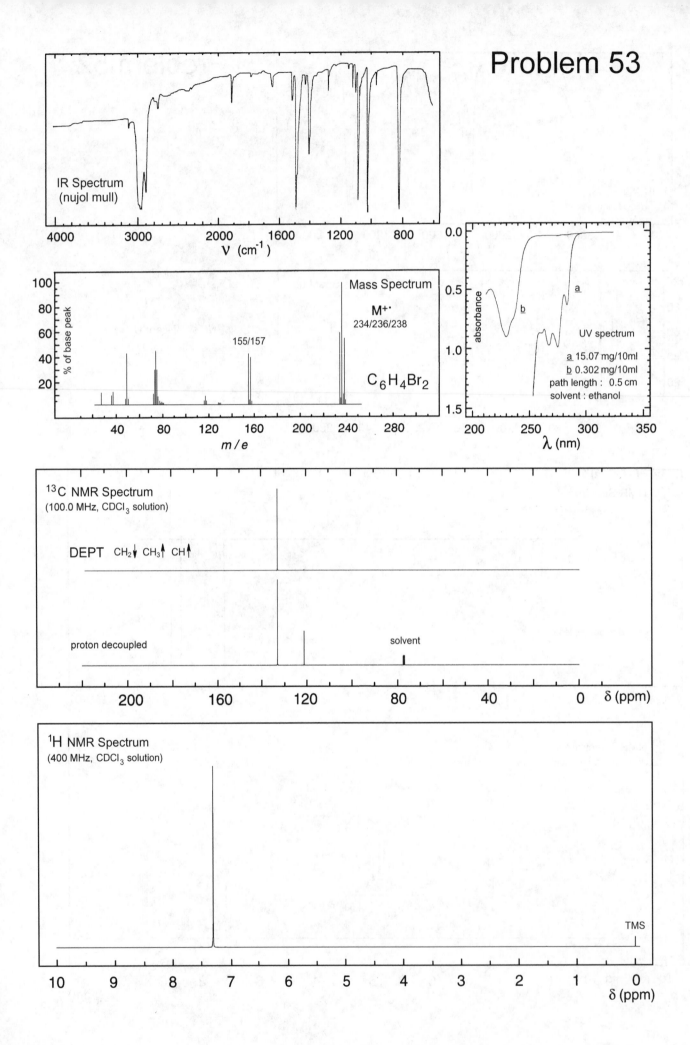

# Problem 53

IR Spectrum
(nujol mull)

$\nu$ (cm$^{-1}$)

Mass Spectrum

M$^{+\cdot}$
234/236/238

155/157

$C_6H_4Br_2$

% of base peak

$m/e$

UV spectrum

a 15.07 mg/10ml
b 0.302 mg/10ml
path length : 0.5 cm
solvent : ethanol

absorbance

$\lambda$ (nm)

$^{13}$C NMR Spectrum
(100.0 MHz, CDCl$_3$ solution)

DEPT  CH$_2$↓ CH$_3$↑ CH↑

proton decoupled

solvent

$\delta$ (ppm)

$^1$H NMR Spectrum
(400 MHz, CDCl$_3$ solution)

TMS

$\delta$ (ppm)

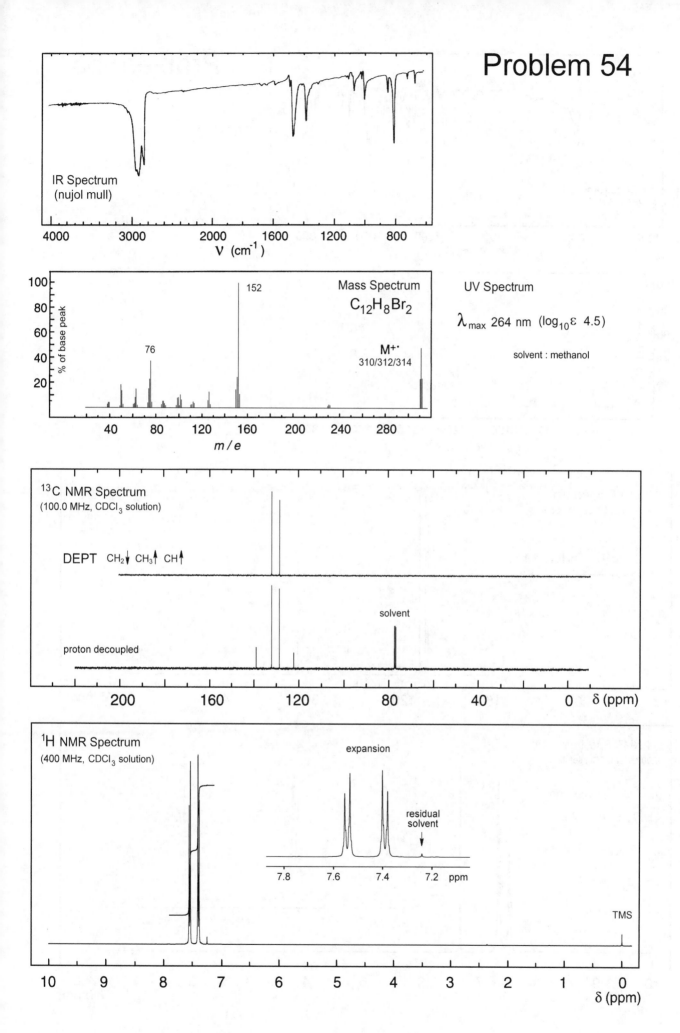

IR Spectrum
(nujol mull)

ν (cm⁻¹)

Mass Spectrum
$C_{12}H_8Br_2$

% of base peak

152

76

M⁺˙
310/312/314

m / e

UV Spectrum

$\lambda_{max}$ 264 nm  ($\log_{10}\varepsilon$  4.5)

solvent : methanol

$^{13}C$ NMR Spectrum
(100.0 MHz, CDCl₃ solution)

DEPT   CH₂↓ CH₃↑ CH↑

solvent

proton decoupled

δ (ppm)

$^1H$ NMR Spectrum
(400 MHz, CDCl₃ solution)

expansion

residual
solvent

7.8   7.6   7.4   7.2  ppm

TMS

δ (ppm)

165

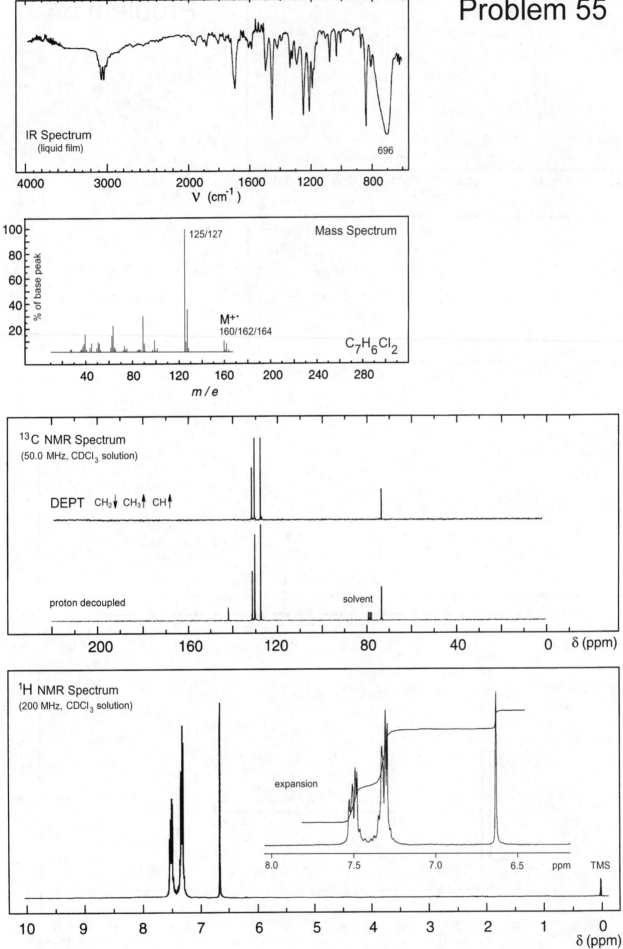

# Problem 55

**IR Spectrum**
(liquid film)

696

ν (cm⁻¹)

$\nu$ (cm$^{-1}$)

**Mass Spectrum**

% of base peak

125/127

M⁺·
160/162/164

$C_7H_6Cl_2$

m/e

**¹³C NMR Spectrum**
(50.0 MHz, CDCl₃ solution)

DEPT   CH₂↓ CH₃↑ CH↑

proton decoupled                                    solvent

δ (ppm)

**¹H NMR Spectrum**
(200 MHz, CDCl₃ solution)

expansion

8.0        7.5        7.0        6.5      ppm      TMS

δ (ppm)

# Problem 56

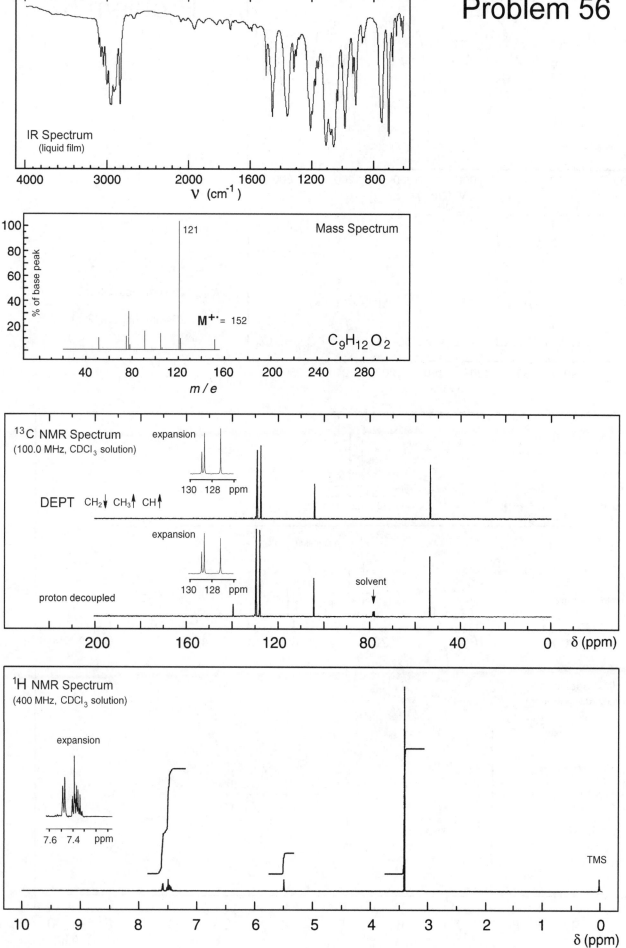

IR Spectrum
(liquid film)

$\nu$ (cm$^{-1}$)

Mass Spectrum

% of base peak

121

$M^{+\cdot} = 152$

$C_9H_{12}O_2$

$m/e$

$^{13}$C NMR Spectrum
(100.0 MHz, CDCl$_3$ solution)

expansion

DEPT   CH$_2\downarrow$ CH$_3\uparrow$ CH$\uparrow$

expansion

proton decoupled

solvent

$\delta$ (ppm)

$^1$H NMR Spectrum
(400 MHz, CDCl$_3$ solution)

expansion

TMS

$\delta$ (ppm)

# Problem 57

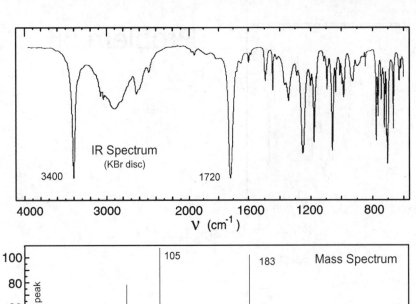

IR Spectrum
(KBr disc)

3400

1720

ν (cm⁻¹)

4000   3000   2000   1600   1200   800

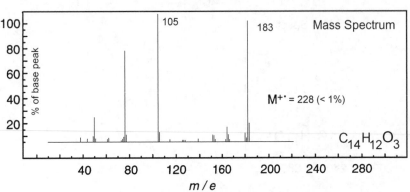

Mass Spectrum

% of base peak

100
80
60
40
20

105

183

M⁺˙ = 228 (< 1%)

$C_{14}H_{12}O_3$

m/e

40   80   120   160   200   240   280

UV Spectrum

$\lambda_{max}$ 253 nm (log₁₀ε 2.6)

$\lambda_{max}$ 259 nm (log₁₀ε 2.7)

$\lambda_{max}$ 264 nm (log₁₀ε 2.5)

solvent : ethanol

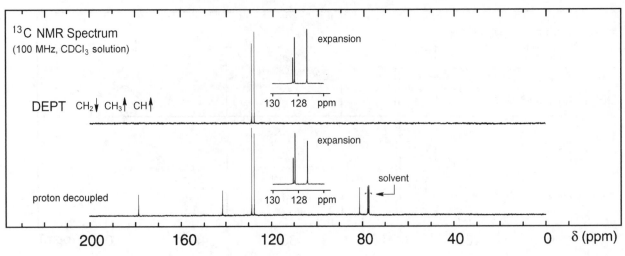

¹³C NMR Spectrum
(100 MHz, CDCl₃ solution)

expansion

DEPT   CH₂↓ CH₃↑ CH↑

130  128  ppm

expansion

solvent

proton decoupled

130  128  ppm

δ (ppm)

200   160   120   80   40   0

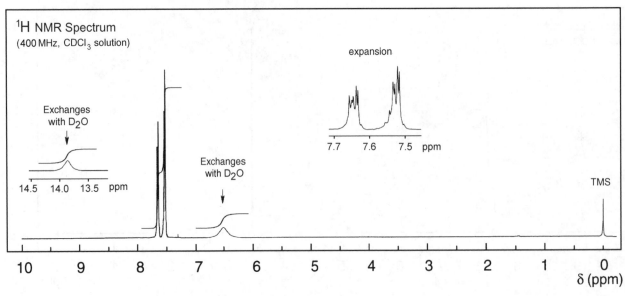

¹H NMR Spectrum
(400 MHz, CDCl₃ solution)

Exchanges
with D₂O

expansion

14.5  14.0  13.5  ppm

7.7   7.6   7.5  ppm

Exchanges
with D₂O

TMS

δ (ppm)

10   9   8   7   6   5   4   3   2   1   0

# Problem 58

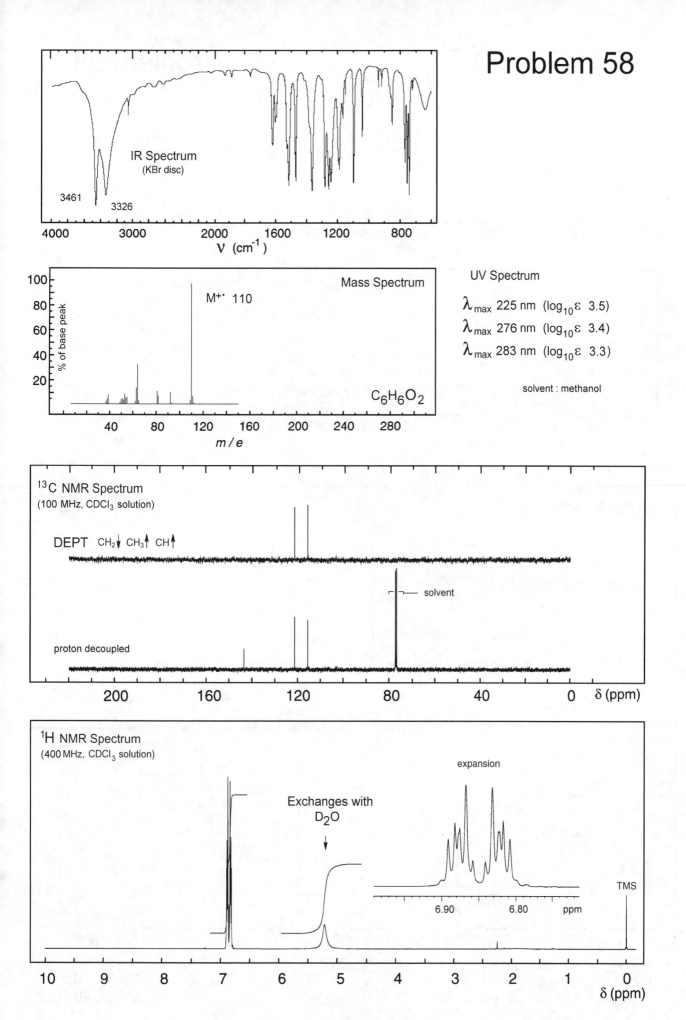

IR Spectrum
(KBr disc)

3461
3326

ν (cm⁻¹)

Mass Spectrum

M⁺˙ 110

% of base peak

m/e

C₆H₆O₂

UV Spectrum

$\lambda_{max}$ 225 nm (log₁₀ε 3.5)
$\lambda_{max}$ 276 nm (log₁₀ε 3.4)
$\lambda_{max}$ 283 nm (log₁₀ε 3.3)

solvent : methanol

¹³C NMR Spectrum
(100 MHz, CDCl₃ solution)

DEPT  CH₂↓ CH₃↑ CH↑

solvent

proton decoupled

δ (ppm)

¹H NMR Spectrum
(400 MHz, CDCl₃ solution)

expansion

Exchanges with
D₂O

6.90    6.80  ppm

TMS

δ (ppm)

169

# Problem 59

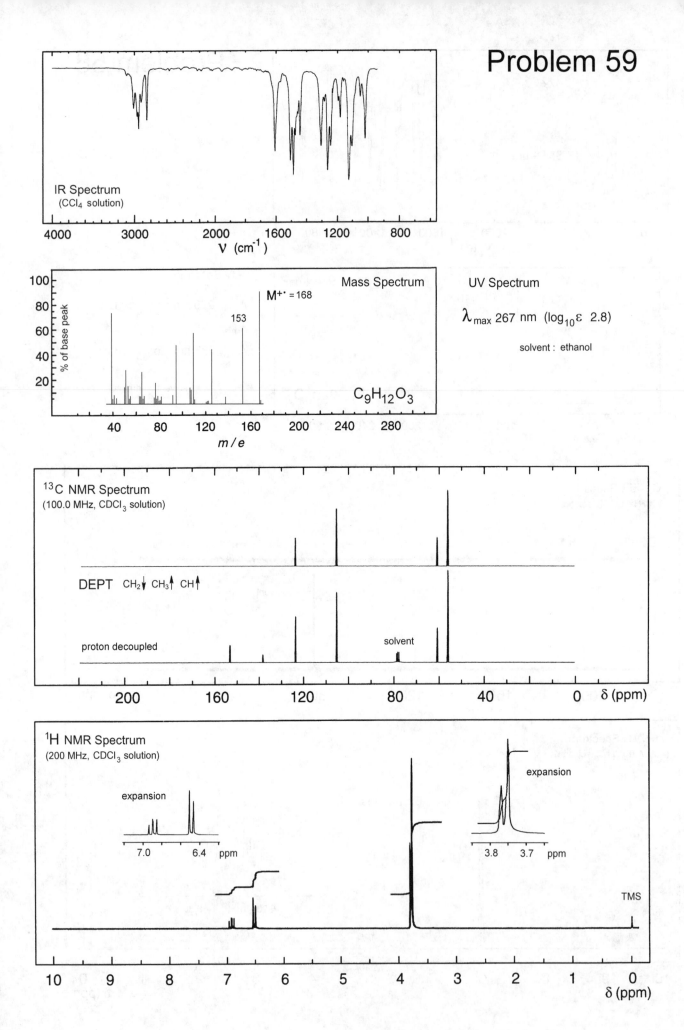

IR Spectrum
(CCl$_4$ solution)

ν (cm$^{-1}$)

Mass Spectrum

M$^{+\cdot}$ = 168

153

% of base peak

m/e

C$_9$H$_{12}$O$_3$

UV Spectrum

λ$_{max}$ 267 nm (log$_{10}$ε 2.8)

solvent : ethanol

$^{13}$C NMR Spectrum
(100.0 MHz, CDCl$_3$ solution)

DEPT  CH$_2$↓ CH$_3$↑ CH↑

proton decoupled

solvent

δ (ppm)

$^1$H NMR Spectrum
(200 MHz, CDCl$_3$ solution)

expansion

7.0  6.4  ppm

expansion

3.8  3.7  ppm

TMS

δ (ppm)

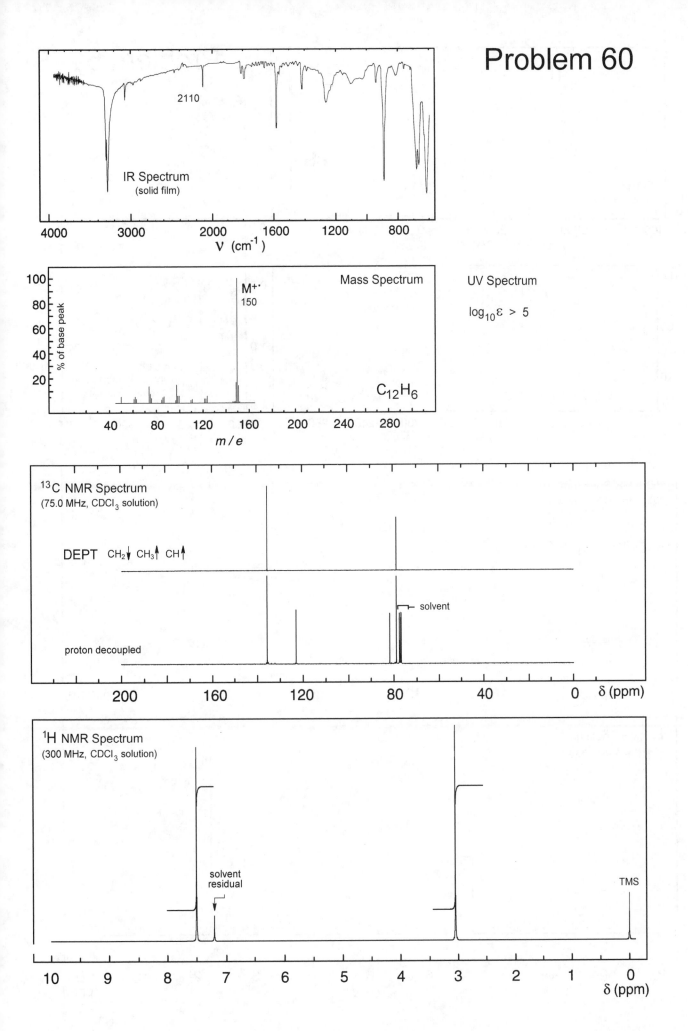

# Problem 60

**IR Spectrum** (solid film)

2110

ν (cm⁻¹)

**Mass Spectrum**

M⁺· 150

% of base peak

m/e

$C_{12}H_6$

**UV Spectrum**

$\log_{10}\varepsilon > 5$

**¹³C NMR Spectrum** (75.0 MHz, CDCl₃ solution)

DEPT  CH₂↓ CH₃↑ CH↑

solvent

proton decoupled

δ (ppm)

**¹H NMR Spectrum** (300 MHz, CDCl₃ solution)

solvent residual

TMS

δ (ppm)

# Problem 61

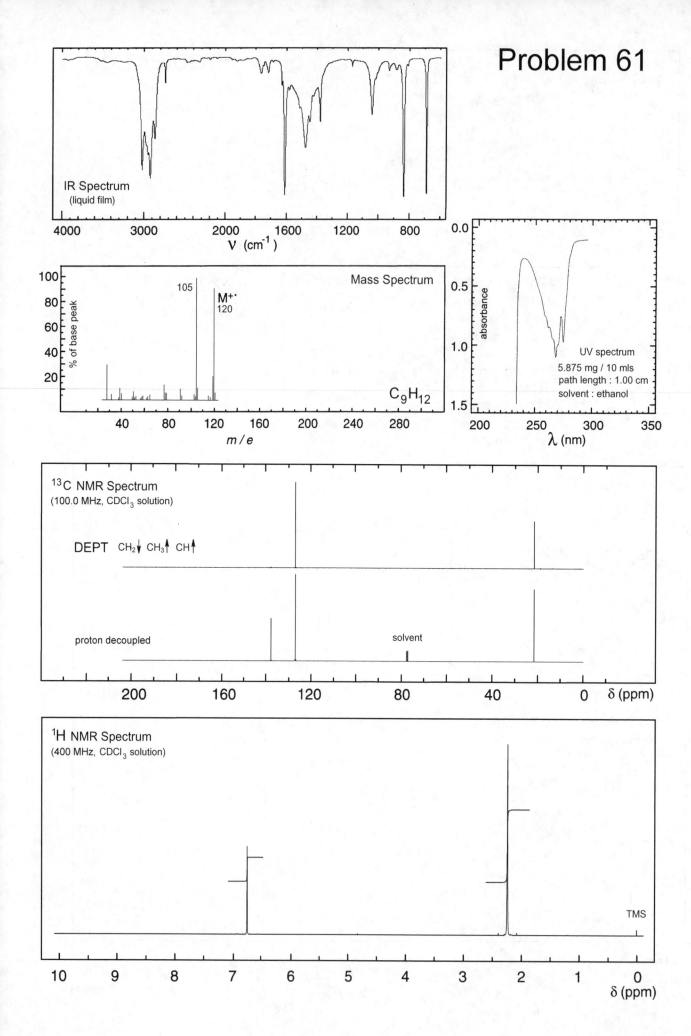

IR Spectrum
(liquid film)

$\nu$ (cm$^{-1}$)

Mass Spectrum

105

M$^{+\cdot}$
120

% of base peak

$m/e$

C$_9$H$_{12}$

absorbance

UV spectrum
5.875 mg / 10 mls
path length : 1.00 cm
solvent : ethanol

$\lambda$ (nm)

$^{13}$C NMR Spectrum
(100.0 MHz, CDCl$_3$ solution)

DEPT  CH$_2\downarrow$ CH$_3\uparrow$ CH$\uparrow$

proton decoupled

solvent

$\delta$ (ppm)

$^1$H NMR Spectrum
(400 MHz, CDCl$_3$ solution)

TMS

$\delta$ (ppm)

# Problem 62

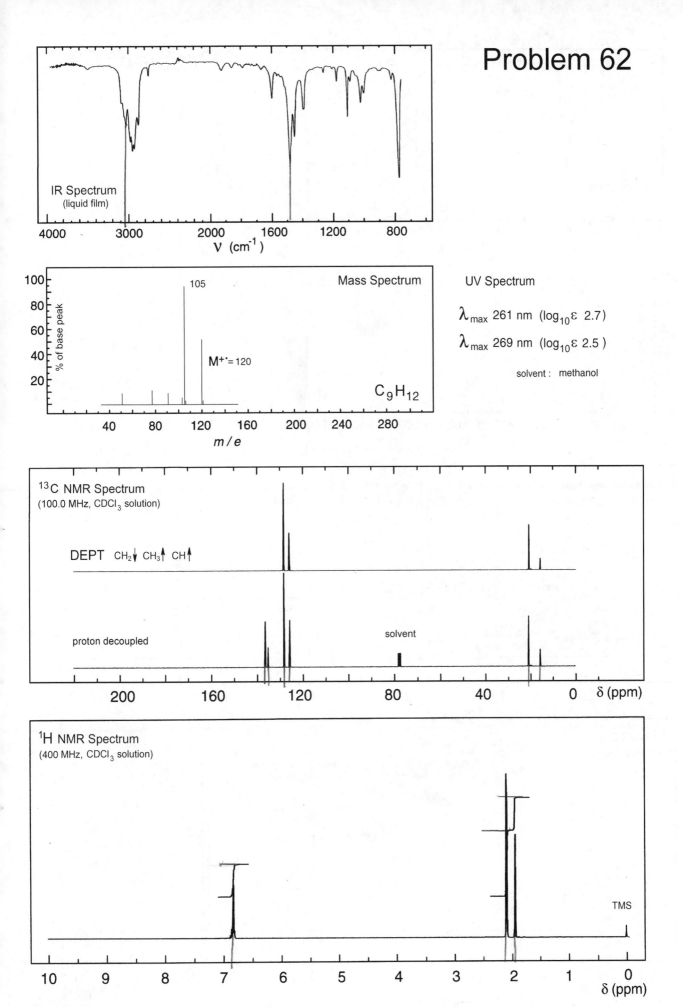

IR Spectrum
(liquid film)

$\nu$ (cm$^{-1}$)

Mass Spectrum

105

% of base peak

M$^{+ \cdot}$= 120

$C_9H_{12}$

$m/e$

UV Spectrum

$\lambda_{max}$ 261 nm (log$_{10}\varepsilon$ 2.7)

$\lambda_{max}$ 269 nm (log$_{10}\varepsilon$ 2.5)

solvent : methanol

$^{13}$C NMR Spectrum
(100.0 MHz, CDCl$_3$ solution)

DEPT  CH$_2\downarrow$ CH$_3\uparrow$ CH$\uparrow$

proton decoupled

solvent

$\delta$ (ppm)

$^1$H NMR Spectrum
(400 MHz, CDCl$_3$ solution)

TMS

$\delta$ (ppm)

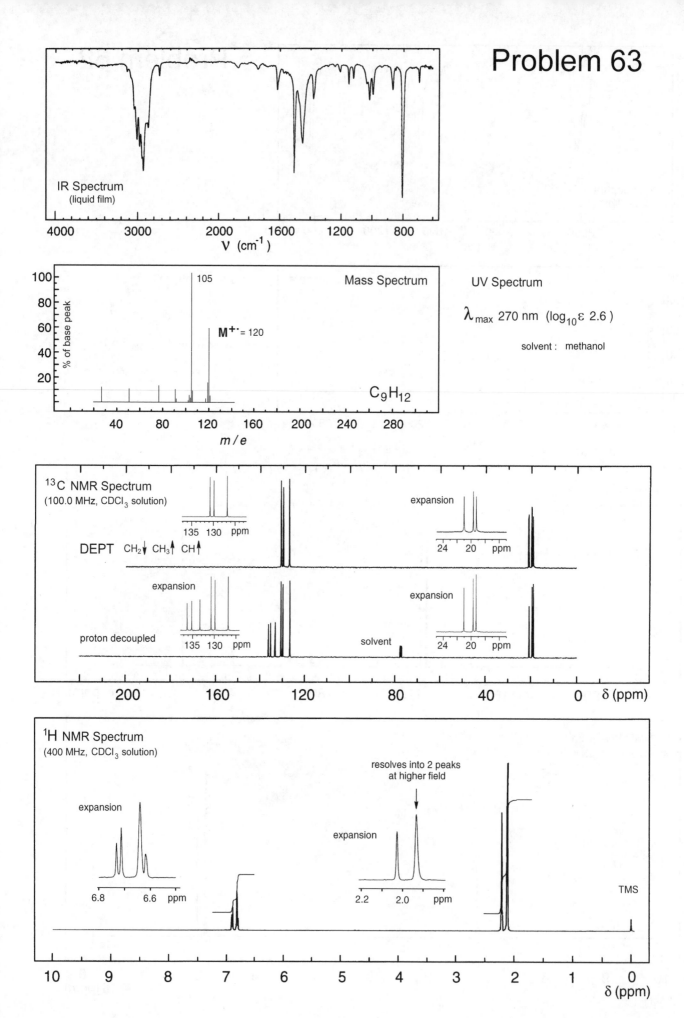

# Problem 63

**IR Spectrum**
(liquid film)

ν (cm⁻¹)

**Mass Spectrum**

105

M⁺· = 120

C₉H₁₂

m/e

**UV Spectrum**

λ_max 270 nm (log₁₀ε 2.6)

solvent : methanol

¹³C NMR Spectrum
(100.0 MHz, CDCl₃ solution)

expansion

135  130  ppm

DEPT   CH₂↓ CH₃↑ CH↑

expansion

24  20  ppm

expansion

135  130  ppm

proton decoupled

solvent

expansion

24  20  ppm

δ (ppm)

¹H NMR Spectrum
(400 MHz, CDCl₃ solution)

expansion

6.8   6.6  ppm

resolves into 2 peaks
at higher field

expansion

2.2   2.0  ppm

TMS

δ (ppm)

174

# Problem 64

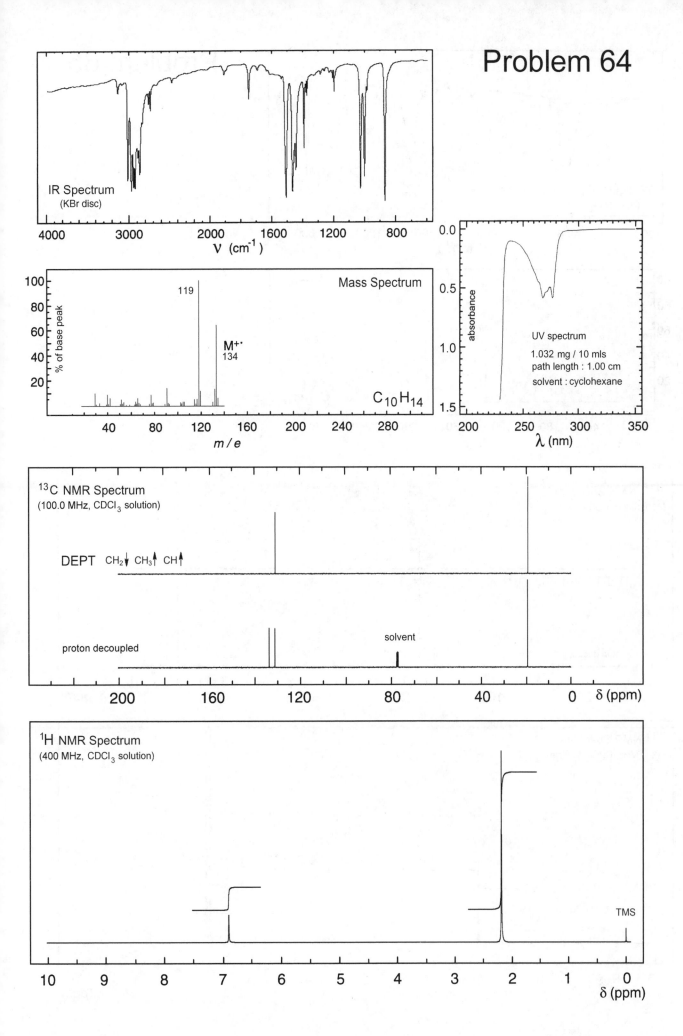

IR Spectrum
(KBr disc)

ν (cm⁻¹)

Mass Spectrum

% of base peak

119

M⁺·
134

C₁₀H₁₄

m/e

UV spectrum

1.032 mg / 10 mls
path length : 1.00 cm
solvent : cyclohexane

absorbance

λ (nm)

¹³C NMR Spectrum
(100.0 MHz, CDCl₃ solution)

DEPT  CH₂↓ CH₃↑ CH↑

proton decoupled

solvent

δ (ppm)

¹H NMR Spectrum
(400 MHz, CDCl₃ solution)

TMS

δ (ppm)

# Problem 65

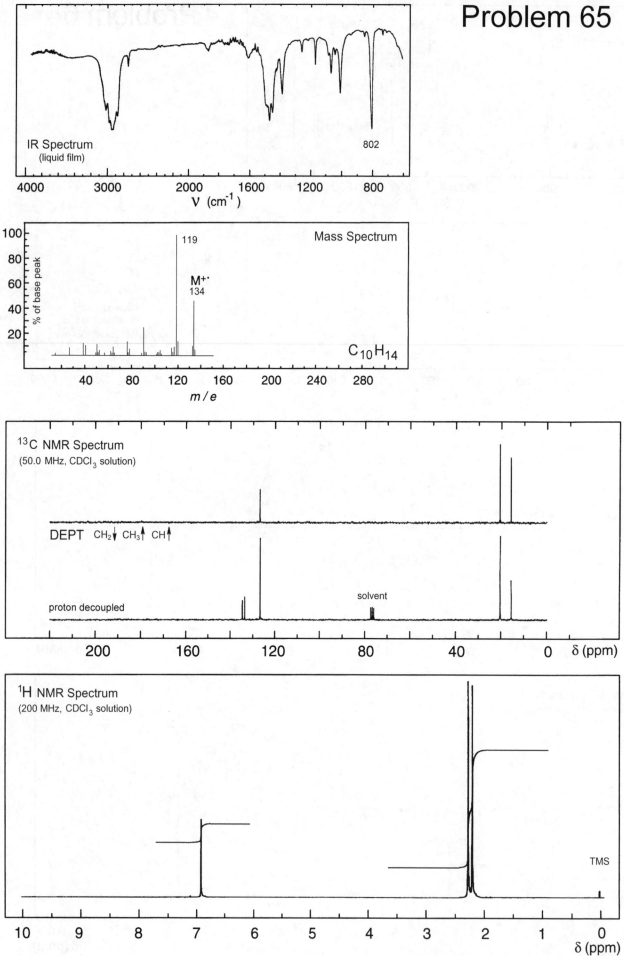

IR Spectrum
(liquid film)

802

ν (cm⁻¹)

Mass Spectrum

119

M⁺·
134

% of base peak

C₁₀H₁₄

m/e

¹³C NMR Spectrum
(50.0 MHz, CDCl₃ solution)

DEPT  CH₂↓ CH₃↑ CH↑

solvent

proton decoupled

δ (ppm)

¹H NMR Spectrum
(200 MHz, CDCl₃ solution)

TMS

δ (ppm)

# Problem 66

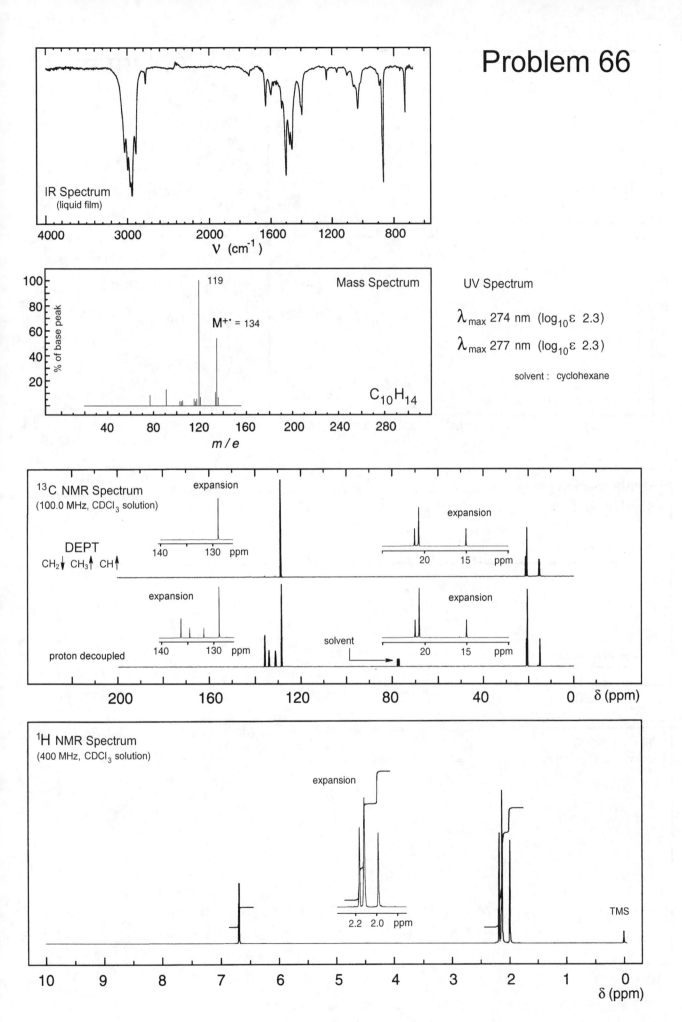

IR Spectrum
(liquid film)

ν (cm⁻¹)

Mass Spectrum

119

M⁺˙ = 134

C₁₀H₁₄

% of base peak

m/e

UV Spectrum

λ_max 274 nm  (log₁₀ε  2.3)

λ_max 277 nm  (log₁₀ε  2.3)

solvent :  cyclohexane

¹³C NMR Spectrum
(100.0 MHz, CDCl₃ solution)

DEPT
CH₂↓ CH₃↑ CH↑

expansion

140    130  ppm

expansion

20    15   ppm

expansion

140    130  ppm

proton decoupled

solvent

expansion

20    15   ppm

δ (ppm)

¹H NMR Spectrum
(400 MHz, CDCl₃ solution)

expansion

2.2   2.0  ppm

TMS

δ (ppm)

# Problem 67

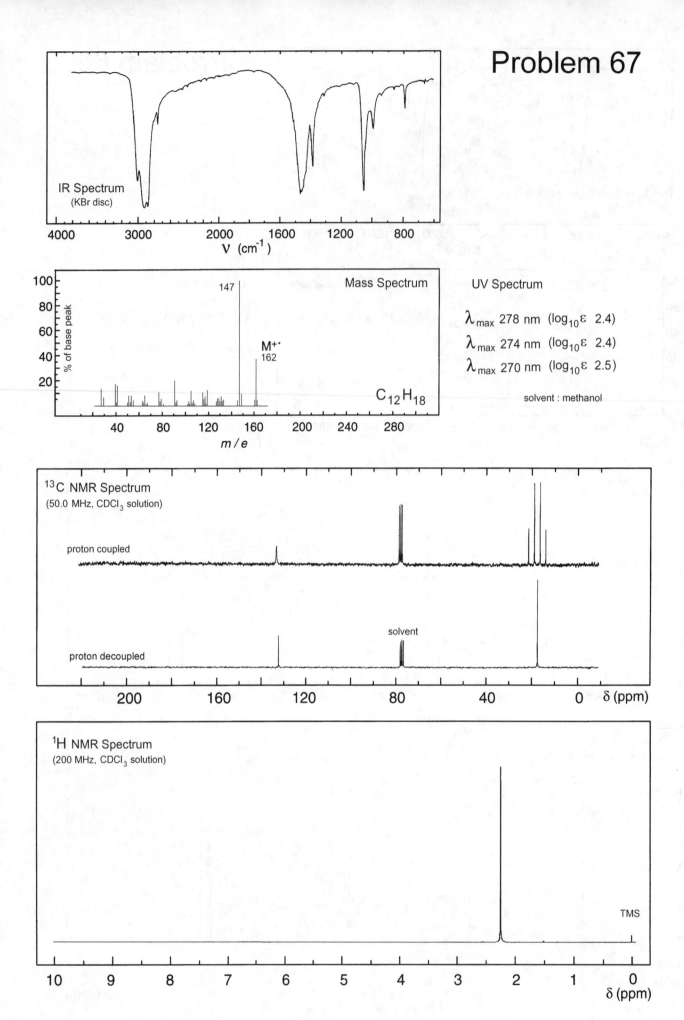

IR Spectrum
(KBr disc)

ν (cm⁻¹)

Mass Spectrum

147

M⁺·
162

C₁₂H₁₈

% of base peak

m/e

UV Spectrum

λ_max 278 nm (log₁₀ε 2.4)

λ_max 274 nm (log₁₀ε 2.4)

λ_max 270 nm (log₁₀ε 2.5)

solvent : methanol

¹³C NMR Spectrum
(50.0 MHz, CDCl₃ solution)

proton coupled

proton decoupled

solvent

δ (ppm)

¹H NMR Spectrum
(200 MHz, CDCl₃ solution)

TMS

δ (ppm)

# Problem 68

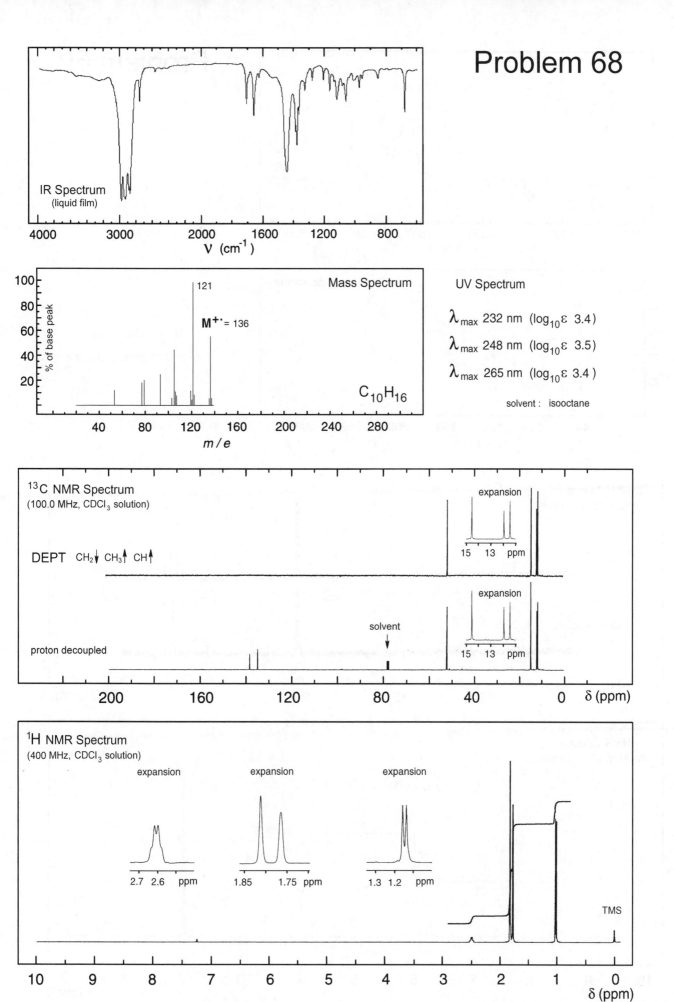

IR Spectrum
(liquid film)

ν (cm⁻¹)

Mass Spectrum

121

M⁺· = 136

C₁₀H₁₆

% of base peak

m/e

UV Spectrum

λₘₐₓ 232 nm (log₁₀ε 3.4)

λₘₐₓ 248 nm (log₁₀ε 3.5)

λₘₐₓ 265 nm (log₁₀ε 3.4)

solvent : isooctane

¹³C NMR Spectrum
(100.0 MHz, CDCl₃ solution)

DEPT  CH₂↓ CH₃↑ CH↑

expansion

15   13   ppm

proton decoupled

solvent

expansion

15   13   ppm

δ (ppm)

¹H NMR Spectrum
(400 MHz, CDCl₃ solution)

expansion

2.7  2.6  ppm

expansion

1.85  1.75  ppm

expansion

1.3  1.2  ppm

TMS

δ (ppm)

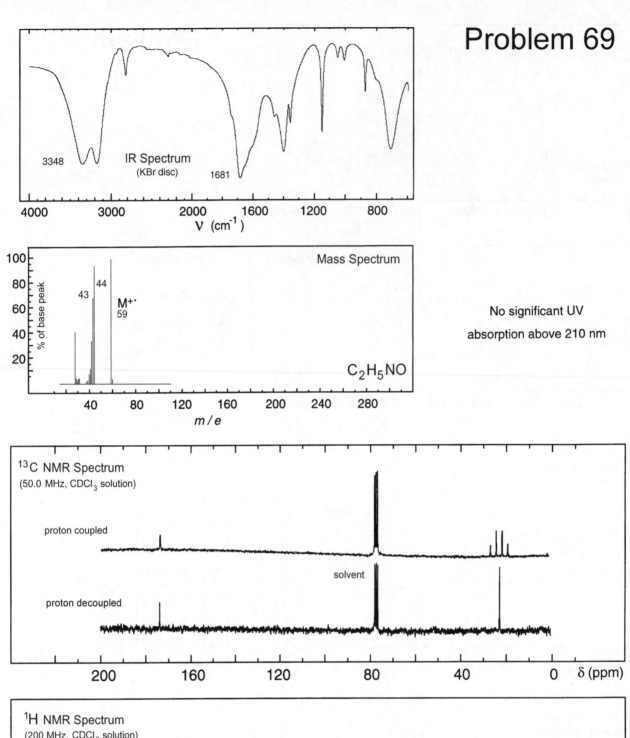

# Problem 69

**IR Spectrum**
(KBr disc)

3348

1681

**Mass Spectrum**

43

44

M⁺·
59

$C_2H_5NO$

No significant UV
absorption above 210 nm

**¹³C NMR Spectrum**
(50.0 MHz, CDCl₃ solution)

proton coupled

proton decoupled

solvent

δ (ppm)

**¹H NMR Spectrum**
(200 MHz, CDCl₃ solution)

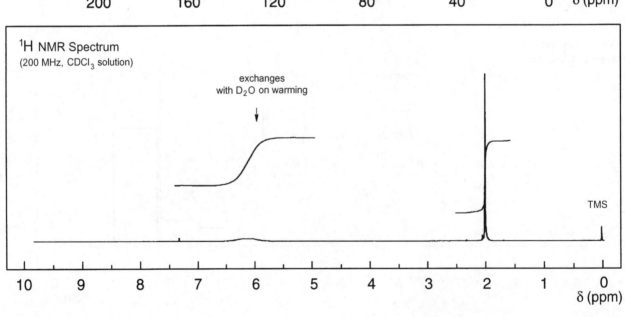

exchanges
with D₂O on warming

TMS

δ (ppm)

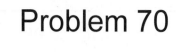

# Problem 70

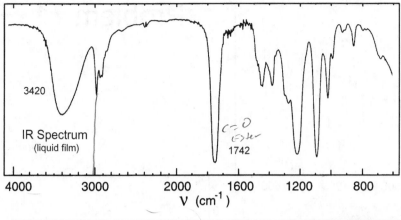

IR Spectrum
(liquid film)

3420

C=O
Ester
1742

ν (cm⁻¹)

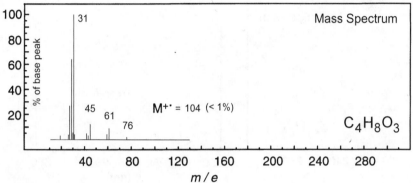

Mass Spectrum

31

45

61

76

M⁺˙ = 104 (< 1%)

C₄H₈O₃

% of base peak

m/e

No significant UV
absorption above 220 nm

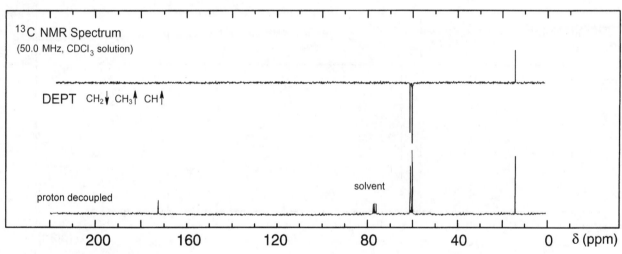

¹³C NMR Spectrum
(50.0 MHz, CDCl₃ solution)

DEPT  CH₂↓ CH₃↑ CH↑

solvent

proton decoupled

δ (ppm)

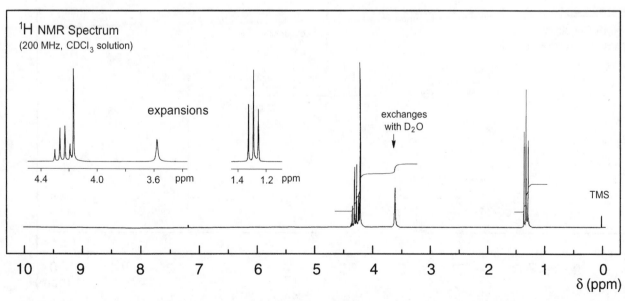

¹H NMR Spectrum
(200 MHz, CDCl₃ solution)

expansions

exchanges
with D₂O

TMS

δ (ppm)

# Problem 71

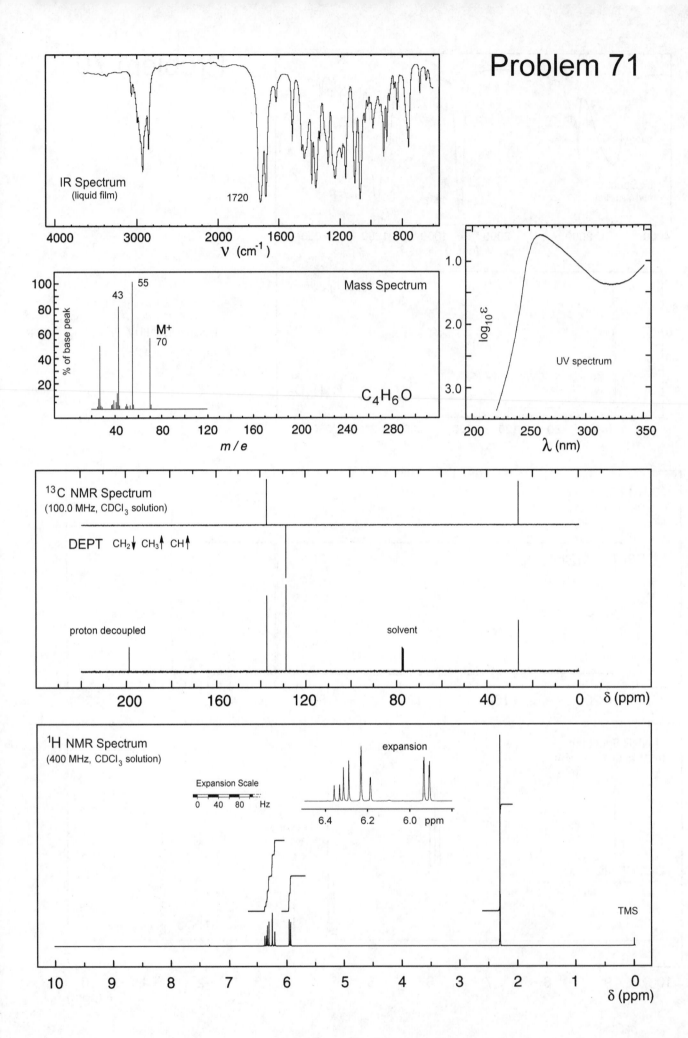

IR Spectrum
(liquid film)
1720

Mass Spectrum

43
55
M+
70

% of base peak

m/e

$C_4H_6O$

UV spectrum

$\log_{10} \varepsilon$

$\lambda$ (nm)

$^{13}C$ NMR Spectrum
(100.0 MHz, $CDCl_3$ solution)

DEPT  $CH_2\downarrow$ $CH_3\uparrow$ $CH\uparrow$

proton decoupled

solvent

$\delta$ (ppm)

$^1H$ NMR Spectrum
(400 MHz, $CDCl_3$ solution)

expansion

Expansion Scale
0   40   80   Hz

6.4      6.2      6.0  ppm

TMS

$\delta$ (ppm)

# Problem 72

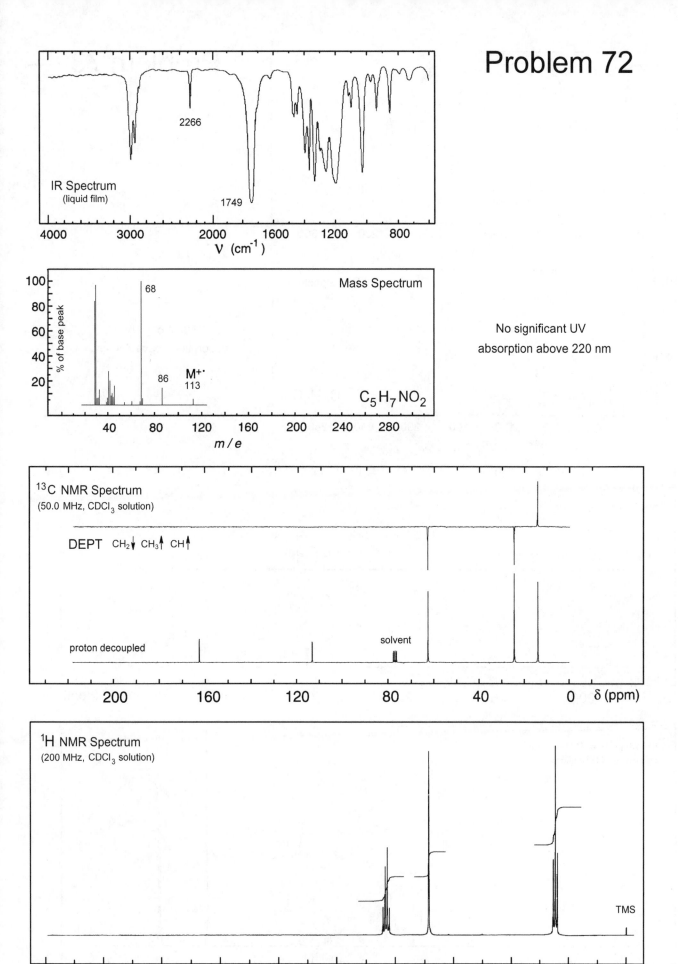

IR Spectrum
(liquid film)

2266

1749

ν (cm⁻¹)

Mass Spectrum

% of base peak

68

86

M⁺·
113

C₅H₇NO₂

No significant UV
absorption above 220 nm

m / e

¹³C NMR Spectrum
(50.0 MHz, CDCl₃ solution)

DEPT   CH₂↓ CH₃↑ CH↑

proton decoupled

solvent

δ (ppm)

¹H NMR Spectrum
(200 MHz, CDCl₃ solution)

TMS

δ (ppm)

183

# Problem 73

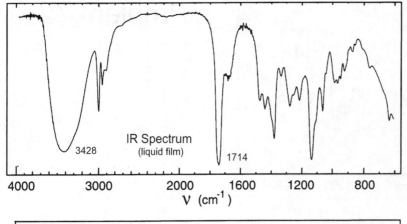

IR Spectrum
(liquid film)

3428
1714

ν (cm⁻¹)

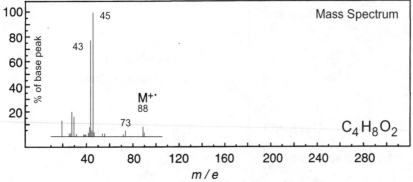

Mass Spectrum

45
43
M⁺·
88
73

% of base peak

m/e

No strong UV
absorption above 220 nm

C₄H₈O₂

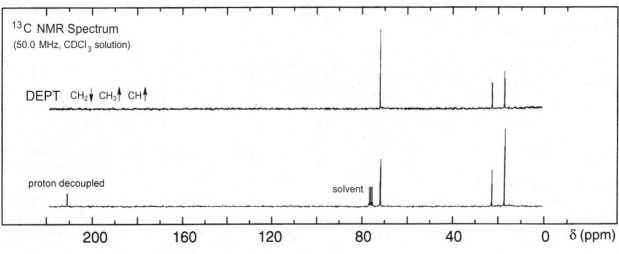

¹³C NMR Spectrum
(50.0 MHz, CDCl₃ solution)

DEPT   CH₂↓ CH₃↑ CH↑

proton decoupled

solvent

δ (ppm)

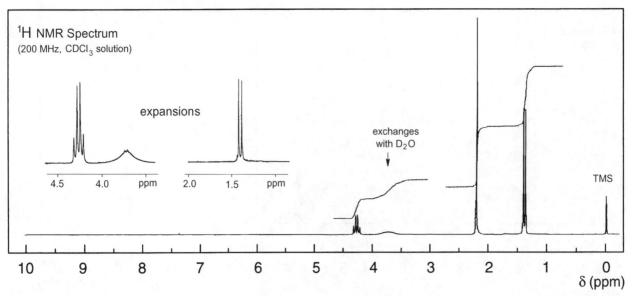

¹H NMR Spectrum
(200 MHz, CDCl₃ solution)

expansions

4.5   4.0   ppm        2.0   1.5   ppm

exchanges
with D₂O

TMS

δ (ppm)

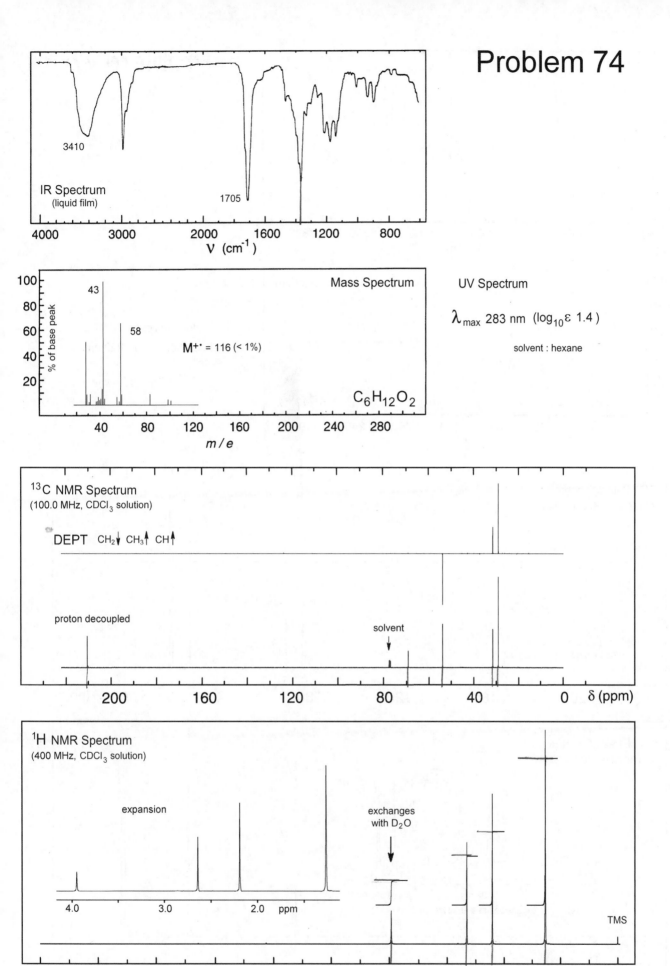

# Problem 74

**IR Spectrum**
(liquid film)

3410

1705

$\nu$ (cm$^{-1}$)

4000   3000   2000   1600   1200   800

**Mass Spectrum**

43

58

M$^{+\cdot}$ = 116 (< 1%)

$C_6H_{12}O_2$

% of base peak

40   80   120   160   200   240   280

$m/e$

**UV Spectrum**

$\lambda_{max}$ 283 nm (log$_{10}\varepsilon$ 1.4)

solvent : hexane

**$^{13}$C NMR Spectrum**
(100.0 MHz, CDCl$_3$ solution)

DEPT   CH$_2\downarrow$ CH$_3\uparrow$ CH$\uparrow$

proton decoupled

solvent

200   160   120   80   40   0   $\delta$ (ppm)

**$^1$H NMR Spectrum**
(400 MHz, CDCl$_3$ solution)

expansion

exchanges
with D$_2$O

4.0   3.0   2.0   ppm

TMS

10   9   8   7   6   5   4   3   2   1   0

$\delta$ (ppm)

185

# Problem 75

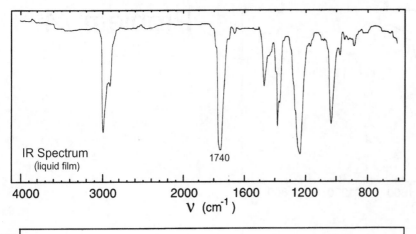

IR Spectrum
(liquid film)

1740

ν (cm⁻¹)

$$\nu \ (cm^{-1})$$

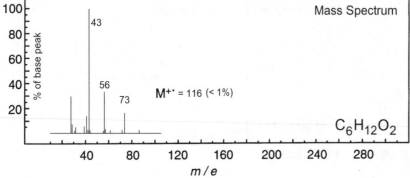

Mass Spectrum

100
80
60
40
20

% of base peak

43

56

73

M⁺˙ = 116 (< 1%)

$$M^{+\bullet} = 116 \ (< 1\%)$$

$$C_6H_{12}O_2$$

m/e

No significant UV
absorption above 220 nm

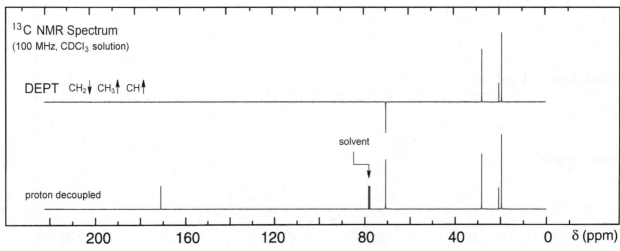

¹³C NMR Spectrum
(100 MHz, CDCl₃ solution)

DEPT  CH₂↓ CH₃↑ CH↑

solvent

proton decoupled

200   160   120   80   40   0   δ (ppm)

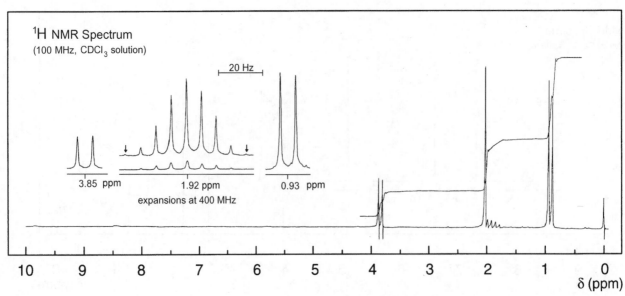

¹H NMR Spectrum
(100 MHz, CDCl₃ solution)

20 Hz

3.85 ppm

1.92 ppm

0.93 ppm

expansions at 400 MHz

10   9   8   7   6   5   4   3   2   1   0   δ (ppm)

# Problem 76

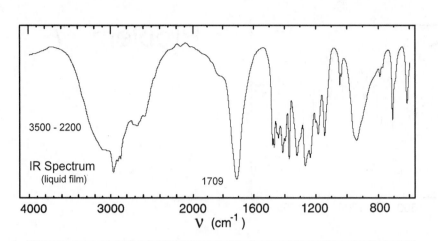

IR Spectrum
(liquid film)

3500 - 2200

1709

ν (cm⁻¹)

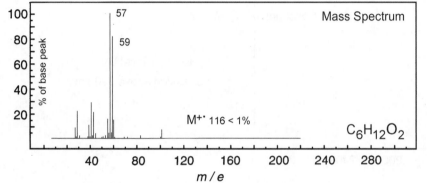

Mass Spectrum

% of base peak

57

59

M⁺ᐧ 116 < 1%

$C_6H_{12}O_2$

m / e

No significant UV
absorption above 220 nm

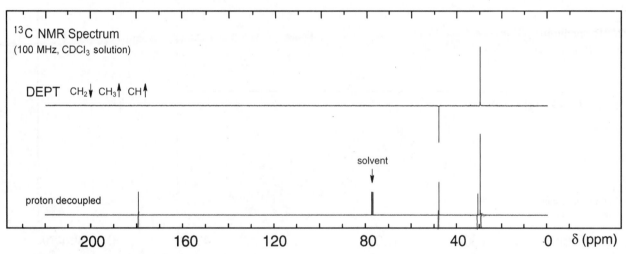

¹³C NMR Spectrum
(100 MHz, CDCl₃ solution)

DEPT   CH₂↓ CH₃↑ CH↑

solvent

proton decoupled

δ (ppm)

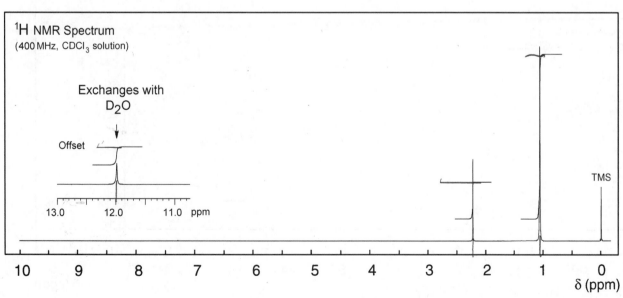

¹H NMR Spectrum
(400 MHz, CDCl₃ solution)

Exchanges with
D₂O

Offset

13.0   12.0   11.0   ppm

TMS

δ (ppm)

187

# Problem 77

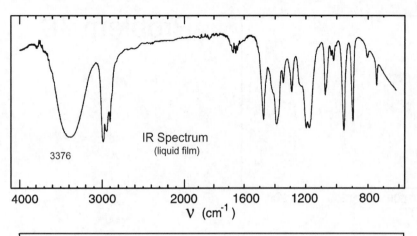

IR Spectrum
(liquid film)

3376

ν (cm⁻¹)

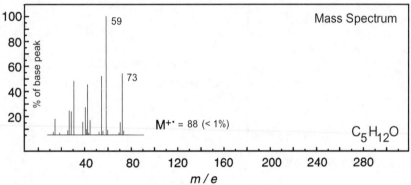

Mass Spectrum

59

73

M⁺˙ = 88 (< 1%)

$C_5H_{12}O$

% of base peak

m / e

No significant UV
absorption above 220 nm

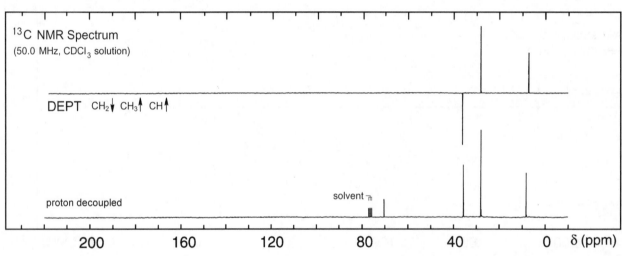

¹³C NMR Spectrum
(50.0 MHz, CDCl₃ solution)

DEPT   CH₂↓ CH₃↑ CH↑

proton decoupled

solvent

δ (ppm)

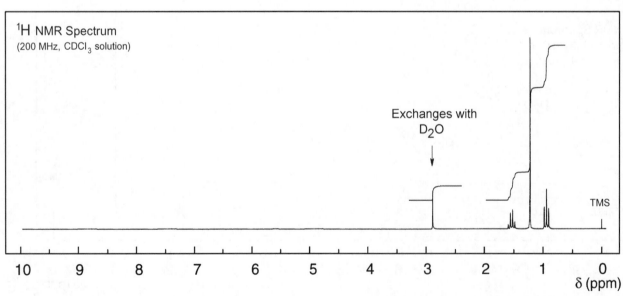

¹H NMR Spectrum
(200 MHz, CDCl₃ solution)

Exchanges with
D₂O

TMS

δ (ppm)

# Problem 78

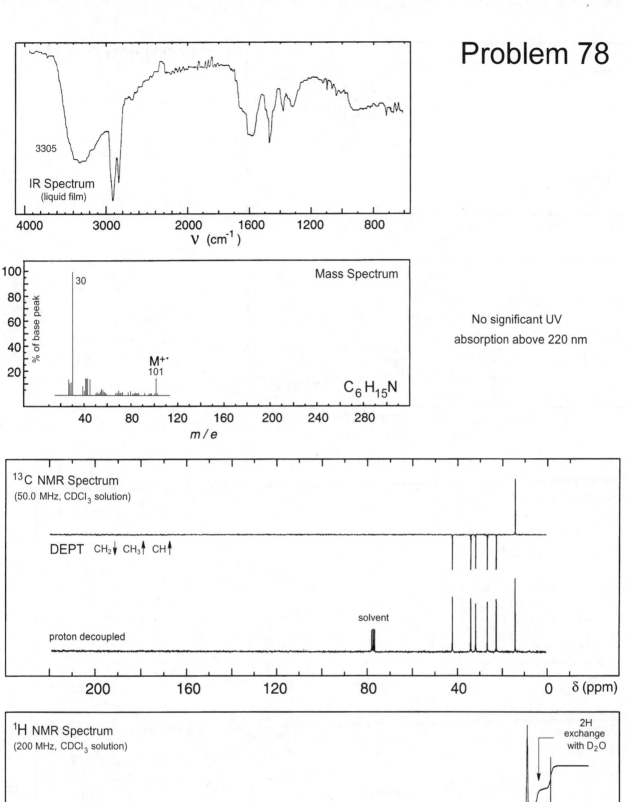

IR Spectrum
(liquid film)

3305

$\nu$ (cm$^{-1}$)

Mass Spectrum

% of base peak

30

M$^{+\bullet}$
101

C$_6$H$_{15}$N

m/e

No significant UV
absorption above 220 nm

$^{13}$C NMR Spectrum
(50.0 MHz, CDCl$_3$ solution)

DEPT  CH$_2\downarrow$ CH$_3\uparrow$ CH$\uparrow$

solvent

proton decoupled

$\delta$ (ppm)

$^1$H NMR Spectrum
(200 MHz, CDCl$_3$ solution)

2H
exchange
with D$_2$O

TMS

$\delta$ (ppm)

# Problem 79

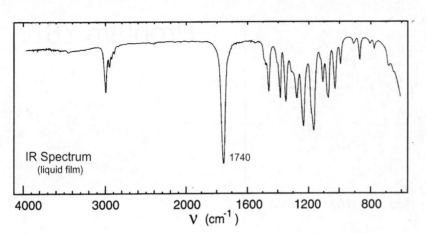

IR Spectrum
(liquid film)

1740

V (cm⁻¹)

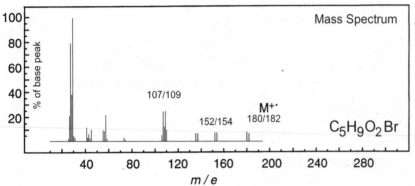

Mass Spectrum

% of base peak

107/109

152/154

M⁺·
180/182

$C_5H_9O_2Br$

m/e

No significant UV
absorption above 220 nm

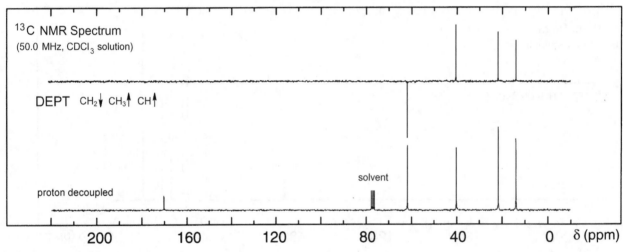

¹³C NMR Spectrum
(50.0 MHz, CDCl₃ solution)

DEPT   CH₂↓ CH₃↑ CH↑

solvent

proton decoupled

δ (ppm)

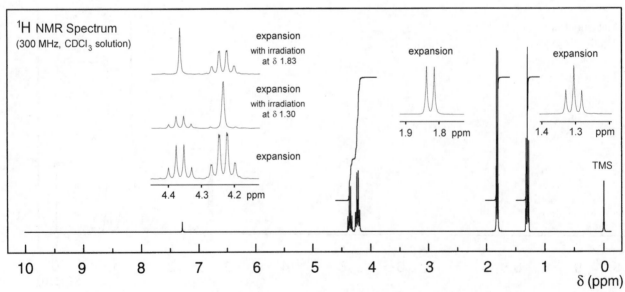

¹H NMR Spectrum
(300 MHz, CDCl₃ solution)

expansion
with irradiation
at δ 1.83

expansion
with irradiation
at δ 1.30

expansion

expansion

expansion

4.4    4.3    4.2 ppm

1.9    1.8 ppm

1.4    1.3   ppm

TMS

δ (ppm)

# Problem 80

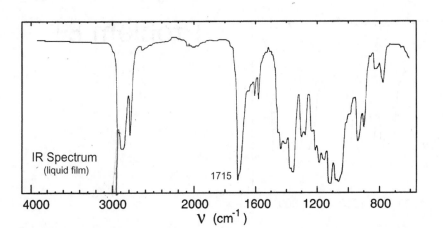

IR Spectrum
(liquid film)

1715

ν (cm⁻¹)

$\nu$ (cm$^{-1}$)

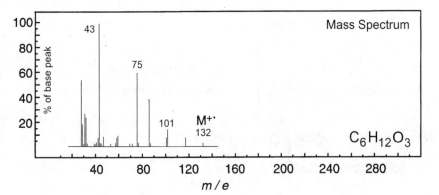

Mass Spectrum

% of base peak

43

75

101

M⁺·
132

$M^{+\cdot}$

No significant UV
absorption above 220 nm

$C_6H_{12}O_3$

m/e

$m/e$

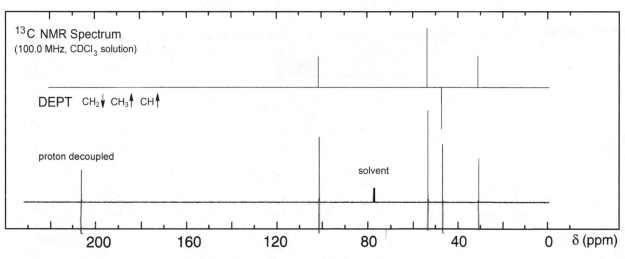

$^{13}C$ NMR Spectrum
(100.0 MHz, CDCl₃ solution)

DEPT  CH₂↓ CH₃↑ CH↑

proton decoupled

solvent

δ (ppm)

$\delta$ (ppm)

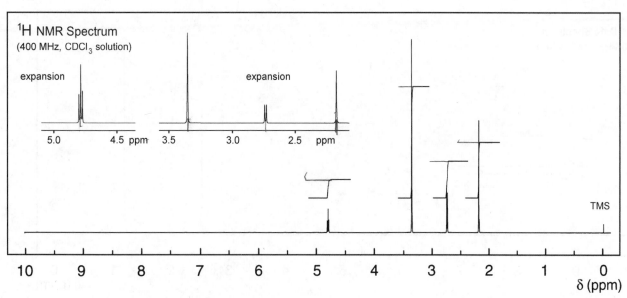

$^1H$ NMR Spectrum
(400 MHz, CDCl₃ solution)

expansion

expansion

5.0    4.5  ppm  3.5    3.0    2.5   ppm

TMS

δ (ppm)

$\delta$ (ppm)

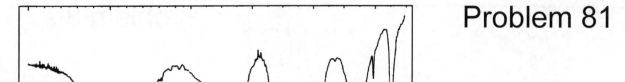

# Problem 81

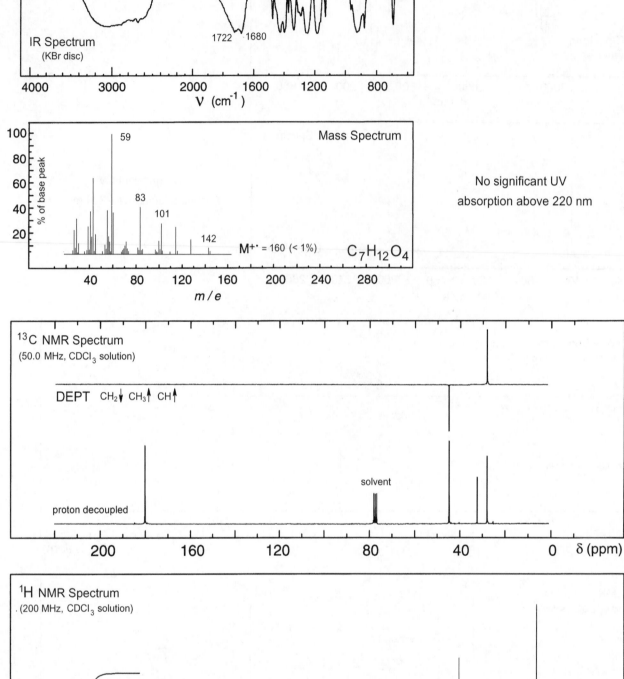

IR Spectrum
(KBr disc)

1722  1680

ν (cm⁻¹)

Mass Spectrum

% of base peak

59

83

101

142

M⁺˙ = 160 (< 1%)

$C_7H_{12}O_4$

m/e

No significant UV
absorption above 220 nm

¹³C NMR Spectrum
(50.0 MHz, CDCl₃ solution)

DEPT  CH₂↓ CH₃↑ CH↑

proton decoupled

solvent

δ (ppm)

¹H NMR Spectrum
(200 MHz, CDCl₃ solution)

exchanges
with D₂O

11.0        10.0  ppm

TMS

δ (ppm)

# Problem 82

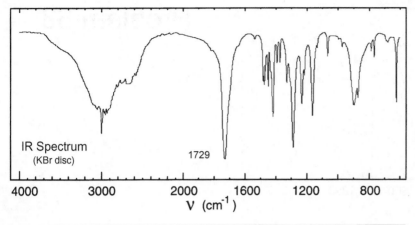

IR Spectrum
(KBr disc)

1729

ν (cm⁻¹)

$\nu$ (cm$^{-1}$)

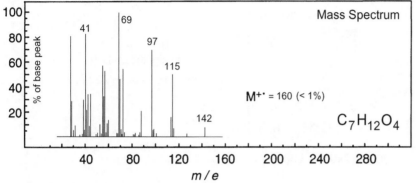

Mass Spectrum

% of base peak

41
69
97
115
142

M⁺˙ = 160 (< 1%)

$M^{+\cdot} = 160$ (< 1%)

$C_7H_{12}O_4$

No significant UV
absorption above 220 nm

m / e

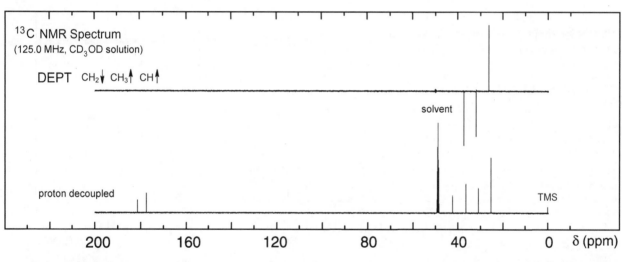

¹³C NMR Spectrum
(125.0 MHz, CD₃OD solution)

$^{13}C$ NMR Spectrum
(125.0 MHz, CD$_3$OD solution)

DEPT   CH₂↓ CH₃↑ CH↑

solvent

proton decoupled

TMS

δ (ppm)

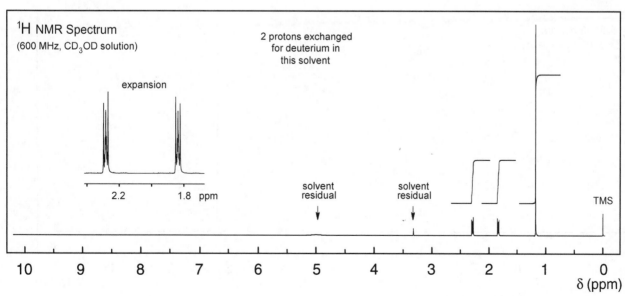

¹H NMR Spectrum
(600 MHz, CD₃OD solution)

$^1H$ NMR Spectrum
(600 MHz, CD$_3$OD solution)

2 protons exchanged
for deuterium in
this solvent

expansion

2.2      1.8  ppm

solvent
residual

solvent
residual

TMS

δ (ppm)

# Problem 83

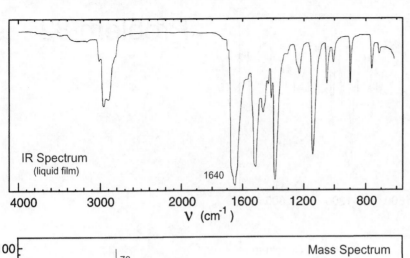

IR Spectrum
(liquid film)

1640

$\nu$ (cm$^{-1}$)

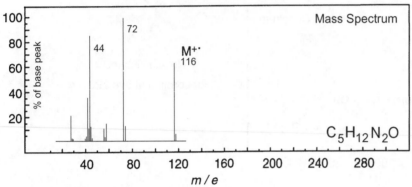

Mass Spectrum

% of base peak

44

72

M$^{+\cdot}$
116

$C_5H_{12}N_2O$

No significant UV
absorption above 220 nm

$m/e$

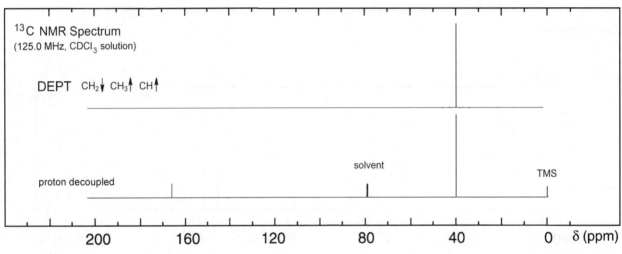

$^{13}$C NMR Spectrum
(125.0 MHz, CDCl$_3$ solution)

DEPT   CH$_2\downarrow$ CH$_3\uparrow$ CH$\uparrow$

proton decoupled

solvent

TMS

$\delta$ (ppm)

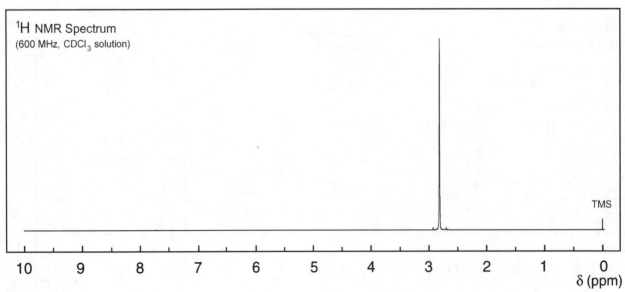

$^1$H NMR Spectrum
(600 MHz, CDCl$_3$ solution)

TMS

$\delta$ (ppm)

194

# Problem 84

IR Spectrum
(liquid film)

ν (cm$^{-1}$)

M$^{+\cdot}$ = 88

Mass Spectrum

% of base peak

No significant UV
absorption above 220 nm

$C_4H_8O_2$

m / e

$^{13}$C NMR Spectrum
(100.0 MHz, CDCl$_3$ solution)

DEPT  CH$_2$↓ CH$_3$↑ CH↑

solvent

proton decoupled

δ (ppm)

$^1$H NMR Spectrum
(400 MHz, CDCl$_3$ solution)

expansion

expansion

3.8    3.6  ppm

1.8    1.6  ppm

TMS

δ (ppm)

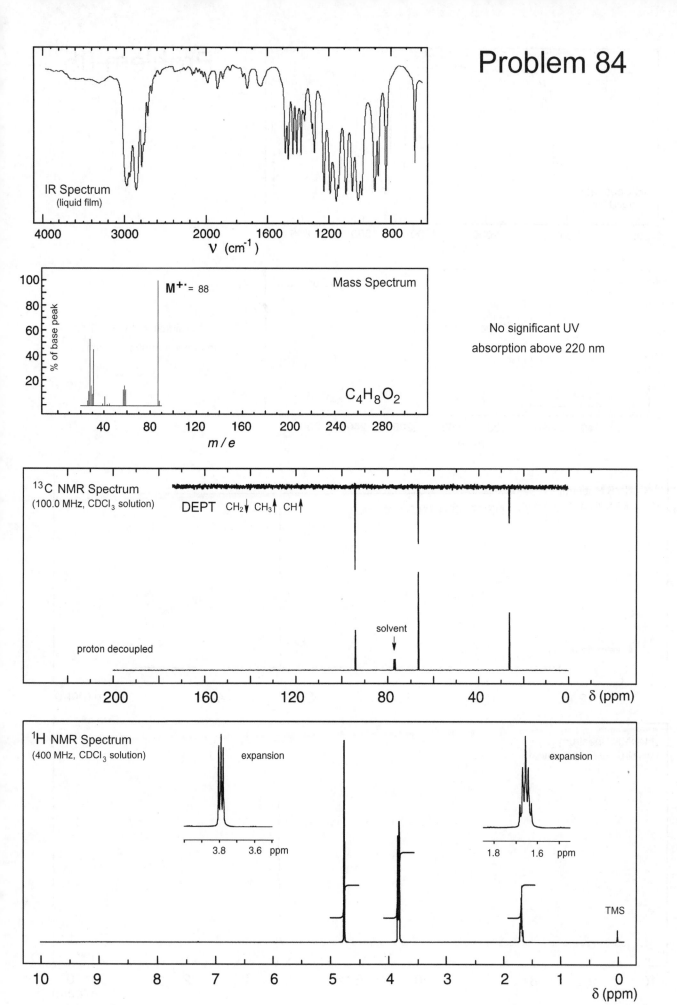

195

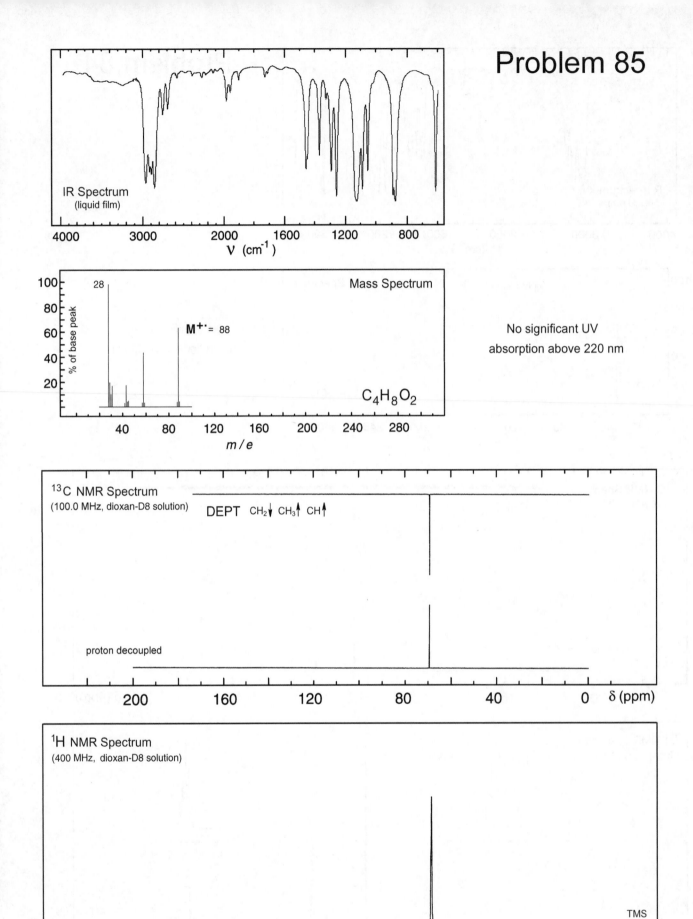

# Problem 85

**IR Spectrum**
(liquid film)

4000  3000  2000  1600  1200  800

$\nu$ (cm$^{-1}$)

Mass Spectrum

% of base peak

28

**M$^{+\cdot}$**= 88

$C_4H_8O_2$

40  80  120  160  200  240  280

*m/e*

No significant UV
absorption above 220 nm

**$^{13}$C NMR Spectrum**
(100.0 MHz, dioxan-D8 solution)

DEPT  CH$_2$↓ CH$_3$↑ CH↑

proton decoupled

200  160  120  80  40  0  $\delta$ (ppm)

**$^1$H NMR Spectrum**
(400 MHz, dioxan-D8 solution)

TMS

10  9  8  7  6  5  4  3  2  1  0

$\delta$ (ppm)

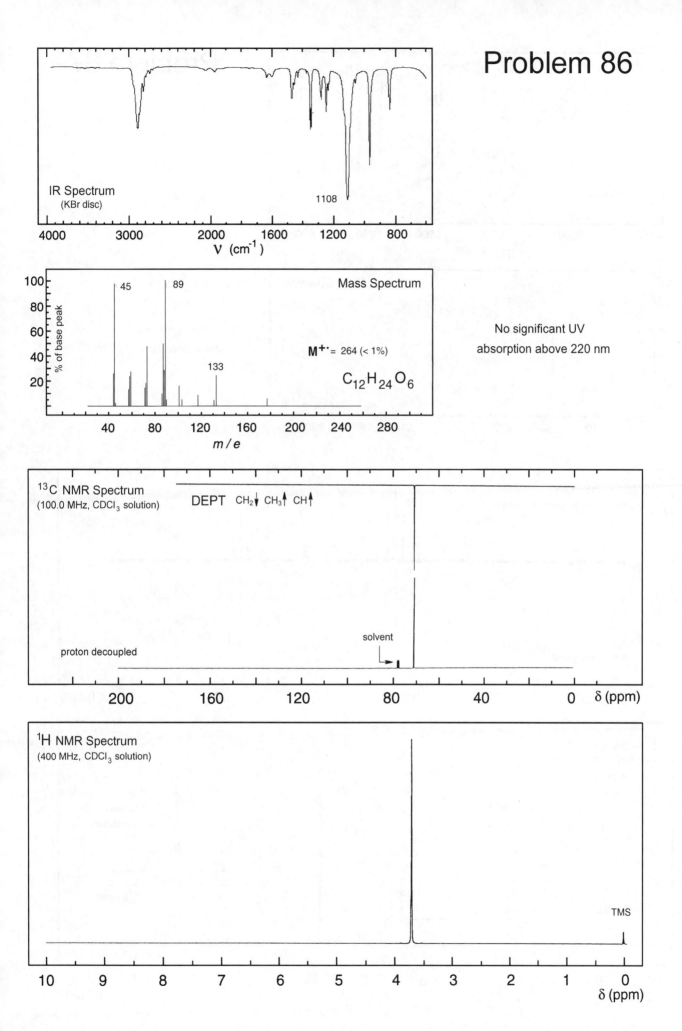

# Problem 86

**IR Spectrum**
(KBr disc)

1108

ν (cm⁻¹)

100  80  60  40  20

% of base peak

45

89

**Mass Spectrum**

M⁺· = 264 (< 1%)

$C_{12}H_{24}O_6$

133

m / e

No significant UV
absorption above 220 nm

**¹³C NMR Spectrum**
(100.0 MHz, CDCl₃ solution)

DEPT  CH₂↓ CH₃↑ CH↑

solvent

proton decoupled

δ (ppm)

**¹H NMR Spectrum**
(400 MHz, CDCl₃ solution)

TMS

δ (ppm)

197

# Problem 87

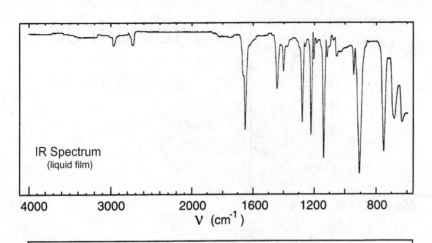

**IR Spectrum**
(liquid film)

ν (cm⁻¹)

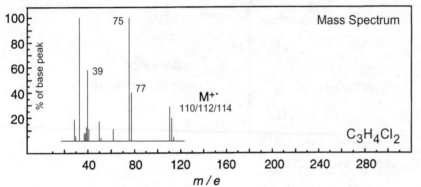

Mass Spectrum

% of base peak

M⁺˙
110/112/114

C₃H₄Cl₂

No significant UV
absorption above 220 nm

m/e

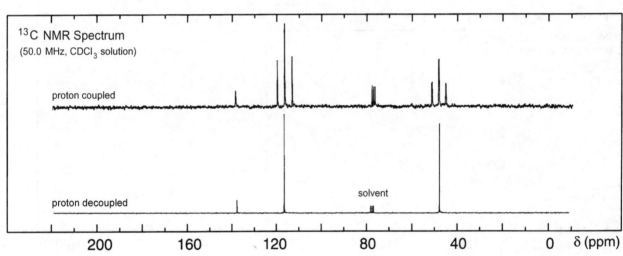

¹³C NMR Spectrum
(50.0 MHz, CDCl₃ solution)

proton coupled

proton decoupled

solvent

δ (ppm)

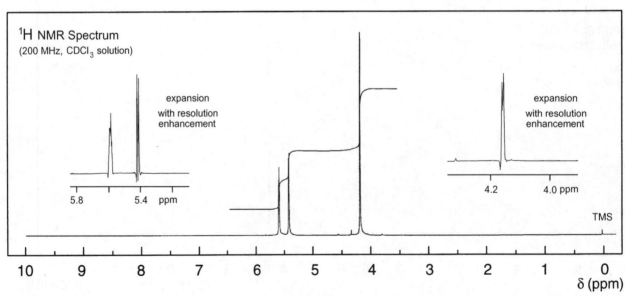

¹H NMR Spectrum
(200 MHz, CDCl₃ solution)

expansion
with resolution
enhancement

5.8      5.4   ppm

expansion
with resolution
enhancement

4.2        4.0 ppm

TMS

δ (ppm)

# Problem 88

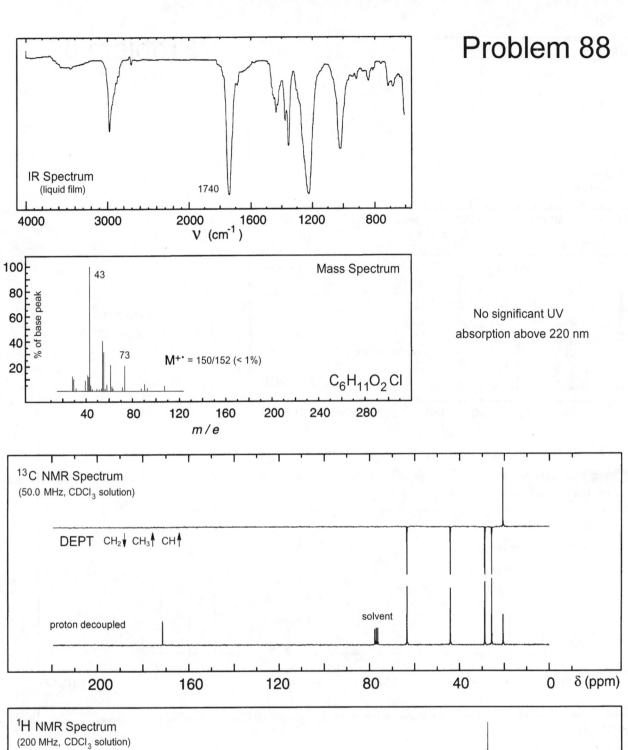

IR Spectrum
(liquid film)

1740

ν (cm⁻¹)

Mass Spectrum

43

73

M⁺˙ = 150/152 (< 1%)

$C_6H_{11}O_2Cl$

No significant UV
absorption above 220 nm

% of base peak

*m / e*

¹³C NMR Spectrum
(50.0 MHz, CDCl₃ solution)

DEPT   CH₂↓  CH₃↑  CH↑

proton decoupled

solvent

δ (ppm)

¹H NMR Spectrum
(200 MHz, CDCl₃ solution)

TMS

δ (ppm)

# Problem 89

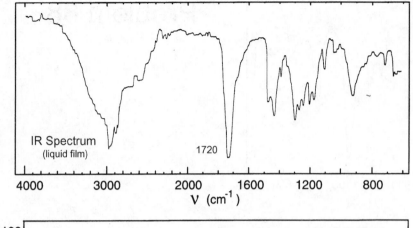

IR Spectrum
(liquid film)

1720

ν (cm⁻¹)

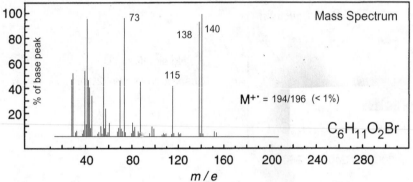

Mass Spectrum

73

138 140

115

M⁺· = 194/196 (< 1%)

$C_6H_{11}O_2Br$

% of base peak

m/e

No significant UV
absorption above 220 nm

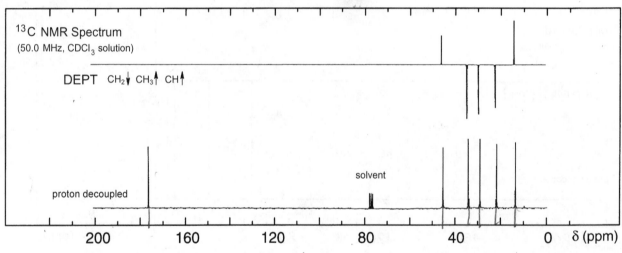

$^{13}C$ NMR Spectrum
(50.0 MHz, CDCl₃ solution)

DEPT   CH₂↓ CH₃↑ CH↑

solvent

proton decoupled

δ (ppm)

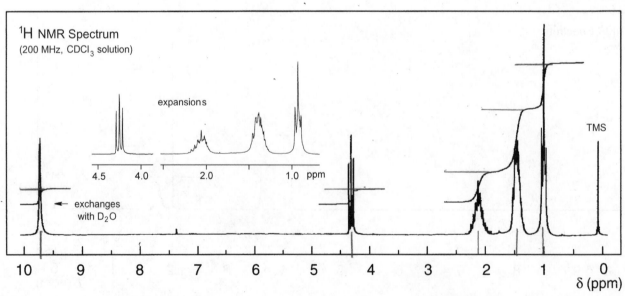

$^1H$ NMR Spectrum
(200 MHz, CDCl₃ solution)

expansions

4.5    4.0         2.0         1.0  ppm

← exchanges
with D₂O

TMS

δ (ppm)

# Problem 90

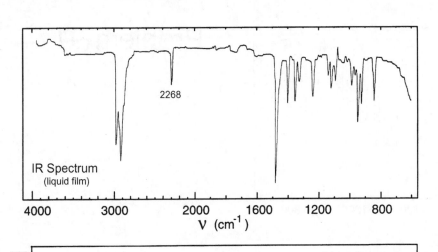

IR Spectrum
(liquid film)

2268

ν (cm⁻¹) — $\nu \ (cm^{-1})$

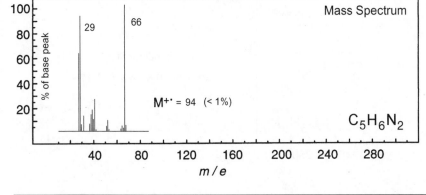

Mass Spectrum

29

66

$M^{+\bullet} = 94$  (< 1%)

$C_5H_6N_2$

% of base peak

m/e

No significant UV
absorption above 220 nm

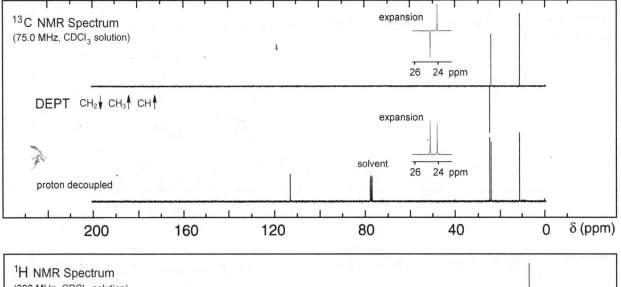

¹³C NMR Spectrum
(75.0 MHz, CDCl₃ solution)

expansion

26   24 ppm

DEPT   CH₂↓ CH₃↑ CH↑

expansion

26   24 ppm

solvent

proton decoupled

δ (ppm)

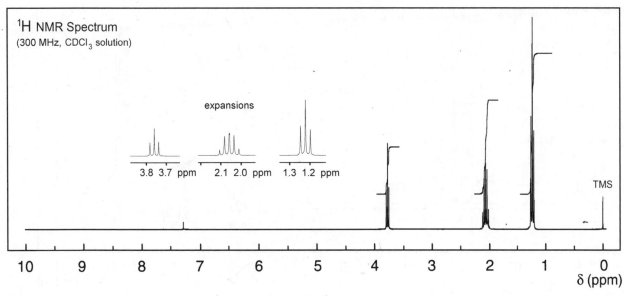

¹H NMR Spectrum
(300 MHz, CDCl₃ solution)

expansions

3.8  3.7 ppm    2.1  2.0 ppm    1.3  1.2 ppm

TMS

δ (ppm)

# Problem 91

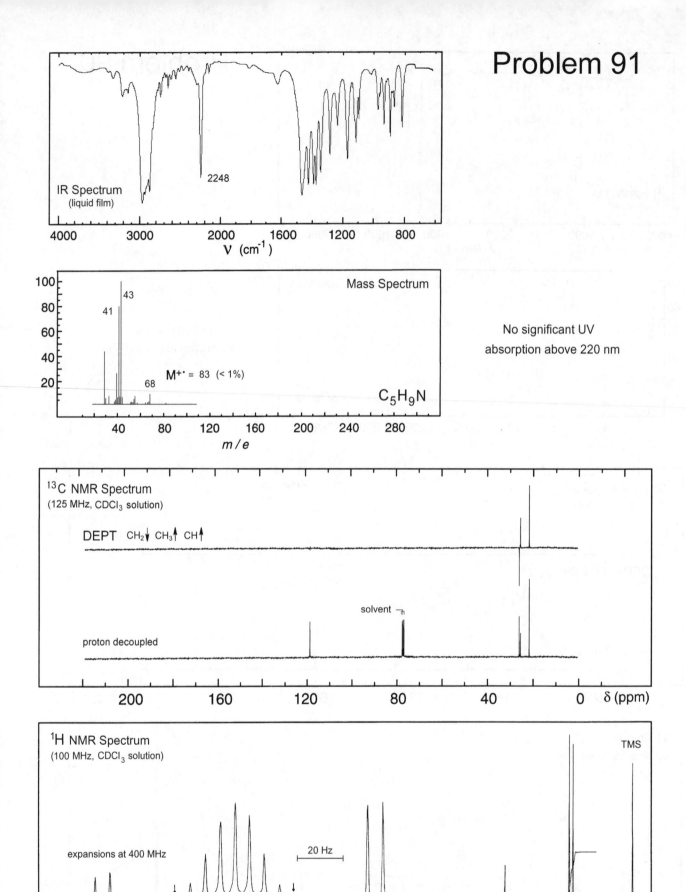

IR Spectrum
(liquid film)

2248

ν (cm⁻¹)

Mass Spectrum

No significant UV
absorption above 220 nm

41
43
68
M⁺· = 83 (< 1%)

C₅H₉N

m/e

¹³C NMR Spectrum
(125 MHz, CDCl₃ solution)

DEPT  CH₂↓ CH₃↑ CH↑

proton decoupled

solvent

δ (ppm)

¹H NMR Spectrum
(100 MHz, CDCl₃ solution)

TMS

expansions at 400 MHz

20 Hz

2.26

2.03

1.07

δ (ppm)

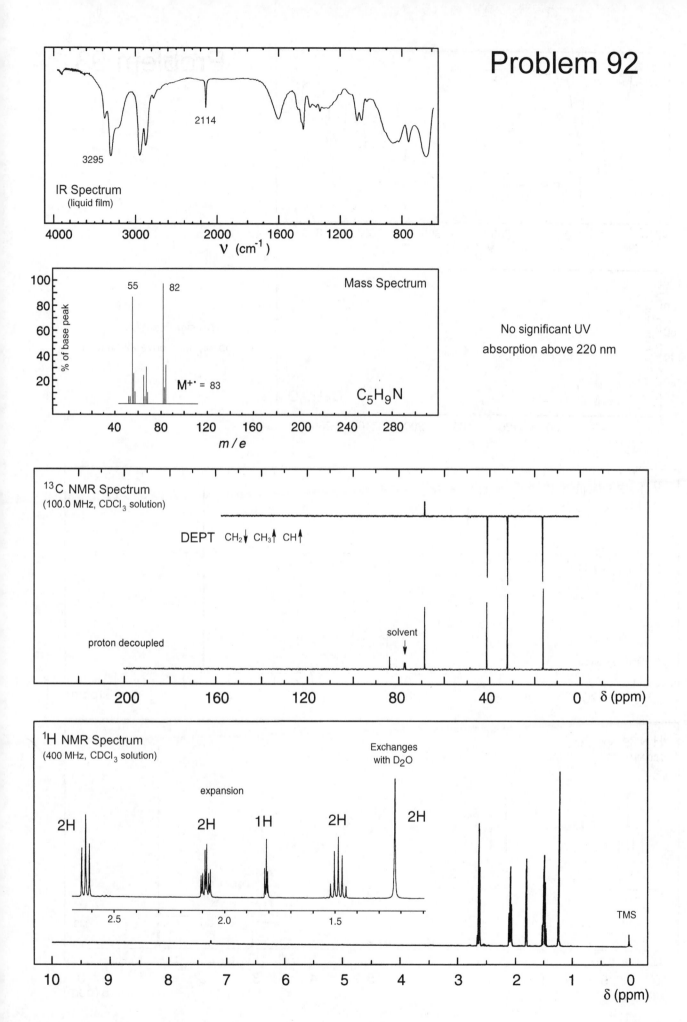

# Problem 92

**IR Spectrum** (liquid film)

3295
2114

**Mass Spectrum**

55
82
$M^{+\cdot}$ = 83

$C_5H_9N$

No significant UV absorption above 220 nm

**$^{13}$C NMR Spectrum** (100.0 MHz, CDCl$_3$ solution)

DEPT  CH$_2$↓ CH$_3$↑ CH↑

proton decoupled

solvent

**$^1$H NMR Spectrum** (400 MHz, CDCl$_3$ solution)

Exchanges with D$_2$O

expansion

2H
2H
1H
2H
2H

TMS

# Problem 93

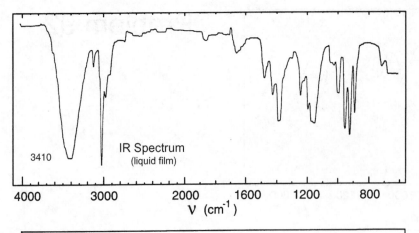

3410

IR Spectrum
(liquid film)

$\nu$ (cm$^{-1}$)

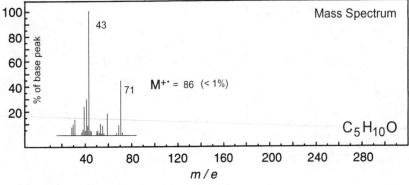

Mass Spectrum

43

71

$M^{+\cdot} = 86$ (< 1%)

% of base peak

$m/e$

No significant UV
absorption above 220 nm

$C_5H_{10}O$

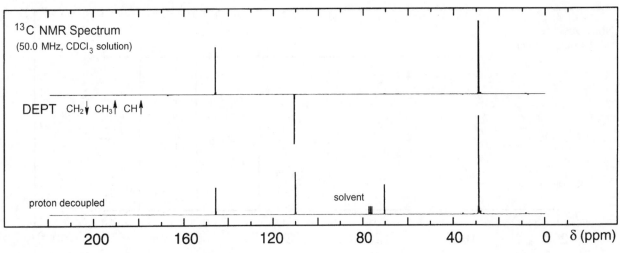

$^{13}C$ NMR Spectrum
(50.0 MHz, CDCl$_3$ solution)

DEPT  CH$_2\downarrow$ CH$_3\uparrow$ CH$\uparrow$

proton decoupled

solvent

$\delta$ (ppm)

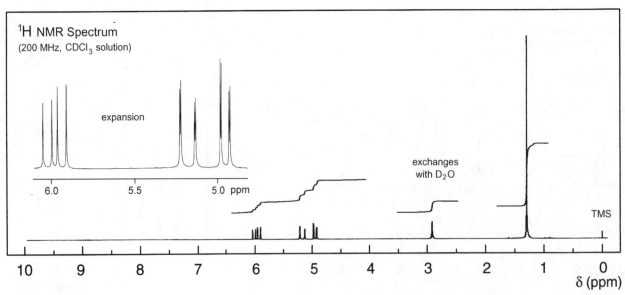

$^1H$ NMR Spectrum
(200 MHz, CDCl$_3$ solution)

expansion

6.0    5.5    5.0 ppm

exchanges
with D$_2$O

TMS

$\delta$ (ppm)

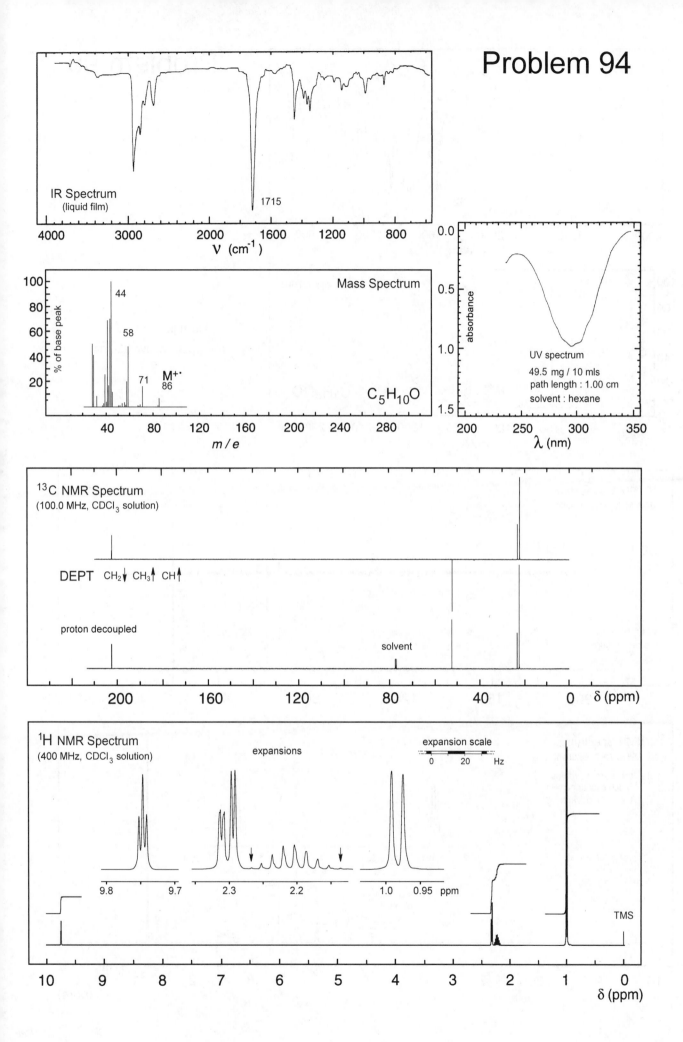

IR Spectrum
(liquid film)

1715

ν (cm⁻¹)

Mass Spectrum

% of base peak

44

58

71

M⁺˙
86

C₅H₁₀O

m/e

UV spectrum

49.5 mg / 10 mls
path length : 1.00 cm
solvent : hexane

λ (nm)

absorbance

¹³C NMR Spectrum
(100.0 MHz, CDCl₃ solution)

DEPT    CH₂↓ CH₃↑ CH↑

proton decoupled

solvent

δ (ppm)

¹H NMR Spectrum
(400 MHz, CDCl₃ solution)

expansions

expansion scale

0    20    Hz

9.8    9.7

2.3    2.2

1.0    0.95 ppm

TMS

δ (ppm)

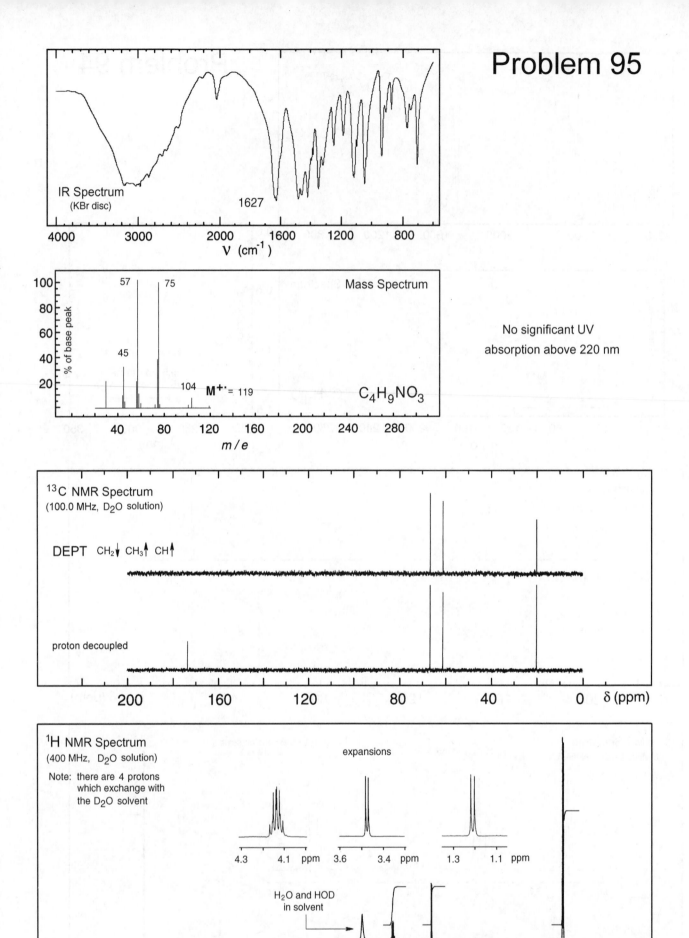

# Problem 95

IR Spectrum
(KBr disc)

1627

ν (cm⁻¹)

$\nu$ (cm$^{-1}$)

Mass Spectrum

No significant UV
absorption above 220 nm

100
80
60
40
20

% of base peak

57
75
45
104

M⁺˙= 119

$M^{+\cdot}$= 119

$C_4H_9NO_3$

40   80   120   160   200   240   280

m/e

¹³C NMR Spectrum
(100.0 MHz, D₂O solution)

$^{13}C$ NMR Spectrum
(100.0 MHz, D$_2$O solution)

DEPT   CH₂↓ CH₃↑ CH↑

proton decoupled

200   160   120   80   40   0   δ (ppm)

¹H NMR Spectrum
(400 MHz, D₂O solution)

$^1H$ NMR Spectrum
(400 MHz, D$_2$O solution)

Note: there are 4 protons
which exchange with
the D₂O solvent

expansions

4.3   4.1  ppm      3.6   3.4  ppm      1.3   1.1  ppm

H₂O and HOD
in solvent

10   9   8   7   6   5   4   3   2   1   0
δ (ppm)

206

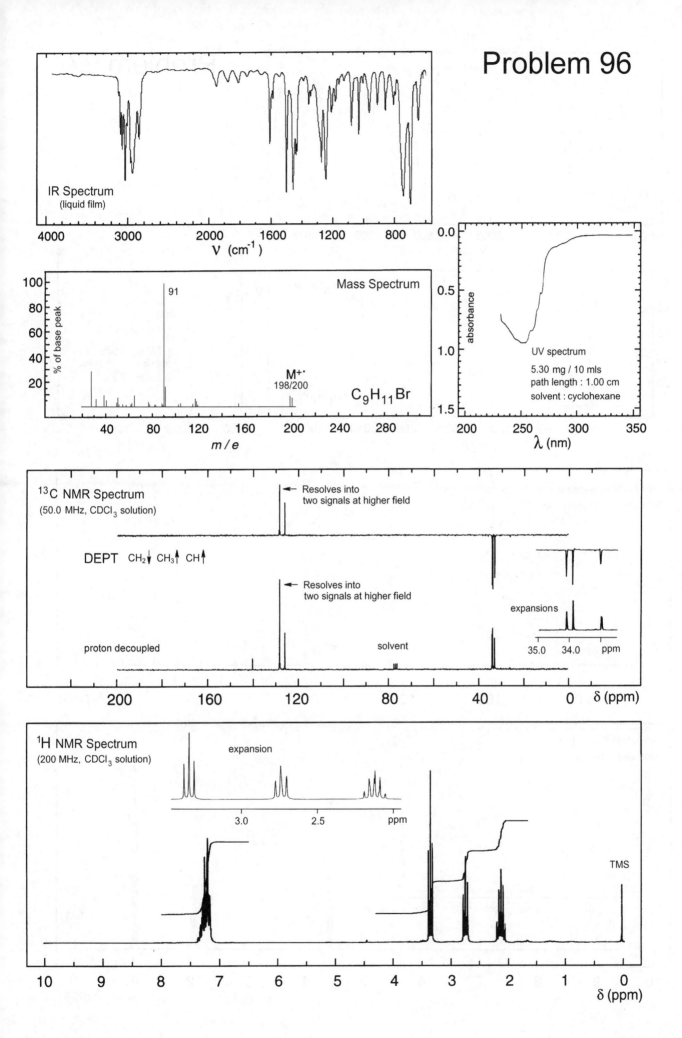

# Problem 96

**IR Spectrum**
(liquid film)

ν (cm⁻¹)

Mass Spectrum

91

M⁺˙
198/200

$C_9H_{11}Br$

m/e

UV spectrum

5.30 mg / 10 mls
path length : 1.00 cm
solvent : cyclohexane

λ (nm)

¹³C NMR Spectrum
(50.0 MHz, CDCl₃ solution)

← Resolves into
two signals at higher field

DEPT  CH₂↓ CH₃↑ CH↑

← Resolves into
two signals at higher field

proton decoupled

solvent

expansions

35.0  34.0  ppm

δ (ppm)

¹H NMR Spectrum
(200 MHz, CDCl₃ solution)

expansion

3.0  2.5  ppm

TMS

δ (ppm)

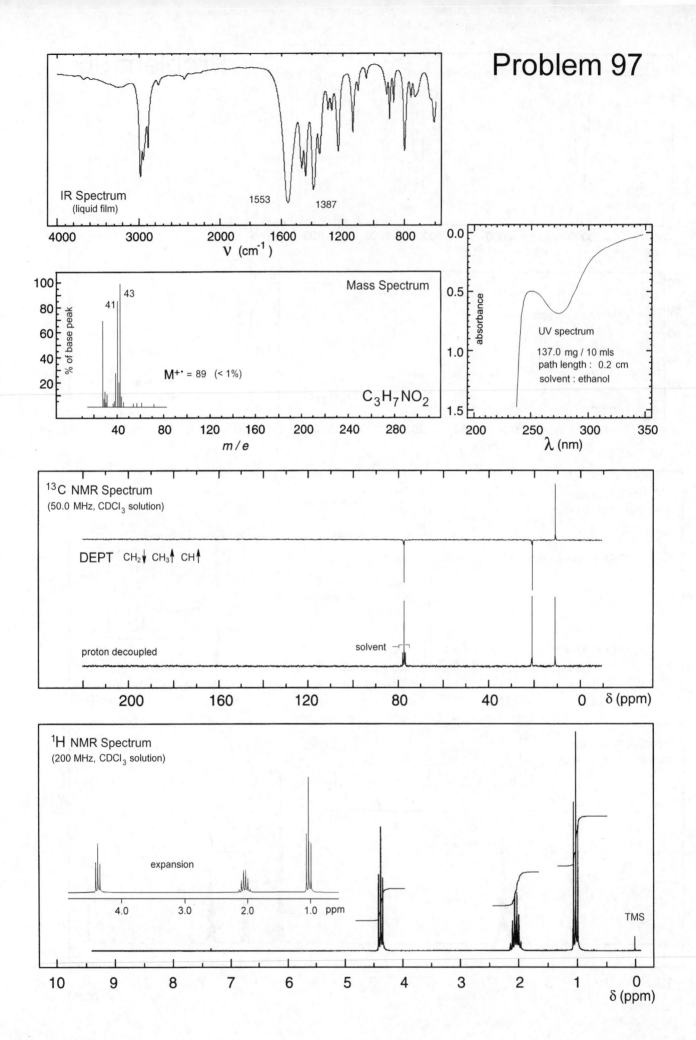

# Problem 97

IR Spectrum
(liquid film)

1553   1387

ν (cm⁻¹)

4000   3000   2000   1600   1200   800

Mass Spectrum

100
80
60
40
20

% of base peak

41   43

M⁺ˑ = 89   (< 1%)

$C_3H_7NO_2$

40   80   120   160   200   240   280

m / e

UV spectrum

137.0 mg / 10 mls
path length :  0.2 cm
solvent : ethanol

absorbance

0.0
0.5
1.0
1.5

200   250   300   350

λ (nm)

¹³C NMR Spectrum
(50.0 MHz, CDCl₃ solution)

DEPT   CH₂↓ CH₃↑ CH↑

proton decoupled

solvent

200   160   120   80   40   0   δ (ppm)

¹H NMR Spectrum
(200 MHz, CDCl₃ solution)

expansion

4.0   3.0   2.0   1.0   ppm

TMS

10   9   8   7   6   5   4   3   2   1   0

δ (ppm)

208

# Problem 98

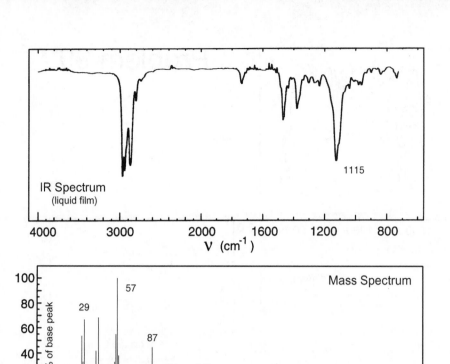

IR Spectrum
(liquid film)

ν (cm⁻¹)

1115

Mass Spectrum

% of base peak

29

57

87

73

101

M⁺·

130

$C_8H_{18}O$

No significant UV
absorption above 220 nm

m/e

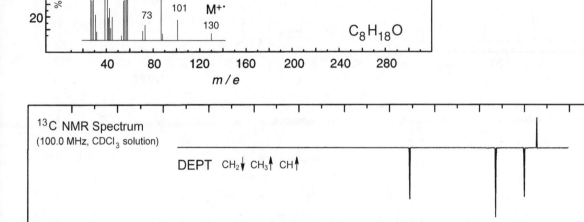

$^{13}C$ NMR Spectrum
(100.0 MHz, CDCl₃ solution)

DEPT  CH₂↓ CH₃↑ CH↑

proton decoupled

solvent

δ (ppm)

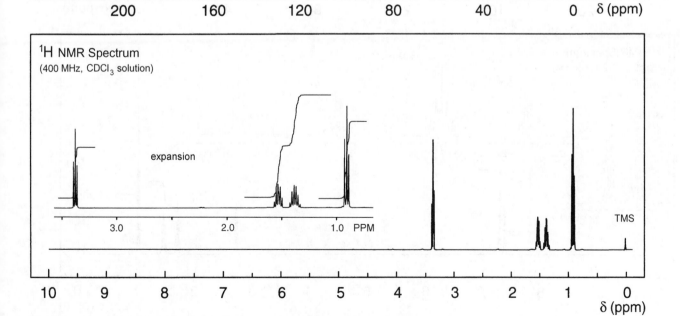

$^1H$ NMR Spectrum
(400 MHz, CDCl₃ solution)

expansion

PPM

TMS

δ (ppm)

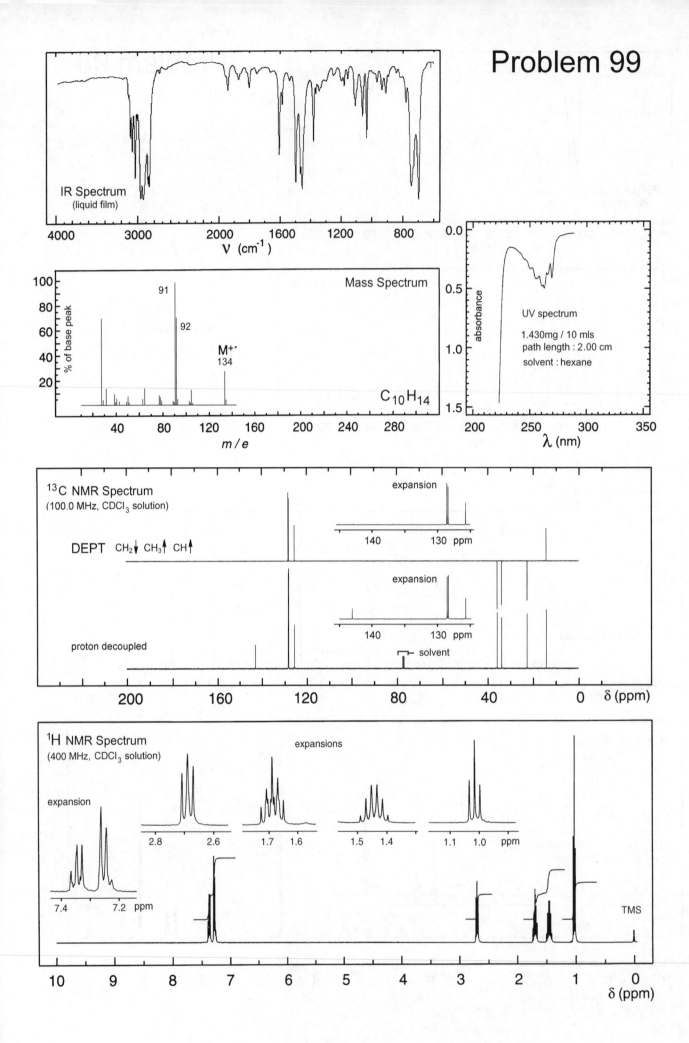

Problem 99

IR Spectrum
(liquid film)

ν (cm⁻¹)

Mass Spectrum

91

92

M⁺·
134

$C_{10}H_{14}$

% of base peak

m/e

UV spectrum

1.430mg / 10 mls
path length : 2.00 cm
solvent : hexane

absorbance

λ (nm)

¹³C NMR Spectrum
(100.0 MHz, CDCl₃ solution)

expansion

DEPT   CH₂↓ CH₃↑ CH↑

expansion

proton decoupled

solvent

δ (ppm)

¹H NMR Spectrum
(400 MHz, CDCl₃ solution)

expansions

expansion

2.8    2.6        1.7  1.6        1.5   1.4        1.1  1.0   ppm

7.4    7.2  ppm

TMS

δ (ppm)

# Problem 100

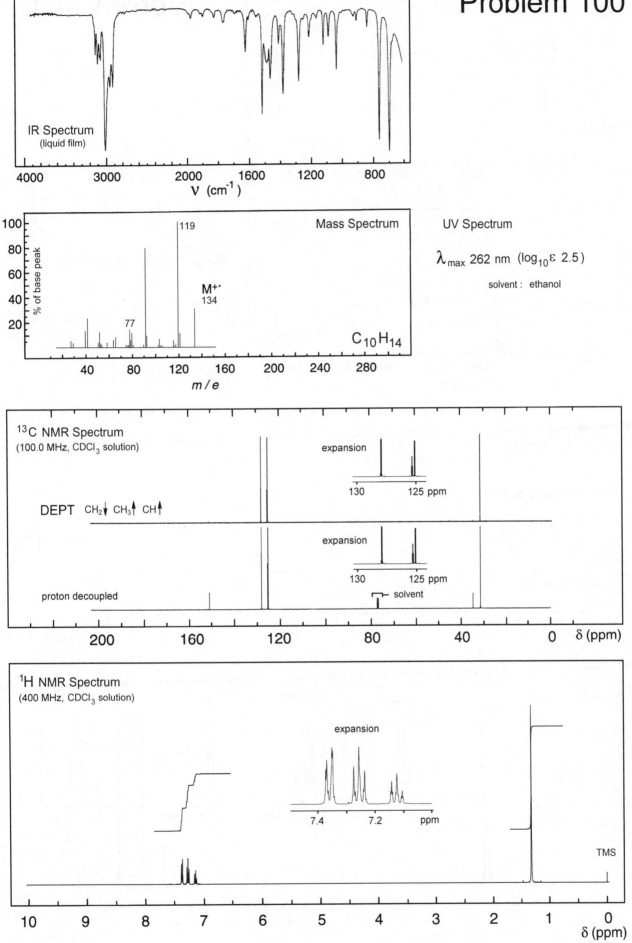

IR Spectrum
(liquid film)

ν (cm⁻¹)

Mass Spectrum

119

77

M⁺·
134

% of base peak

$C_{10}H_{14}$

m/e

UV Spectrum

$\lambda_{max}$ 262 nm ($\log_{10}\varepsilon$ 2.5)

solvent : ethanol

¹³C NMR Spectrum
(100.0 MHz, CDCl₃ solution)

expansion

130      125 ppm

DEPT   CH₂↓ CH₃↑ CH↑

expansion

130      125 ppm

proton decoupled

solvent

δ (ppm)

¹H NMR Spectrum
(400 MHz, CDCl₃ solution)

expansion

7.4      7.2      ppm

TMS

δ (ppm)

# Problem 101

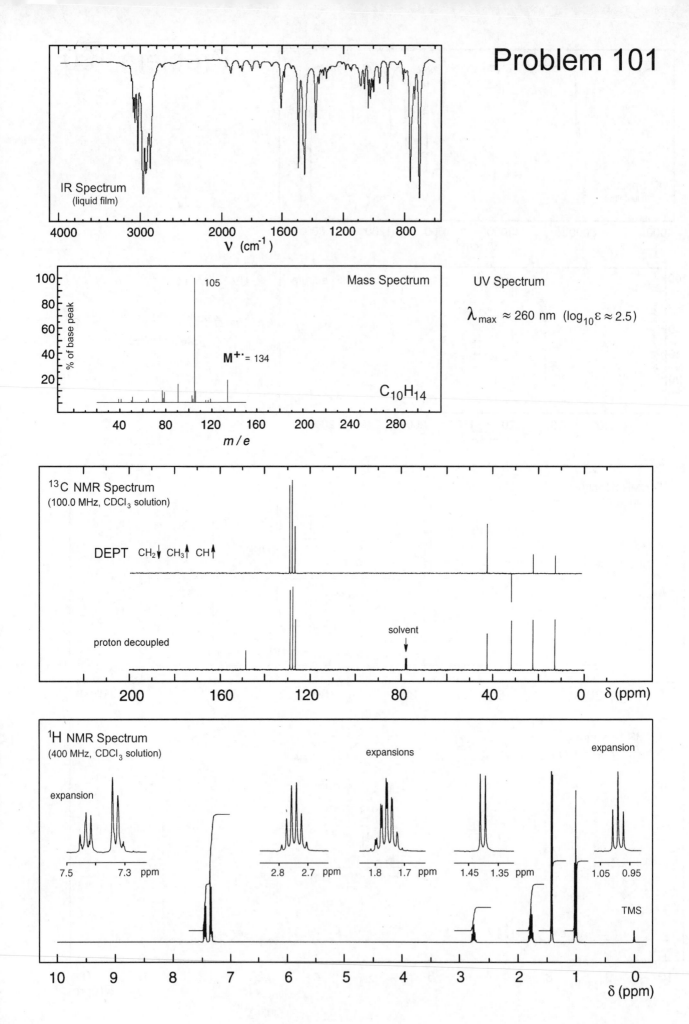

### IR Spectrum
(liquid film)

$\nu$ (cm$^{-1}$)

### Mass Spectrum

105

$M^{+\cdot}$ = 134

$C_{10}H_{14}$

### UV Spectrum

$\lambda_{max} \approx 260$ nm $(\log_{10}\varepsilon \approx 2.5)$

### $^{13}$C NMR Spectrum
(100.0 MHz, CDCl$_3$ solution)

DEPT    CH$_2$↓ CH$_3$↑ CH↑

proton decoupled

solvent

$\delta$ (ppm)

### $^1$H NMR Spectrum
(400 MHz, CDCl$_3$ solution)

expansions

expansion

expansion

7.5    7.3 ppm

2.8    2.7 ppm

1.8    1.7 ppm

1.45    1.35 ppm

1.05    0.95

TMS

$\delta$ (ppm)

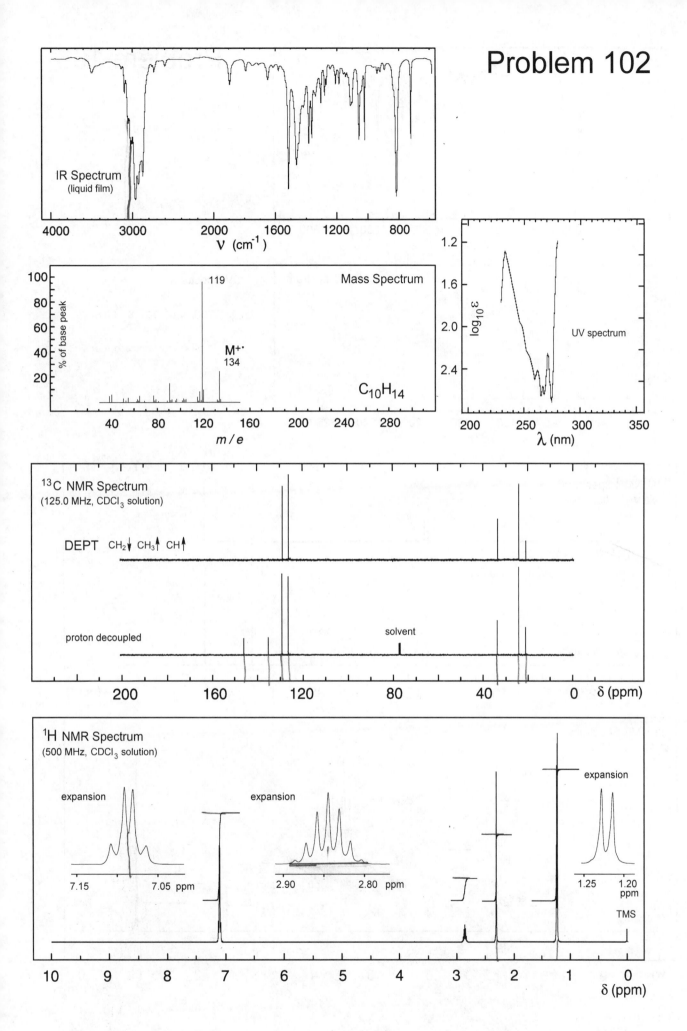

# Problem 102

IR Spectrum
(liquid film)

ν (cm⁻¹)

Mass Spectrum

% of base peak

M⁺·
134

119

$C_{10}H_{14}$

m/e

UV spectrum

λ (nm)

$^{13}C$ NMR Spectrum
(125.0 MHz, CDCl₃ solution)

DEPT   CH₂↓ CH₃↑ CH↑

proton decoupled

solvent

δ (ppm)

$^1H$ NMR Spectrum
(500 MHz, CDCl₃ solution)

expansion

7.15   7.05 ppm

expansion

2.90   2.80 ppm

expansion

1.25  1.20
ppm

TMS

δ (ppm)

# Problem 103

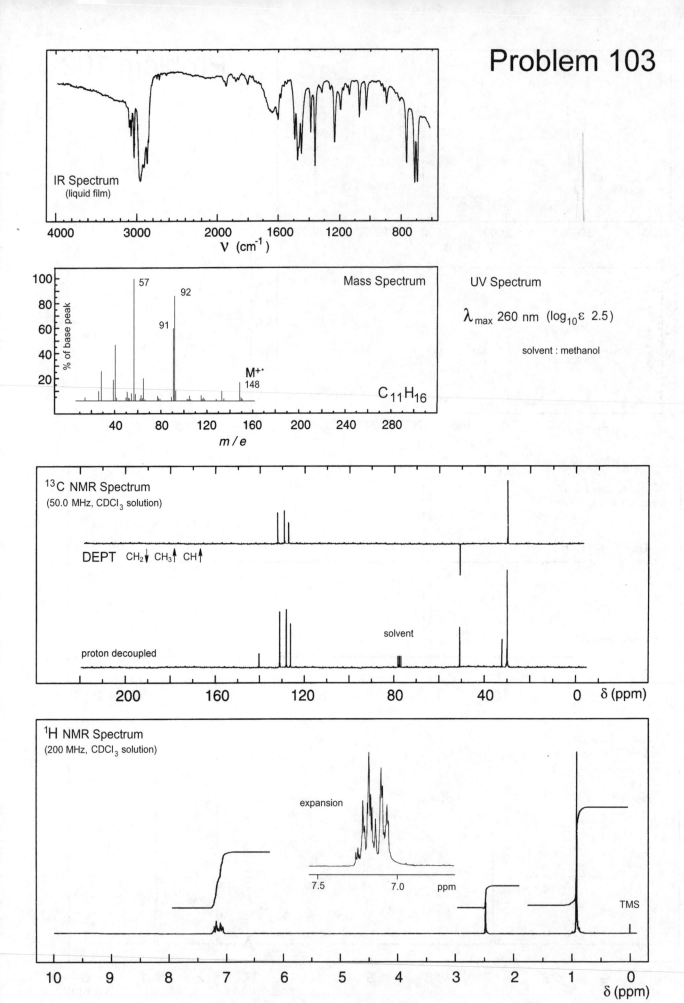

IR Spectrum
(liquid film)

ν (cm⁻¹)
$\nu$ (cm$^{-1}$)

Mass Spectrum

UV Spectrum

$\lambda_{max}$ 260 nm  (log$_{10}$ε 2.5)

solvent : methanol

% of base peak

m/e
$m/e$

M⁺˙
148

$C_{11}H_{16}$

¹³C NMR Spectrum
$^{13}$C NMR Spectrum
(50.0 MHz, CDCl₃ solution)

DEPT  CH₂↓ CH₃↑ CH↑

proton decoupled

solvent

δ (ppm)
$\delta$ (ppm)

¹H NMR Spectrum
$^{1}$H NMR Spectrum
(200 MHz, CDCl₃ solution)

expansion

ppm

TMS

δ (ppm)
$\delta$ (ppm)

# Problem 104

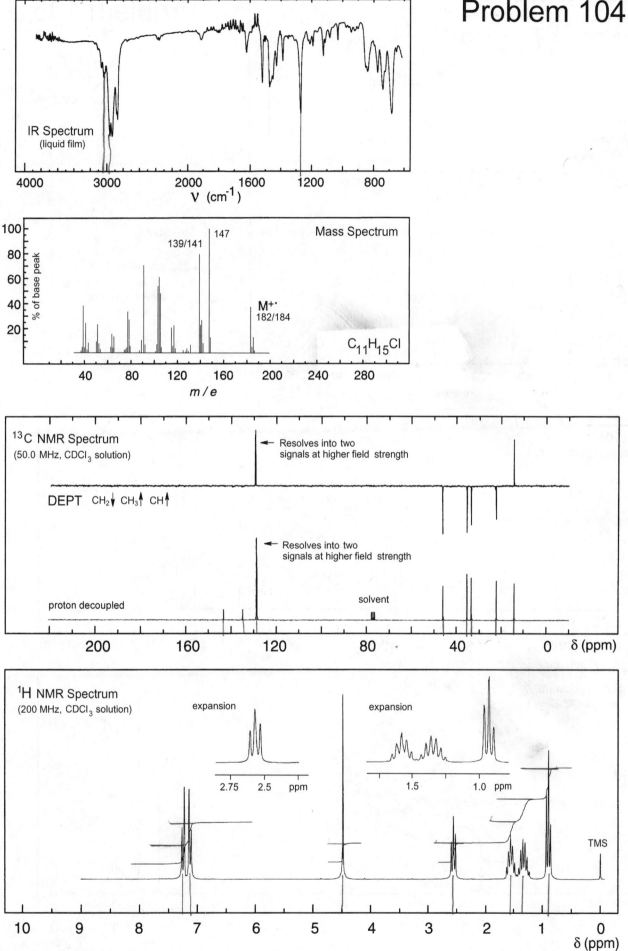

# Problem 105

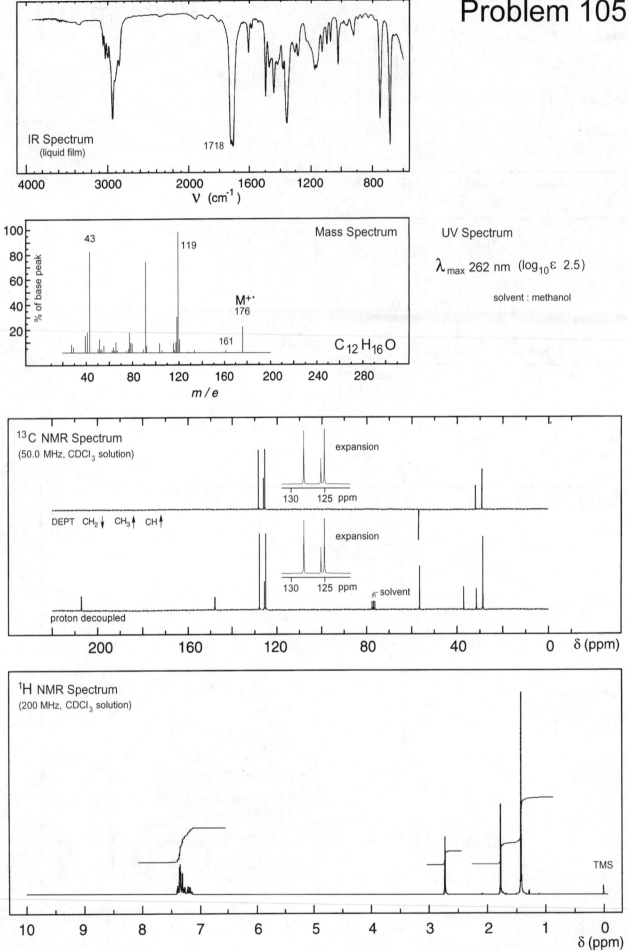

IR Spectrum
(liquid film)

1718

4000    3000    2000    1600    1200    800

ν (cm⁻¹)

Mass Spectrum

43

119

M⁺·
176

161

C₁₂H₁₆O

40    80    120    160    200    240    280

m/e

% of base peak

UV Spectrum

λmax 262 nm  (log₁₀ε 2.5)

solvent : methanol

¹³C NMR Spectrum
(50.0 MHz, CDCl₃ solution)

expansion

130    125  ppm

DEPT  CH₂↓  CH₃↑  CH↑

expansion

130    125  ppm

solvent

proton decoupled

200    160    120    80    40    0    δ (ppm)

¹H NMR Spectrum
(200 MHz, CDCl₃ solution)

TMS

10    9    8    7    6    5    4    3    2    1    0

δ (ppm)

216

# Problem 106

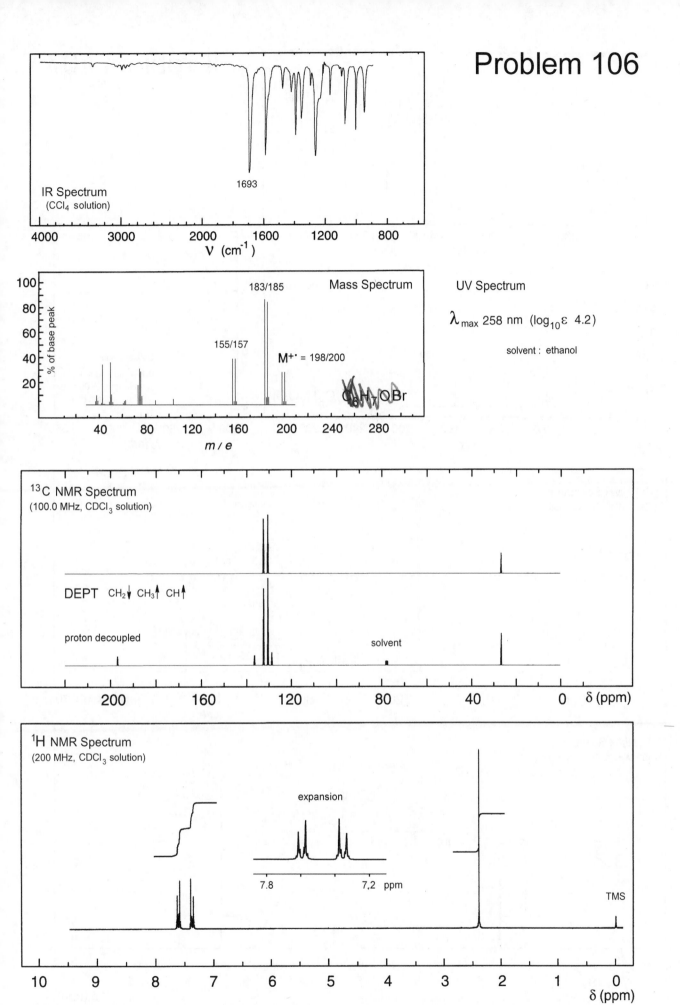

IR Spectrum
(CCl$_4$ solution)

1693

$\nu$ (cm$^{-1}$)

Mass Spectrum

183/185

155/157

M$^{+\cdot}$ = 198/200

% of base peak

m / e

C$_8$H$_7$OBr

UV Spectrum

$\lambda_{max}$ 258 nm (log$_{10}\varepsilon$ 4.2)

solvent : ethanol

$^{13}$C NMR Spectrum
(100.0 MHz, CDCl$_3$ solution)

DEPT   CH$_2\downarrow$  CH$_3\uparrow$  CH$\uparrow$

proton decoupled

solvent

$\delta$ (ppm)

$^1$H NMR Spectrum
(200 MHz, CDCl$_3$ solution)

expansion

7.8          7.2  ppm

TMS

$\delta$ (ppm)

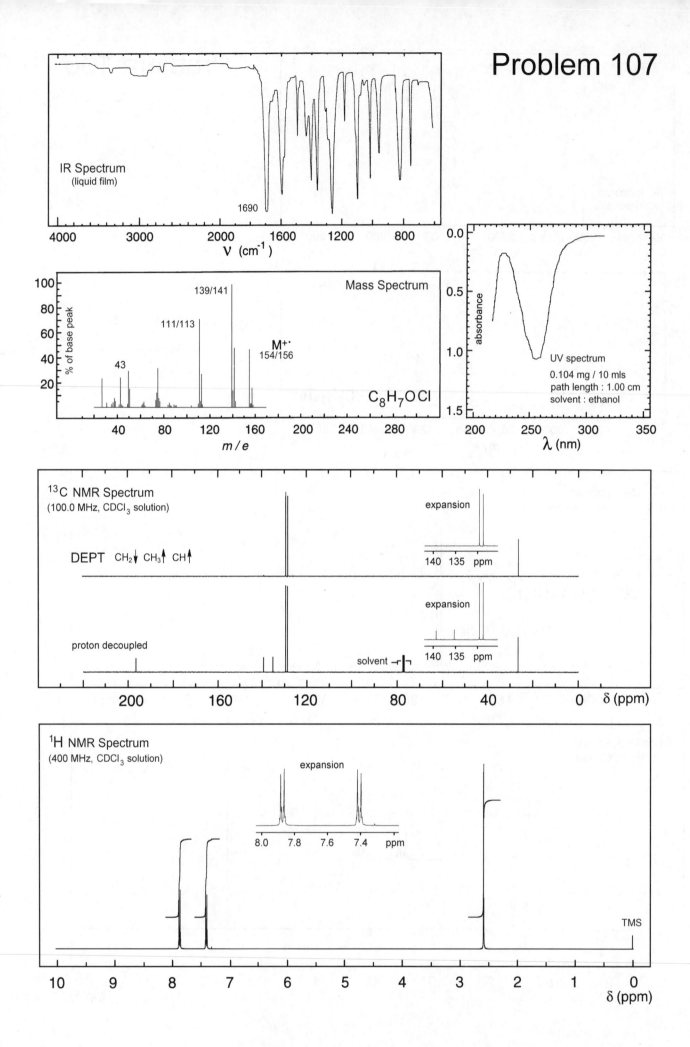

IR Spectrum
(liquid film)

1690

4000    3000    2000  1600    1200    800

ν (cm⁻¹)

Mass Spectrum

100
80
60
40
20

% of base peak

139/141

111/113

43

M⁺˙
154/156

$C_8H_7OCl$

40    80    120    160    200    240    280

m/e

0.0

0.5

1.0

1.5

absorbance

UV spectrum
0.104 mg / 10 mls
path length : 1.00 cm
solvent : ethanol

200    250    300    350

λ (nm)

¹³C NMR Spectrum
(100.0 MHz, CDCl₃ solution)

DEPT  CH₂↓ CH₃↑ CH↑

expansion

140  135  ppm

proton decoupled

solvent

expansion

140  135  ppm

200    160    120    80    40    0    δ (ppm)

¹H NMR Spectrum
(400 MHz, CDCl₃ solution)

expansion

8.0  7.8  7.6  7.4  ppm

TMS

10   9   8   7   6   5   4   3   2   1   0
                                        δ (ppm)

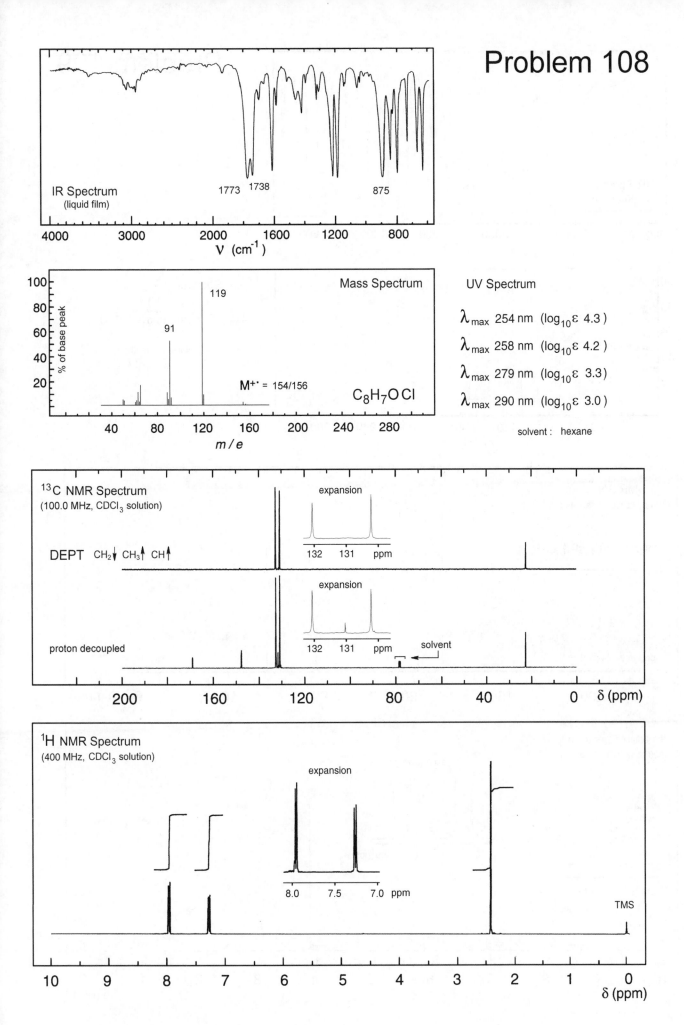

IR Spectrum
(liquid film)
1773 1738 875
ν (cm⁻¹)
4000 3000 2000 1600 1200 800

Mass Spectrum
100
80
60
40
20
% of base peak
119
91
M⁺· = 154/156
$C_8H_7OCl$
40 80 120 160 200 240 280
m/e

UV Spectrum

$\lambda_{max}$ 254 nm (log₁₀ε 4.3)

$\lambda_{max}$ 258 nm (log₁₀ε 4.2)

$\lambda_{max}$ 279 nm (log₁₀ε 3.3)

$\lambda_{max}$ 290 nm (log₁₀ε 3.0)

solvent : hexane

¹³C NMR Spectrum
(100.0 MHz, CDCl₃ solution)

DEPT   CH₂↓ CH₃↑ CH↑

expansion
132 131 ppm

proton decoupled
expansion
132 131 ppm
solvent

200 160 120 80 40 0 δ (ppm)

¹H NMR Spectrum
(400 MHz, CDCl₃ solution)

expansion
8.0 7.5 7.0 ppm

TMS

10 9 8 7 6 5 4 3 2 1 0
δ (ppm)

# Problem 109

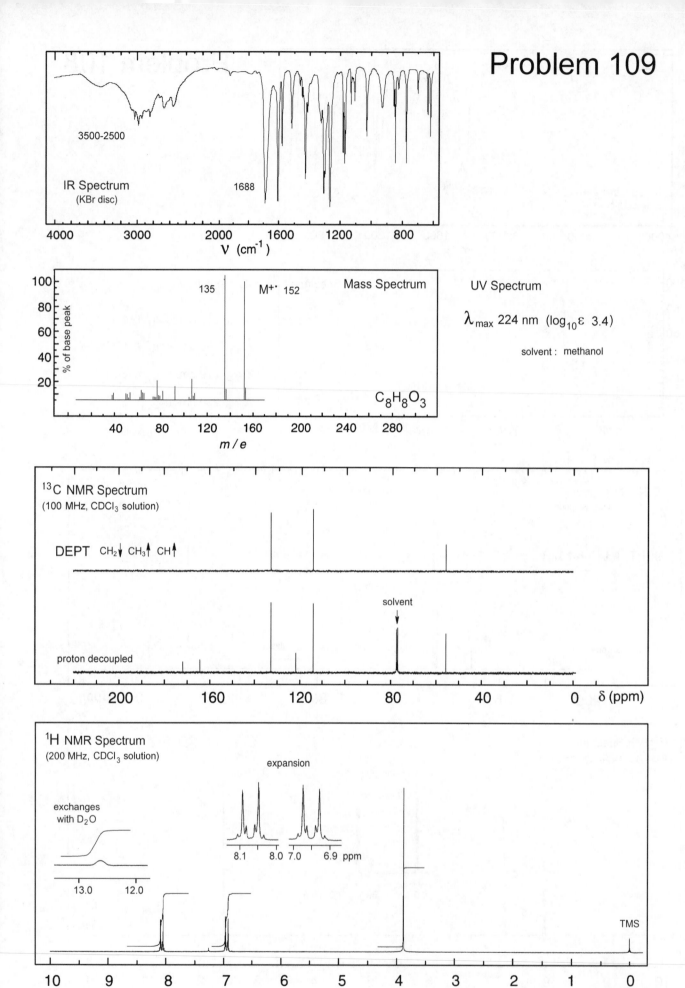

IR Spectrum
(KBr disc)

3500-2500

1688

ν (cm⁻¹)

Mass Spectrum

135    M⁺˙ 152

% of base peak

$C_8H_8O_3$

m/e

UV Spectrum

$\lambda_{max}$ 224 nm  (log₁₀ε  3.4)

solvent :  methanol

¹³C NMR Spectrum
(100 MHz, CDCl₃ solution)

DEPT   CH₂↓ CH₃↑ CH↑

solvent

proton decoupled

δ (ppm)

¹H NMR Spectrum
(200 MHz, CDCl₃ solution)

expansion

exchanges
with D₂O

8.1    8.0 7.0    6.9 ppm

13.0    12.0

TMS

δ (ppm)

# Problem 110

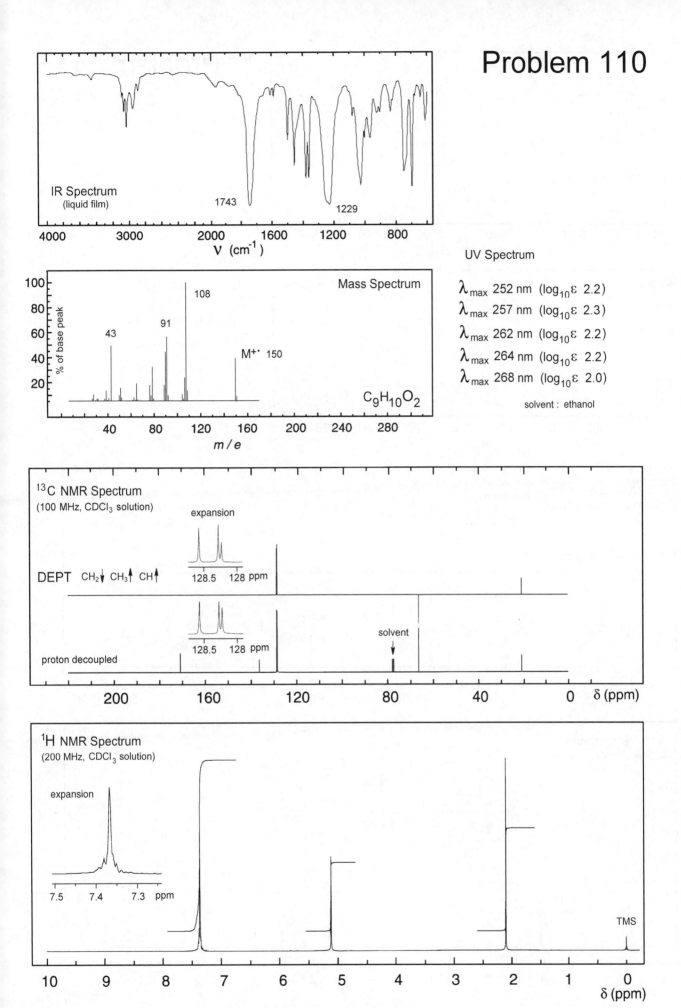

IR Spectrum
(liquid film)

1743

1229

ν (cm⁻¹)

Mass Spectrum

108

91

43

M⁺˙ 150

C₉H₁₀O₂

% of base peak

m/e

UV Spectrum

$\lambda_{max}$ 252 nm (log₁₀ε 2.2)
$\lambda_{max}$ 257 nm (log₁₀ε 2.3)
$\lambda_{max}$ 262 nm (log₁₀ε 2.2)
$\lambda_{max}$ 264 nm (log₁₀ε 2.2)
$\lambda_{max}$ 268 nm (log₁₀ε 2.0)

solvent : ethanol

¹³C NMR Spectrum
(100 MHz, CDCl₃ solution)

expansion

DEPT   CH₂↓ CH₃↑ CH↑

128.5   128 ppm

solvent

proton decoupled

128.5   128 ppm

δ (ppm)

¹H NMR Spectrum
(200 MHz, CDCl₃ solution)

expansion

7.5   7.4   7.3 ppm

TMS

δ (ppm)

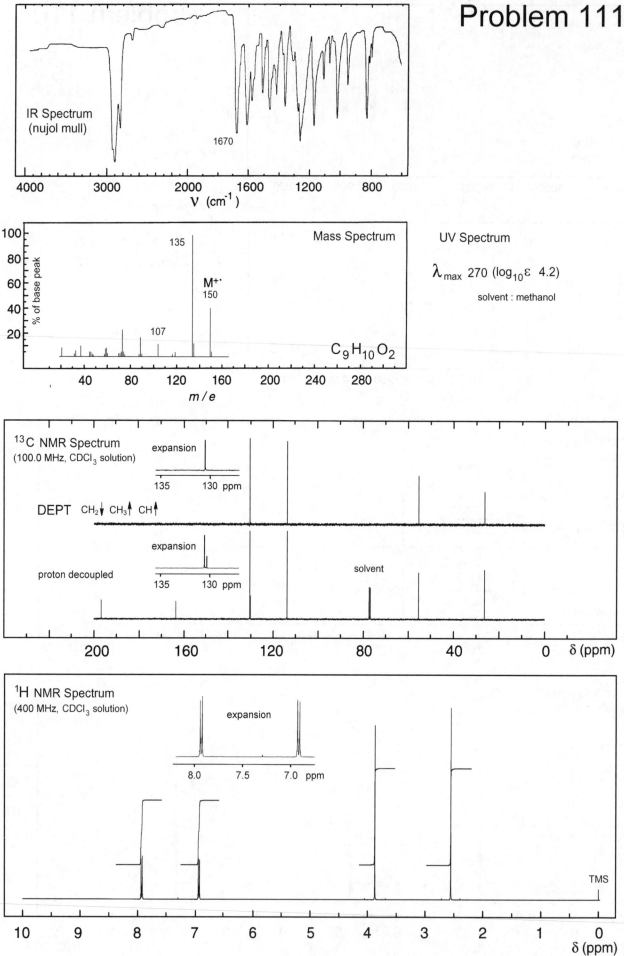

**IR Spectrum (nujol mull)**

1670

ν (cm⁻¹)
4000  3000  2000  1600  1200  800

**Mass Spectrum**

135

M⁺·
150

107

$C_9H_{10}O_2$

% of base peak

m/e
40  80  120  160  200  240  280

**UV Spectrum**

$\lambda_{max}$ 270 ($\log_{10}\varepsilon$ 4.2)

solvent : methanol

**$^{13}$C NMR Spectrum**
(100.0 MHz, CDCl₃ solution)

expansion
135  130 ppm

DEPT  CH₂↓ CH₃↑ CH↑

expansion
135  130 ppm

proton decoupled

solvent

δ (ppm)
200  160  120  80  40  0

**$^1$H NMR Spectrum**
(400 MHz, CDCl₃ solution)

expansion

8.0  7.5  7.0 ppm

TMS

δ (ppm)
10  9  8  7  6  5  4  3  2  1  0

# Problem 112

IR Spectrum
(liquid film)

1760

ν (cm⁻¹)

Mass Spectrum

% of base peak

108

107

M⁺· = 150

C₉H₁₀O₂

m/e

UV Spectrum

$\lambda_{max}$ 265 nm (log₁₀ε 2.6)

$\lambda_{max}$ 271 nm (log₁₀ε 2.6)

solvent : methanol

¹³C NMR Spectrum
(100.0 MHz, CDCl₃ solution)

DEPT   CH₂↓ CH₃↑ CH↑

expansion

24   20  ppm

proton decoupled

solvent

expansion

24   20  ppm

δ (ppm)

¹H NMR Spectrum
(400 MHz, CDCl₃ solution)

expansion

7.2   7.0  ppm

TMS

δ (ppm)

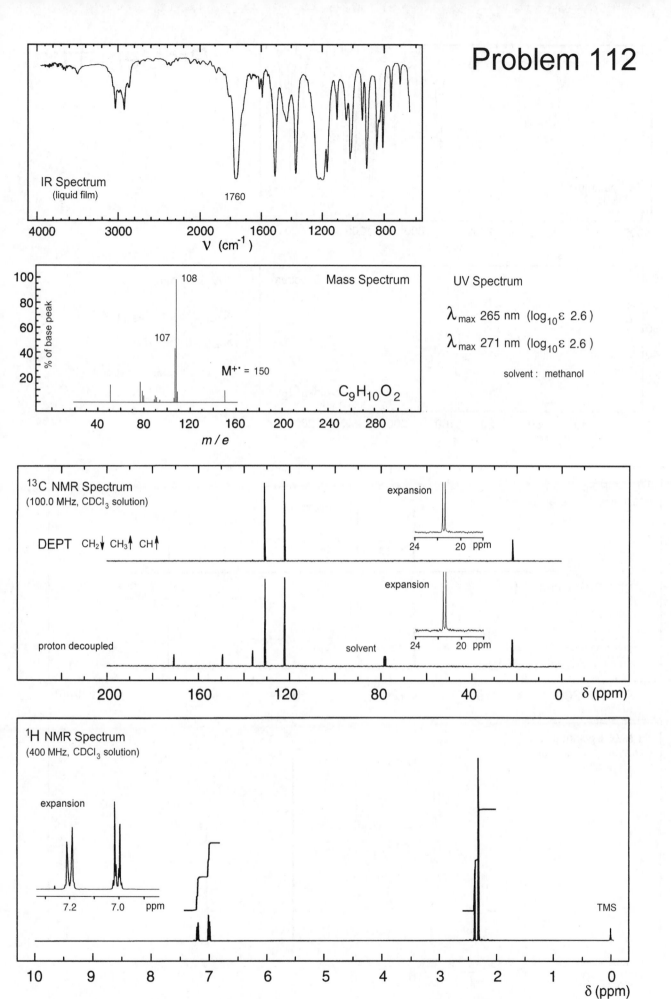

# Problem 113

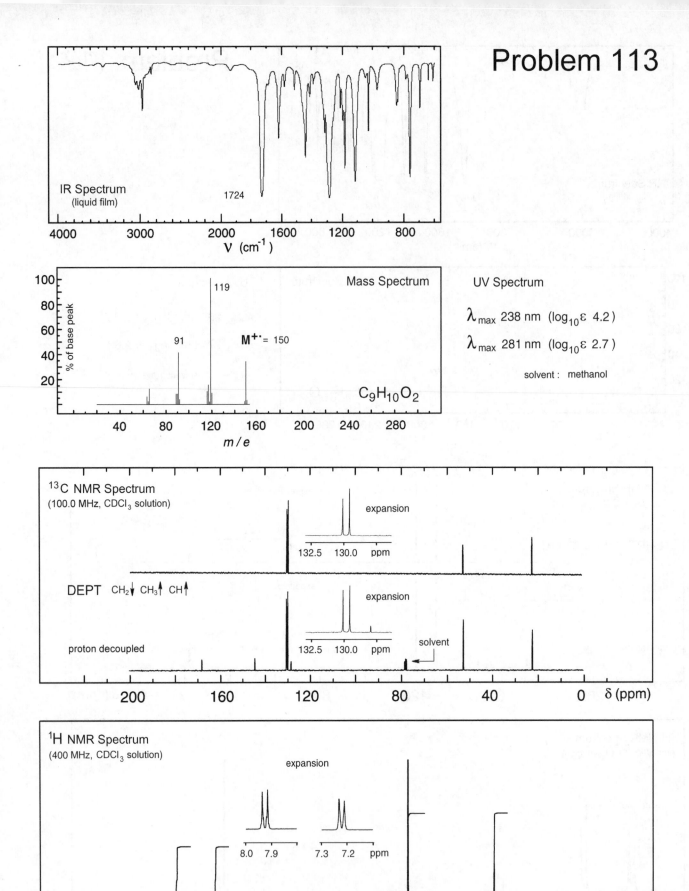

IR Spectrum
(liquid film)

1724

ν (cm⁻¹)

Mass Spectrum

119

91

M⁺· = 150

C₉H₁₀O₂

m/e

UV Spectrum

λ_max 238 nm (log₁₀ε 4.2)

λ_max 281 nm (log₁₀ε 2.7)

solvent: methanol

¹³C NMR Spectrum
(100.0 MHz, CDCl₃ solution)

expansion

DEPT  CH₂↓ CH₃↑ CH↑

proton decoupled

solvent

δ (ppm)

¹H NMR Spectrum
(400 MHz, CDCl₃ solution)

expansion

TMS

δ (ppm)

# Problem 114

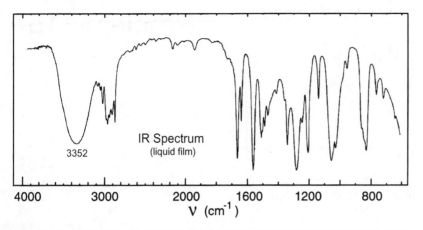

IR Spectrum
(liquid film)

3352

ν (cm⁻¹)

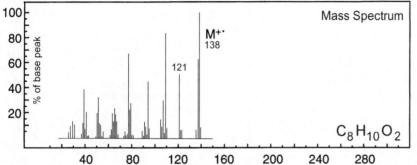

Mass Spectrum

M⁺·
138

121

C₈H₁₀O₂

m/e

UV Spectrum

$\lambda_{max}$ 270 nm (log₁₀ε 3.1)

$\lambda_{max}$ 282 nm (log₁₀ε 3.1)

solvent : hexane

Note: UV spectrum not changed
significantly on addition of base

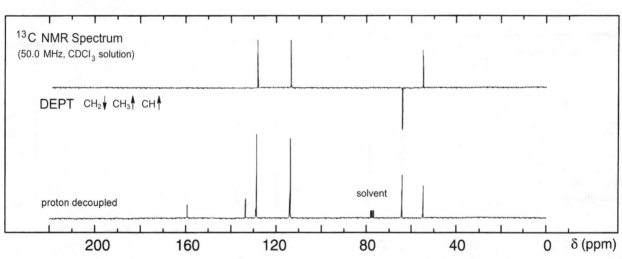

¹³C NMR Spectrum
(50.0 MHz, CDCl₃ solution)

DEPT   CH₂↓ CH₃↑ CH↑

proton decoupled

solvent

δ (ppm)

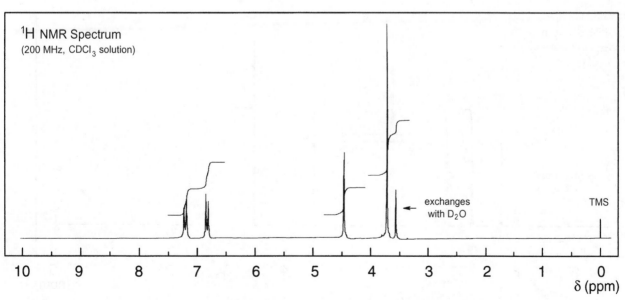

¹H NMR Spectrum
(200 MHz, CDCl₃ solution)

exchanges
with D₂O

TMS

δ (ppm)

# Problem 115

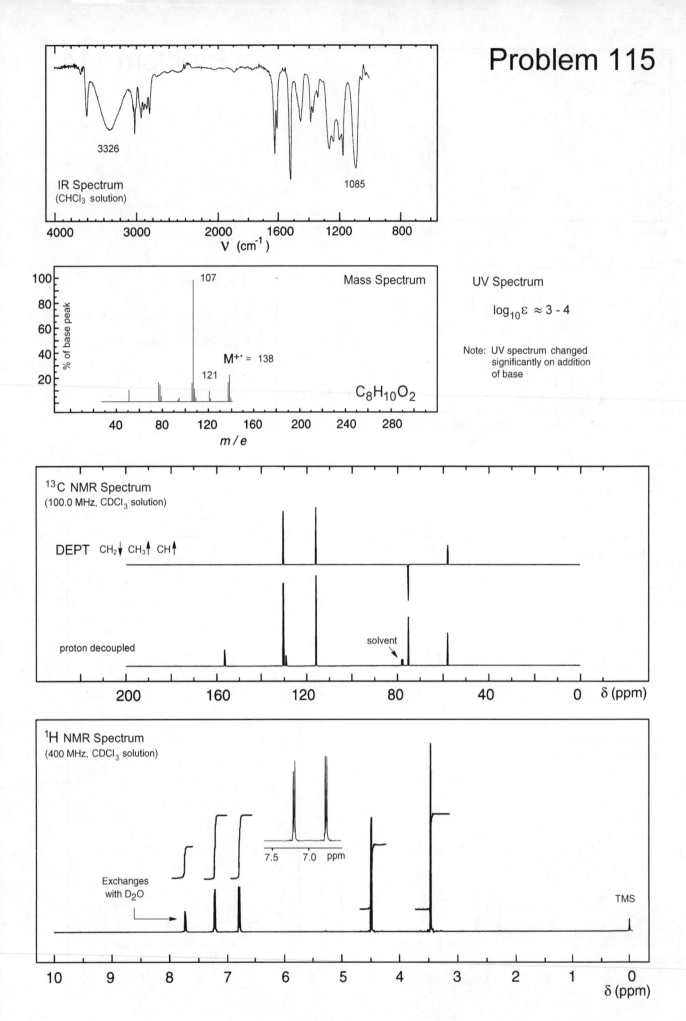

# Problem 116

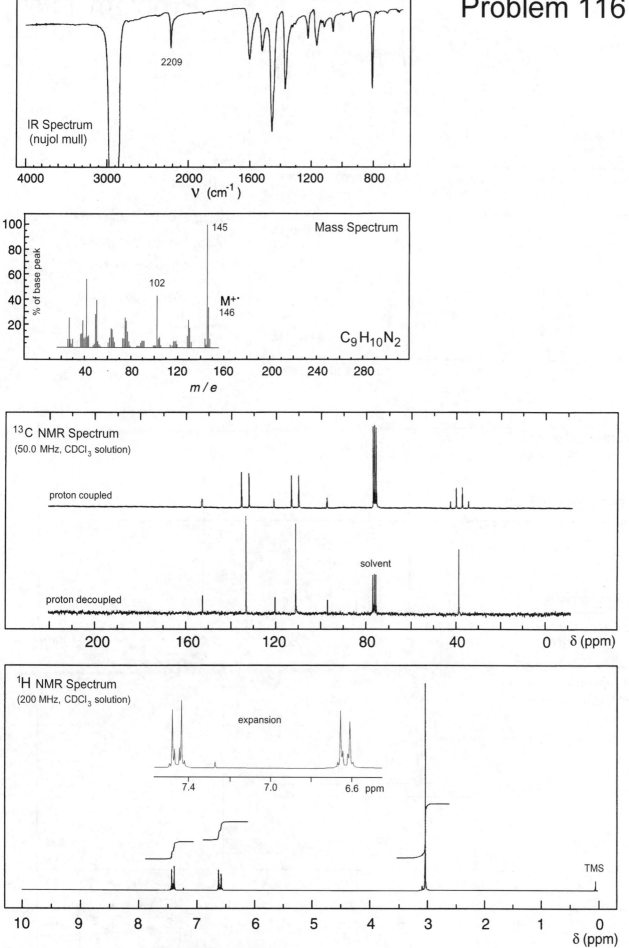

IR Spectrum
(nujol mull)

2209

Mass Spectrum

145

102

M+·
146

$C_9H_{10}N_2$

$^{13}C$ NMR Spectrum
(50.0 MHz, $CDCl_3$ solution)

proton coupled

proton decoupled

solvent

$\delta$ (ppm)

$^1H$ NMR Spectrum
(200 MHz, $CDCl_3$ solution)

expansion

7.4    7.0    6.6 ppm

TMS

$\delta$ (ppm)

227

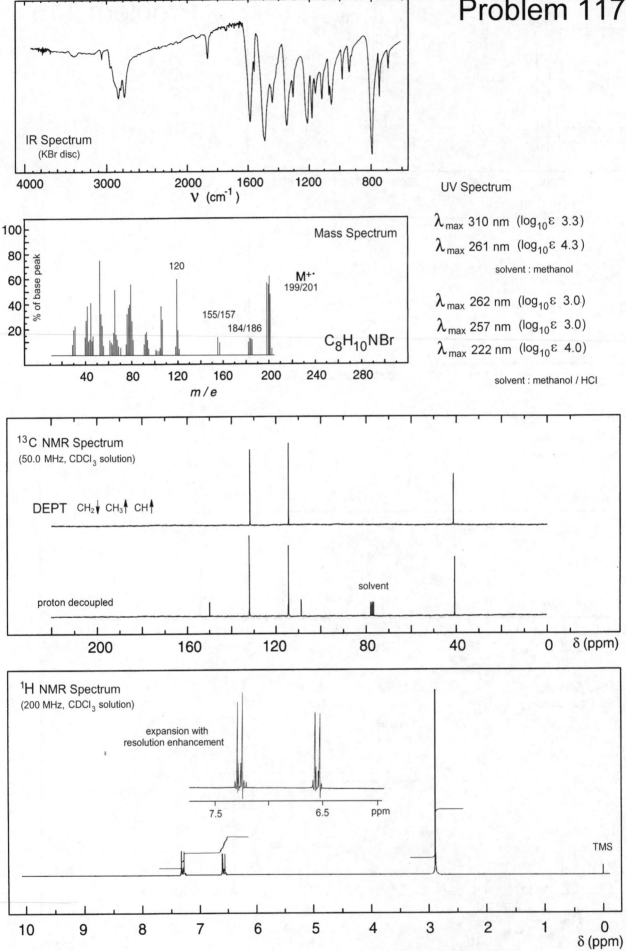

# Problem 117

**IR Spectrum**
(KBr disc)

ν (cm⁻¹)

**Mass Spectrum**

120

M⁺˙
199/201

155/157

184/186

C₈H₁₀NBr

% of base peak

m/e

**UV Spectrum**

λ_max 310 nm (log₁₀ε 3.3)
λ_max 261 nm (log₁₀ε 4.3)
  solvent : methanol

λ_max 262 nm (log₁₀ε 3.0)
λ_max 257 nm (log₁₀ε 3.0)
λ_max 222 nm (log₁₀ε 4.0)

  solvent : methanol / HCl

**¹³C NMR Spectrum**
(50.0 MHz, CDCl₃ solution)

DEPT  CH₂↓ CH₃↑ CH↑

proton decoupled

solvent

δ (ppm)

**¹H NMR Spectrum**
(200 MHz, CDCl₃ solution)

expansion with
resolution enhancement

7.5     6.5     ppm

TMS

δ (ppm)

228

# Problem 118

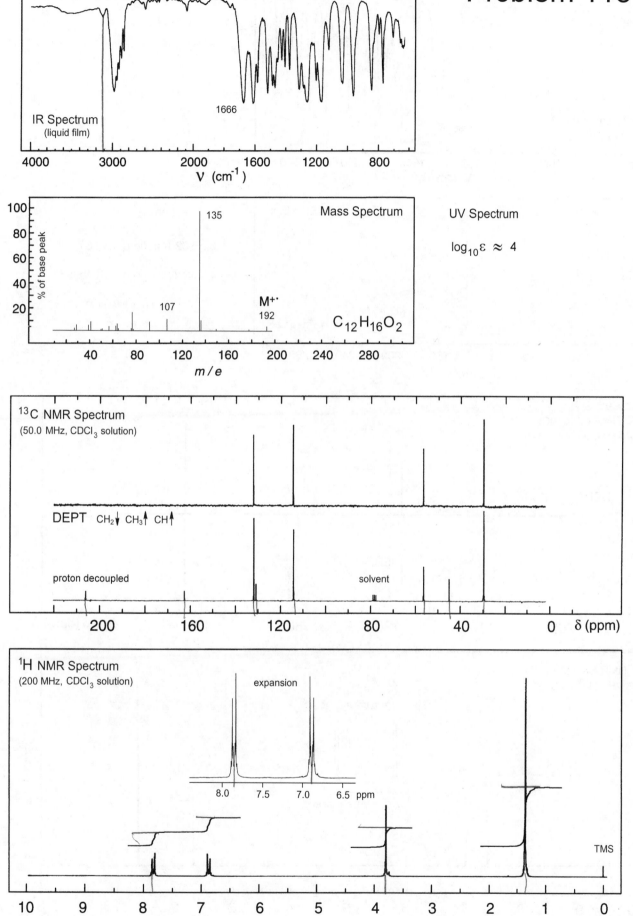

IR Spectrum
(liquid film)

1666

ν (cm⁻¹)

Mass Spectrum

100
80
60
40
20

% of base peak

135

107

M⁺˙
192

C₁₂H₁₆O₂

m/e

UV Spectrum

log₁₀ε ≈ 4

¹³C NMR Spectrum
(50.0 MHz, CDCl₃ solution)

DEPT  CH₂↓ CH₃↑ CH↑

proton decoupled

solvent

δ (ppm)

¹H NMR Spectrum
(200 MHz, CDCl₃ solution)

expansion

8.0    7.5    7.0    6.5 ppm

TMS

δ (ppm)

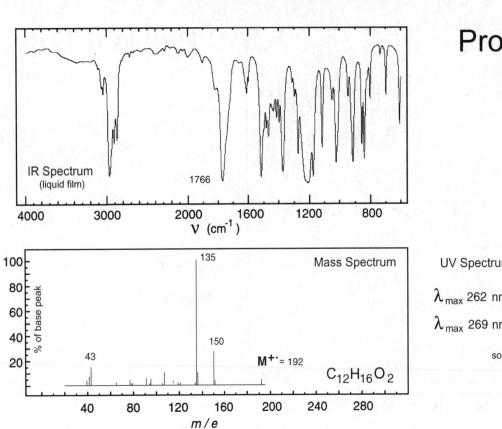

IR Spectrum
(liquid film)

1766

$\nu$ (cm$^{-1}$)

Mass Spectrum

135

150

43

$M^{+\cdot} = 192$

$C_{12}H_{16}O_2$

UV Spectrum

$\lambda_{max}$ 262 nm (log$_{10}\varepsilon$ 2.6)

$\lambda_{max}$ 269 nm (log$_{10}\varepsilon$ 2.6)

solvent : methanol

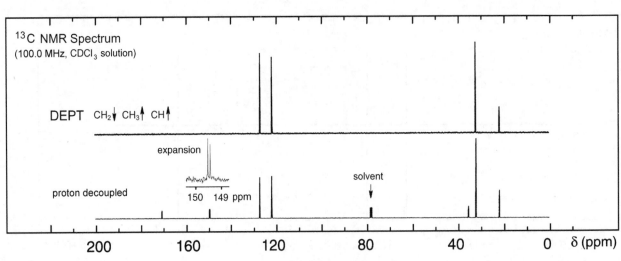

$^{13}$C NMR Spectrum
(100.0 MHz, CDCl$_3$ solution)

DEPT   CH$_2$↓ CH$_3$↑ CH↑

expansion

150   149 ppm

proton decoupled

solvent

$\delta$ (ppm)

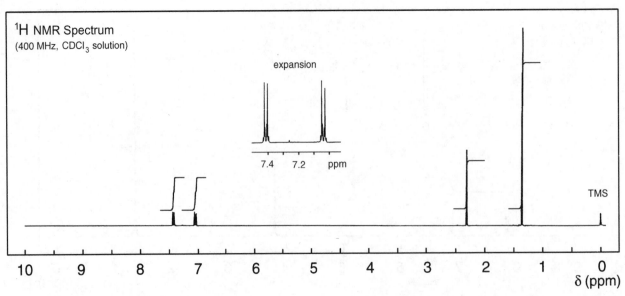

$^1$H NMR Spectrum
(400 MHz, CDCl$_3$ solution)

expansion

7.4   7.2   ppm

TMS

$\delta$ (ppm)

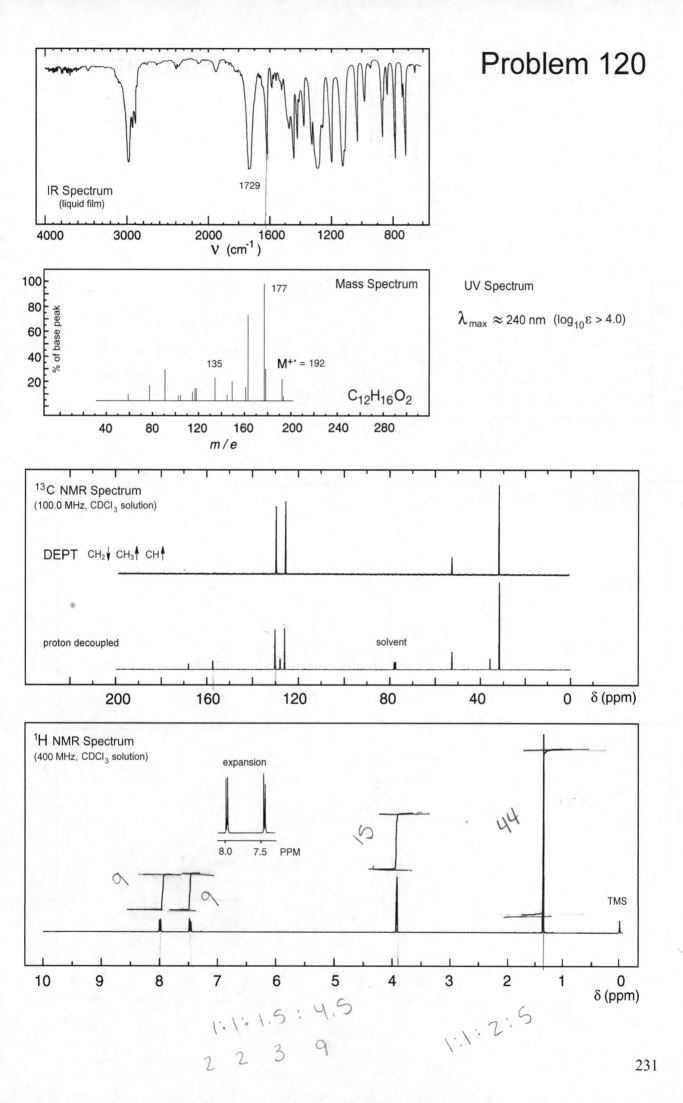

# Problem 120

**IR Spectrum**
(liquid film)

1729

ν (cm⁻¹)

**Mass Spectrum**

177

135

M⁺• = 192

$C_{12}H_{16}O_2$

% of base peak

m / e

**UV Spectrum**

$\lambda_{max} \approx 240$ nm $(\log_{10}\varepsilon > 4.0)$

**¹³C NMR Spectrum**
(100.0 MHz, CDCl₃ solution)

DEPT  CH₂↓ CH₃↑ CH↑

proton decoupled

solvent

δ (ppm)

**¹H NMR Spectrum**
(400 MHz, CDCl₃ solution)

expansion

8.0   7.5  PPM

9

9

15

44

TMS

δ (ppm)

1:1: 1.5 : 4.5
2   2   3    9

1:1: 2: 5

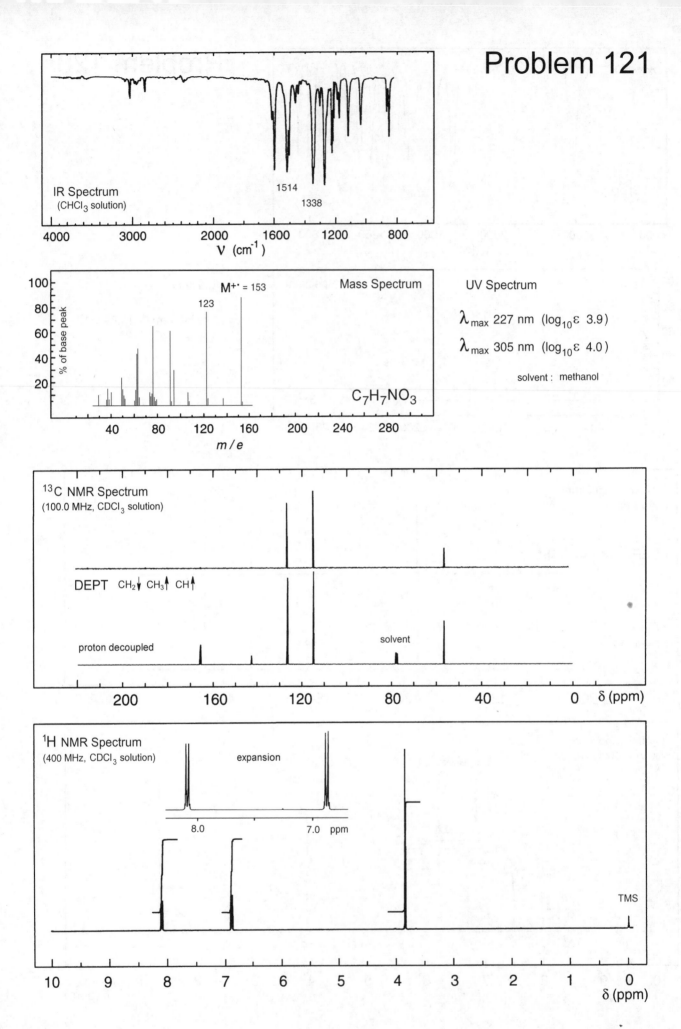

# Problem 121

**IR Spectrum**
(CHCl₃ solution)

1514

1338

ν (cm⁻¹)

M⁺˙ = 153    Mass Spectrum

123

% of base peak

$C_7H_7NO_3$

m/e

**UV Spectrum**

$\lambda_{max}$ 227 nm (log₁₀ε 3.9)

$\lambda_{max}$ 305 nm (log₁₀ε 4.0)

solvent : methanol

¹³C NMR Spectrum
(100.0 MHz, CDCl₃ solution)

DEPT  CH₂↓ CH₃↑ CH↑

proton decoupled

solvent

δ (ppm)

¹H NMR Spectrum
(400 MHz, CDCl₃ solution)

expansion

8.0          7.0   ppm

TMS

δ (ppm)

# Problem 122

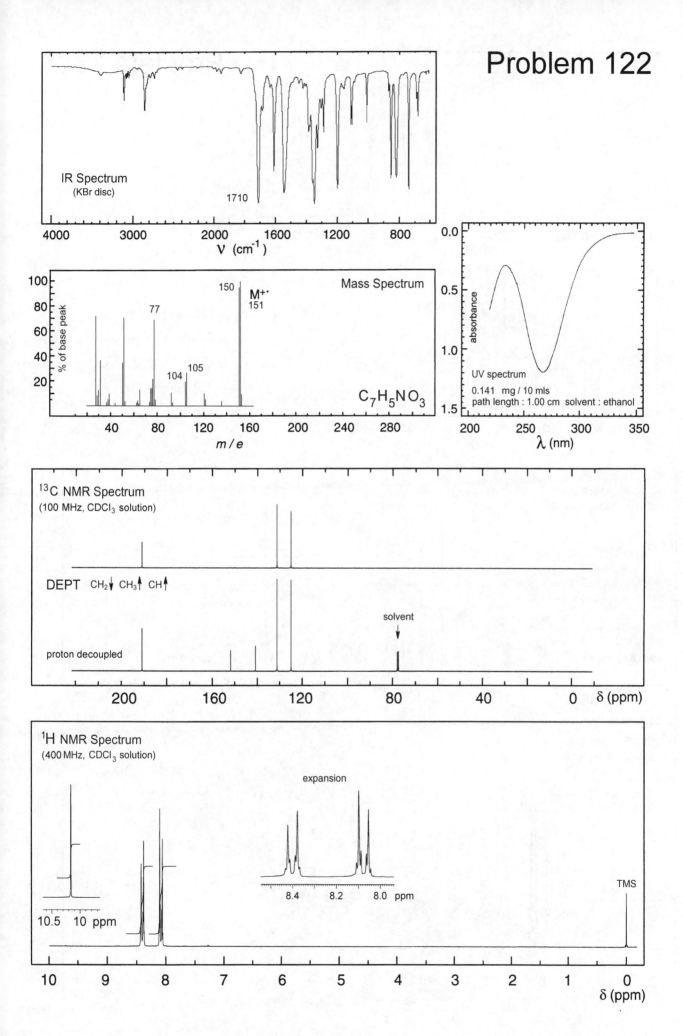

IR Spectrum
(KBr disc)

1710

ν (cm⁻¹)

Mass Spectrum

150

M⁺˙
151

77

104   105

% of base peak

m/e

C₇H₅NO₃

UV spectrum
0.141 mg / 10 mls
path length : 1.00 cm  solvent : ethanol

absorbance

λ (nm)

¹³C NMR Spectrum
(100 MHz, CDCl₃ solution)

DEPT   CH₂↓ CH₃↑ CH↑

proton decoupled

solvent

δ (ppm)

¹H NMR Spectrum
(400 MHz, CDCl₃ solution)

expansion

8.4   8.2   8.0  ppm

TMS

10.5   10  ppm

δ (ppm)

233

# Problem 123

IR Spectrum
(KBr disc)

1699

ν (cm⁻¹)

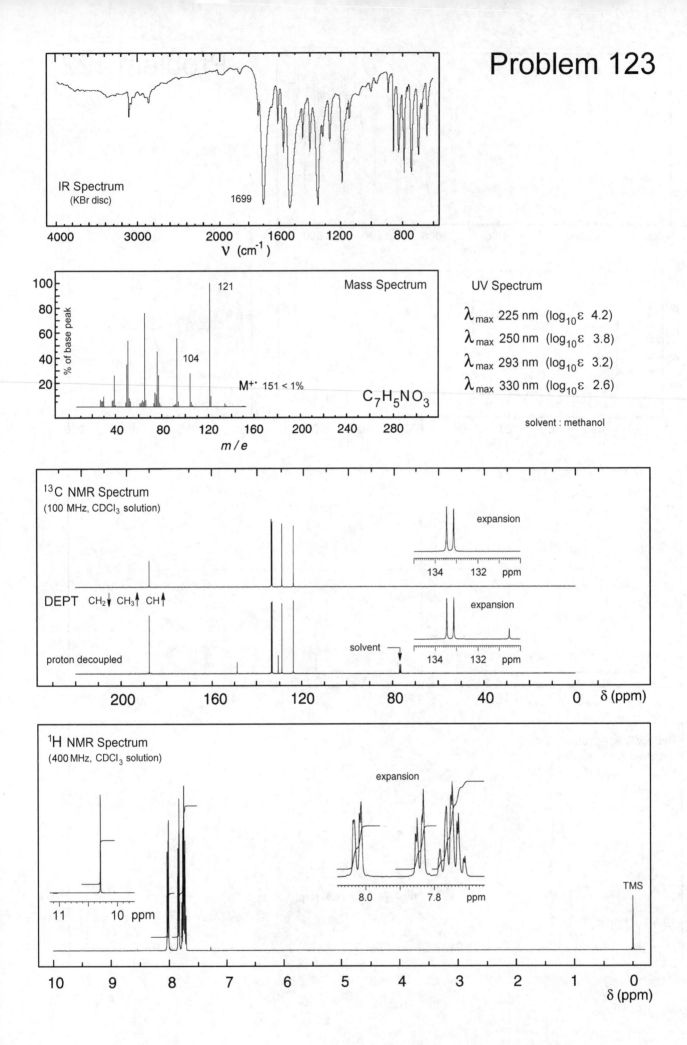

Mass Spectrum

121

104

M⁺˙ 151 < 1%

$C_7H_5NO_3$

% of base peak

m/e

UV Spectrum

$\lambda_{max}$ 225 nm (log₁₀ε 4.2)

$\lambda_{max}$ 250 nm (log₁₀ε 3.8)

$\lambda_{max}$ 293 nm (log₁₀ε 3.2)

$\lambda_{max}$ 330 nm (log₁₀ε 2.6)

solvent : methanol

¹³C NMR Spectrum
(100 MHz, CDCl₃ solution)

expansion

134  132  ppm

DEPT  CH₂↓ CH₃↑ CH↑

expansion

proton decoupled

solvent

134  132  ppm

δ (ppm)

¹H NMR Spectrum
(400 MHz, CDCl₃ solution)

expansion

8.0  7.8  ppm

TMS

11  10  ppm

δ (ppm)

# Problem 124

IR Spectrum
(liquid film)

1698

ν (cm⁻¹)

4000  3000  2000  1600  1200  800

Mass Spectrum

M⁺· = 136

92  107

$C_8H_8O_2$

% of base peak

100
80
60
40
20

m/e

40  80  120  160  200  240  280

UV Spectrum

$\lambda_{max}$ 277 nm (log₁₀ε 4.2)

solvent : methanol

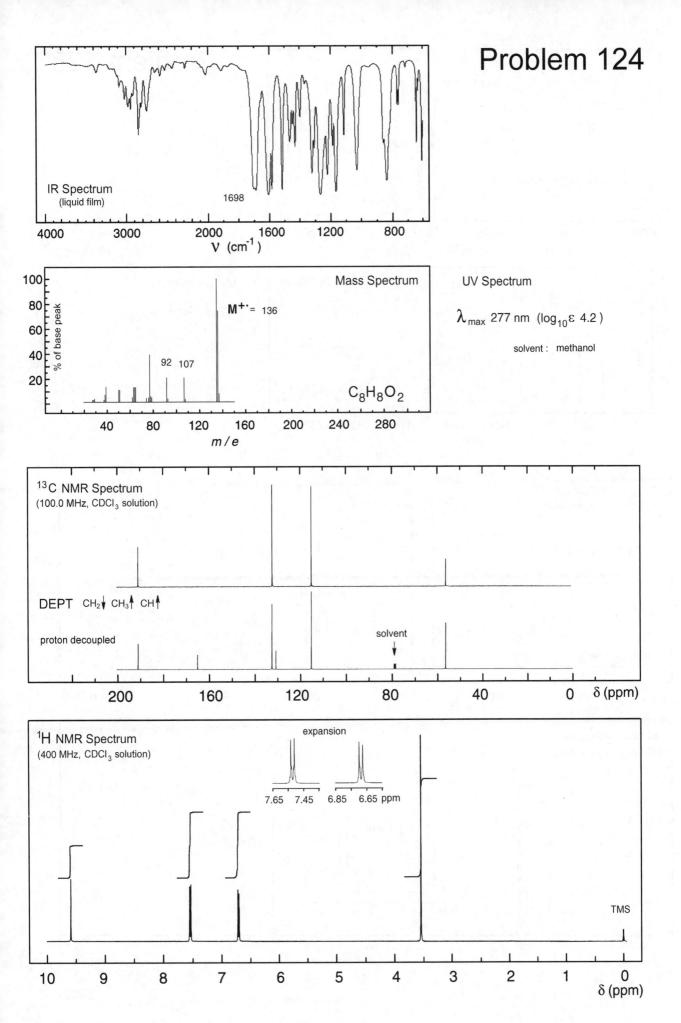

¹³C NMR Spectrum
(100.0 MHz, CDCl₃ solution)

DEPT   CH₂↓ CH₃↑ CH↑

proton decoupled

solvent

200  160  120  80  40  0   δ (ppm)

¹H NMR Spectrum
(400 MHz, CDCl₃ solution)

expansion

7.65  7.45   6.85  6.65 ppm

TMS

10  9  8  7  6  5  4  3  2  1  0   δ (ppm)

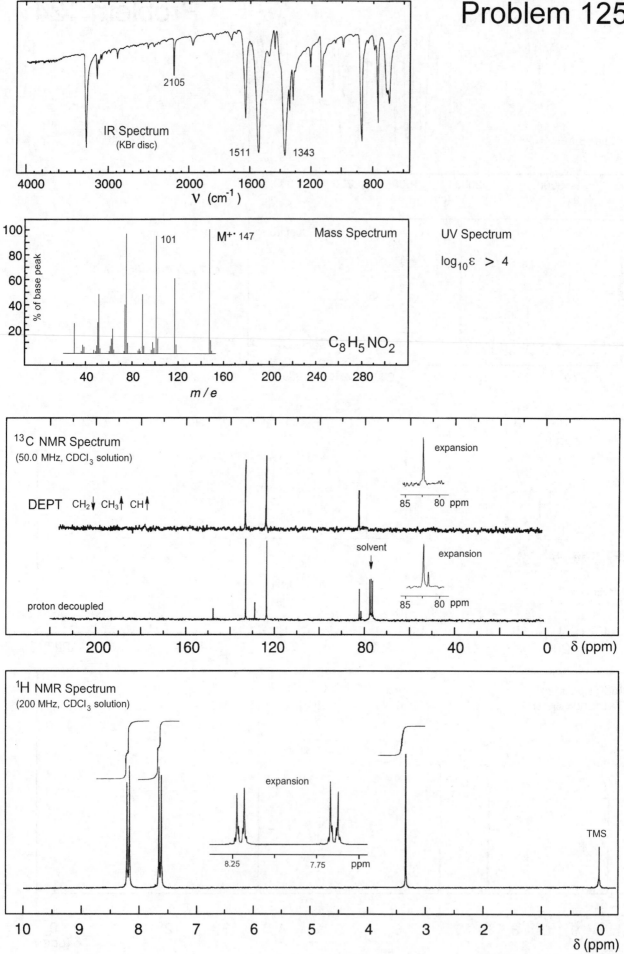

IR Spectrum
(KBr disc)

2105

1511    1343

Mass Spectrum

101    M⁺· 147

% of base peak

C₈H₅NO₂

m/e

UV Spectrum

$\log_{10}\varepsilon > 4$

¹³C NMR Spectrum
(50.0 MHz, CDCl₃ solution)

DEPT  CH₂↓ CH₃↑ CH↑

expansion

proton decoupled

solvent

expansion

δ (ppm)

¹H NMR Spectrum
(200 MHz, CDCl₃ solution)

expansion

TMS

δ (ppm)

# Problem 126

## IR Spectrum
(KBr disc)

1766  1685

ν (cm⁻¹)

## Mass Spectrum

43  121  138

M⁺· = 180

C₉H₈O₄

% of base peak

m/e

## UV Spectrum

λ_max 235 nm (log₁₀ε 4.1)

λ_max 265 nm (log₁₀ε 3.0)

solvent : ethanol

## ¹³C NMR Spectrum
(100.0 MHz, CDCl₃ solution)

DEPT  CH₂↓ CH₃↑ CH↑

proton decoupled

solvent

δ (ppm)

## ¹H NMR Spectrum
(400 MHz, CDCl₃ solution)

exchanges
with D₂O

expansion

11.0  10.0

8.5  8.0  7.5  7.0  ppm

TMS

δ (ppm)

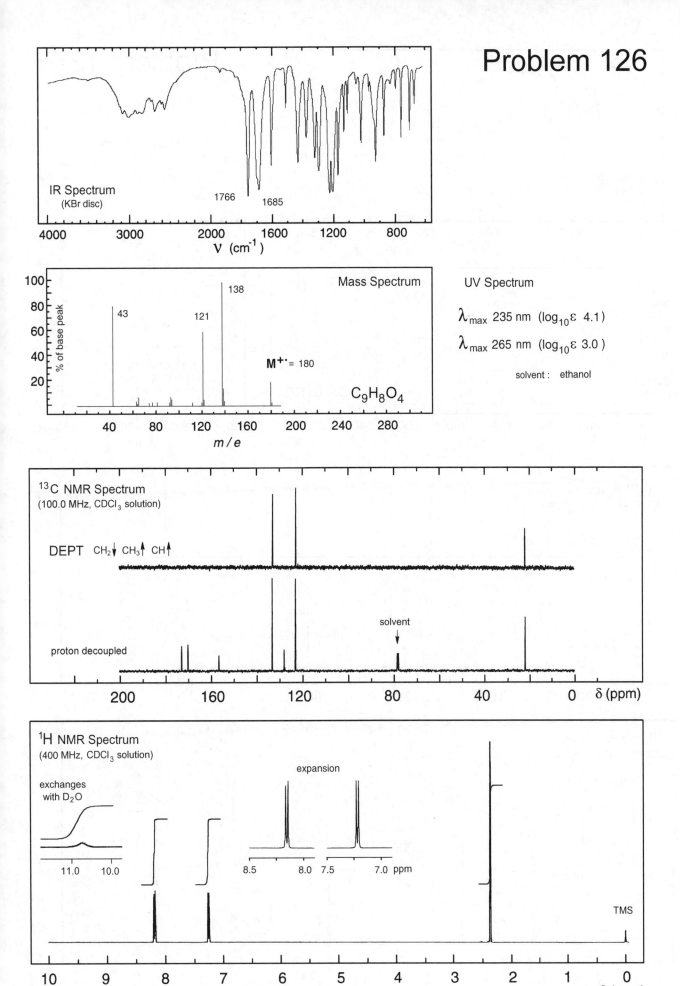

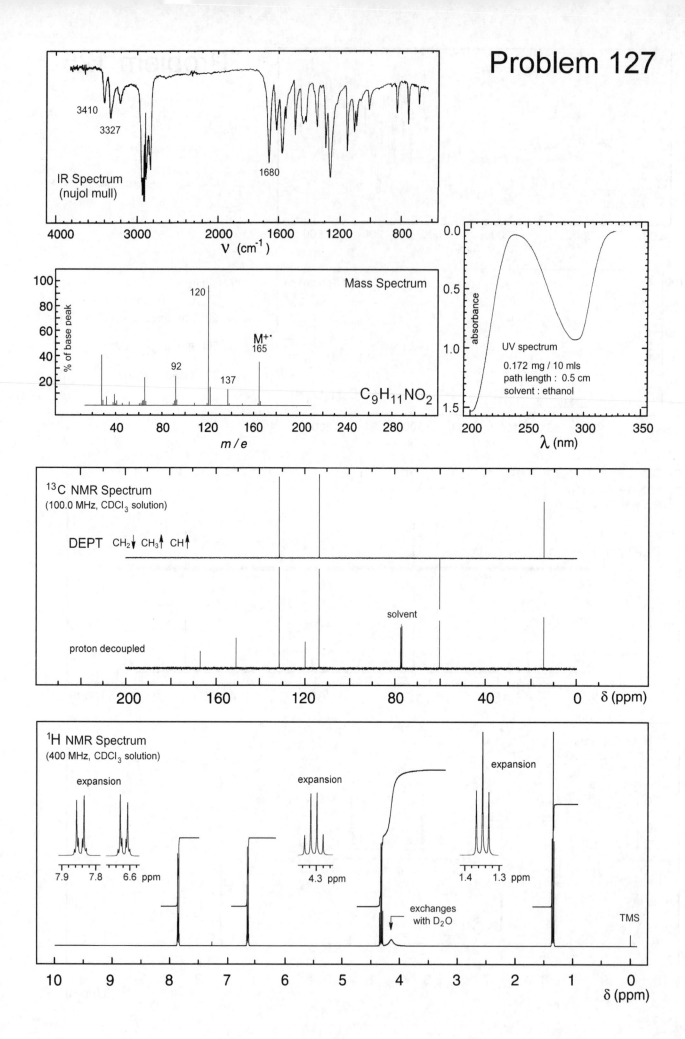

# Problem 127

IR Spectrum
(nujol mull)

3410
3327

1680

ν (cm$^{-1}$)
4000    3000    2000    1600    1200    800

Mass Spectrum

120
92
M$^{+\cdot}$
165
137

% of base peak
100
80
60
40
20

$C_9H_{11}NO_2$

m / e
40    80    120    160    200    240    280

UV spectrum

0.172 mg / 10 mls
path length : 0.5 cm
solvent : ethanol

absorbance
0.0
0.5
1.0
1.5

λ (nm)
200    250    300    350

$^{13}$C NMR Spectrum
(100.0 MHz, CDCl$_3$ solution)

DEPT    CH$_2$↓ CH$_3$↑ CH↑

solvent

proton decoupled

δ (ppm)
200    160    120    80    40    0

$^1$H NMR Spectrum
(400 MHz, CDCl$_3$ solution)

expansion

7.9    7.8    6.6 ppm

expansion

4.3 ppm

expansion

1.4    1.3 ppm

exchanges
with D$_2$O

TMS

δ (ppm)
10    9    8    7    6    5    4    3    2    1    0

238

# Problem 128

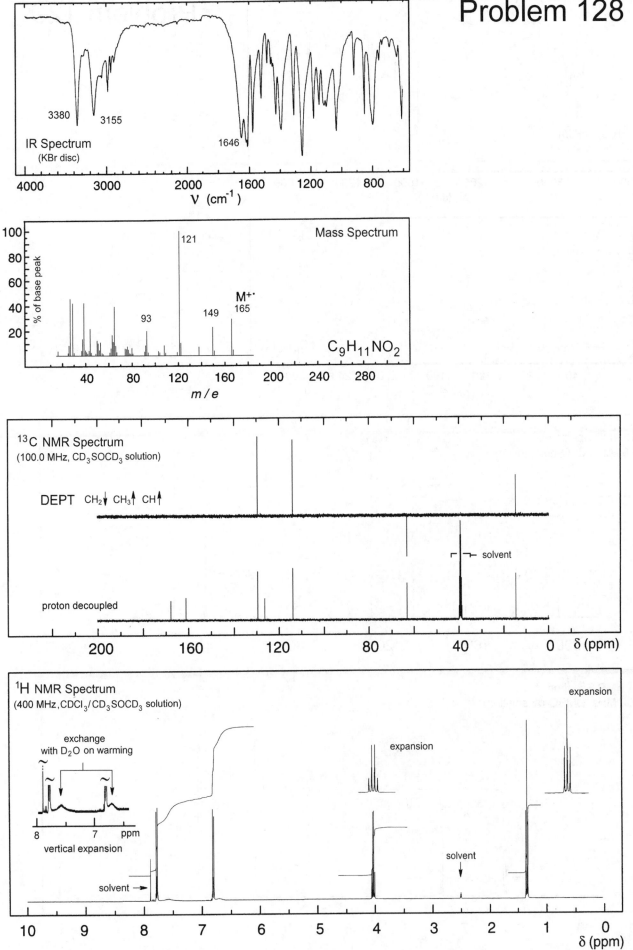

IR Spectrum (KBr disc)

3380  3155  1646

Mass Spectrum

121  93  149  M+· 165

$C_9H_{11}NO_2$

$^{13}C$ NMR Spectrum
(100.0 MHz, $CD_3SOCD_3$ solution)

DEPT  $CH_2\downarrow$  $CH_3\uparrow$  $CH\uparrow$

solvent

proton decoupled

$^1H$ NMR Spectrum
(400 MHz, $CDCl_3$/$CD_3SOCD_3$ solution)

exchange with $D_2O$ on warming

expansion

expansion

vertical expansion

solvent

solvent

expansion

# Problem 129

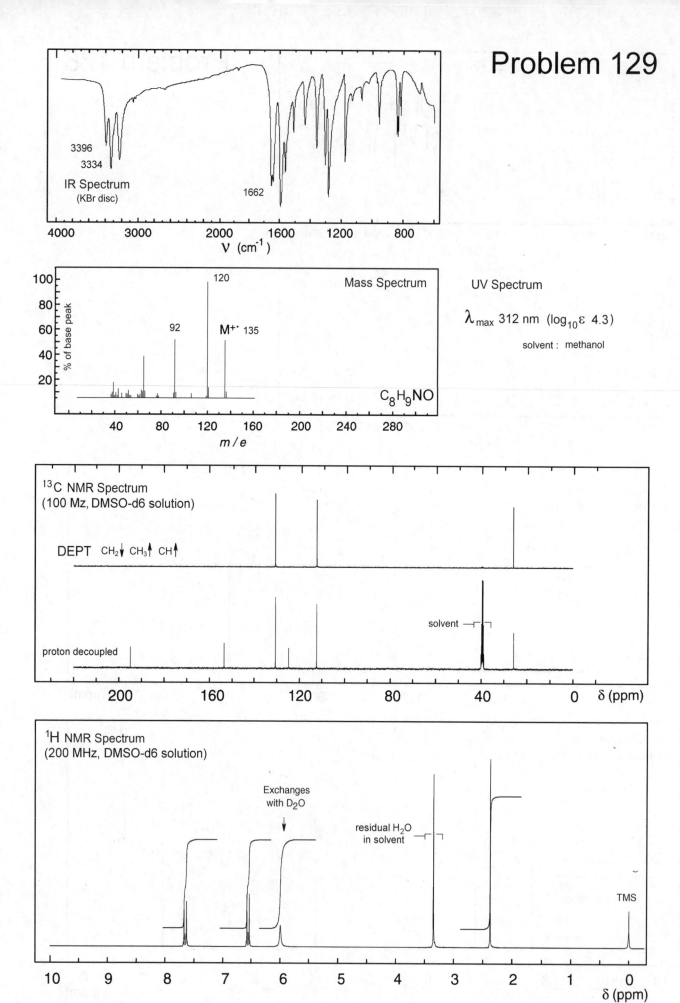

IR Spectrum
(KBr disc)

3396
3334

1662

ν (cm⁻¹)

Mass Spectrum

120
92
M⁺˙ 135

C₈H₉NO

% of base peak

m/e

UV Spectrum

λ_max 312 nm (log₁₀ε 4.3)

solvent : methanol

¹³C NMR Spectrum
(100 Mz, DMSO-d6 solution)

DEPT CH₂↓ CH₃↑ CH↑

proton decoupled

solvent

δ (ppm)

¹H NMR Spectrum
(200 MHz, DMSO-d6 solution)

Exchanges
with D₂O

residual H₂O
in solvent

TMS

δ (ppm)

# Problem 130

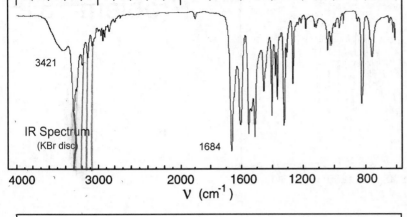

IR Spectrum
(KBr disc)

3421

1684

ν (cm⁻¹)

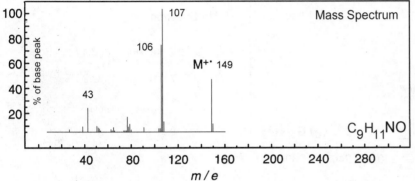

Mass Spectrum

107

106

M⁺˙ 149

43

% of base peak

$C_9H_{11}NO$

m/e

UV Spectrum

$\lambda_{max}$ 246 nm (log₁₀ε 4.2)

$\lambda_{max}$ 280 nm (log₁₀ε 3.1)

solvent : methanol

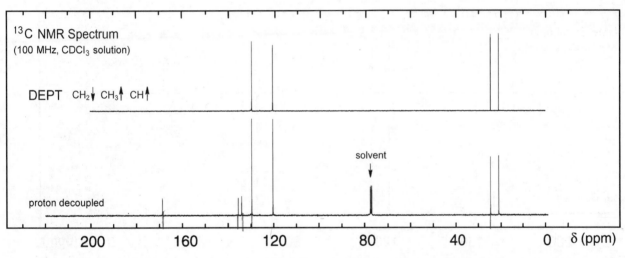

¹³C NMR Spectrum
(100 MHz, CDCl₃ solution)

DEPT   CH₂↓ CH₃↑ CH↑

solvent

proton decoupled

δ (ppm)

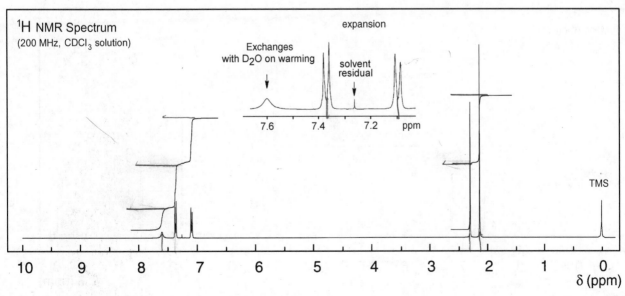

¹H NMR Spectrum
(200 MHz, CDCl₃ solution)

expansion

Exchanges
with D₂O on warming

solvent
residual

7.6   7.4   7.2   ppm

TMS

δ (ppm)

# Problem 131

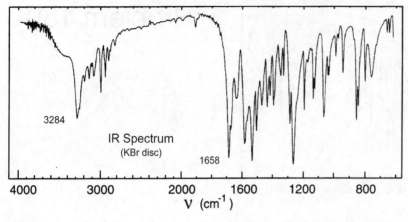

IR Spectrum
(KBr disc)

3284

1658

ν (cm⁻¹)

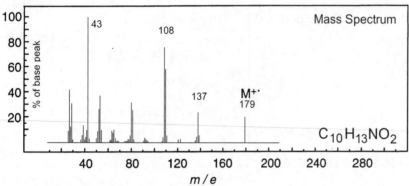

Mass Spectrum

43

108

137

M⁺˙
179

C₁₀H₁₃NO₂

% of base peak

m/e

UV Spectrum

$\lambda_{max}$ 250 nm (log₁₀ε 3.1)

$\lambda_{max}$ 287 nm (log₁₀ε 2.2)

solvent : chloroform

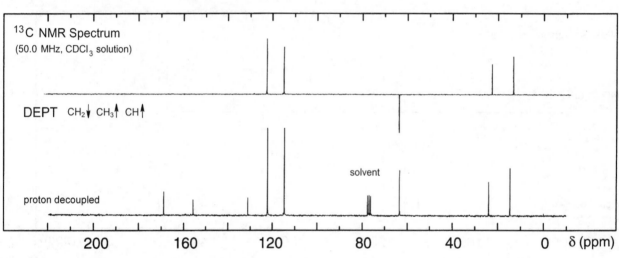

¹³C NMR Spectrum
(50.0 MHz, CDCl₃ solution)

DEPT  CH₂↓ CH₃↑ CH↑

solvent

proton decoupled

δ (ppm)

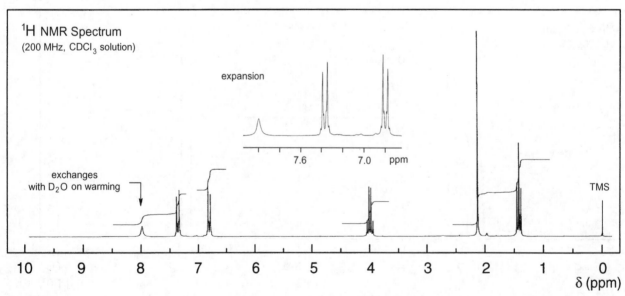

¹H NMR Spectrum
(200 MHz, CDCl₃ solution)

expansion

7.6    7.0    ppm

exchanges
with D₂O on warming

TMS

δ (ppm)

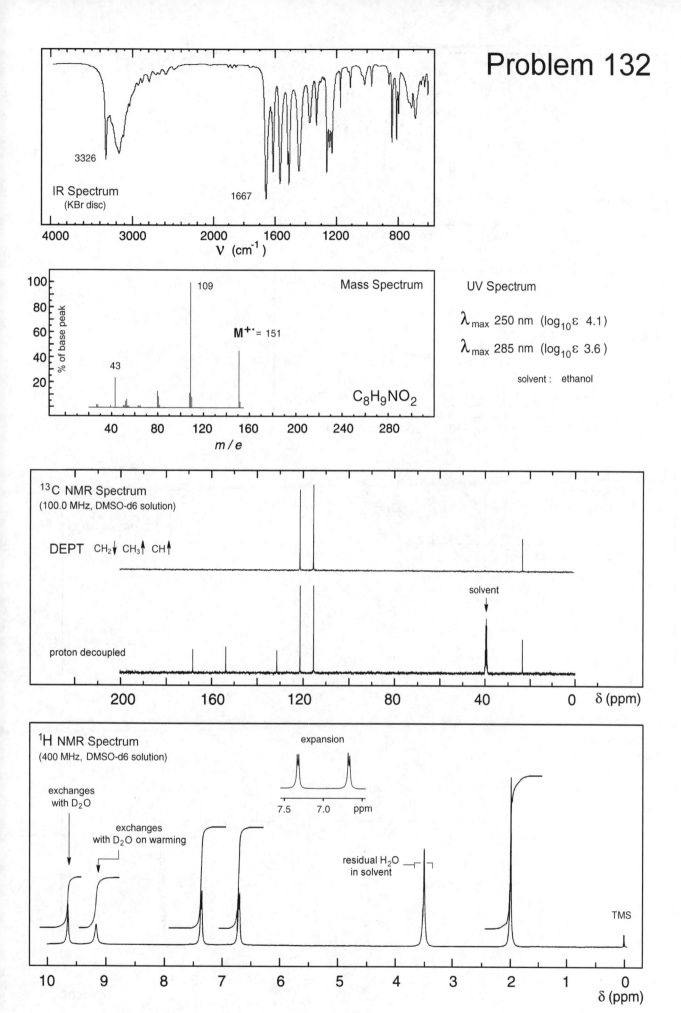

# Problem 132

**IR Spectrum**
(KBr disc)

3326

1667

ν (cm⁻¹)

**Mass Spectrum**

109

**M⁺· = 151**

43

C₈H₉NO₂

**UV Spectrum**

$\lambda_{max}$ 250 nm (log₁₀ε 4.1)

$\lambda_{max}$ 285 nm (log₁₀ε 3.6)

solvent : ethanol

**¹³C NMR Spectrum**
(100.0 MHz, DMSO-d6 solution)

DEPT   CH₂↓ CH₃↑ CH↑

solvent

proton decoupled

δ (ppm)

**¹H NMR Spectrum**
(400 MHz, DMSO-d6 solution)

expansion

7.5    7.0    ppm

exchanges
with D₂O

exchanges
with D₂O on warming

residual H₂O
in solvent

TMS

δ (ppm)

# Problem 133

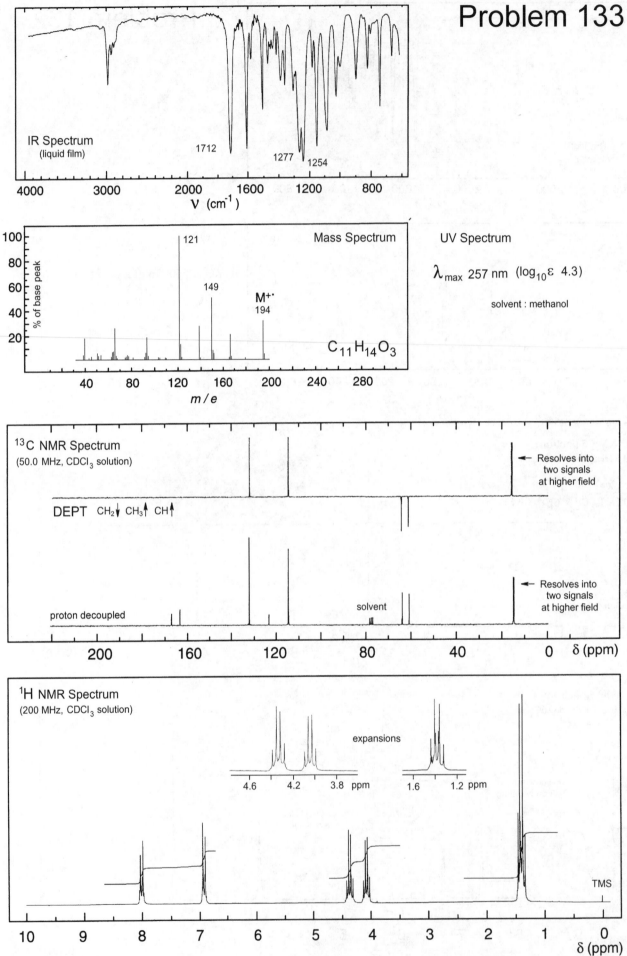

IR Spectrum
(liquid film)

1712

1277 1254

ν (cm⁻¹)

Mass Spectrum

121

149

M⁺˙
194

% of base peak

$C_{11}H_{14}O_3$

m/e

UV Spectrum

$\lambda_{max}$ 257 nm  ($\log_{10}\varepsilon$  4.3)

solvent : methanol

¹³C NMR Spectrum
(50.0 MHz, CDCl₃ solution)

← Resolves into
two signals
at higher field

DEPT   CH₂↓ CH₃↑ CH↑

← Resolves into
two signals
at higher field

proton decoupled

solvent

δ (ppm)

¹H NMR Spectrum
(200 MHz, CDCl₃ solution)

expansions

4.6   4.2   3.8  ppm

1.6   1.2  ppm

TMS

δ (ppm)

244

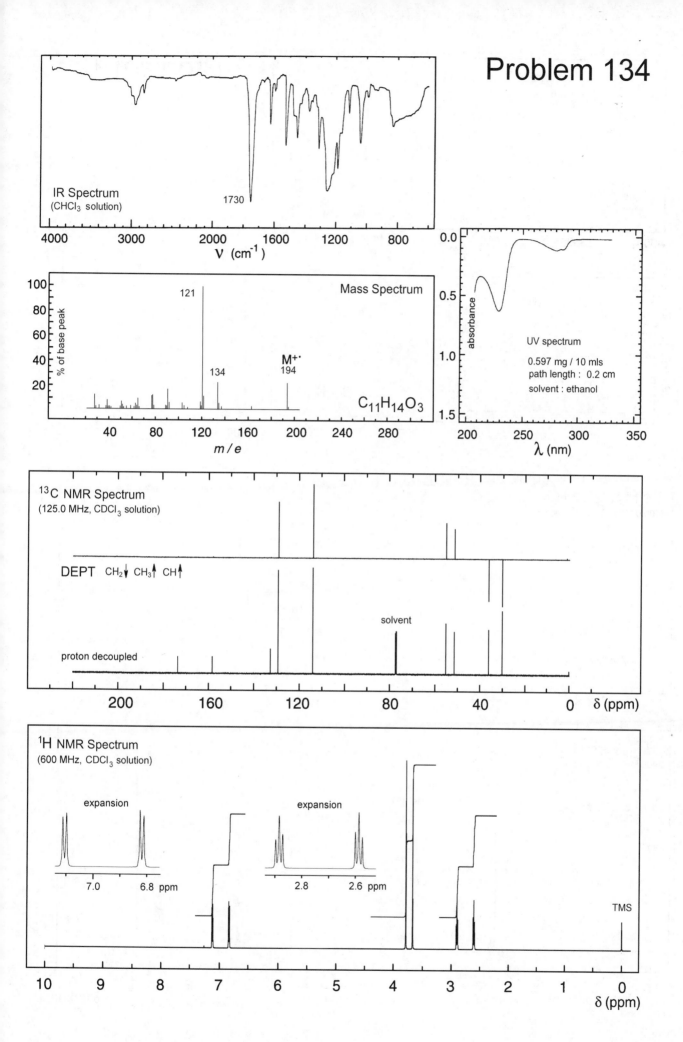

# Problem 134

**IR Spectrum**
(CHCl₃ solution)

1730

ν (cm⁻¹)

**Mass Spectrum**

121

134

M⁺·
194

C₁₁H₁₄O₃

% of base peak

m/e

**UV spectrum**

0.597 mg / 10 mls
path length : 0.2 cm
solvent : ethanol

λ (nm)

**¹³C NMR Spectrum**
(125.0 MHz, CDCl₃ solution)

DEPT   CH₂↓ CH₃↑ CH↑

solvent

proton decoupled

δ (ppm)

**¹H NMR Spectrum**
(600 MHz, CDCl₃ solution)

expansion

7.0      6.8 ppm

expansion

2.8      2.6 ppm

TMS

δ (ppm)

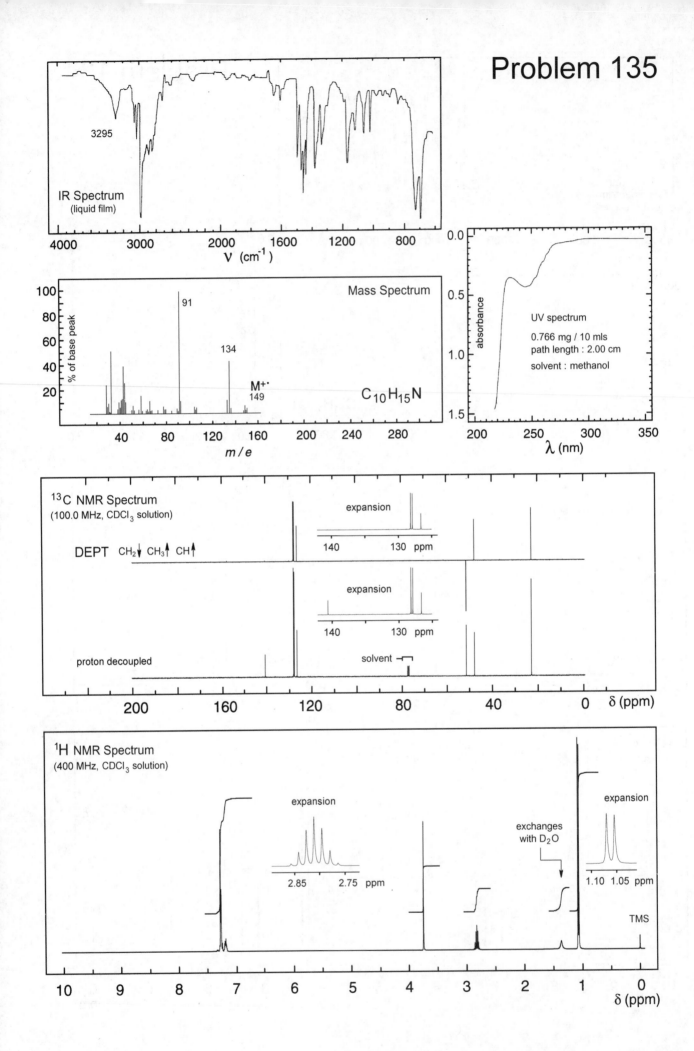

# Problem 135

IR Spectrum
(liquid film)

3295

$\nu$ (cm$^{-1}$)

Mass Spectrum

% of base peak

91

134

M$^{+\cdot}$
149

$C_{10}H_{15}N$

$m/e$

UV spectrum

0.766 mg / 10 mls
path length : 2.00 cm

solvent : methanol

absorbance

$\lambda$ (nm)

$^{13}C$ NMR Spectrum
(100.0 MHz, CDCl$_3$ solution)

DEPT  CH$_2\downarrow$ CH$_3\uparrow$ CH$\uparrow$

expansion

140   130  ppm

expansion

140   130  ppm

proton decoupled

solvent

$\delta$ (ppm)

$^1H$ NMR Spectrum
(400 MHz, CDCl$_3$ solution)

expansion

2.85   2.75  ppm

exchanges
with D$_2$O

expansion

1.10  1.05  ppm

TMS

$\delta$ (ppm)

# Problem 136

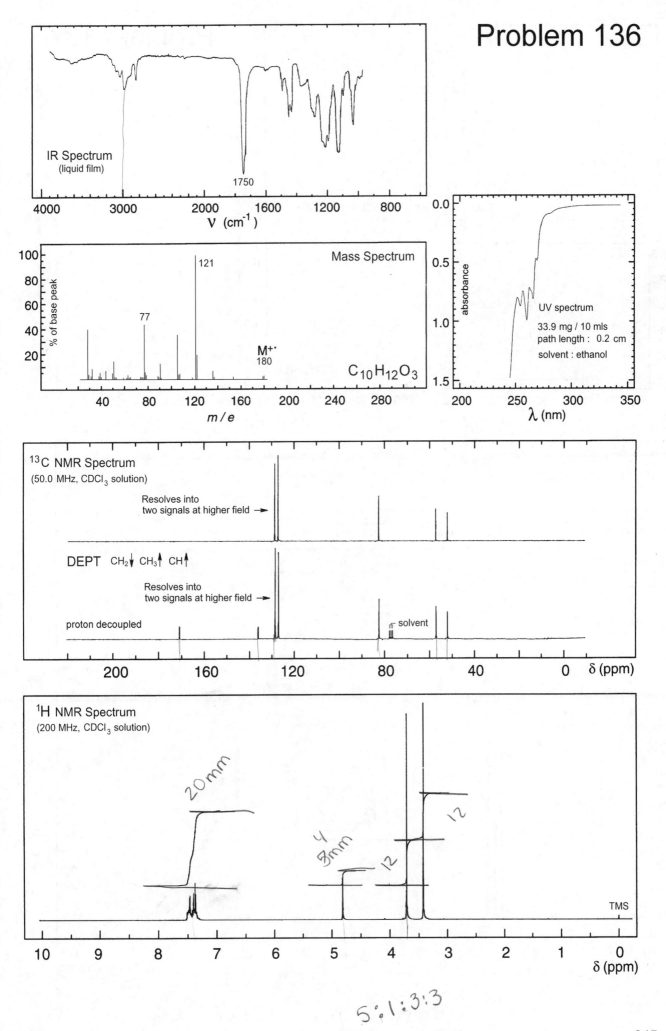

IR Spectrum
(liquid film)

1750

ν (cm⁻¹)

Mass Spectrum

121

77

M⁺·
180

$C_{10}H_{12}O_3$

% of base peak

m/e

UV spectrum

33.9 mg / 10 mls
path length :  0.2 cm

solvent : ethanol

absorbance

λ (nm)

$^{13}$C NMR Spectrum
(50.0 MHz, CDCl$_3$ solution)

Resolves into
two signals at higher field →

DEPT  CH$_2$↓ CH$_3$↑ CH↑

Resolves into
two signals at higher field →

proton decoupled

solvent

δ (ppm)

$^1$H NMR Spectrum
(200 MHz, CDCl$_3$ solution)

20mm

4
3mm

12

12

TMS

δ (ppm)

5:1:3:3

247

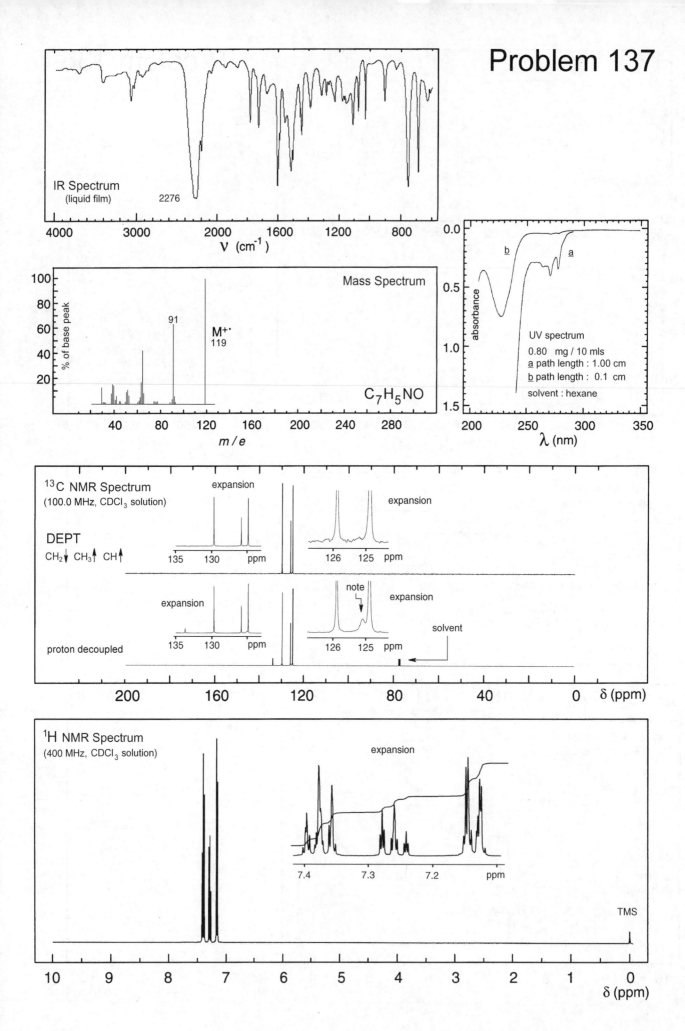

# Problem 137

**IR Spectrum** (liquid film)
2276

ν (cm⁻¹)

**Mass Spectrum**

91

M⁺•
119

$C_7H_5NO$

% of base peak

m/e

**UV spectrum**
0.80 mg / 10 mls
a path length : 1.00 cm
b path length : 0.1 cm
solvent : hexane

absorbance

λ (nm)

**¹³C NMR Spectrum**
(100.0 MHz, CDCl₃ solution)

**DEPT**
CH₂↓ CH₃↑ CH↑

expansion

expansion

note

expansion

expansion

solvent

proton decoupled

δ (ppm)

**¹H NMR Spectrum**
(400 MHz, CDCl₃ solution)

expansion

TMS

δ (ppm)

# Problem 138

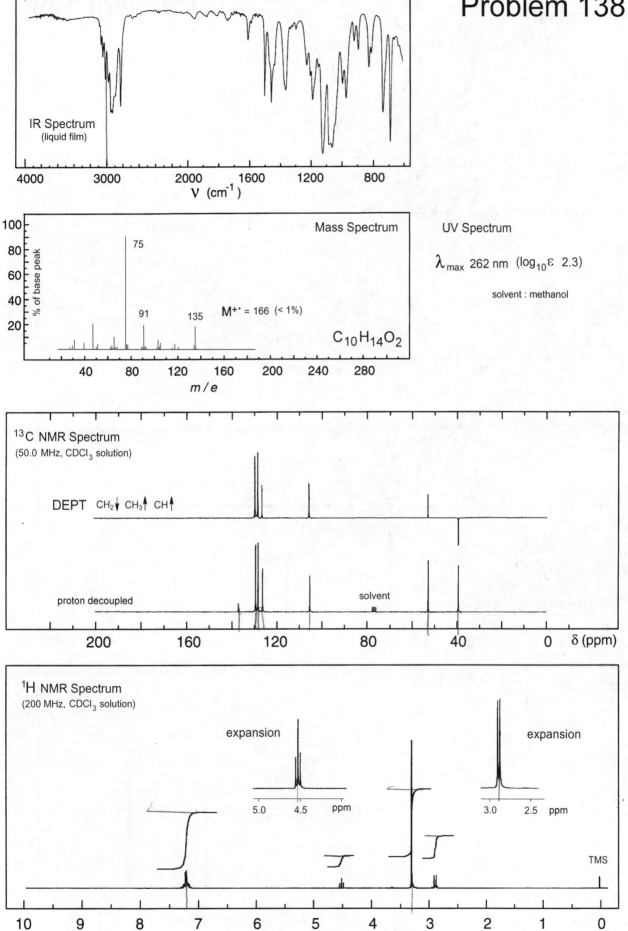

IR Spectrum
(liquid film)

$\nu$ (cm$^{-1}$)

Mass Spectrum

75

91

135

M$^{+\bullet}$ = 166 (< 1%)

$C_{10}H_{14}O_2$

% of base peak

$m/e$

UV Spectrum

$\lambda_{max}$ 262 nm (log$_{10}\varepsilon$ 2.3)

solvent : methanol

$^{13}$C NMR Spectrum
(50.0 MHz, CDCl$_3$ solution)

DEPT    CH$_2\downarrow$ CH$_3\uparrow$ CH$\uparrow$

proton decoupled

solvent

$\delta$ (ppm)

$^1$H NMR Spectrum
(200 MHz, CDCl$_3$ solution)

expansion

5.0    4.5    ppm

expansion

3.0    2.5    ppm

TMS

$\delta$ (ppm)

# Problem 139

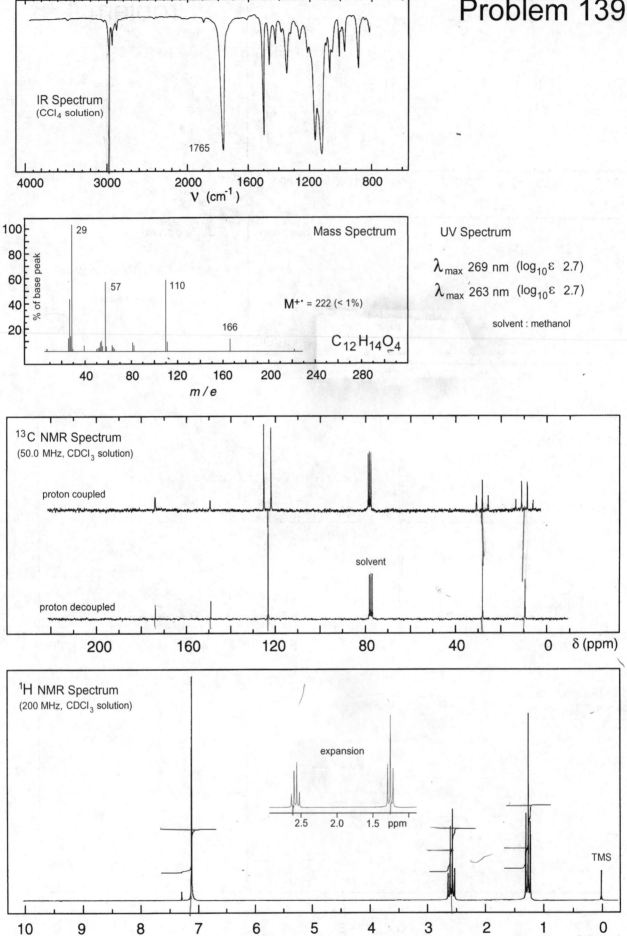

IR Spectrum
(CCl₄ solution)

1765

ν (cm⁻¹)

Mass Spectrum

29

57    110

M⁺· = 222 (< 1%)

166

$C_{12}H_{14}O_4$

% of base peak

m/e

UV Spectrum

$\lambda_{max}$ 269 nm  (log₁₀ε  2.7)

$\lambda_{max}$ 263 nm  (log₁₀ε  2.7)

solvent : methanol

¹³C NMR Spectrum
(50.0 MHz, CDCl₃ solution)

proton coupled

solvent

proton decoupled

δ (ppm)

¹H NMR Spectrum
(200 MHz, CDCl₃ solution)

expansion

2.5    2.0    1.5   ppm

TMS

δ (ppm)

250

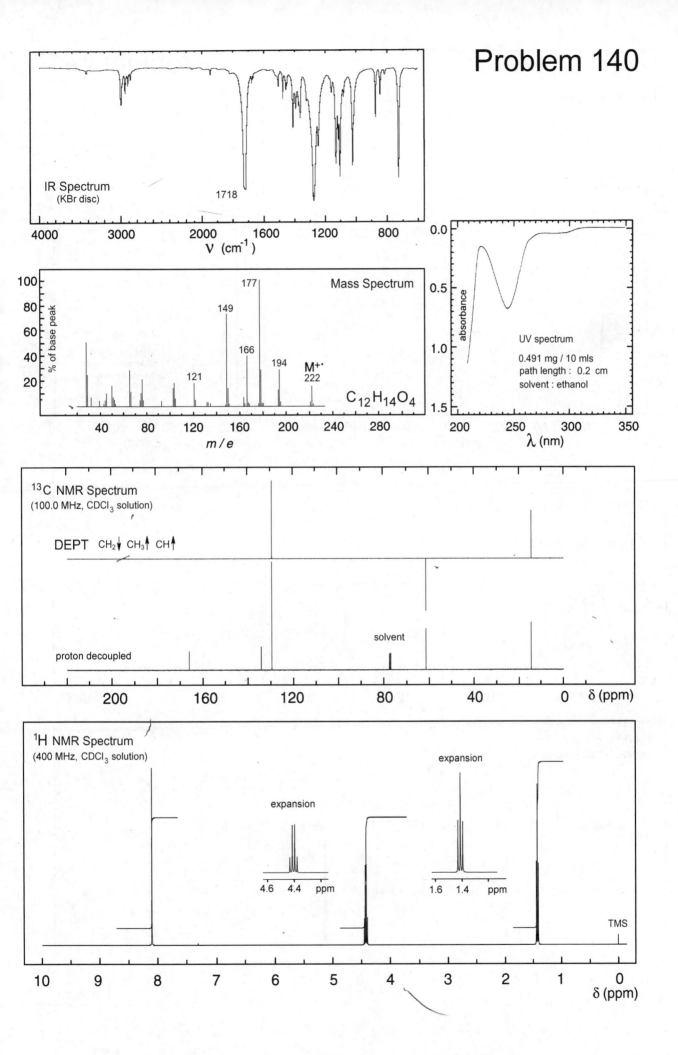

# Problem 140

**IR Spectrum** (KBr disc)

1718

ν (cm⁻¹)

$C_{12}H_{14}O_4$

Mass Spectrum

177
149
166
121
194
M⁺·
222

% of base peak

m / e

**UV spectrum**

0.491 mg / 10 mls
path length : 0.2 cm
solvent : ethanol

absorbance

λ (nm)

**¹³C NMR Spectrum**
(100.0 MHz, CDCl₃ solution)

DEPT  CH₂↓ CH₃↑ CH↑

proton decoupled

solvent

δ (ppm)

**¹H NMR Spectrum**
(400 MHz, CDCl₃ solution)

expansion

4.6   4.4   ppm

expansion

1.6   1.4   ppm

TMS

δ (ppm)

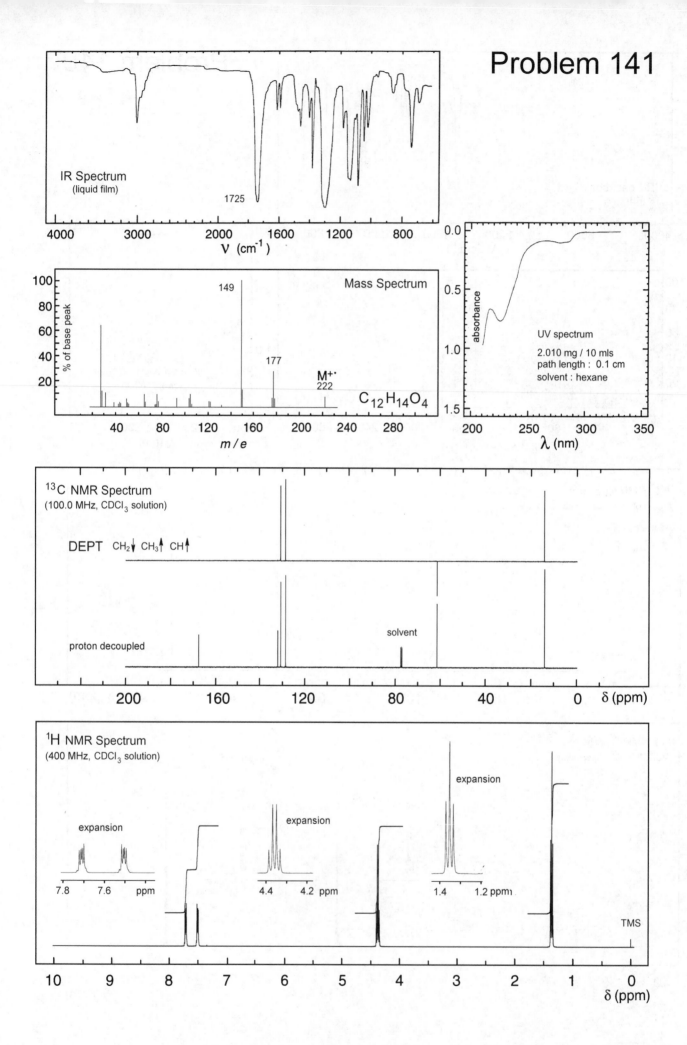

# Problem 141

**IR Spectrum** (liquid film)

1725

ν (cm⁻¹)

**Mass Spectrum**

149

177

M⁺·
222

C₁₂H₁₄O₄

m / e

**UV spectrum**

2.010 mg / 10 mls
path length : 0.1 cm
solvent : hexane

λ (nm)

**¹³C NMR Spectrum**
(100.0 MHz, CDCl₃ solution)

DEPT  CH₂↓ CH₃↑ CH↑

proton decoupled

solvent

δ (ppm)

**¹H NMR Spectrum**
(400 MHz, CDCl₃ solution)

expansion

7.8   7.6   ppm

expansion

4.4   4.2  ppm

expansion

1.4   1.2 ppm

TMS

δ (ppm)

# Problem 142

IR Spectrum
(liquid film)

1724

ν (cm⁻¹)

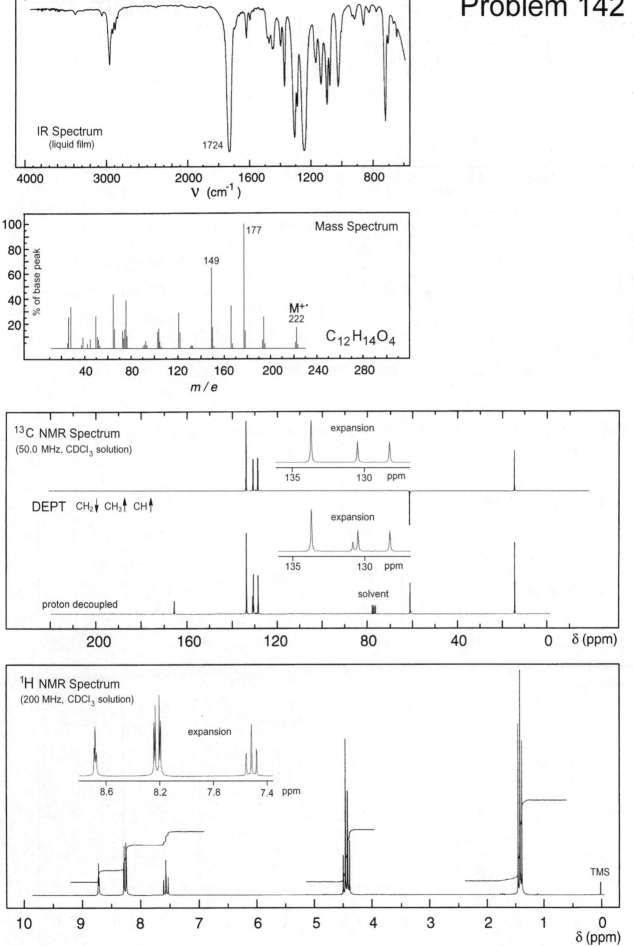

Mass Spectrum

177

149

M⁺˙
222

$C_{12}H_{14}O_4$

% of base peak

m / e

¹³C NMR Spectrum
(50.0 MHz, CDCl₃ solution)

expansion

135   130   ppm

DEPT   CH₂↓ CH₃↑ CH↑

expansion

135   130   ppm

proton decoupled

solvent

δ (ppm)

¹H NMR Spectrum
(200 MHz, CDCl₃ solution)

expansion

8.6   8.2   7.8   7.4   ppm

TMS

δ (ppm)

# Problem 143

IR Spectrum
(liquid film)

1765

4000   3000   2000   1600   1200   800

$\nu$ (cm$^{-1}$)

100
80
60
40
20

% of base peak

110

166

$M^{+\cdot}$ = 222   Mass Spectrum

$C_{12}H_{14}O_4$

40   80   120   160   200   240   280

$m/e$

UV Spectrum

$\lambda_{max}$ 220 nm (log$_{10}\varepsilon$ 3.7)

$\lambda_{max}$ 274 nm (log$_{10}\varepsilon$ 3.3)

solvent : ethanol

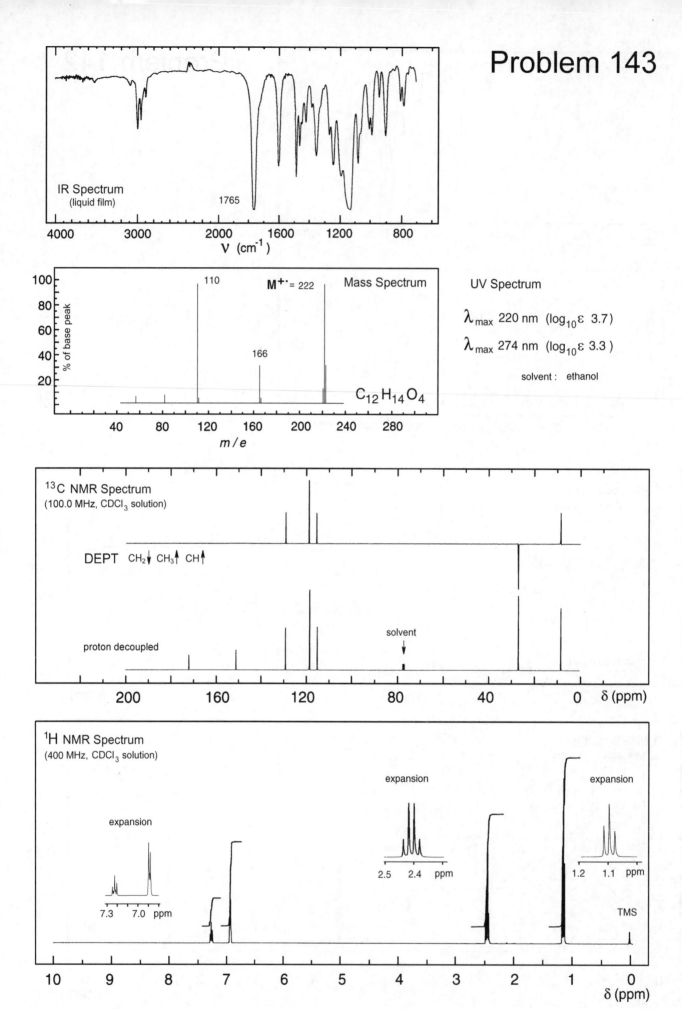

$^{13}$C NMR Spectrum
(100.0 MHz, CDCl$_3$ solution)

DEPT   CH$_2\downarrow$ CH$_3\uparrow$ CH$\uparrow$

proton decoupled

solvent

200   160   120   80   40   0   $\delta$ (ppm)

$^1$H NMR Spectrum
(400 MHz, CDCl$_3$ solution)

expansion

expansion

expansion

7.3   7.0   ppm

2.5   2.4   ppm

1.2   1.1   ppm

TMS

10   9   8   7   6   5   4   3   2   1   0

$\delta$ (ppm)

# Problem 144

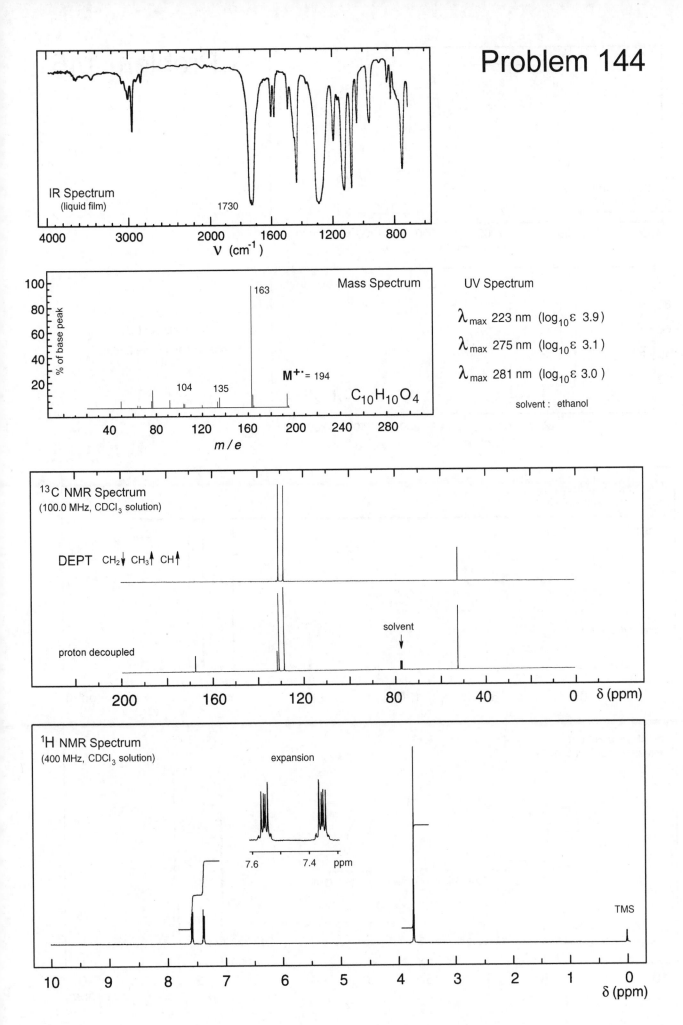

IR Spectrum
(liquid film)
1730

Mass Spectrum

% of base peak

163
104
135
$M^{+ \cdot} = 194$

$C_{10}H_{10}O_4$

UV Spectrum

$\lambda_{max}$ 223 nm ($\log_{10}\varepsilon$ 3.9)

$\lambda_{max}$ 275 nm ($\log_{10}\varepsilon$ 3.1)

$\lambda_{max}$ 281 nm ($\log_{10}\varepsilon$ 3.0 )

solvent : ethanol

$^{13}C$ NMR Spectrum
(100.0 MHz, CDCl$_3$ solution)

DEPT  CH$_2\downarrow$ CH$_3\uparrow$ CH$\uparrow$

solvent

proton decoupled

$^1H$ NMR Spectrum
(400 MHz, CDCl$_3$ solution)

expansion

7.6        7.4    ppm

TMS

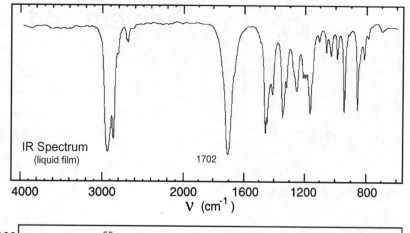

IR Spectrum
(liquid film)

1702

4000    3000    2000    1600    1200    800

ν (cm⁻¹)

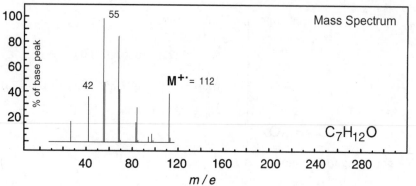

Mass Spectrum

100
80
60
40
20

% of base peak

55

42

M⁺· = 112

C₇H₁₂O

40    80    120    160    200    240    280

m/e

No significant UV
absorption above 220 nm

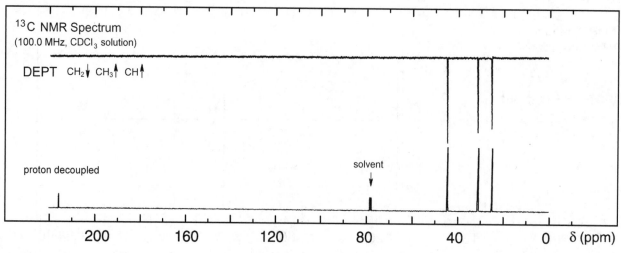

¹³C NMR Spectrum
(100.0 MHz, CDCl₃ solution)

DEPT   CH₂↓  CH₃↑  CH↑

proton decoupled

solvent

200    160    120    80    40    0    δ (ppm)

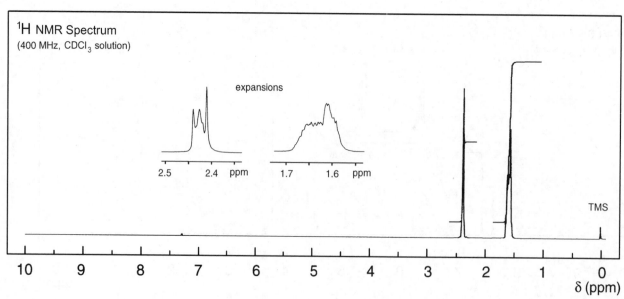

¹H NMR Spectrum
(400 MHz, CDCl₃ solution)

expansions

2.5    2.4  ppm       1.7    1.6  ppm

TMS

10    9    8    7    6    5    4    3    2    1    0    δ (ppm)

# Problem 146

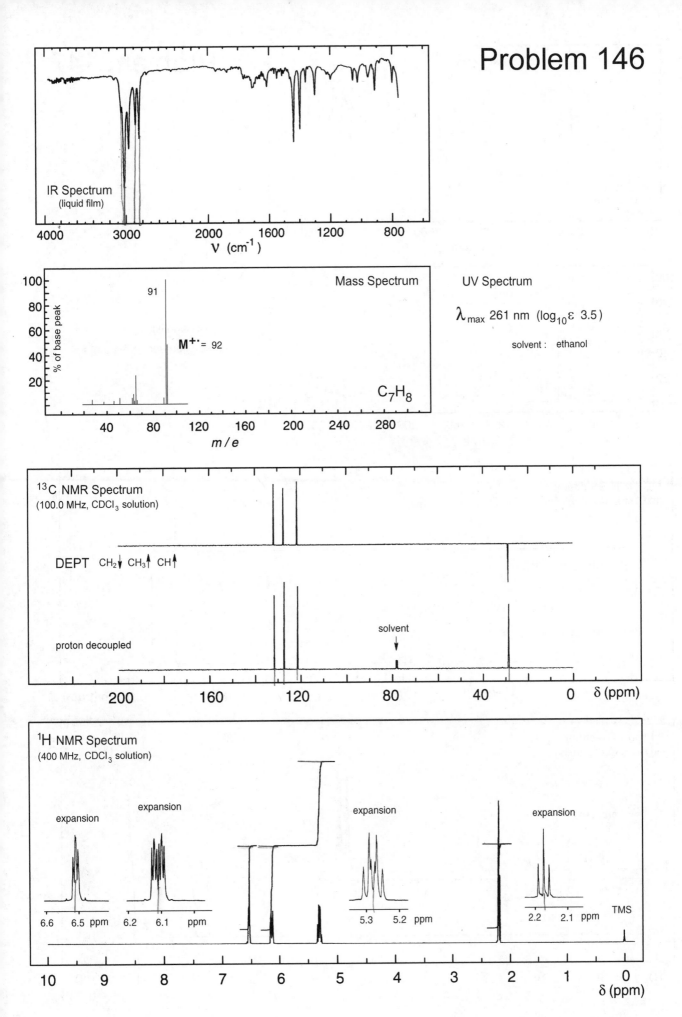

IR Spectrum
(liquid film)

ν (cm⁻¹)

Mass Spectrum

UV Spectrum

$\lambda_{max}$ 261 nm $(\log_{10}\varepsilon\ 3.5)$

solvent : ethanol

91

$M^{+\cdot} = 92$

% of base peak

$C_7H_8$

m/e

$^{13}C$ NMR Spectrum
(100.0 MHz, CDCl₃ solution)

DEPT   CH₂↓ CH₃↑ CH↑

proton decoupled

solvent

δ (ppm)

$^1H$ NMR Spectrum
(400 MHz, CDCl₃ solution)

expansion

expansion

6.6   6.5 ppm

6.2   6.1 ppm

expansion

5.3   5.2 ppm

expansion

2.2   2.1 ppm

TMS

δ (ppm)

# Problem 147

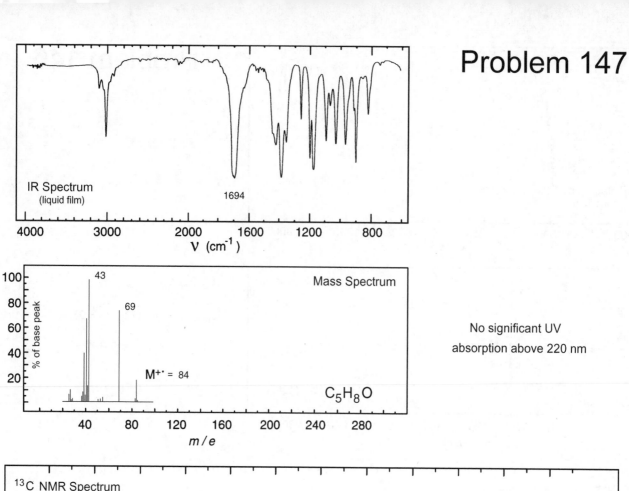

IR Spectrum
(liquid film)

1694

ν (cm⁻¹)

Mass Spectrum

No significant UV
absorption above 220 nm

M⁺˙ = 84

C₅H₈O

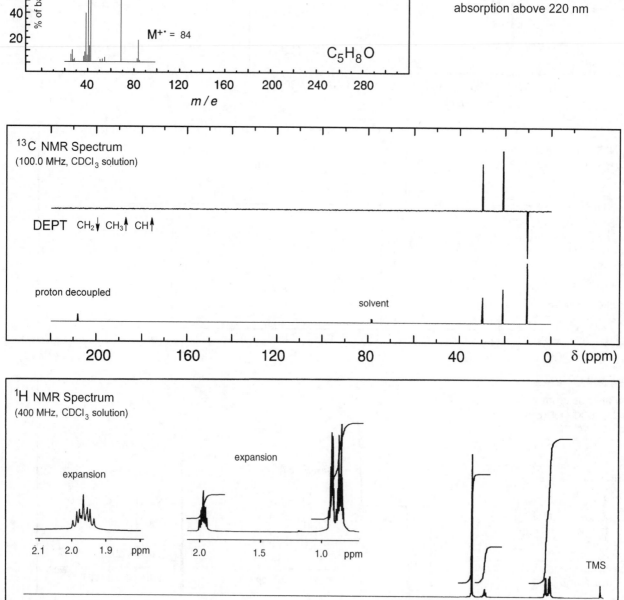

¹³C NMR Spectrum
(100.0 MHz, CDCl₃ solution)

DEPT   CH₂↓ CH₃↑ CH↑

proton decoupled

solvent

δ (ppm)

¹H NMR Spectrum
(400 MHz, CDCl₃ solution)

expansion

expansion

TMS

δ (ppm)

# Problem 148

IR Spectrum
(liquid film)

1696

ν (cm⁻¹)

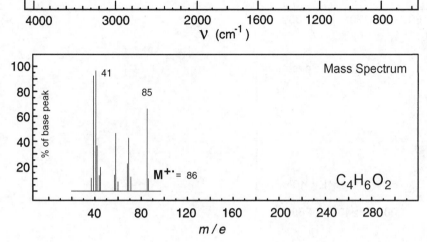

Mass Spectrum

41

85

M⁺· = 86

C₄H₆O₂

% of base peak

m/e

No significant UV
absorption above 220 nm

¹³C NMR Spectrum
(100.0 MHz, CDCl₃ solution)

DEPT   CH₂↓ CH₃↑ CH↑

solvent

proton decoupled

δ (ppm)

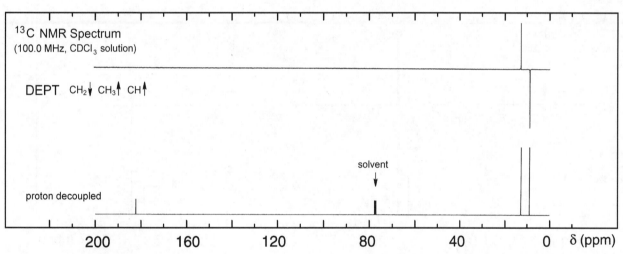

¹H NMR Spectrum
(400 MHz, CDCl₃ solution)

expansion

expansion

Exchanges
with D₂O

12.0      11.5   ppm

1.7     1.5   ppm

1.1      0.9   ppm

TMS

δ (ppm)

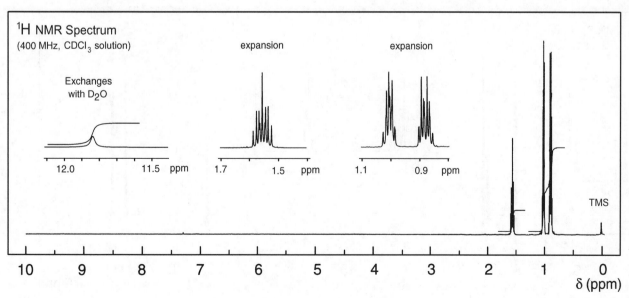

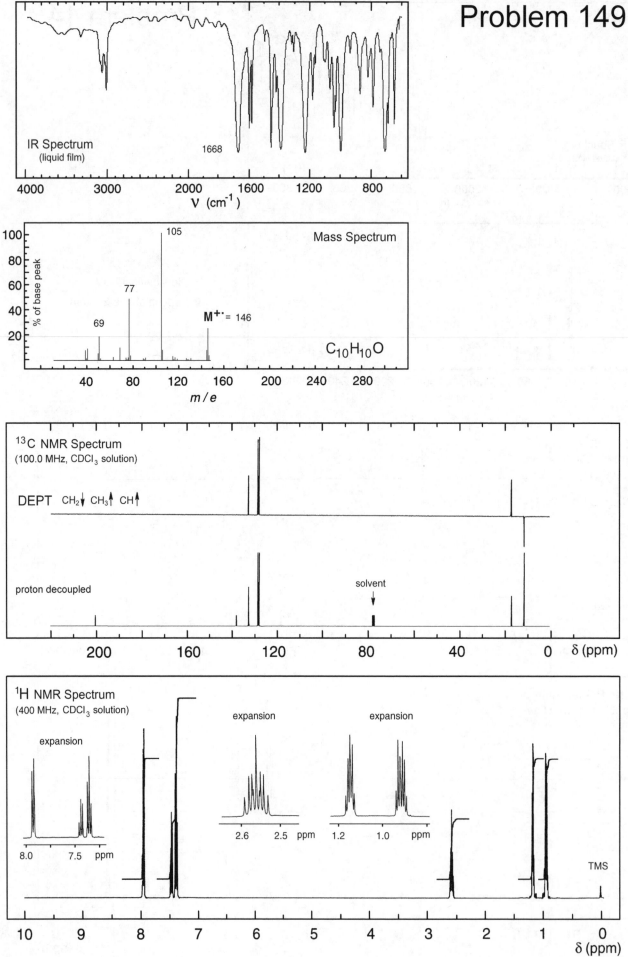

# Problem 149

**IR Spectrum** (liquid film)

1668

ν (cm⁻¹)

**Mass Spectrum**

105

77

69

**M⁺·** = 146

$C_{10}H_{10}O$

m/e

**¹³C NMR Spectrum** (100.0 MHz, CDCl₃ solution)

DEPT   CH₂↓ CH₃↑ CH↑

proton decoupled

solvent

δ (ppm)

**¹H NMR Spectrum** (400 MHz, CDCl₃ solution)

expansion

expansion

expansion

TMS

δ (ppm)

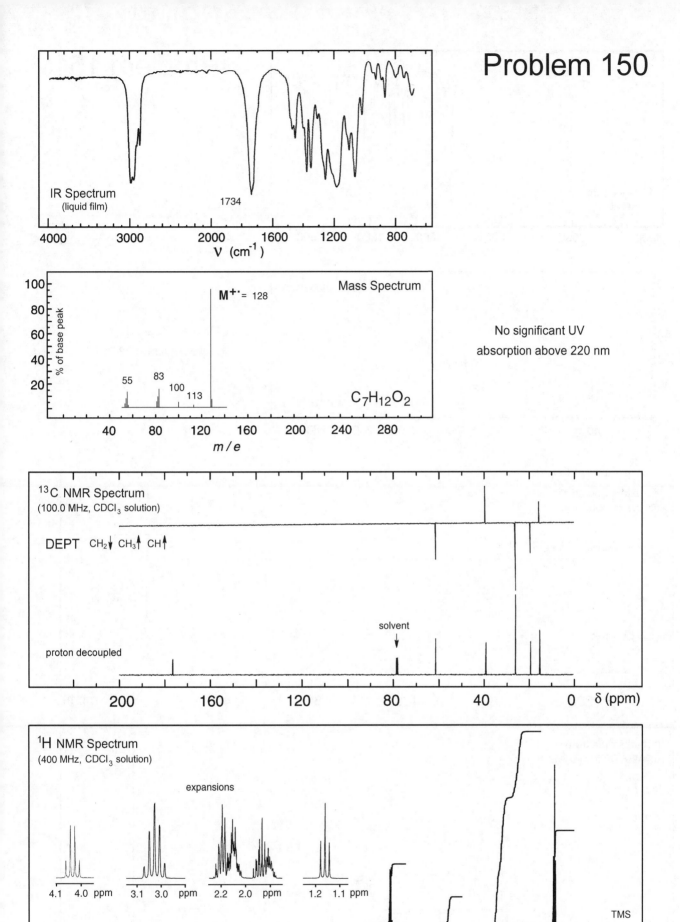

# Problem 150

**IR Spectrum**
(liquid film)

1734

$\nu$ (cm$^{-1}$)

**Mass Spectrum**

$M^{+\cdot}$ = 128

% of base peak

55   83   100   113

$m/e$

No significant UV
absorption above 220 nm

$C_7H_{12}O_2$

**$^{13}$C NMR Spectrum**
(100.0 MHz, CDCl$_3$ solution)

DEPT   CH$_2\downarrow$  CH$_3\uparrow$  CH$\uparrow$

solvent

proton decoupled

$\delta$ (ppm)

**$^1$H NMR Spectrum**
(400 MHz, CDCl$_3$ solution)

expansions

4.1   4.0 ppm        3.1   3.0 ppm        2.2   2.0 ppm        1.2   1.1 ppm

TMS

$\delta$ (ppm)

261

# Problem 151

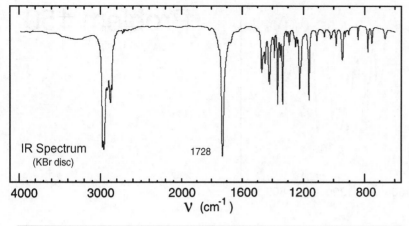

IR Spectrum
(KBr disc)

1728

$\nu$ (cm$^{-1}$)

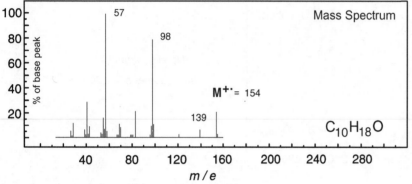

Mass Spectrum

100

% of base peak

80

60

40

20

57

98

$M^{+\cdot}$ = 154

139

$C_{10}H_{18}O$

40   80   120   160   200   240   280

$m/e$

No significant UV
absorption above 220 nm

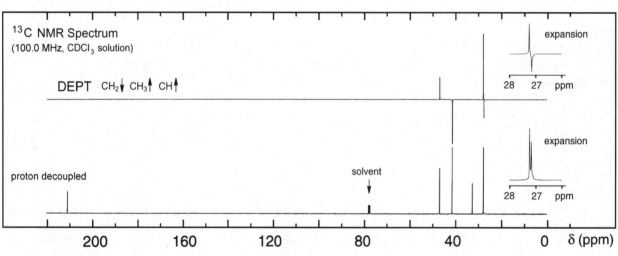

$^{13}C$ NMR Spectrum
(100.0 MHz, CDCl$_3$ solution)

DEPT   CH$_2\downarrow$ CH$_3\uparrow$ CH$\uparrow$

expansion

28   27   ppm

expansion

proton decoupled

solvent

28   27   ppm

200   160   120   80   40   0   $\delta$ (ppm)

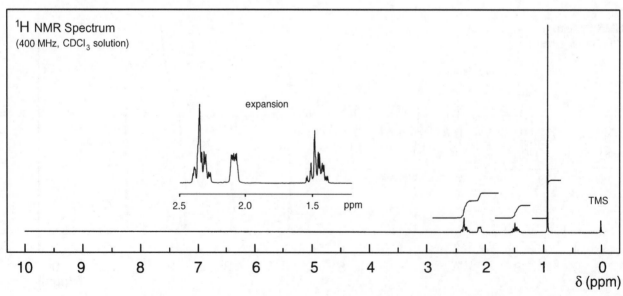

$^1H$ NMR Spectrum
(400 MHz, CDCl$_3$ solution)

expansion

2.5   2.0   1.5   ppm

TMS

10   9   8   7   6   5   4   3   2   1   0
$\delta$ (ppm)

# Problem 152

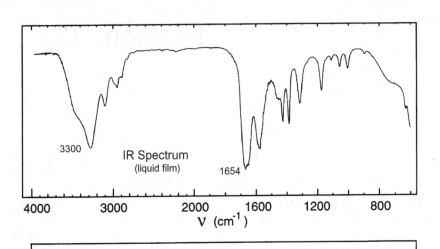

IR Spectrum
(liquid film)

3300

1654

ν (cm⁻¹)

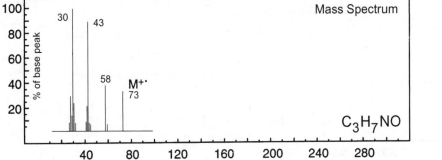

Mass Spectrum

% of base peak

30

43

58

M⁺·
73

C₃H₇NO

m/e

No significant UV
absorption above 220 nm

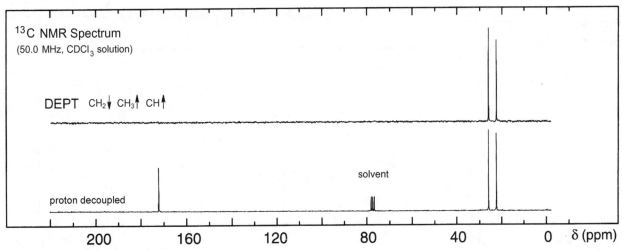

¹³C NMR Spectrum
(50.0 MHz, CDCl₃ solution)

DEPT  CH₂↓ CH₃↑ CH↑

solvent

proton decoupled

δ (ppm)

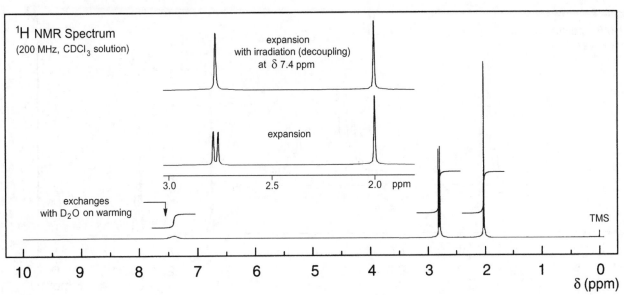

¹H NMR Spectrum
(200 MHz, CDCl₃ solution)

expansion
with irradiation (decoupling)
at δ 7.4 ppm

expansion

3.0        2.5        2.0   ppm

exchanges
with D₂O on warming

TMS

δ (ppm)

# Problem 153

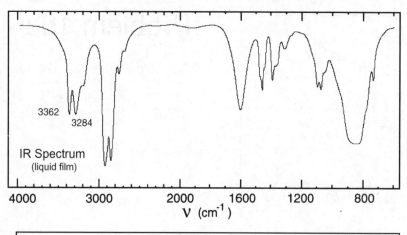

IR Spectrum
(liquid film)

3362
3284

$\nu$ (cm$^{-1}$)

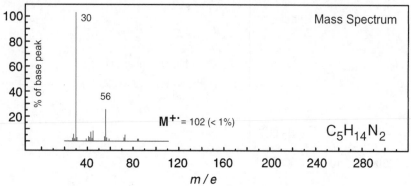

Mass Spectrum

% of base peak

30

56

$M^{+\cdot}$ = 102 (< 1%)

$C_5H_{14}N_2$

$m/e$

No significant UV
absorption above 220 nm

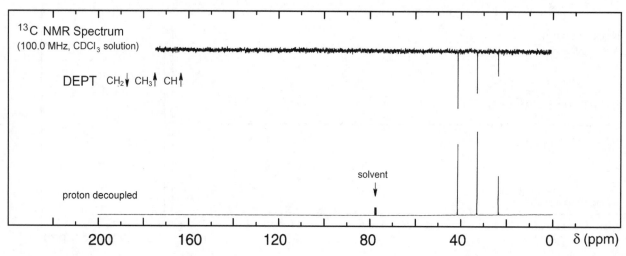

$^{13}$C NMR Spectrum
(100.0 MHz, CDCl$_3$ solution)

DEPT   CH$_2\downarrow$  CH$_3\uparrow$  CH$\uparrow$

solvent

proton decoupled

$\delta$ (ppm)

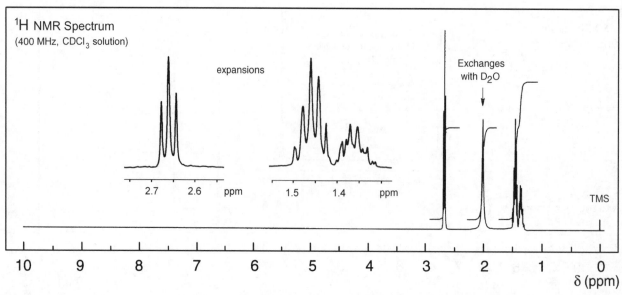

$^1$H NMR Spectrum
(400 MHz, CDCl$_3$ solution)

expansions

Exchanges
with D$_2$O

2.7     2.6    ppm

1.5     1.4    ppm

TMS

$\delta$ (ppm)

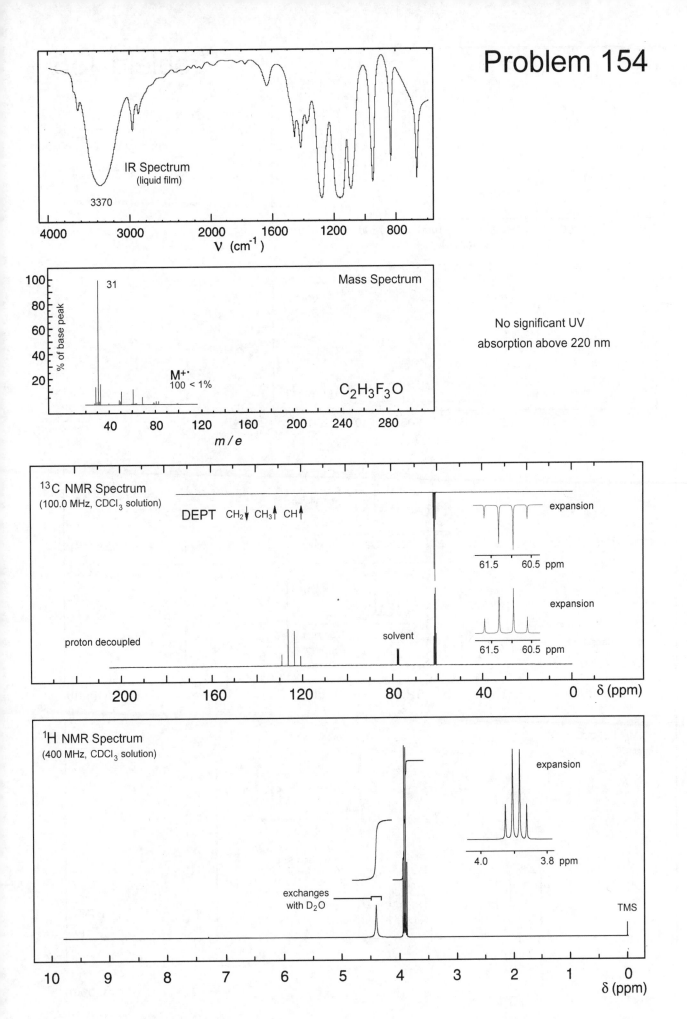

# Problem 154

IR Spectrum
(liquid film)

3370

ν (cm⁻¹)

Mass Spectrum

No significant UV
absorption above 220 nm

$C_2H_3F_3O$

$M^{+\cdot}$
100 < 1%

m/e

¹³C NMR Spectrum
(100.0 MHz, CDCl₃ solution)

DEPT  CH₂↓ CH₃↑ CH↑

expansion

61.5    60.5 ppm

expansion

61.5    60.5 ppm

proton decoupled

solvent

δ (ppm)

¹H NMR Spectrum
(400 MHz, CDCl₃ solution)

expansion

4.0    3.8 ppm

exchanges
with D₂O

TMS

δ (ppm)

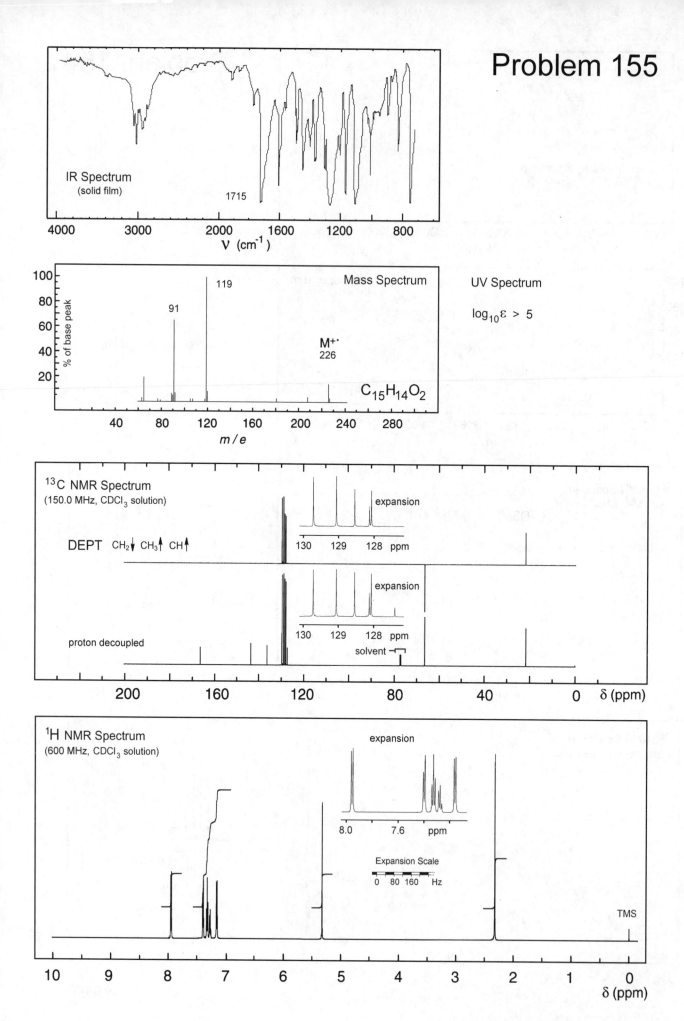

# Problem 155

## IR Spectrum
(solid film)

1715

ν (cm⁻¹)

## Mass Spectrum

119

91

M⁺·
226

C₁₅H₁₄O₂

% of base peak

m/e

## ¹³C NMR Spectrum
(150.0 MHz, CDCl₃ solution)

DEPT   CH₂↓ CH₃↑ CH↑

expansion

130   129   128   ppm

expansion

130   129   128   ppm

proton decoupled

solvent

δ (ppm)

## ¹H NMR Spectrum
(600 MHz, CDCl₃ solution)

expansion

8.0   7.6   ppm

Expansion Scale

0   80  160   Hz

TMS

δ (ppm)

## UV Spectrum

$\log_{10} \varepsilon > 5$

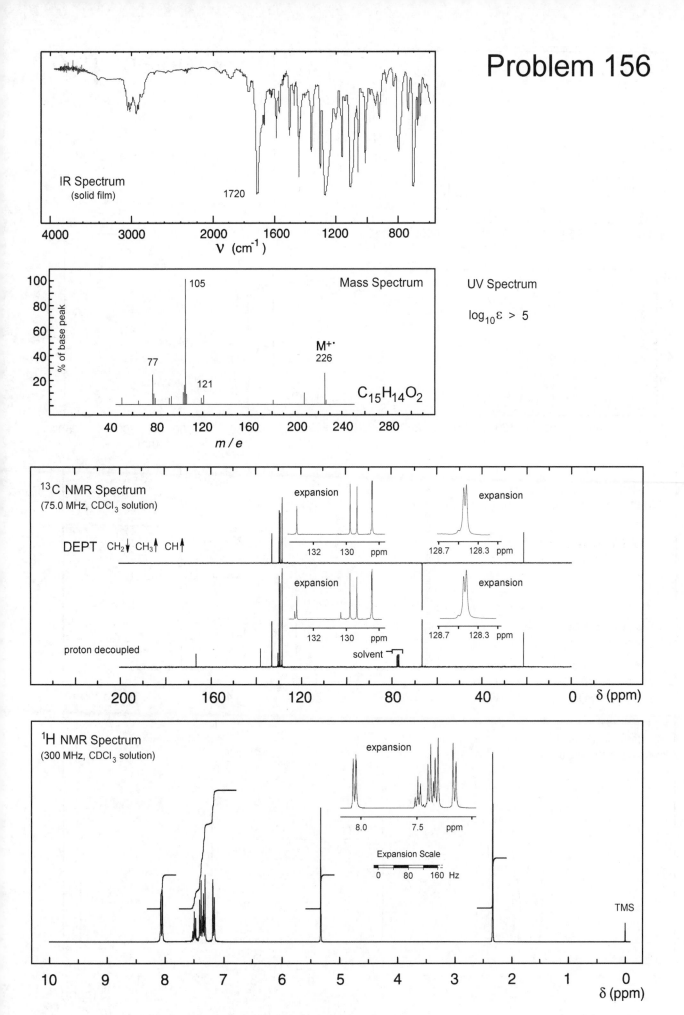

Problem 156

IR Spectrum
(solid film)

1720

ν (cm⁻¹)

Mass Spectrum

105

77

121

M⁺·
226

$C_{15}H_{14}O_2$

% of base peak

m/e

UV Spectrum

$\log_{10}\varepsilon > 5$

¹³C NMR Spectrum
(75.0 MHz, CDCl₃ solution)

DEPT  CH₂↓ CH₃↑ CH↑

expansion

expansion

132    130    ppm

128.7    128.3    ppm

proton decoupled

solvent

δ (ppm)

¹H NMR Spectrum
(300 MHz, CDCl₃ solution)

expansion

8.0    7.5    ppm

Expansion Scale

0    80    160  Hz

TMS

δ (ppm)

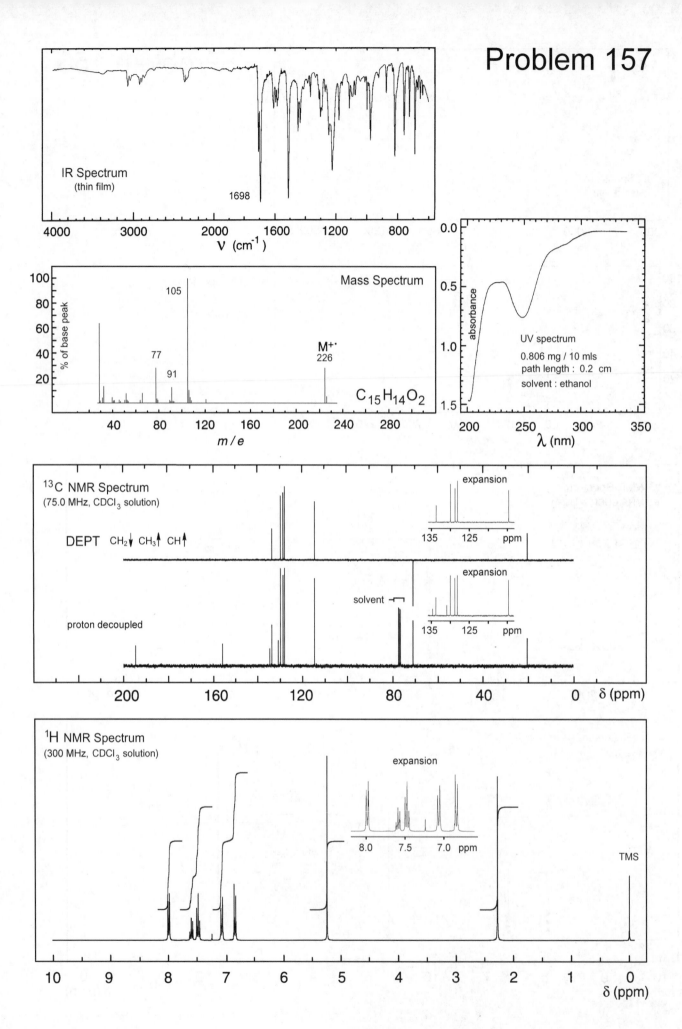

# Problem 157

IR Spectrum
(thin film)

1698

$\nu$ (cm$^{-1}$)

Mass Spectrum

105

77

91

M$^{+\cdot}$
226

$C_{15}H_{14}O_2$

% of base peak

$m/e$

UV spectrum
0.806 mg / 10 mls
path length : 0.2 cm
solvent : ethanol

absorbance

$\lambda$ (nm)

$^{13}$C NMR Spectrum
(75.0 MHz, CDCl$_3$ solution)

DEPT    CH$_2\downarrow$ CH$_3\uparrow$ CH$\uparrow$

expansion

proton decoupled

solvent

expansion

$\delta$ (ppm)

$^1$H NMR Spectrum
(300 MHz, CDCl$_3$ solution)

expansion

TMS

$\delta$ (ppm)

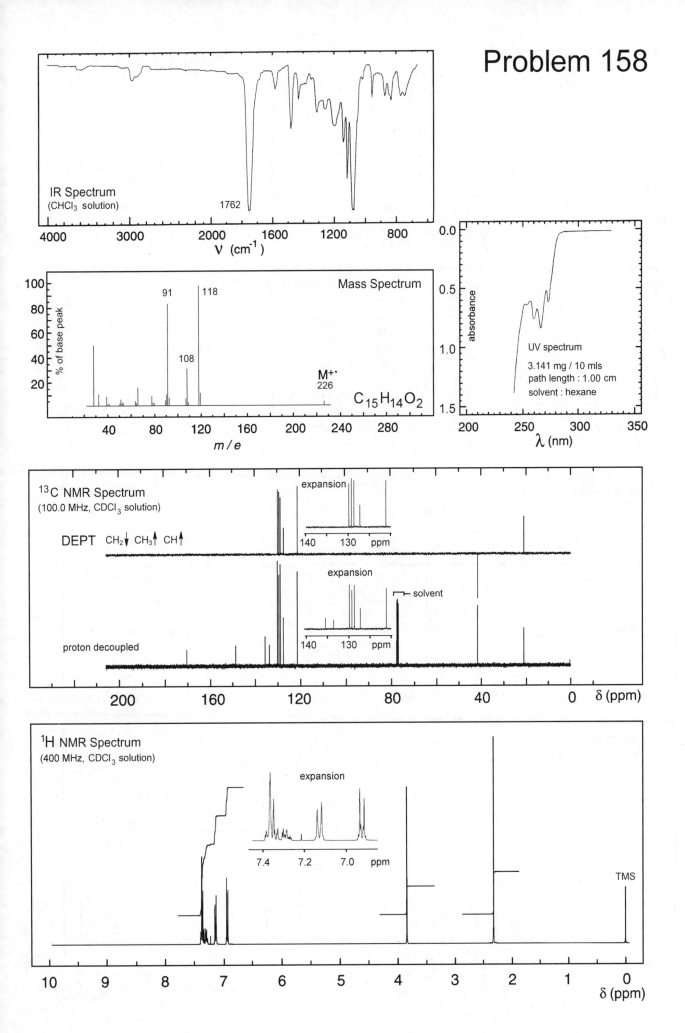

# Problem 158

**IR Spectrum**
(CHCl₃ solution)

1762

ν (cm⁻¹)

**Mass Spectrum**

% of base peak

91
118
108
M⁺˙
226

$C_{15}H_{14}O_2$

m/e

**UV spectrum**

3.141 mg / 10 mls
path length : 1.00 cm
solvent : hexane

absorbance

λ (nm)

**¹³C NMR Spectrum**
(100.0 MHz, CDCl₃ solution)

DEPT  CH₂↓ CH₃↑ CH↑

expansion

140  130  ppm

expansion

solvent

140  130  ppm

proton decoupled

δ (ppm)

**¹H NMR Spectrum**
(400 MHz, CDCl₃ solution)

expansion

7.4  7.2  7.0  ppm

TMS

δ (ppm)

# Problem 159

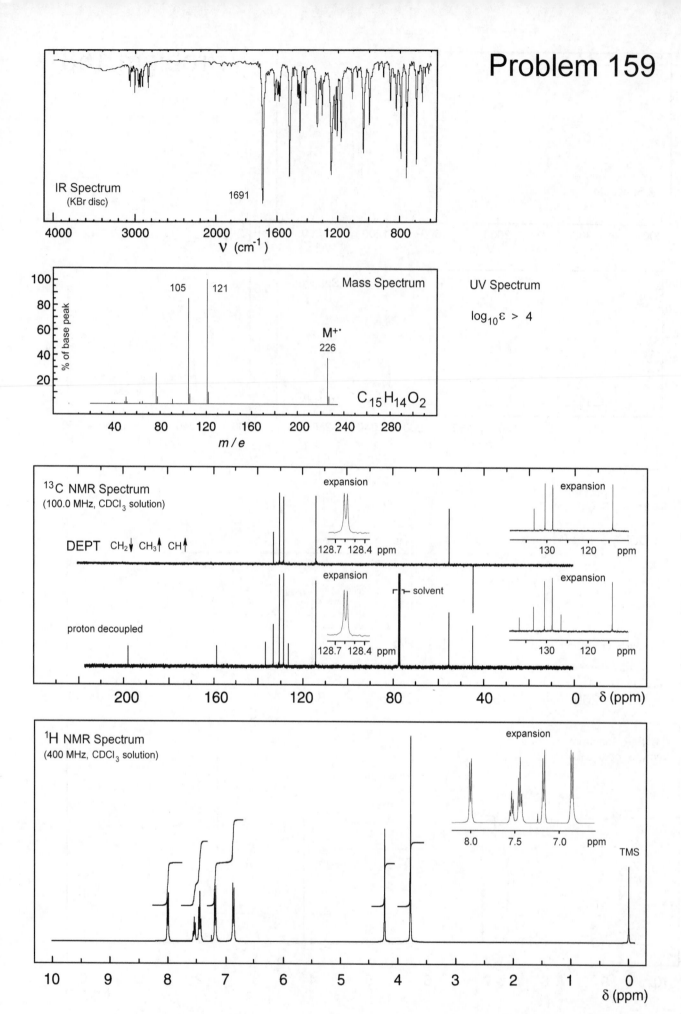

IR Spectrum
(KBr disc)

1691

$\nu$ (cm$^{-1}$)

Mass Spectrum

105

121

M$^{+\cdot}$
226

$C_{15}H_{14}O_2$

% of base peak

m/e

UV Spectrum

$\log_{10}\varepsilon > 4$

$^{13}$C NMR Spectrum
(100.0 MHz, CDCl$_3$ solution)

DEPT   CH$_2\downarrow$ CH$_3\uparrow$ CH$\uparrow$

expansion

128.7   128.4 ppm

expansion

130   120 ppm

proton decoupled

expansion

128.7   128.4 ppm

solvent

expansion

130   120 ppm

$\delta$ (ppm)

$^1$H NMR Spectrum
(400 MHz, CDCl$_3$ solution)

expansion

8.0   7.5   7.0 ppm

TMS

$\delta$ (ppm)

# Problem 160

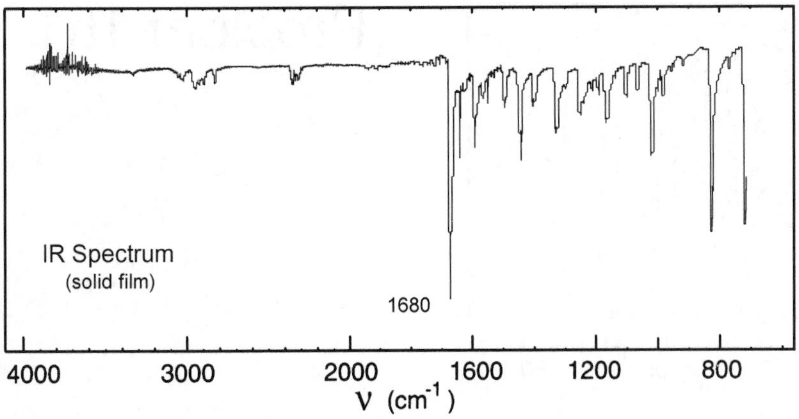

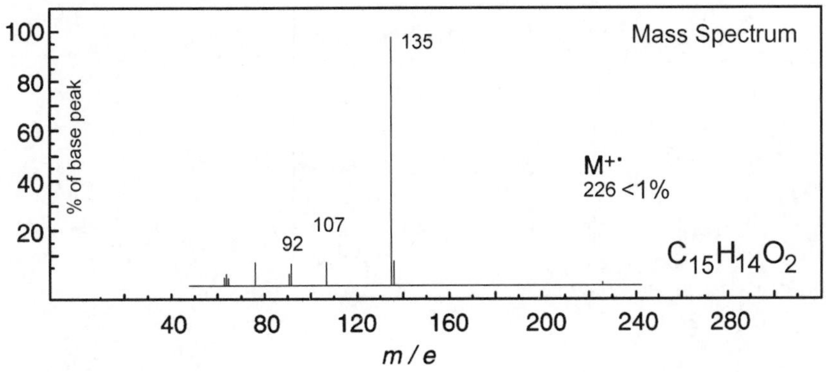

IR Spectrum
(solid film)

1680

ν (cm⁻¹)

$V \ (cm^{-1})$

Mass Spectrum

% of base peak

135

92
107

M⁺˙
226 <1%

$C_{15}H_{14}O_2$

m/e

$m/e$

UV Spectrum

$log_{10}ε > 4$

¹³C NMR Spectrum
(75.0 MHz, CDCl₃ solution)

DEPT   CH₂↓ CH₃↑ CH↑

expansion

134   130   ppm

expansion

134   130   ppm

proton decoupled

solvent

δ (ppm)

¹H NMR Spectrum
(300 MHz, CDCl₃ solution)

expansion

8.0   7.5   7.0   ppm

TMS

δ (ppm)

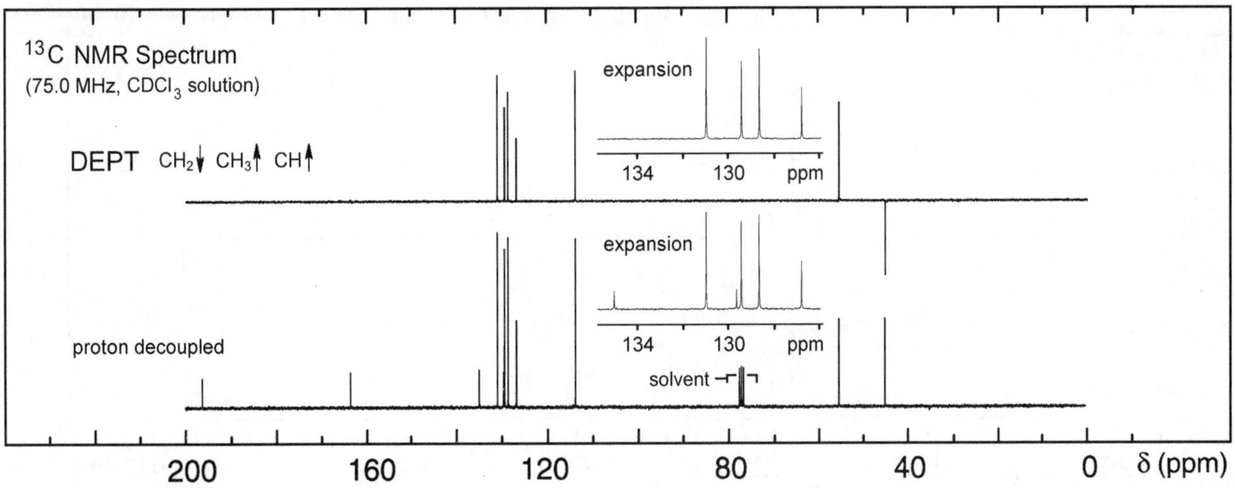

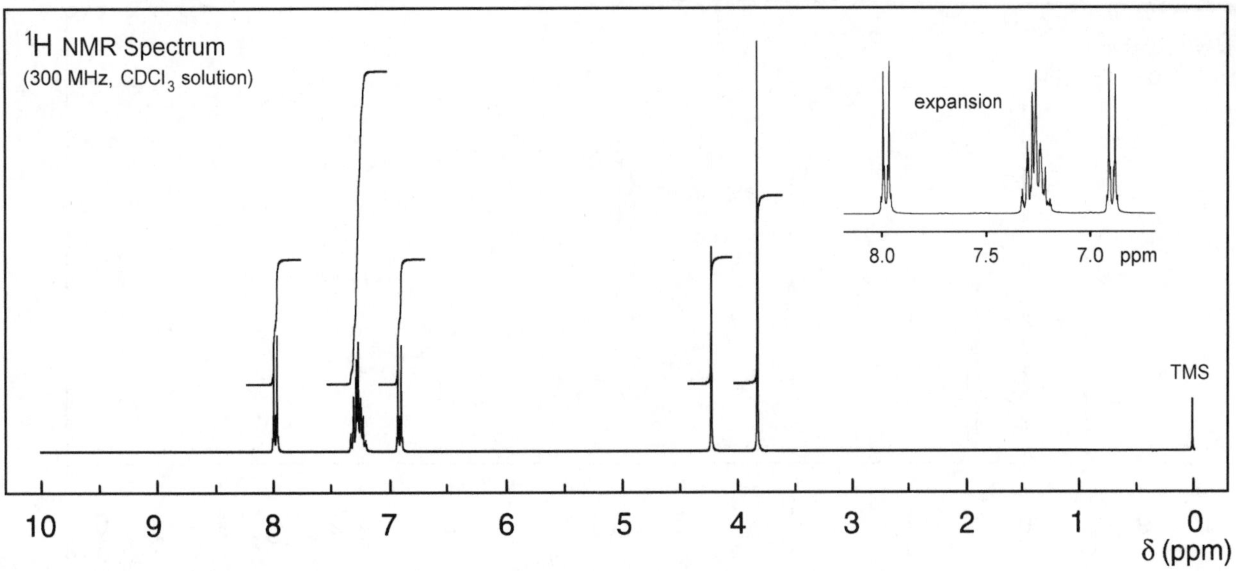

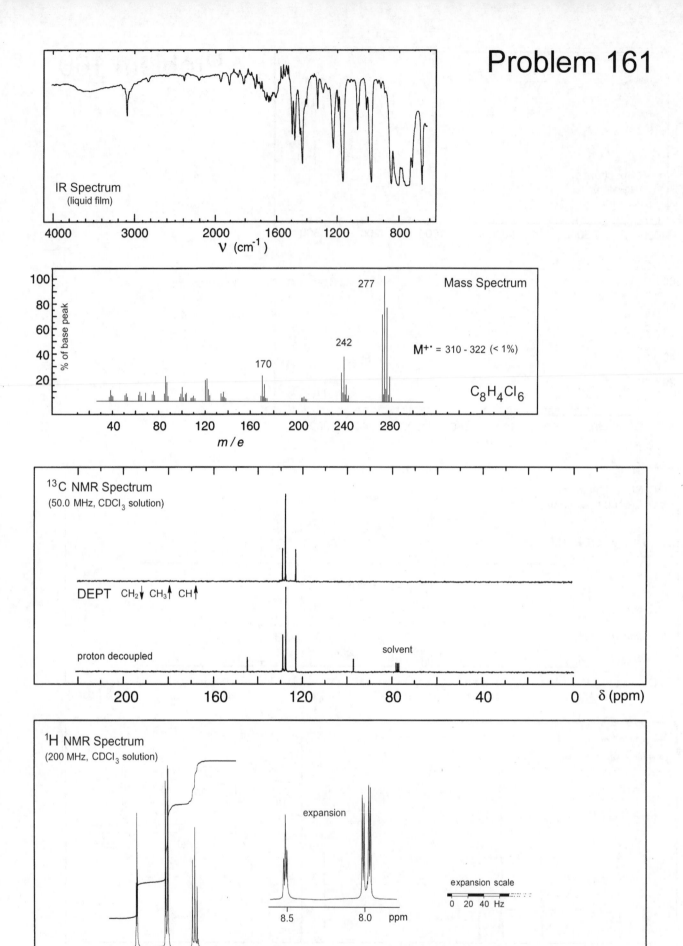

# Problem 161

**IR Spectrum**
(liquid film)

ν (cm⁻¹)

**Mass Spectrum**

277

242

170

M⁺˙ = 310 - 322 (< 1%)

$C_8H_4Cl_6$

% of base peak

m / e

**¹³C NMR Spectrum**
(50.0 MHz, CDCl₃ solution)

DEPT   CH₂↓ CH₃↑ CH↑

proton decoupled

solvent

δ (ppm)

**¹H NMR Spectrum**
(200 MHz, CDCl₃ solution)

expansion

expansion scale

0   20   40 Hz

8.5        8.0    ppm

δ (ppm)

# Problem 162

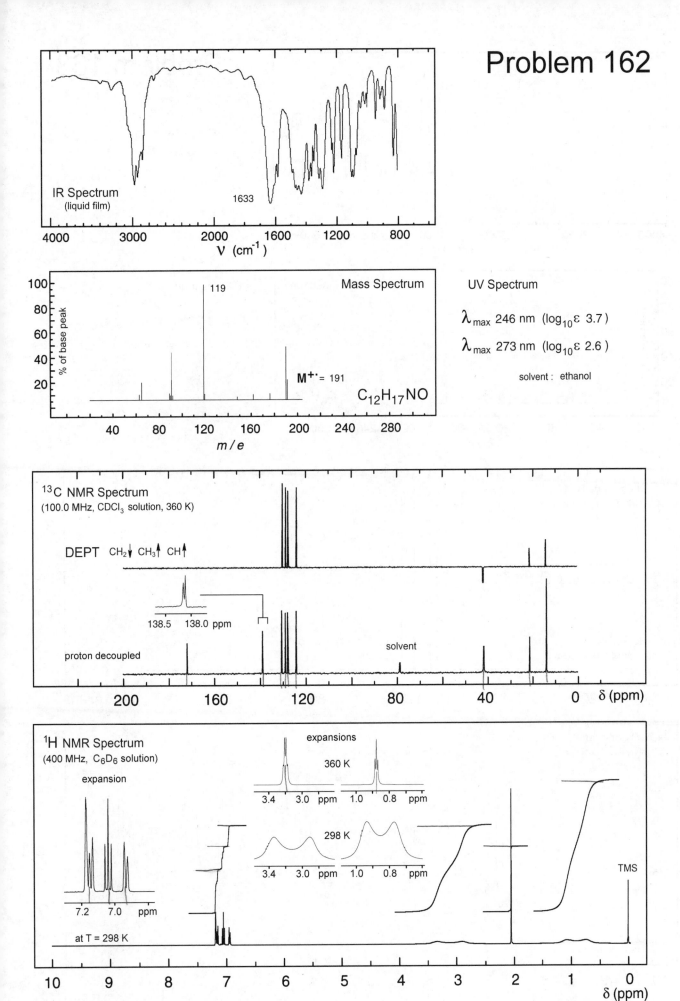

IR Spectrum
(liquid film)

1633

ν (cm⁻¹)

Mass Spectrum

119

% of base peak

M⁺· = 191

C₁₂H₁₇NO

m/e

UV Spectrum

λ_max 246 nm (log₁₀ε 3.7)

λ_max 273 nm (log₁₀ε 2.6)

solvent : ethanol

¹³C NMR Spectrum
(100.0 MHz, CDCl₃ solution, 360 K)

DEPT  CH₂↓ CH₃↑ CH↑

138.5   138.0 ppm

proton decoupled

solvent

δ (ppm)

¹H NMR Spectrum
(400 MHz, C₆D₆ solution)

expansions

360 K

3.4   3.0 ppm   1.0   0.8 ppm

expansion

298 K

3.4   3.0 ppm   1.0   0.8 ppm

7.2   7.0 ppm

at T = 298 K

TMS

δ (ppm)

# Problem 163

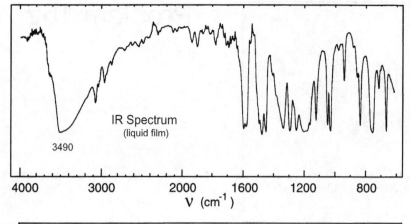

IR Spectrum
(liquid film)

3490

ν (cm⁻¹)

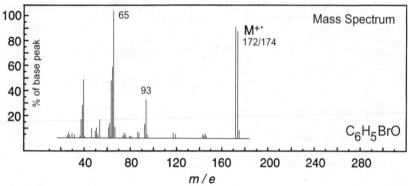

Mass Spectrum

65

93

M⁺˙
172/174

$C_6H_5BrO$

% of base peak

m/e

UV Spectrum

$\lambda_{max}$ 277 nm (log$_{10}\varepsilon$ 3.4)

solvent : methanol

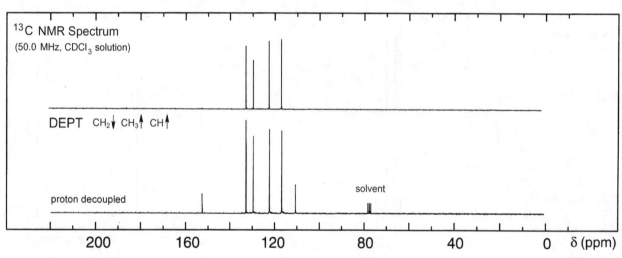

¹³C NMR Spectrum
(50.0 MHz, CDCl₃ solution)

DEPT CH₂↓ CH₃↑ CH↑

proton decoupled

solvent

δ (ppm)

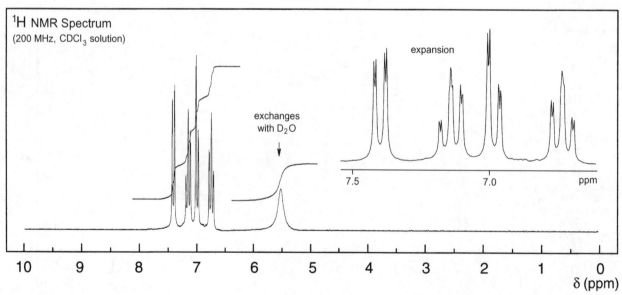

¹H NMR Spectrum
(200 MHz, CDCl₃ solution)

exchanges
with D₂O

expansion

δ (ppm)

# Problem 164

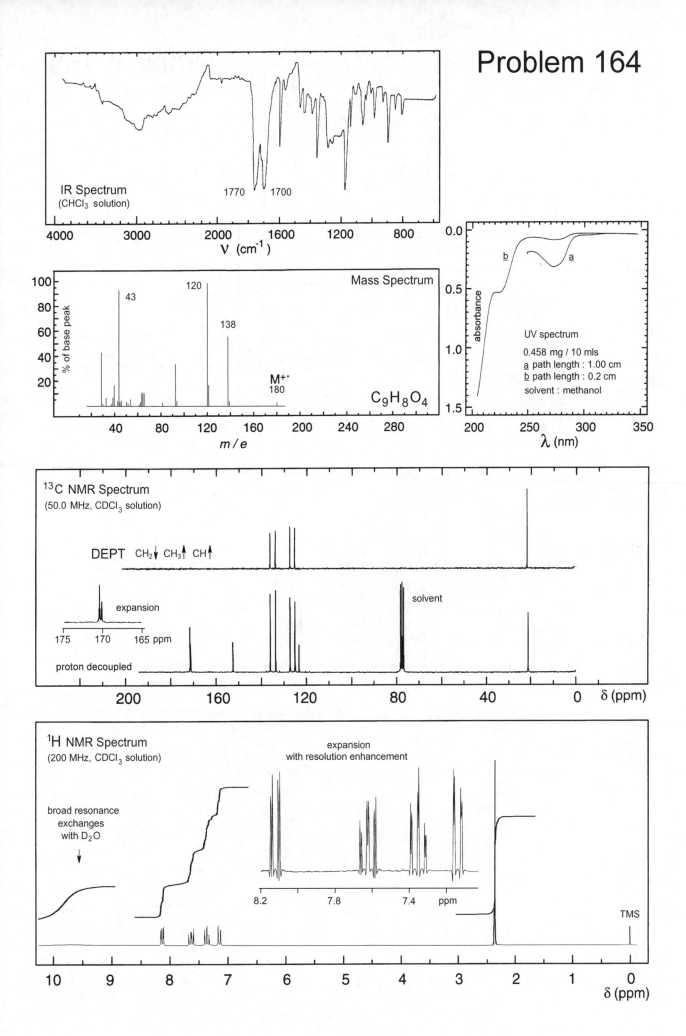

IR Spectrum
(CHCl$_3$ solution)
1770   1700

$\nu$ (cm$^{-1}$)

Mass Spectrum

43
120
138

M$^{+\cdot}$
180

C$_9$H$_8$O$_4$

% of base peak

$m/e$

UV spectrum

0.458 mg / 10 mls
a path length : 1.00 cm
b path length : 0.2 cm
solvent : methanol

absorbance

$\lambda$ (nm)

$^{13}$C NMR Spectrum
(50.0 MHz, CDCl$_3$ solution)

DEPT   CH$_2\downarrow$ CH$_3\uparrow$ CH$\uparrow$

expansion

175   170   165 ppm

solvent

proton decoupled

$\delta$ (ppm)

$^{1}$H NMR Spectrum
(200 MHz, CDCl$_3$ solution)

expansion
with resolution enhancement

broad resonance
exchanges
with D$_2$O

8.2   7.8   7.4   ppm

TMS

$\delta$ (ppm)

# Problem 165

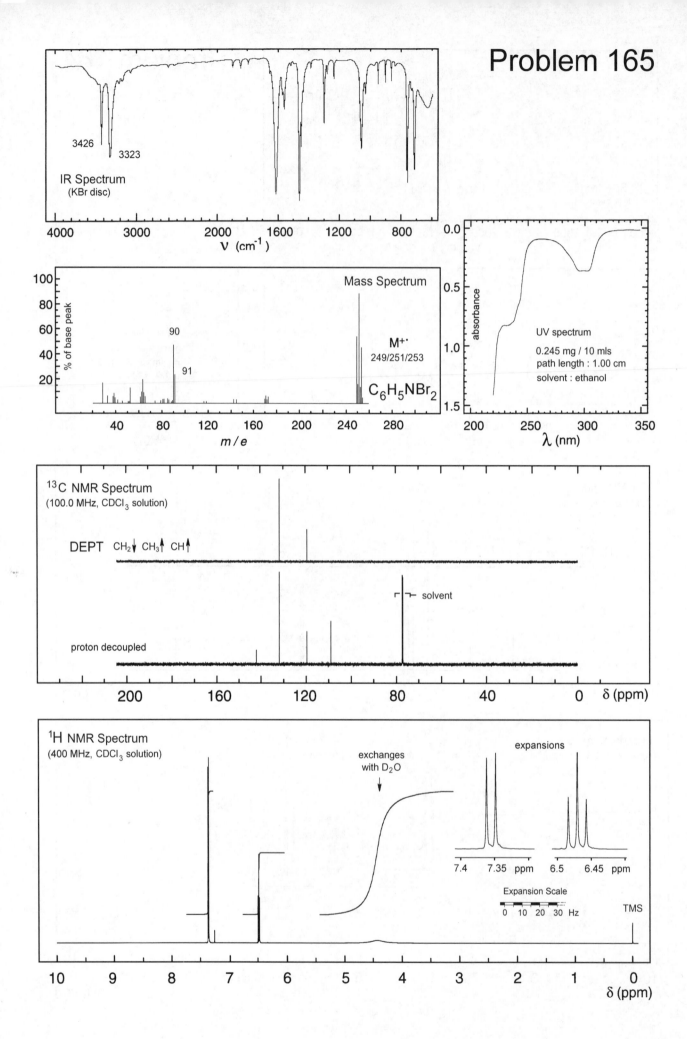

IR Spectrum
(KBr disc)

3426
3323

ν (cm⁻¹)

Mass Spectrum

90
91

M⁺·
249/251/253

C₆H₅NBr₂

% of base peak

m/e

UV spectrum

0.245 mg / 10 mls
path length : 1.00 cm
solvent : ethanol

absorbance

λ (nm)

¹³C NMR Spectrum
(100.0 MHz, CDCl₃ solution)

DEPT  CH₂↓ CH₃↑ CH↑

solvent

proton decoupled

δ (ppm)

¹H NMR Spectrum
(400 MHz, CDCl₃ solution)

exchanges
with D₂O

expansions

7.4   7.35  ppm

6.5   6.45 ppm

Expansion Scale

0  10  20  30  Hz

TMS

δ (ppm)

# Problem 166

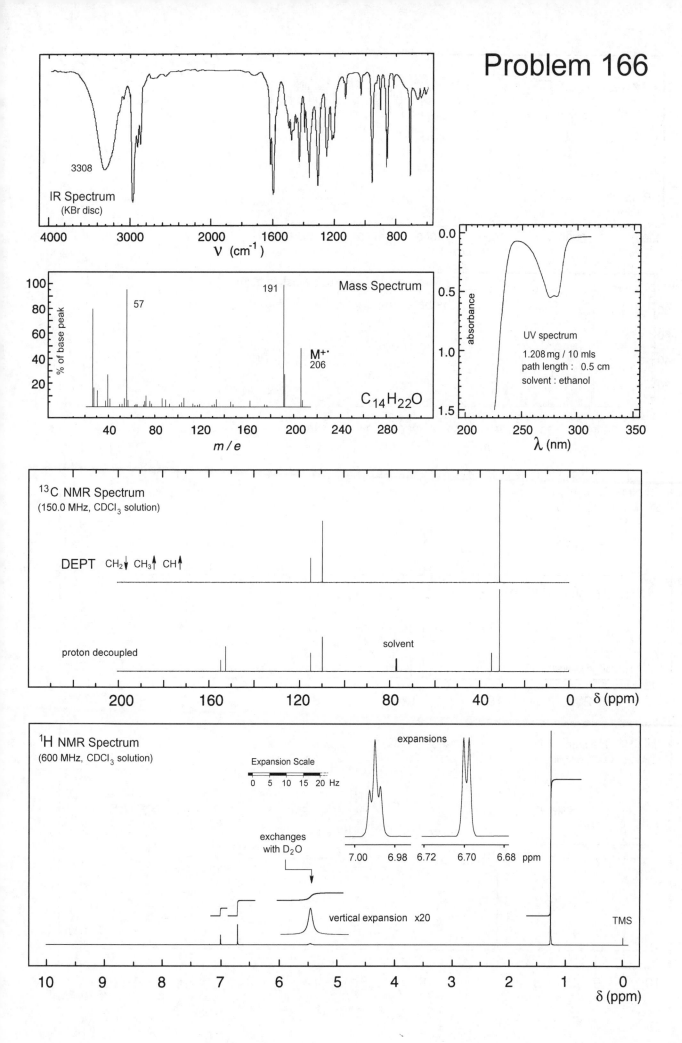

IR Spectrum
(KBr disc)

3308

ν (cm⁻¹)

Mass Spectrum

191

57

M⁺˙
206

C₁₄H₂₂O

% of base peak

m/e

UV spectrum

1.208 mg / 10 mls
path length : 0.5 cm
solvent : ethanol

absorbance

λ (nm)

¹³C NMR Spectrum
(150.0 MHz, CDCl₃ solution)

DEPT   CH₂↓ CH₃↑ CH↑

proton decoupled

solvent

δ (ppm)

¹H NMR Spectrum
(600 MHz, CDCl₃ solution)

Expansion Scale

0   5   10   15   20  Hz

expansions

7.00    6.98         6.72    6.70    6.68  ppm

exchanges
with D₂O

vertical expansion  x20

TMS

δ (ppm)

# Problem 167

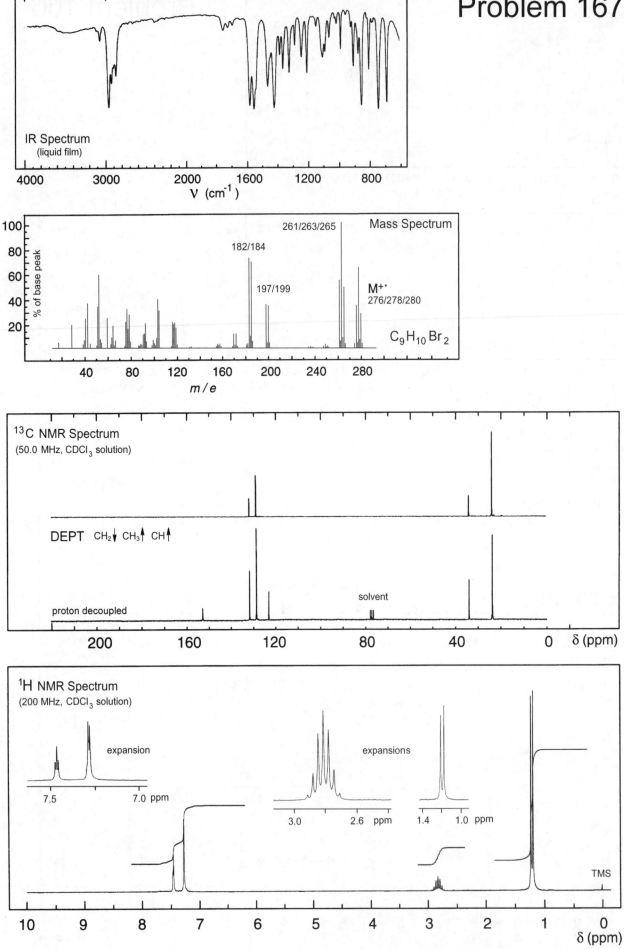

IR Spectrum
(liquid film)

4000  3000  2000  1600  1200  800
ν (cm⁻¹)

Mass Spectrum

261/263/265

182/184

197/199

M⁺˙
276/278/280

C₉H₁₀Br₂

% of base peak

100
80
60
40
20

40  80  120  160  200  240  280
m/e

¹³C NMR Spectrum
(50.0 MHz, CDCl₃ solution)

DEPT   CH₂↓ CH₃↑ CH↑

solvent

proton decoupled

200  160  120  80  40  0  δ (ppm)

¹H NMR Spectrum
(200 MHz, CDCl₃ solution)

expansion

7.5  7.0 ppm

expansions

3.0  2.6 ppm

1.4  1.0 ppm

TMS

10  9  8  7  6  5  4  3  2  1  0
δ (ppm)

278

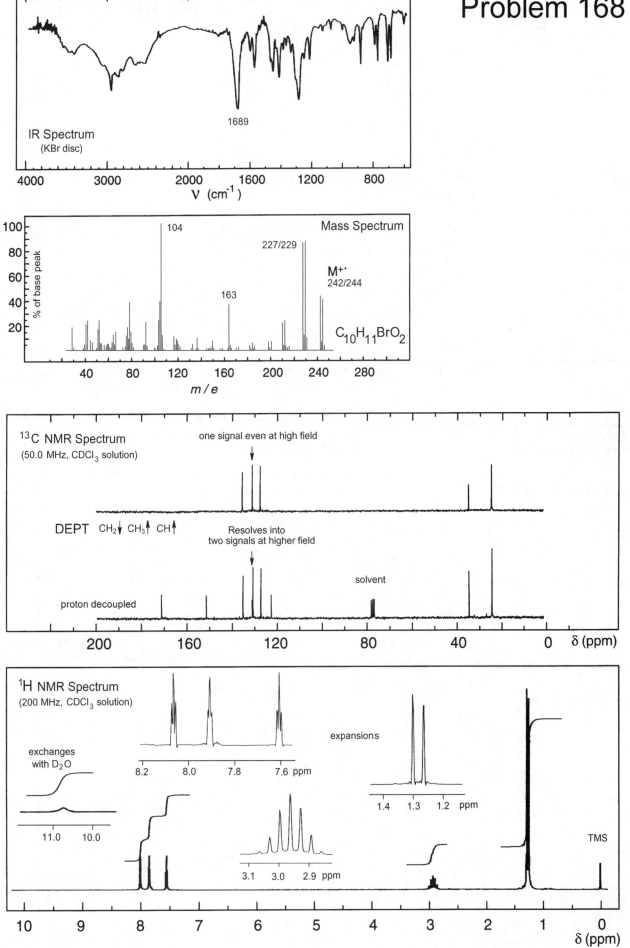

# Problem 168

**IR Spectrum**
(KBr disc)

1689

ν (cm⁻¹)

**Mass Spectrum**

104

227/229

M⁺·
242/244

163

C₁₀H₁₁BrO₂

% of base peak

m/e

**¹³C NMR Spectrum**
(50.0 MHz, CDCl₃ solution)

one signal even at high field

**DEPT**  CH₂↓ CH₃↑ CH↑

Resolves into
two signals at higher field

solvent

proton decoupled

δ (ppm)

**¹H NMR Spectrum**
(200 MHz, CDCl₃ solution)

expansions

exchanges
with D₂O

8.2   8.0   7.8   7.6 ppm

1.4   1.3   1.2   ppm

11.0   10.0

3.1   3.0   2.9 ppm

TMS

δ (ppm)

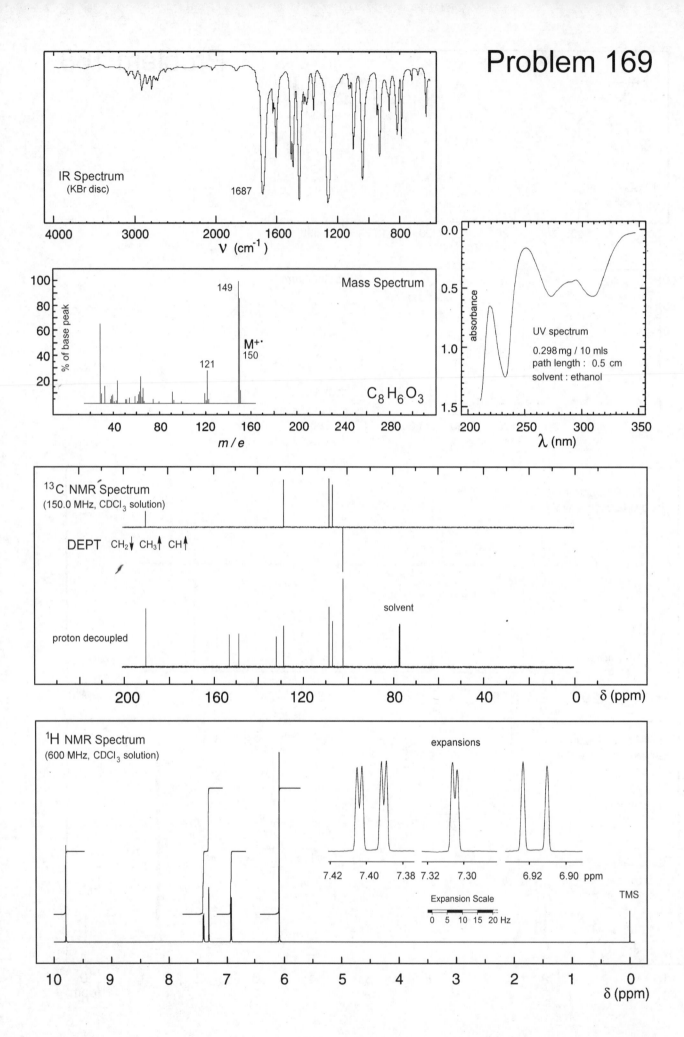

IR Spectrum
(KBr disc)

1687

ν (cm⁻¹)

Mass Spectrum

149

121

M⁺·
150

C₈H₆O₃

% of base peak

m/e

UV spectrum

0.298 mg / 10 mls
path length : 0.5 cm
solvent : ethanol

absorbance

λ (nm)

¹³C NMR Spectrum
(150.0 MHz, CDCl₃ solution)

DEPT   CH₂↓ CH₃↑ CH↑

proton decoupled

solvent

δ (ppm)

¹H NMR Spectrum
(600 MHz, CDCl₃ solution)

expansions

7.42  7.40  7.38   7.32  7.30      6.92  6.90 ppm

Expansion Scale

0  5  10  15  20 Hz

TMS

δ (ppm)

# Problem 170

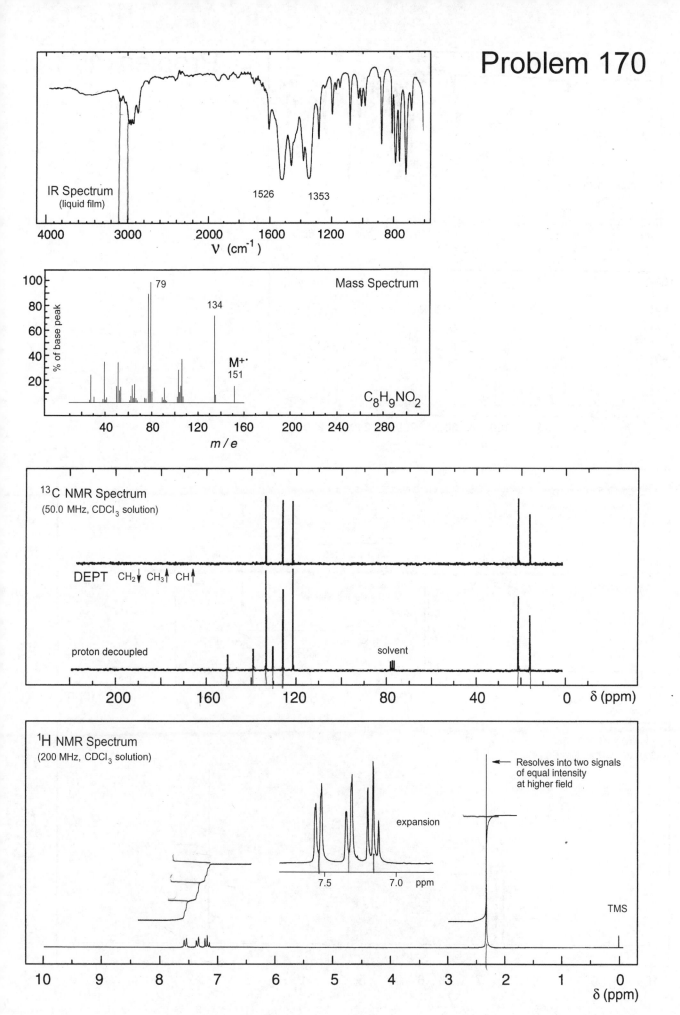

IR Spectrum
(liquid film)

1526    1353

ν (cm⁻¹)

Mass Spectrum

% of base peak

79
134

M⁺·
151

C₈H₉NO₂

m/e

¹³C NMR Spectrum
(50.0 MHz, CDCl₃ solution)

DEPT    CH₂↓ CH₃↑ CH↑

proton decoupled          solvent

δ (ppm)

¹H NMR Spectrum
(200 MHz, CDCl₃ solution)

Resolves into two signals
of equal intensity
at higher field

expansion

7.5        7.0    ppm

TMS

δ (ppm)

# Problem 171

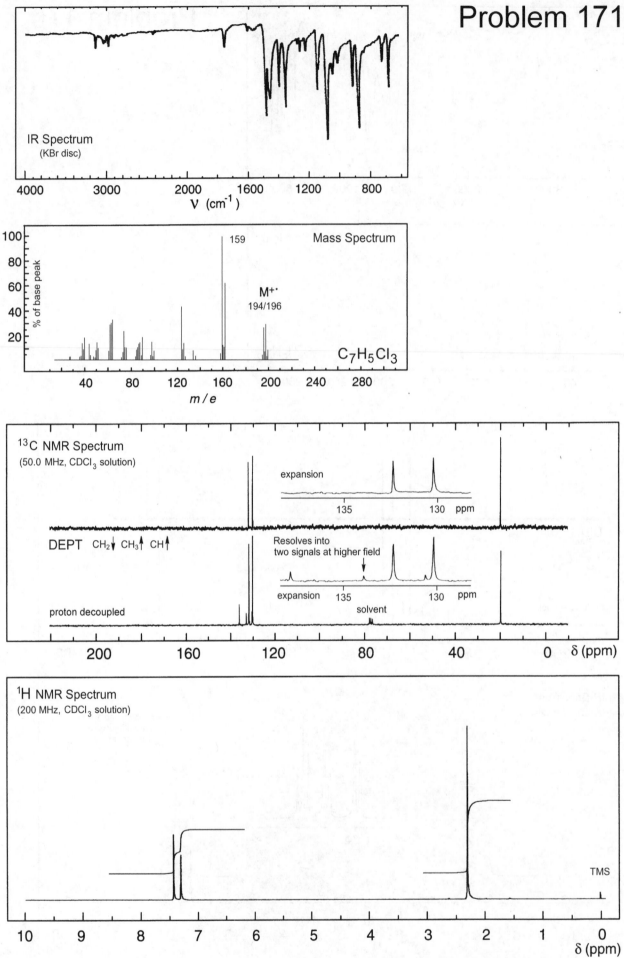

IR Spectrum
(KBr disc)

$\nu$ (cm$^{-1}$)

Mass Spectrum

% of base peak

159

M$^{+\cdot}$
194/196

C$_7$H$_5$Cl$_3$

$m/e$

$^{13}$C NMR Spectrum
(50.0 MHz, CDCl$_3$ solution)

expansion

DEPT   CH$_2$↓ CH$_3$↑ CH↑

Resolves into
two signals at higher field

expansion

proton decoupled

solvent

$\delta$ (ppm)

$^1$H NMR Spectrum
(200 MHz, CDCl$_3$ solution)

TMS

$\delta$ (ppm)

# Problem 172

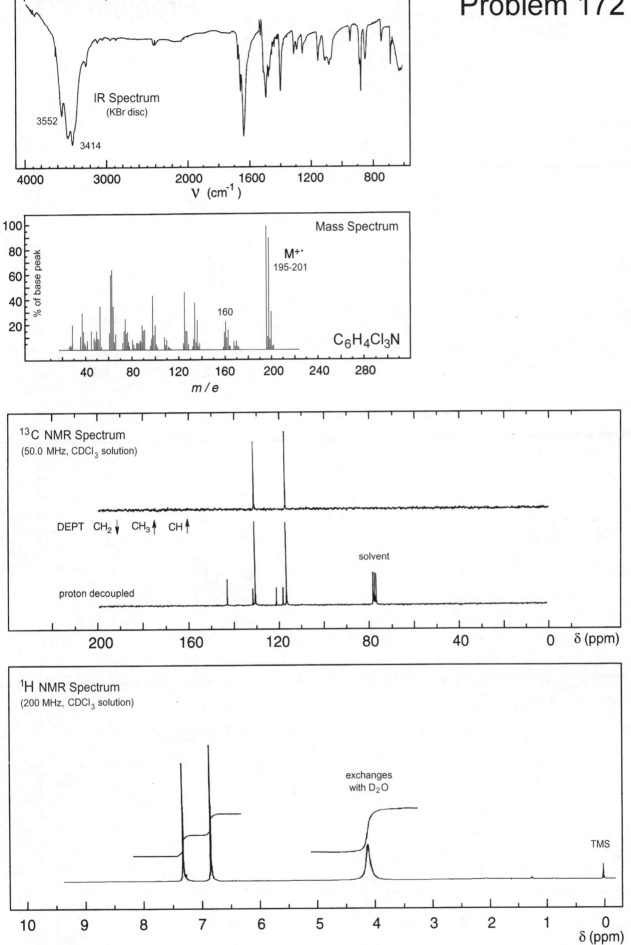

IR Spectrum
(KBr disc)

3552

3414

Mass Spectrum

% of base peak

M$^{+\cdot}$
195-201

160

$C_6H_4Cl_3N$

$^{13}$C NMR Spectrum
(50.0 MHz, CDCl$_3$ solution)

DEPT   CH$_2$↓   CH$_3$↑   CH↑

solvent

proton decoupled

δ (ppm)

$^1$H NMR Spectrum
(200 MHz, CDCl$_3$ solution)

exchanges
with D$_2$O

TMS

δ (ppm)

# Problem 173

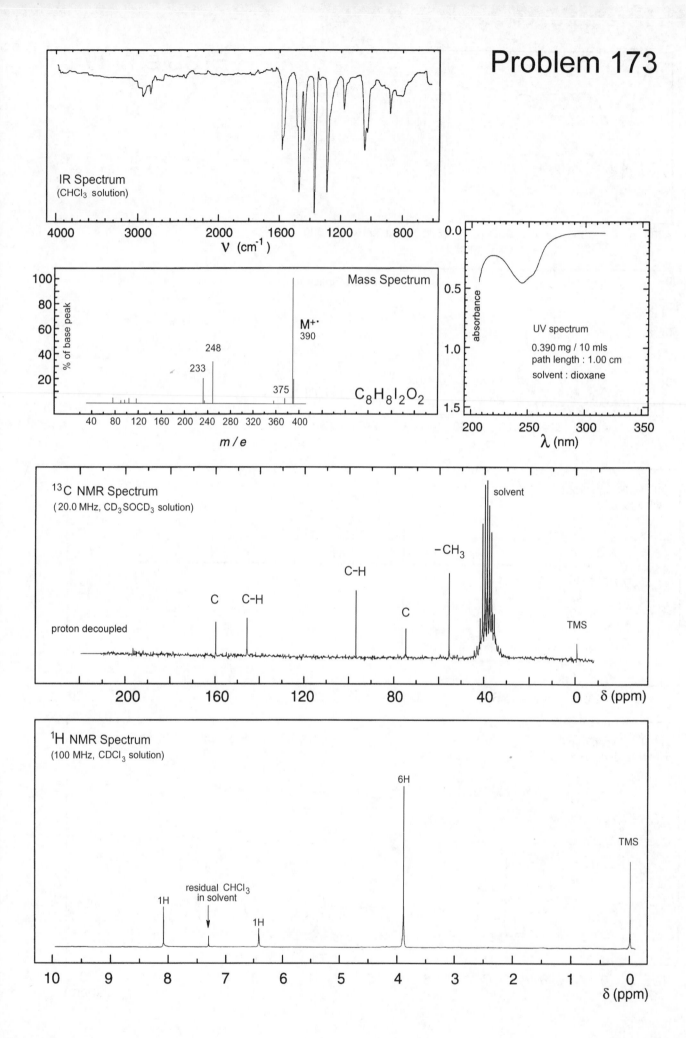

IR Spectrum
(CHCl₃ solution)

ν (cm⁻¹)

Mass Spectrum

M⁺·
390

248

233

375

C₈H₈I₂O₂

% of base peak

m/e

UV spectrum

0.390 mg / 10 mls
path length : 1.00 cm

solvent : dioxane

absorbance

λ (nm)

¹³C NMR Spectrum
( 20.0 MHz, CD₃SOCD₃ solution)

solvent

−CH₃

C−H

C

C−H

C

proton decoupled

TMS

δ (ppm)

¹H NMR Spectrum
(100 MHz, CDCl₃ solution)

6H

residual CHCl₃
in solvent

1H

1H

TMS

δ (ppm)

# Problem 174

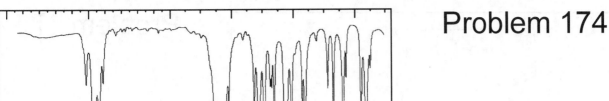

IR Spectrum
(liquid film)

1685

ν (cm⁻¹)

$\nu$ (cm$^{-1}$)

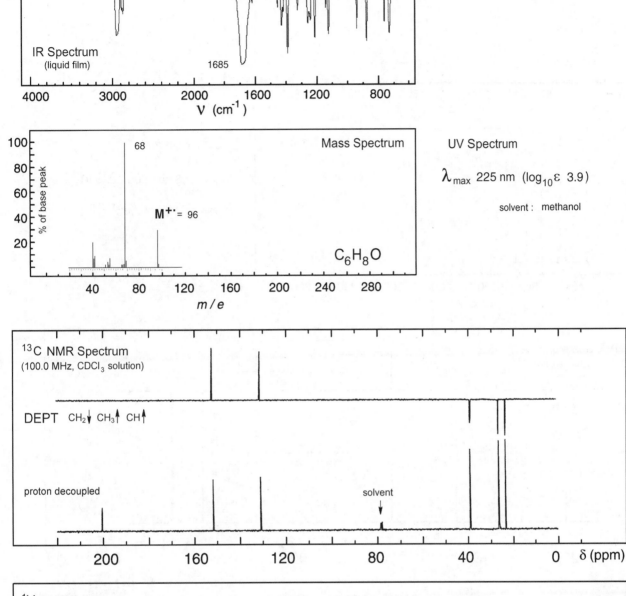

Mass Spectrum

68

**M⁺˙** = 96

$C_6H_8O$

% of base peak

m/e

$m/e$

UV Spectrum

$\lambda_{max}$ 225 nm (log₁₀ε 3.9)

$\lambda_{max}$ 225 nm ($\log_{10}\varepsilon$ 3.9)

solvent : methanol

¹³C NMR Spectrum
(100.0 MHz, CDCl₃ solution)

DEPT  CH₂↓ CH₃↑ CH↑

proton decoupled

solvent

δ (ppm)

$\delta$ (ppm)

¹H NMR Spectrum
(400 MHz, CDCl₃ solution)

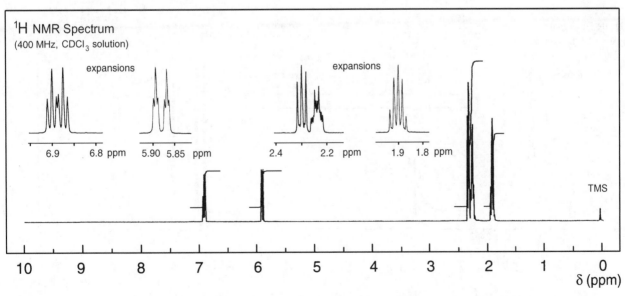

expansions

expansions

6.9    6.8 ppm     5.90 5.85 ppm     2.4    2.2 ppm     1.9    1.8 ppm

TMS

δ (ppm)

$\delta$ (ppm)

# Problem 175

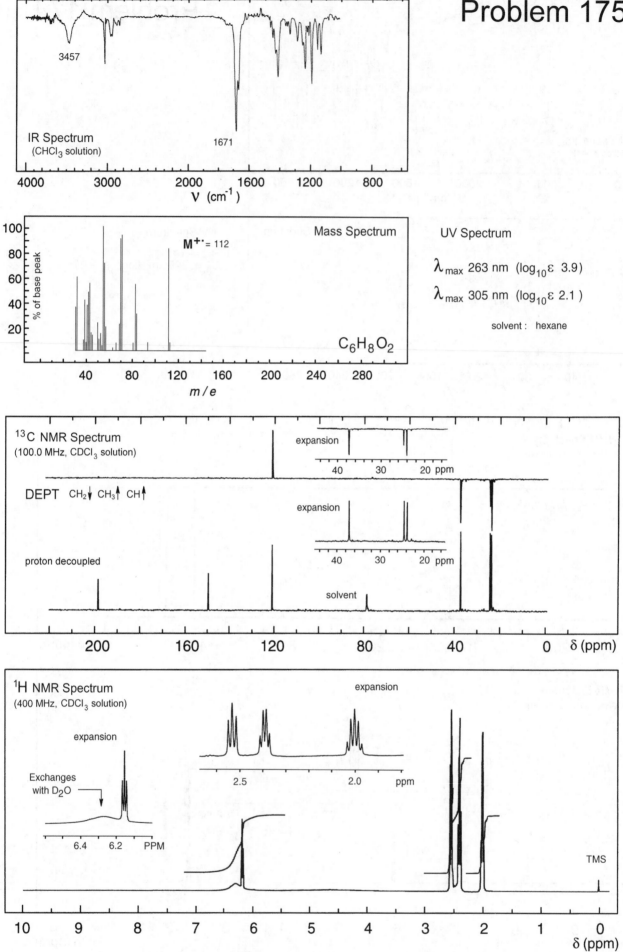

IR Spectrum
(CHCl₃ solution)
3457
1671
ν (cm⁻¹)

Mass Spectrum
M⁺˙ = 112
% of base peak
m/e
$C_6H_8O_2$

UV Spectrum
$\lambda_{max}$ 263 nm (log₁₀ε 3.9)
$\lambda_{max}$ 305 nm (log₁₀ε 2.1)
solvent : hexane

¹³C NMR Spectrum
(100.0 MHz, CDCl₃ solution)
expansion
DEPT  CH₂↓ CH₃↑ CH↑
expansion
proton decoupled
solvent
δ (ppm)

¹H NMR Spectrum
(400 MHz, CDCl₃ solution)
expansion
expansion
Exchanges with D₂O
TMS
δ (ppm)

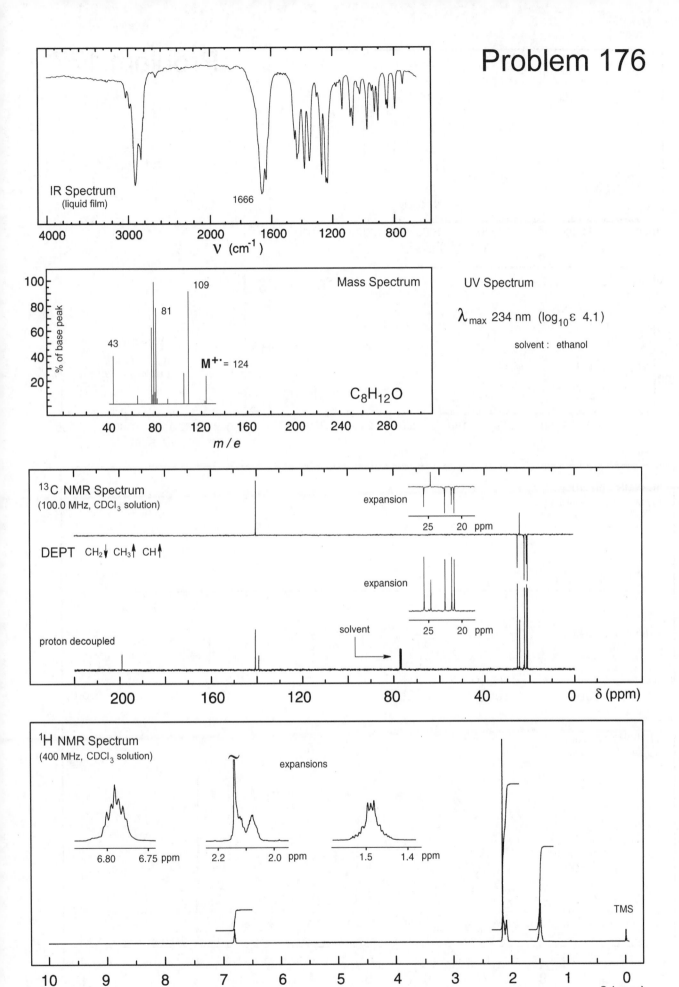

IR Spectrum
(liquid film)
1666

ν (cm⁻¹)

Mass Spectrum

UV Spectrum

$\lambda_{max}$ 234 nm (log$_{10}\varepsilon$ 4.1)

solvent : ethanol

100
80
60
40
20
% of base peak

43
81
109
M⁺˙ = 124

$C_8H_{12}O$

m/e

¹³C NMR Spectrum
(100.0 MHz, CDCl₃ solution)

expansion
25    20  ppm

DEPT   CH₂↓ CH₃↑ CH↑

expansion
25    20  ppm

proton decoupled

solvent

δ (ppm)

¹H NMR Spectrum
(400 MHz, CDCl₃ solution)

expansions

6.80    6.75 ppm

2.2    2.0  ppm

1.5    1.4  ppm

TMS

δ (ppm)

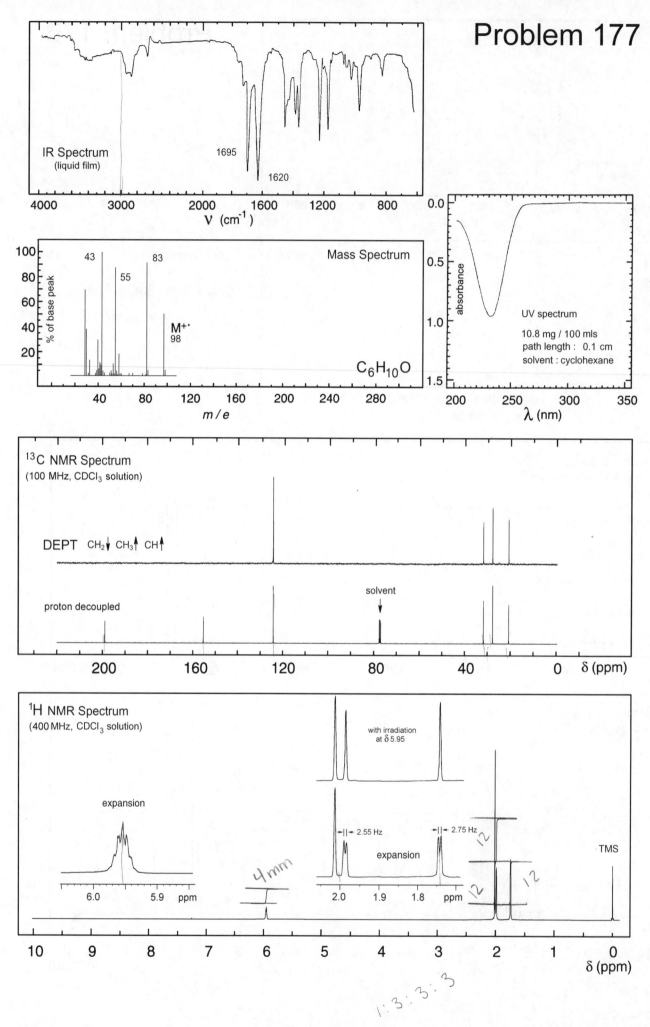

# Problem 177

**IR Spectrum** (liquid film)

1695
1620

ν (cm⁻¹)

**Mass Spectrum**

43
55
83
M⁺·
98

C₆H₁₀O

m/e

**UV spectrum**

10.8 mg / 100 mls
path length : 0.1 cm
solvent : cyclohexane

λ (nm)

**¹³C NMR Spectrum** (100 MHz, CDCl₃ solution)

DEPT CH₂↓ CH₃↑ CH↑

proton decoupled

solvent

δ (ppm)

**¹H NMR Spectrum** (400 MHz, CDCl₃ solution)

expansion

with irradiation at δ 5.95

4 mm

2.55 Hz

2.75 Hz

expansion

TMS

δ (ppm)

1 : 3 : 3 : 3

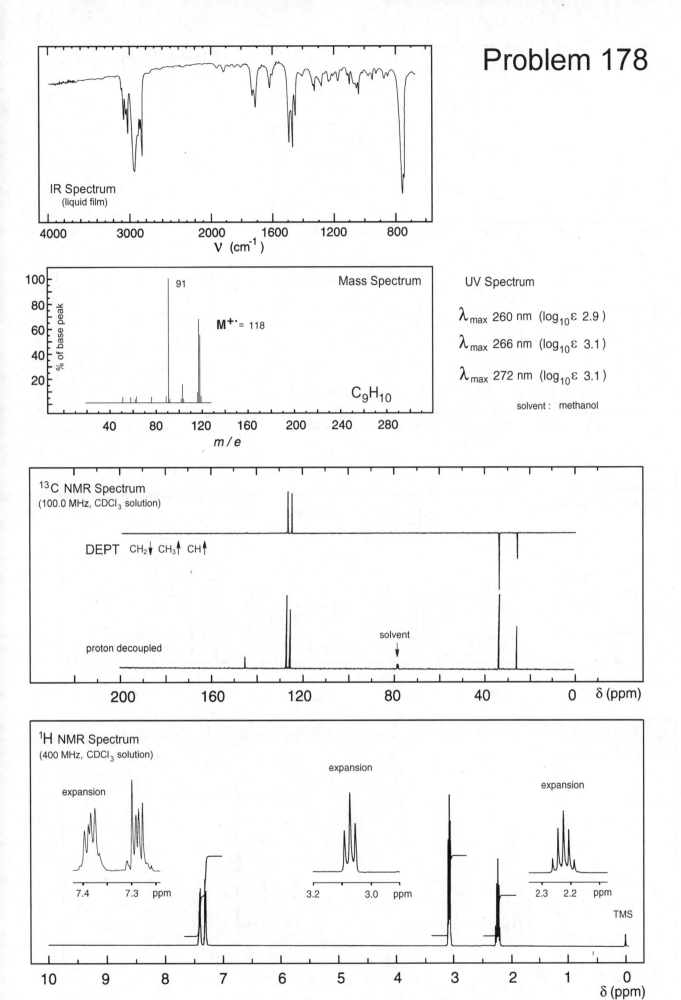

# Problem 178

**IR Spectrum**
(liquid film)

ν (cm⁻¹) → $\nu$ (cm$^{-1}$)

Mass Spectrum

91

$M^{+\cdot} = 118$

$C_9H_{10}$

m/e

UV Spectrum

$\lambda_{max}$ 260 nm (log$_{10}\varepsilon$ 2.9)

$\lambda_{max}$ 266 nm (log$_{10}\varepsilon$ 3.1)

$\lambda_{max}$ 272 nm (log$_{10}\varepsilon$ 3.1)

solvent : methanol

$^{13}$C NMR Spectrum
(100.0 MHz, CDCl$_3$ solution)

DEPT  CH$_2$↓ CH$_3$↑ CH↑

proton decoupled

solvent

δ (ppm)

$^1$H NMR Spectrum
(400 MHz, CDCl$_3$ solution)

expansion

7.4   7.3  ppm

expansion

3.2   3.0  ppm

expansion

2.3   2.2  ppm

TMS

δ (ppm)

289

# Problem 179

IR Spectrum
(CCl₄ Solution)

1719

ν (cm⁻¹)

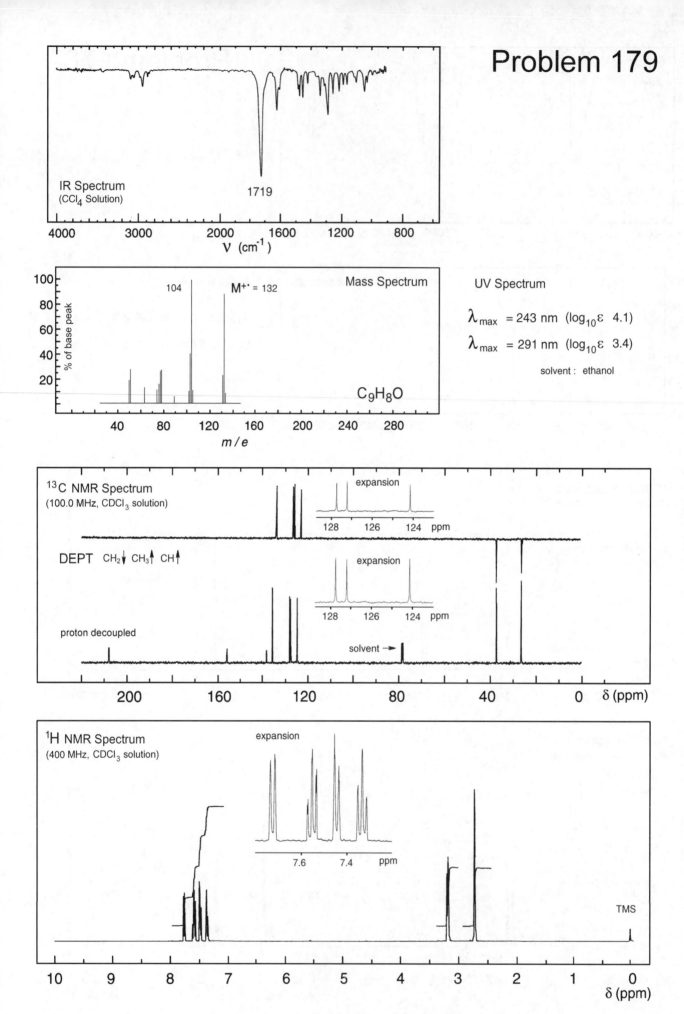

104    M⁺˙ = 132    Mass Spectrum

% of base peak

C₉H₈O

m / e

UV Spectrum

$\lambda_{max}$ = 243 nm (log₁₀ε  4.1)

$\lambda_{max}$ = 291 nm (log₁₀ε  3.4)

solvent : ethanol

¹³C NMR Spectrum
(100.0 MHz, CDCl₃ solution)

expansion

DEPT  CH₂↓ CH₃↑ CH↑

expansion

proton decoupled

solvent →

δ (ppm)

¹H NMR Spectrum
(400 MHz, CDCl₃ solution)

expansion

TMS

δ (ppm)

# Problem 180

IR Spectrum
(CHCl$_3$ solution)

1751

$\nu$ (cm$^{-1}$)

Mass Spectrum

104

% of base peak

M$^{+\cdot}$=132

C$_9$H$_8$O

$m/e$

UV Spectrum

$\lambda_{max}$ 268 nm (log$_{10}\varepsilon$ 3.1)

$\lambda_{max}$ 275 nm (log$_{10}\varepsilon$ 3.1)

solvent : ethanol

$^{13}$C NMR Spectrum
(100.0 MHz, CDCl$_3$ solution)

DEPT  CH$_2\downarrow$ CH$_3\uparrow$ CH$\uparrow$

proton decoupled

solvent

$\delta$ (ppm)

$^1$H NMR Spectrum
(400 MHz, CDCl$_3$ solution)

TMS

$\delta$ (ppm)

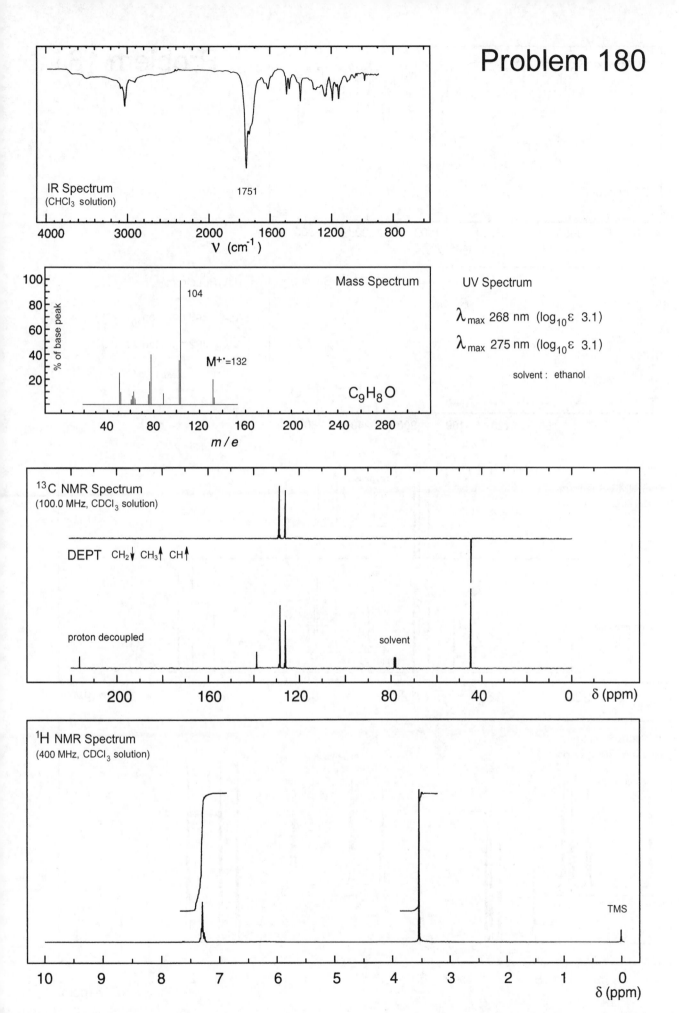

# Problem 181

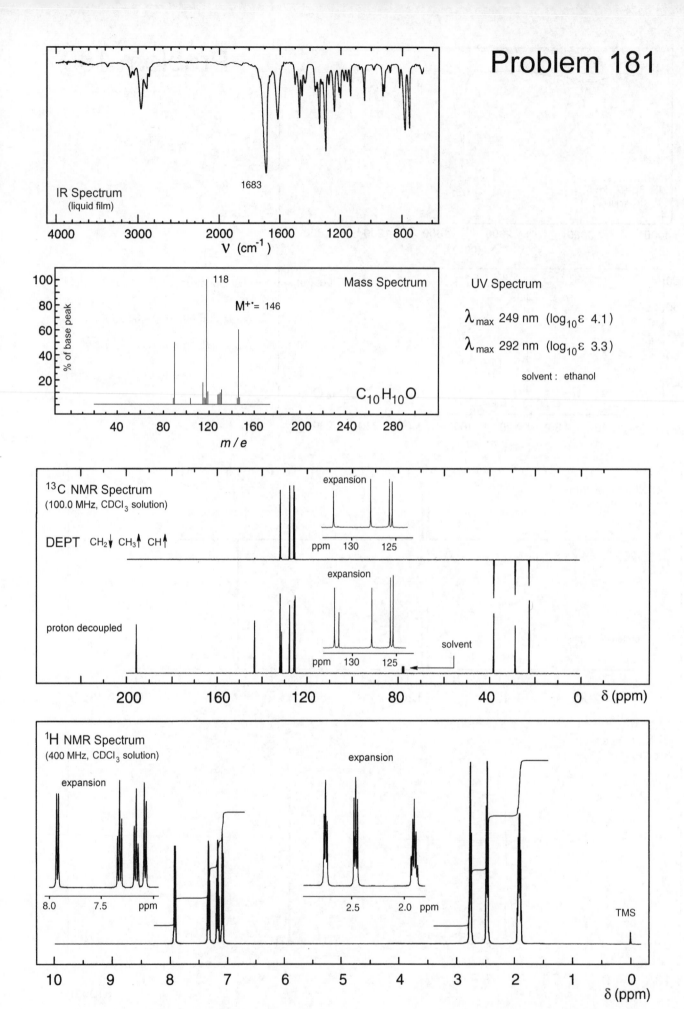

IR Spectrum
(liquid film)

1683

Mass Spectrum

118

M+• = 146

C₁₀H₁₀O

UV Spectrum

$\lambda_{max}$ 249 nm (log₁₀ε 4.1)

$\lambda_{max}$ 292 nm (log₁₀ε 3.3)

solvent : ethanol

¹³C NMR Spectrum
(100.0 MHz, CDCl₃ solution)

DEPT   CH₂↓ CH₃↑ CH↑

expansion

proton decoupled

expansion

solvent

¹H NMR Spectrum
(400 MHz, CDCl₃ solution)

expansion

expansion

TMS

# Problem 182

IR Spectrum
(liquid film)

1716

$\nu$ (cm$^{-1}$)

Mass Spectrum

104

$M^{+ \cdot}$ = 146

$C_{10}H_{10}O$

% of base peak

$m/e$

UV Spectrum

$\lambda_{max}$ 265 nm (log$_{10}\varepsilon$ 3.0)

$\lambda_{max}$ 272 nm (log$_{10}\varepsilon$ 3.1)

$\lambda_{max}$ 294 nm (log$_{10}\varepsilon$ 1.9)

solvent : ethanol

$^{13}$C NMR Spectrum
(100.0 MHz, CDCl$_3$ solution)

expansion

129   128   ppm

DEPT   CH$_2\downarrow$ CH$_3\uparrow$ CH$\uparrow$

expansion

129   128   ppm

proton decoupled

solvent

$\delta$ (ppm)

$^1$H NMR Spectrum
(400 MHz, CDCl$_3$ solution)

expansion

expansion

3.2   3.0  ppm

expansion

2.6   2.5  ppm

expansion

7.3   7.2   ppm

TMS

$\delta$ (ppm)

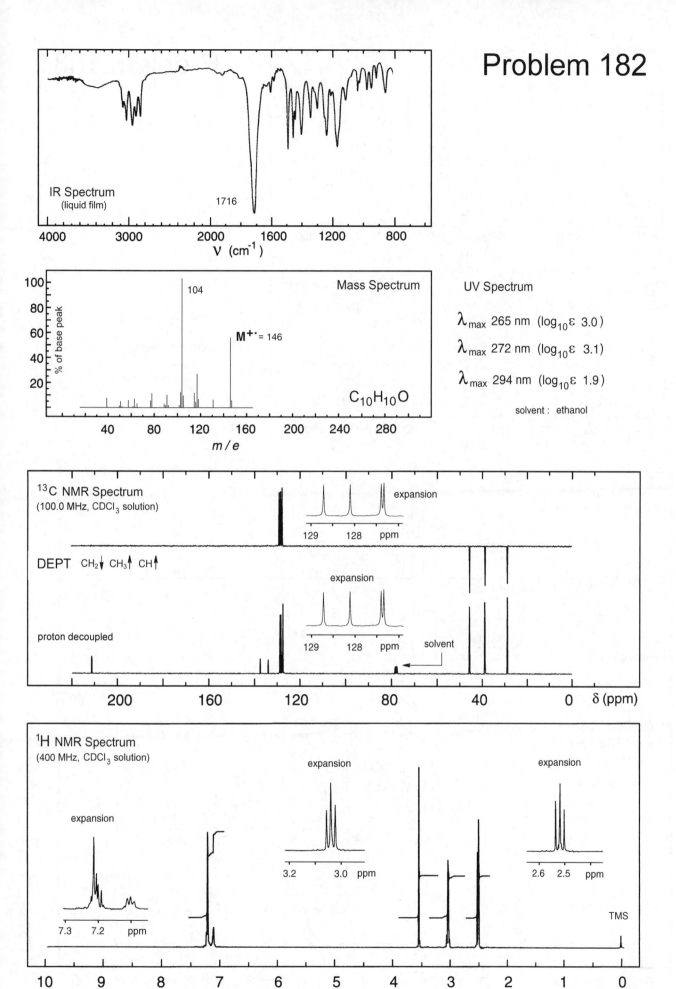

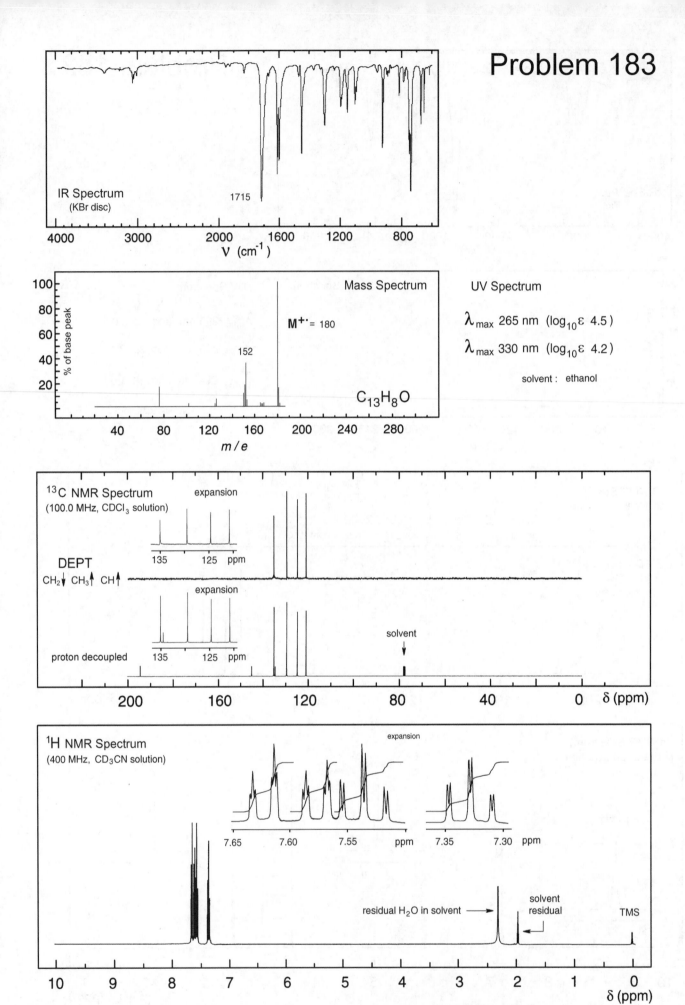

# Problem 183

IR Spectrum
(KBr disc)

1715

$\nu$ (cm$^{-1}$)

Mass Spectrum

M$^{+\cdot}$ = 180

152

% of base peak

$m/e$

$C_{13}H_8O$

UV Spectrum

$\lambda_{max}$ 265 nm (log$_{10}\varepsilon$ 4.5)

$\lambda_{max}$ 330 nm (log$_{10}\varepsilon$ 4.2)

solvent : ethanol

$^{13}C$ NMR Spectrum
(100.0 MHz, CDCl$_3$ solution)

expansion

DEPT
CH$_2\downarrow$ CH$_3\uparrow$ CH$\uparrow$

expansion

proton decoupled

solvent

$\delta$ (ppm)

$^1H$ NMR Spectrum
(400 MHz, CD$_3$CN solution)

expansion

residual H$_2$O in solvent

solvent
residual

TMS

$\delta$ (ppm)

# Problem 184

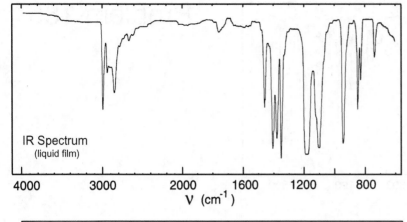

IR Spectrum
(liquid film)

$\nu$ (cm$^{-1}$)

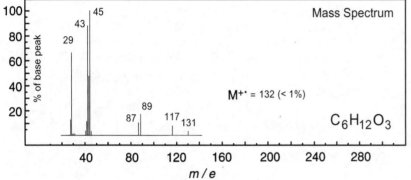

Mass Spectrum

% of base peak

29
43
45
87
89
117 131

M$^{+\cdot}$ = 132 (< 1%)

$C_6H_{12}O_3$

$m/e$

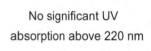

No significant UV
absorption above 220 nm

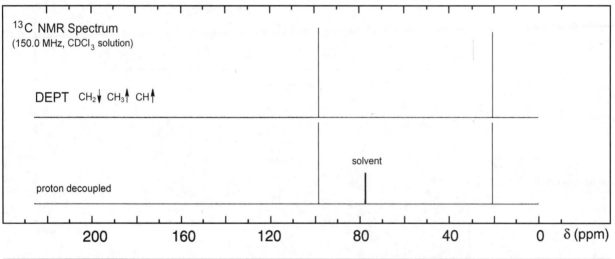

$^{13}$C NMR Spectrum
(150.0 MHz, CDCl$_3$ solution)

DEPT  CH$_2$↓ CH$_3$↑ CH↑

solvent

proton decoupled

$\delta$ (ppm)

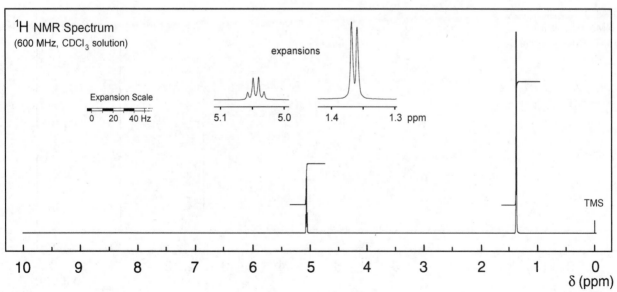

$^1$H NMR Spectrum
(600 MHz, CDCl$_3$ solution)

expansions

Expansion Scale

0  20  40 Hz

5.1  5.0

1.4  1.3 ppm

TMS

$\delta$ (ppm)

# Problem 185

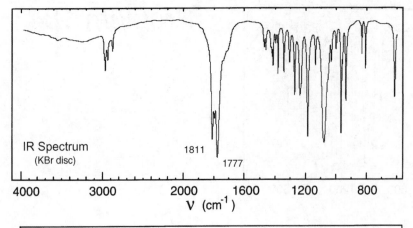

IR Spectrum
(KBr disc)

1811

1777

ν (cm⁻¹)

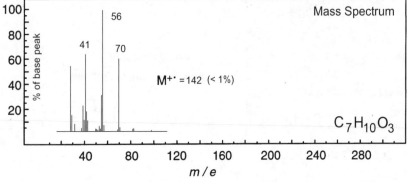

Mass Spectrum

56

41

70

$M^{+\cdot} = 142$ (< 1%)

% of base peak

$C_7H_{10}O_3$

m / e

No significant UV
absorption above 220 nm

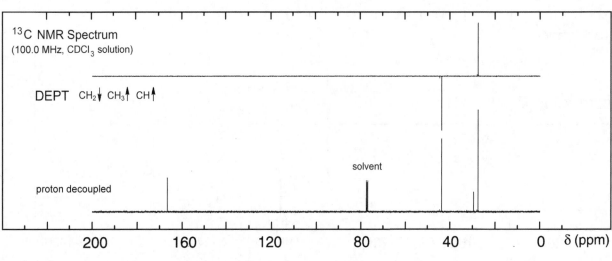

¹³C NMR Spectrum
(100.0 MHz, CDCl₃ solution)

DEPT  CH₂↓ CH₃↑ CH↑

solvent

proton decoupled

δ (ppm)

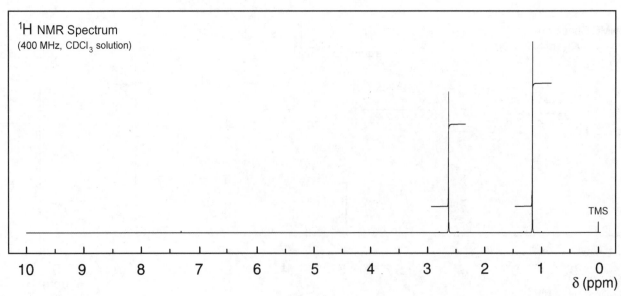

¹H NMR Spectrum
(400 MHz, CDCl₃ solution)

TMS

δ (ppm)

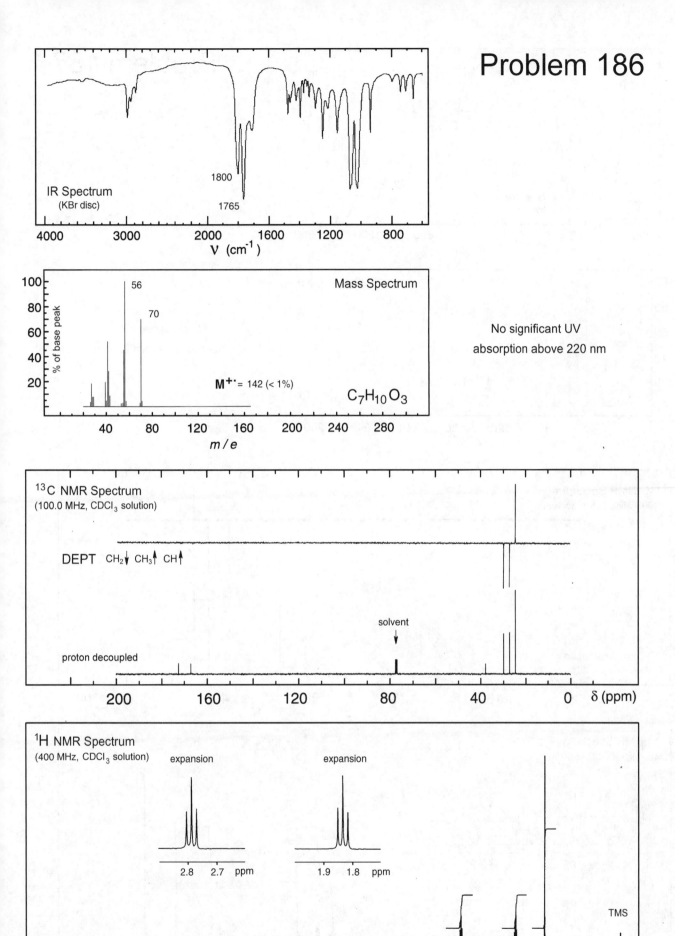

IR Spectrum
(KBr disc)

1800

1765

ν (cm⁻¹)

Mass Spectrum

% of base peak

56

70

**M⁺·** = 142 (< 1%)

$C_7H_{10}O_3$

m / e

No significant UV
absorption above 220 nm

¹³C NMR Spectrum
(100.0 MHz, CDCl₃ solution)

DEPT   CH₂↓ CH₃↑ CH↑

solvent

proton decoupled

δ (ppm)

¹H NMR Spectrum
(400 MHz, CDCl₃ solution)

expansion

expansion

2.8    2.7  ppm

1.9    1.8  ppm

TMS

δ (ppm)

# Problem 187

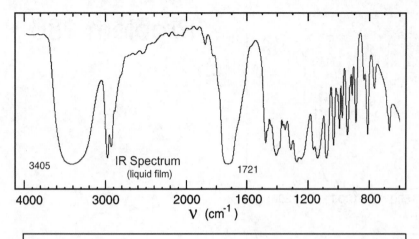

IR Spectrum
(liquid film)

3405

1721

$\nu$ (cm$^{-1}$)

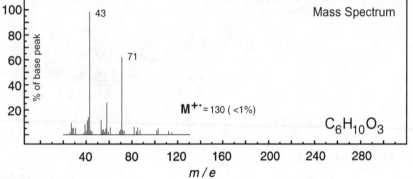

Mass Spectrum

43

71

M$^{+\cdot}$ = 130 ( <1%)

$C_6H_{10}O_3$

% of base peak

m / e

No significant UV
absorption above 220 nm

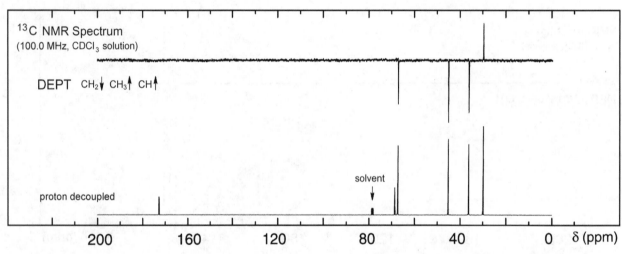

$^{13}$C NMR Spectrum
(100.0 MHz, CDCl$_3$ solution)

DEPT   CH$_2$↓ CH$_3$↑ CH↑

solvent

proton decoupled

$\delta$ (ppm)

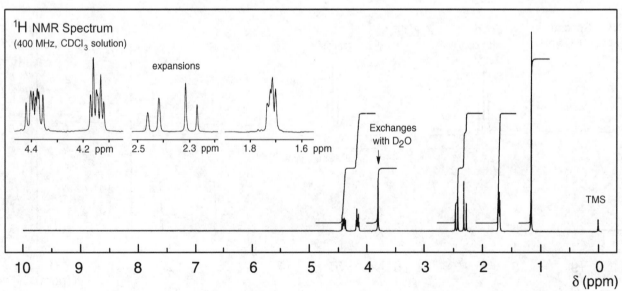

$^1$H NMR Spectrum
(400 MHz, CDCl$_3$ solution)

expansions

4.4   4.2 ppm      2.5   2.3 ppm      1.8   1.6 ppm

Exchanges
with D$_2$O

TMS

$\delta$ (ppm)

# Problem 188

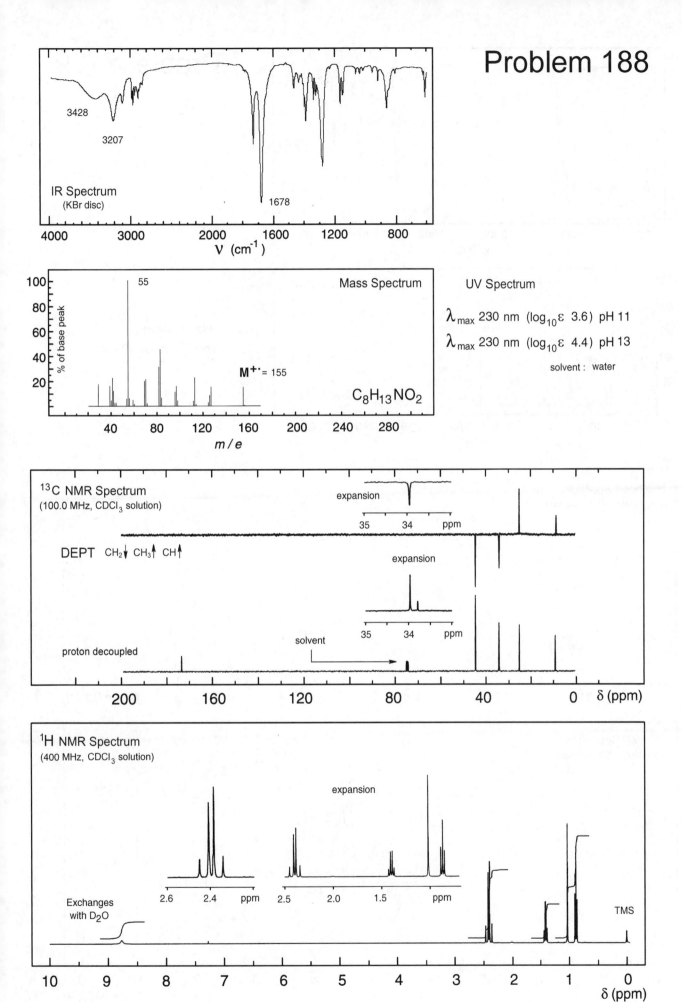

IR Spectrum
(KBr disc)

3428
3207
1678

ν (cm⁻¹)

Mass Spectrum

55

M⁺· = 155

C₈H₁₃NO₂

% of base peak

m/e

UV Spectrum

λ_max 230 nm (log₁₀ε 3.6) pH 11

λ_max 230 nm (log₁₀ε 4.4) pH 13

solvent : water

¹³C NMR Spectrum
(100.0 MHz, CDCl₃ solution)

expansion

DEPT  CH₂↓ CH₃↑ CH↑

expansion

proton decoupled

solvent

δ (ppm)

¹H NMR Spectrum
(400 MHz, CDCl₃ solution)

expansion

Exchanges
with D₂O

TMS

δ (ppm)

299

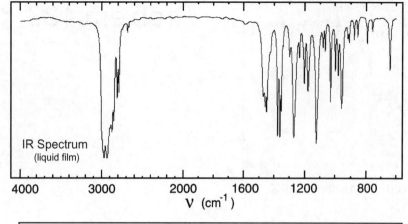

IR Spectrum
(liquid film)

ν (cm⁻¹)

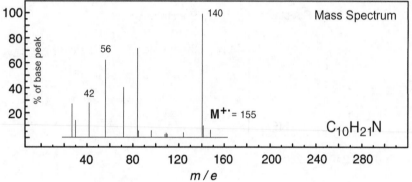

Mass Spectrum

No significant UV
absorption above 220 nm

$M^{+\cdot} = 155$

$C_{10}H_{21}N$

m/e

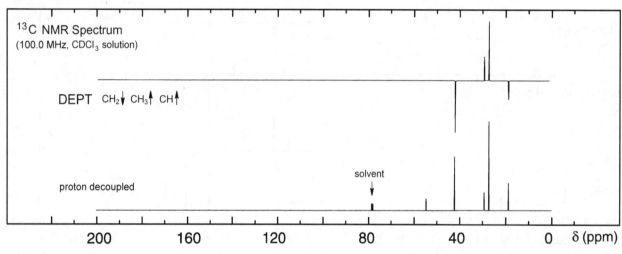

¹³C NMR Spectrum
(100.0 MHz, CDCl₃ solution)

DEPT CH₂↓ CH₃↑ CH↑

proton decoupled

solvent

δ (ppm)

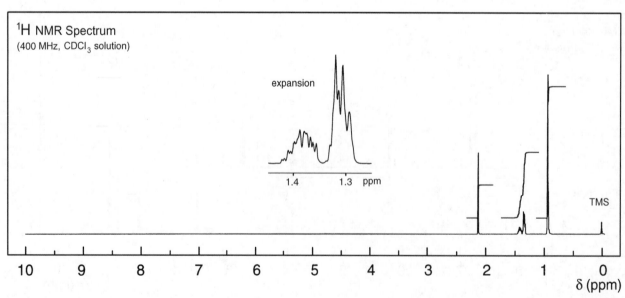

¹H NMR Spectrum
(400 MHz, CDCl₃ solution)

expansion

1.4      1.3   ppm

TMS

δ (ppm)

# Problem 190

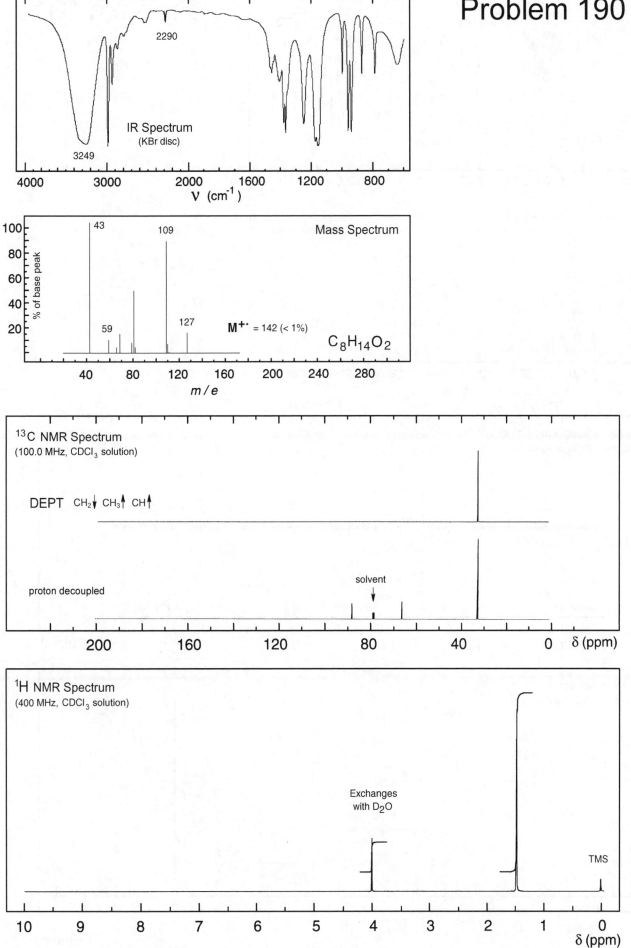

## IR Spectrum
(KBr disc)

3249
2290

$\nu$ (cm$^{-1}$)

## Mass Spectrum

% of base peak

43
109
59
127

$M^{+\cdot}$ = 142 (< 1%)

$C_8H_{14}O_2$

m/e

## $^{13}$C NMR Spectrum
(100.0 MHz, CDCl$_3$ solution)

DEPT   CH$_2\downarrow$ CH$_3\uparrow$ CH$\uparrow$

proton decoupled

solvent

$\delta$ (ppm)

## $^1$H NMR Spectrum
(400 MHz, CDCl$_3$ solution)

Exchanges
with D$_2$O

TMS

$\delta$ (ppm)

# Problem 191

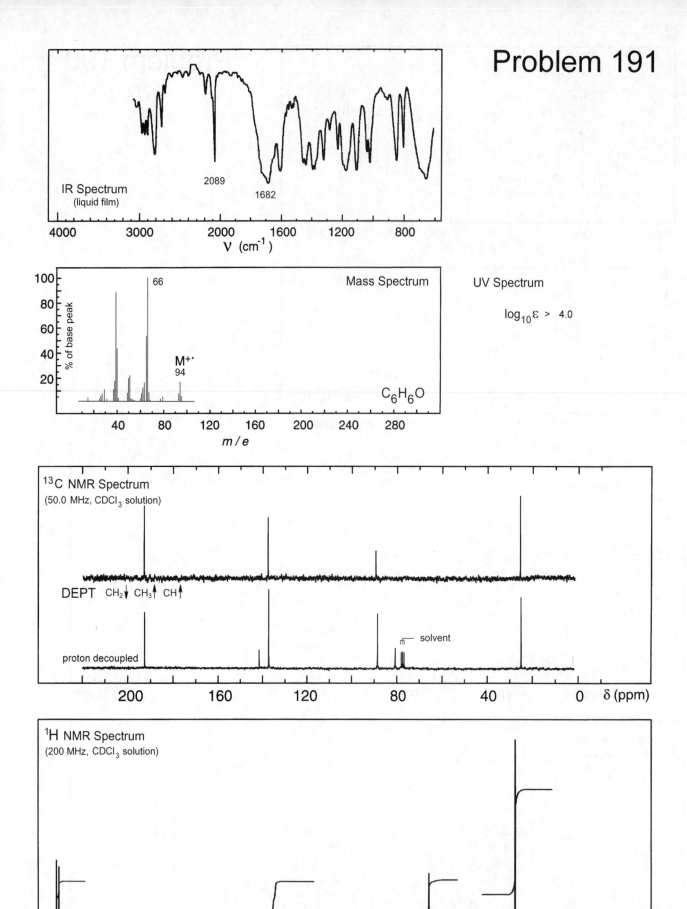

IR Spectrum
(liquid film)

2089
1682

Mass Spectrum

UV Spectrum

$\log_{10}\varepsilon \ > \ 4.0$

66

M+·
94

$C_6H_6O$

% of base peak

$m/e$

$^{13}C$ NMR Spectrum
(50.0 MHz, CDCl$_3$ solution)

DEPT    CH$_2\downarrow$  CH$_3\uparrow$  CH$\uparrow$

proton decoupled

solvent

$\delta$ (ppm)

$^1H$ NMR Spectrum
(200 MHz, CDCl$_3$ solution)

TMS

$\delta$ (ppm)

302

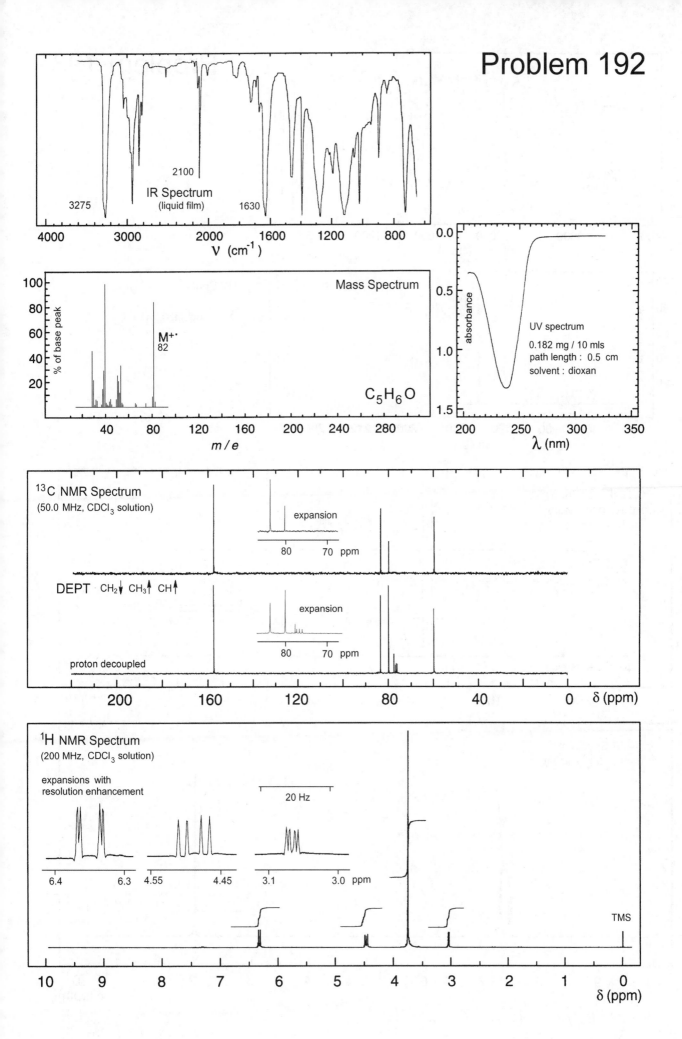

# Problem 192

**IR Spectrum** (liquid film)

3275  2100  1630

ν (cm⁻¹)

**Mass Spectrum**

% of base peak

M⁺˙
82

$C_5H_6O$

m/e

**UV spectrum**

0.182 mg / 10 mls
path length : 0.5 cm
solvent : dioxan

absorbance

λ (nm)

**¹³C NMR Spectrum**
(50.0 MHz, CDCl₃ solution)

expansion

80  70  ppm

**DEPT**  CH₂↓ CH₃↑ CH↑

expansion

80  70  ppm

proton decoupled

δ (ppm)

**¹H NMR Spectrum**
(200 MHz, CDCl₃ solution)

expansions with
resolution enhancement

20 Hz

6.4  6.3  4.55  4.45  3.1  3.0  ppm

TMS

δ (ppm)

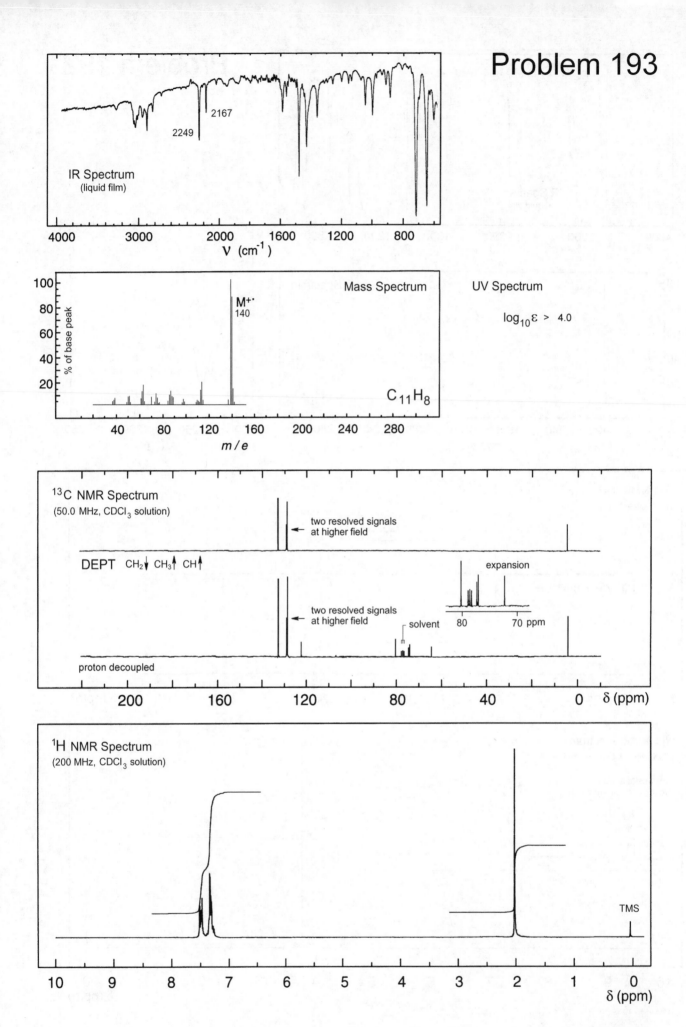

# Problem 193

**IR Spectrum**
(liquid film)

2249
2167

**Mass Spectrum**

M⁺˙
140

C₁₁H₈

**UV Spectrum**

$\log_{10}\varepsilon > 4.0$

**¹³C NMR Spectrum**
(50.0 MHz, CDCl₃ solution)

two resolved signals at higher field

**DEPT** CH₂↓ CH₃↑ CH↑

two resolved signals at higher field

solvent

expansion

80    70 ppm

proton decoupled

**¹H NMR Spectrum**
(200 MHz, CDCl₃ solution)

TMS

# Problem 194

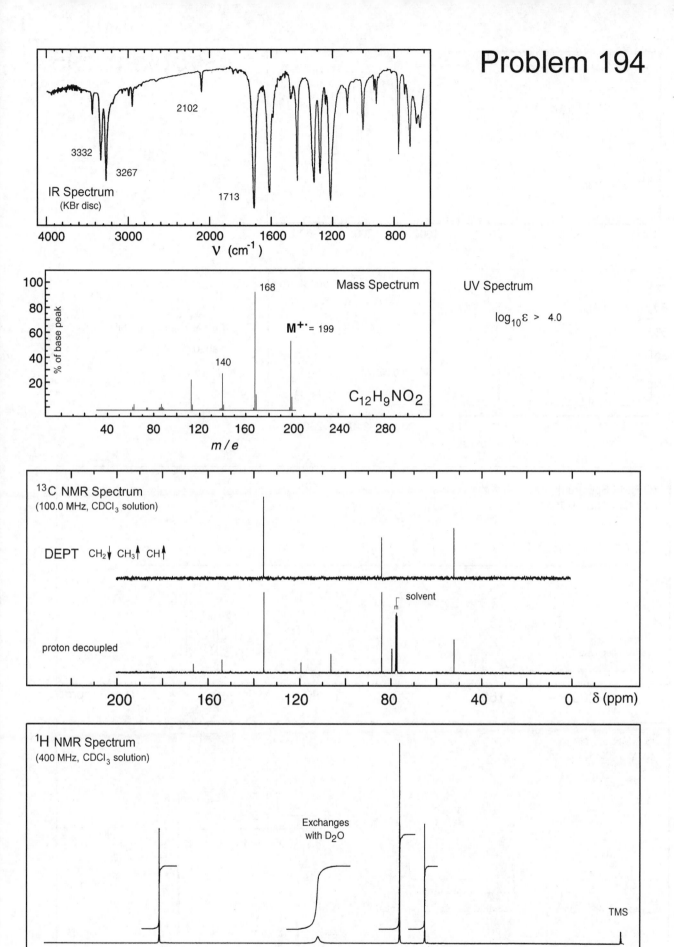

IR Spectrum
(KBr disc)

3332
3267
2102
1713

ν (cm⁻¹)

Mass Spectrum

168
140
M⁺· = 199

% of base peak

m/e

C₁₂H₉NO₂

UV Spectrum

$\log_{10} \varepsilon > 4.0$

¹³C NMR Spectrum
(100.0 MHz, CDCl₃ solution)

DEPT   CH₂↓ CH₃↑ CH↑

solvent

proton decoupled

δ (ppm)

¹H NMR Spectrum
(400 MHz, CDCl₃ solution)

Exchanges
with D₂O

TMS

δ (ppm)

# Problem 195

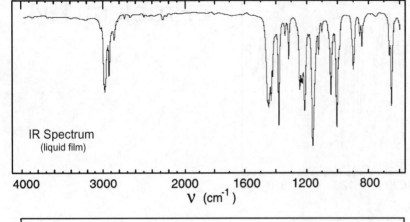

IR Spectrum
(liquid film)

$\nu$ (cm$^{-1}$)

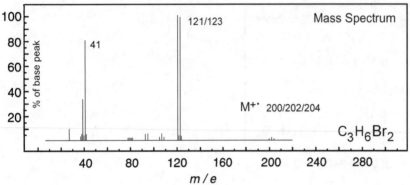

Mass Spectrum

No significant UV
absorption above 220 nm

41

121/123

M$^{+\cdot}$ 200/202/204

$C_3H_6Br_2$

% of base peak

$m/e$

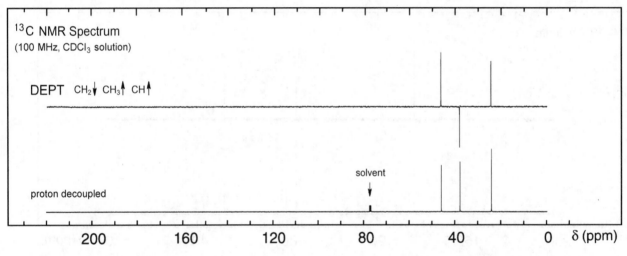

$^{13}$C NMR Spectrum
(100 MHz, CDCl$_3$ solution)

DEPT  CH$_2\downarrow$ CH$_3\uparrow$ CH$\uparrow$

solvent

proton decoupled

$\delta$ (ppm)

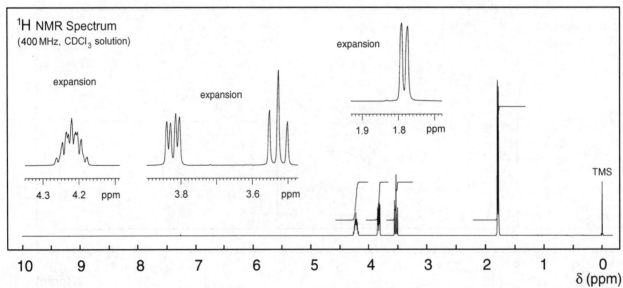

$^1$H NMR Spectrum
(400 MHz, CDCl$_3$ solution)

expansion

expansion

expansion

TMS

$\delta$ (ppm)

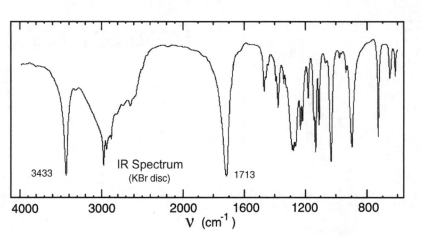

IR Spectrum
(KBr disc)

3433    1713

ν (cm⁻¹)

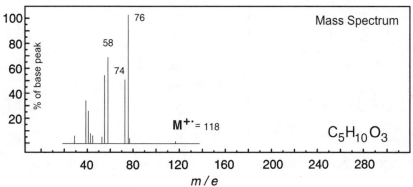

Mass Spectrum

76

58

74

M⁺· = 118

C₅H₁₀O₃

No significant UV
absorption above 220 nm

% of base peak

m/e

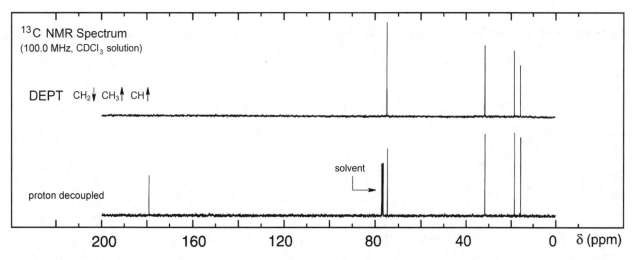

¹³C NMR Spectrum
(100.0 MHz, CDCl₃ solution)

DEPT   CH₂↓ CH₃↑ CH↑

solvent

proton decoupled

δ (ppm)

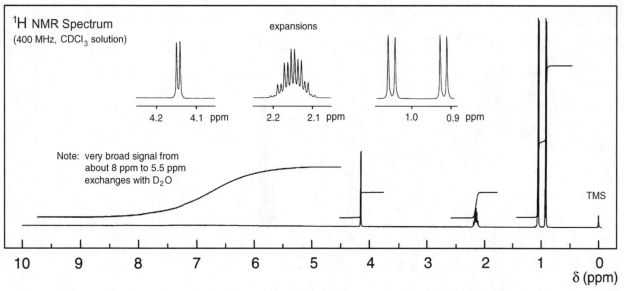

¹H NMR Spectrum
(400 MHz, CDCl₃ solution)

expansions

4.2    4.1  ppm          2.2    2.1  ppm          1.0    0.9  ppm

Note:  very broad signal from
about 8 ppm to 5.5 ppm
exchanges with D₂O

TMS

δ (ppm)

# Problem 197

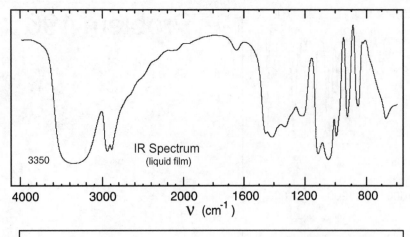

IR Spectrum
(liquid film)

3350

$\nu$ (cm$^{-1}$)

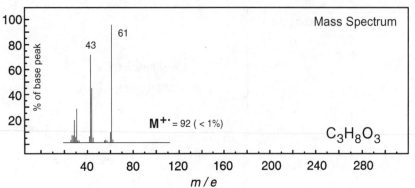

Mass Spectrum

% of base peak

43

61

$M^{+\cdot}$ = 92 ( < 1%)

$C_3H_8O_3$

m/e

No significant UV
absorption above 220 nm

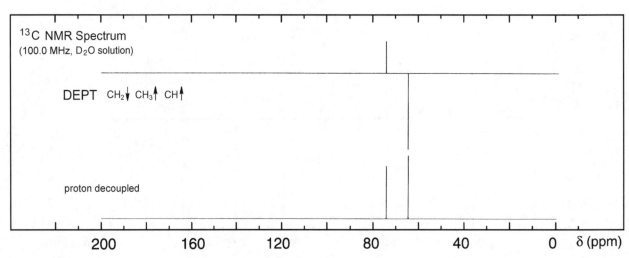

$^{13}$C NMR Spectrum
(100.0 MHz, D$_2$O solution)

DEPT   CH$_2$↓ CH$_3$↑ CH↑

proton decoupled

$\delta$ (ppm)

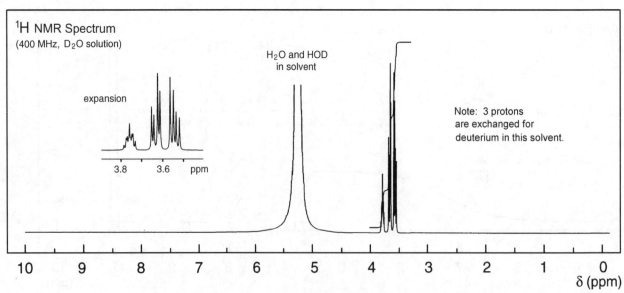

$^1$H NMR Spectrum
(400 MHz, D$_2$O solution)

H$_2$O and HOD
in solvent

expansion

3.8    3.6    ppm

Note:  3 protons
are exchanged for
deuterium in this solvent.

$\delta$ (ppm)

# Problem 198

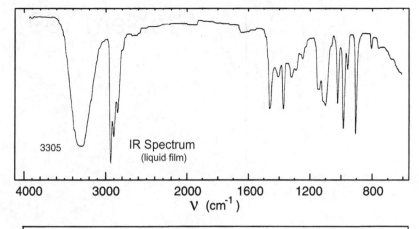

IR Spectrum
(liquid film)

3305

$\nu$ (cm$^{-1}$)

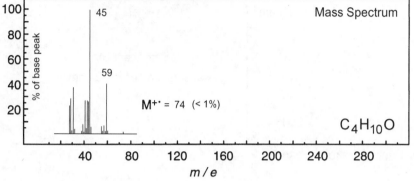

Mass Spectrum

45

59

$M^{+\cdot} = 74$ (< 1%)

$C_4H_{10}O$

% of base peak

$m/e$

No significant UV
absorption above 220 nm

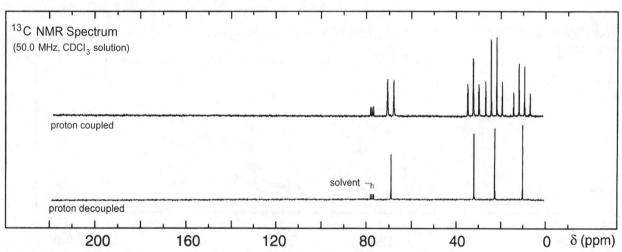

$^{13}$C NMR Spectrum
(50.0 MHz, CDCl$_3$ solution)

proton coupled

solvent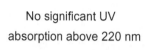

proton decoupled

$\delta$ (ppm)

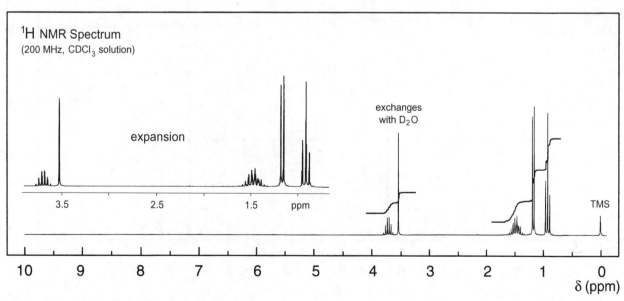

$^1$H NMR Spectrum
(200 MHz, CDCl$_3$ solution)

expansion

exchanges
with D$_2$O

TMS

ppm

$\delta$ (ppm)

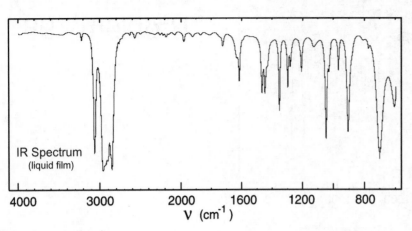

## Problem 199

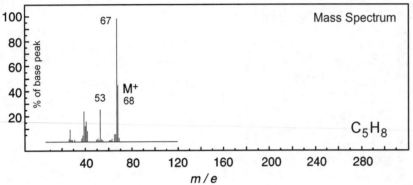

No significant UV
absorption above 220 nm

$C_5H_8$

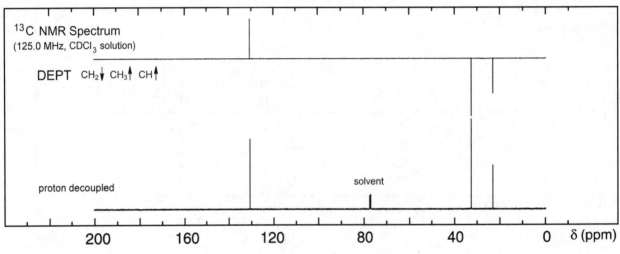

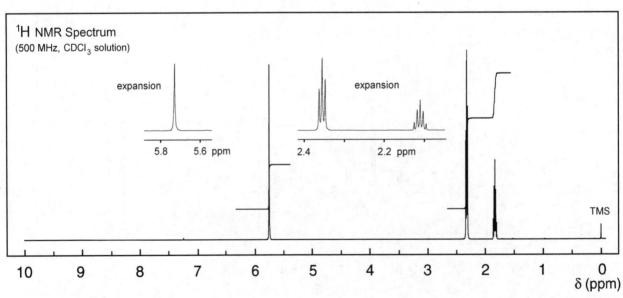

# Problem 200

IR Spectrum
(liquid film)

1709

ν (cm⁻¹)

Mass Spectrum

M⁺·
82

% of base peak

$C_5H_6O$

m/e

UV Spectrum

$\log_{10}\varepsilon \approx 4$

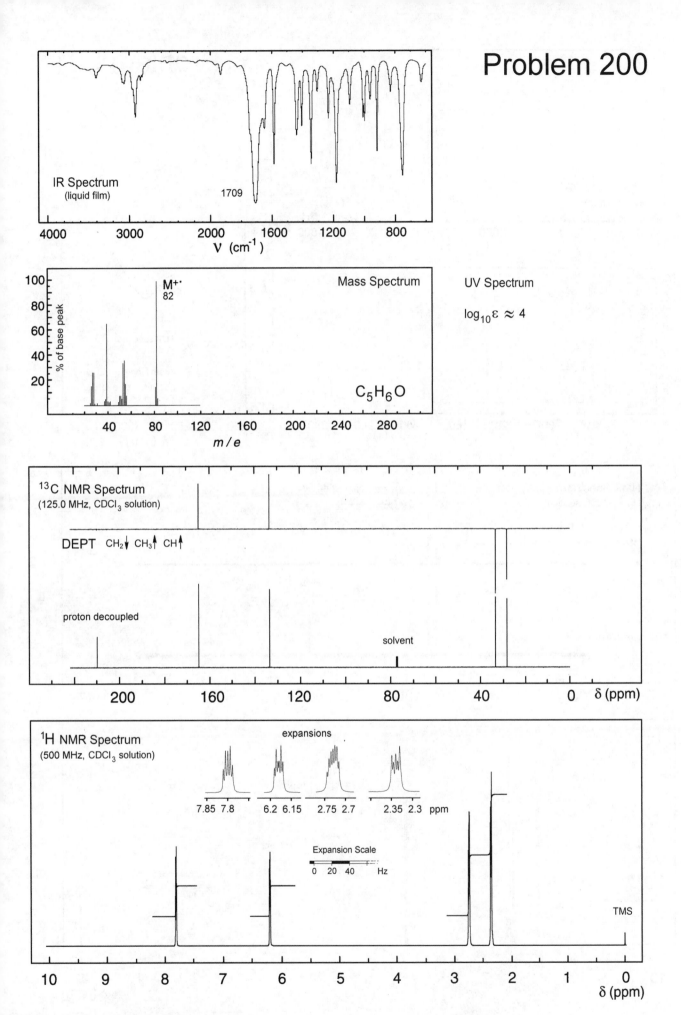

¹³C NMR Spectrum
(125.0 MHz, CDCl₃ solution)

DEPT   CH₂↓ CH₃↑ CH↑

proton decoupled

solvent

δ (ppm)

¹H NMR Spectrum
(500 MHz, CDCl₃ solution)

expansions

7.85 7.8    6.2 6.15    2.75 2.7    2.35 2.3   ppm

Expansion Scale

0  20  40    Hz

TMS

δ (ppm)

# Problem 201

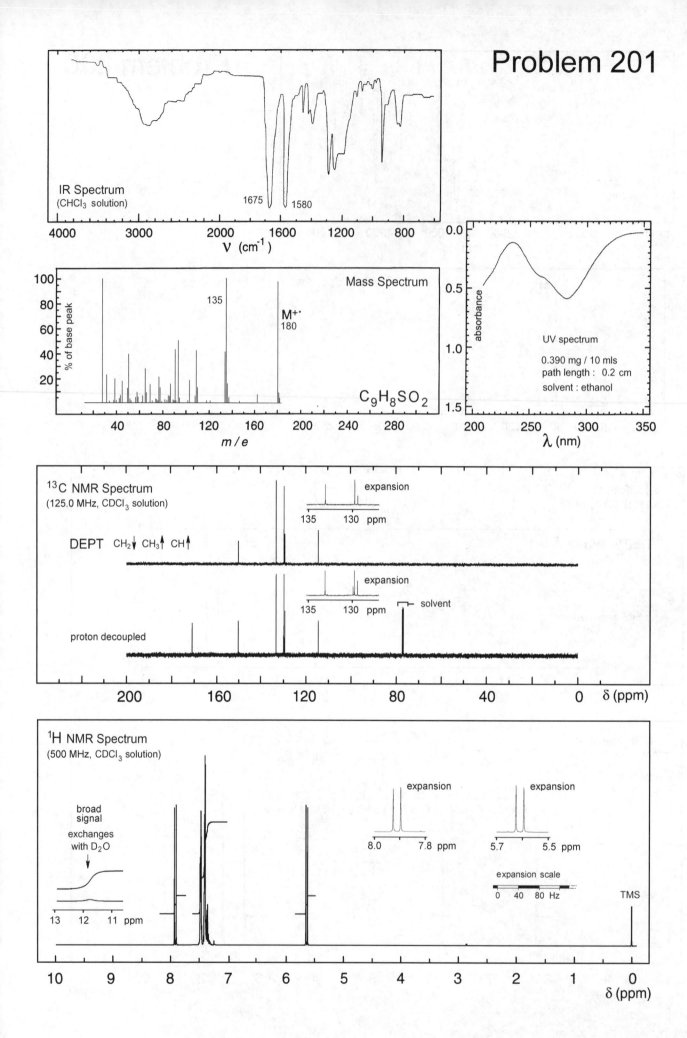

IR Spectrum
(CHCl₃ solution)
1675 1580
ν (cm⁻¹)

Mass Spectrum
135
M⁺˙
180
$C_9H_8SO_2$
% of base peak
m/e

UV spectrum
0.390 mg / 10 mls
path length : 0.2 cm
solvent : ethanol
absorbance
λ (nm)

¹³C NMR Spectrum
(125.0 MHz, CDCl₃ solution)
expansion
135  130  ppm
DEPT   CH₂↓ CH₃↑ CH↑
expansion
135  130  ppm
solvent
proton decoupled
δ (ppm)

¹H NMR Spectrum
(500 MHz, CDCl₃ solution)
broad signal
exchanges with D₂O
13  12  11  ppm
expansion
8.0  7.8  ppm
expansion
5.7  5.5  ppm
expansion scale
0  40  80  Hz
TMS
δ (ppm)

# Problem 202

IR Spectrum
(CHCl₃ solution)

1350

1176

ν (cm⁻¹)

Mass Spectrum

91

155

172

M⁺·
200

$C_9H_{12}O_3S$

% of base peak

m/e

absorbance

UV spectrum

7.260 mg / 10 mls
path length : 0.5 cm
solvent : ethanol

λ (nm)

¹³C NMR Spectrum
(150.0 MHz, CDCl₃ solution)

DEPT  CH₂↓ CH₃↑ CH↑

solvent

proton decoupled

δ (ppm)

¹H NMR Spectrum
(600 MHz, CDCl₃ solution)

expansions

expansions

7.85   7.75 7.40   7.30 ppm

4.15   3.95 1.35   1.25 ppm

TMS

δ (ppm)

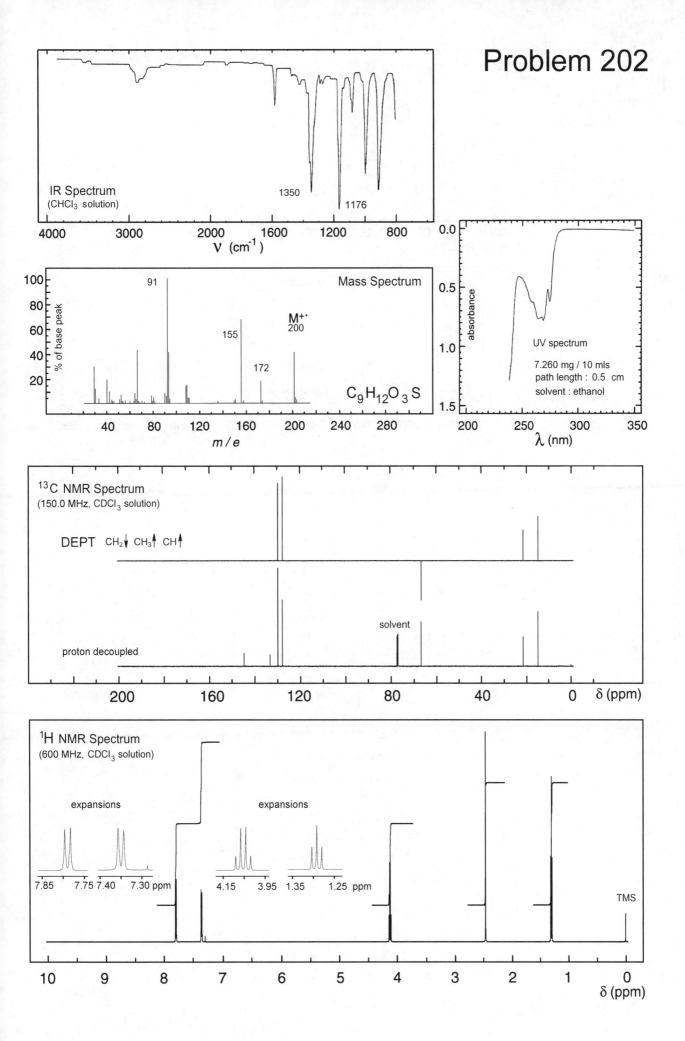

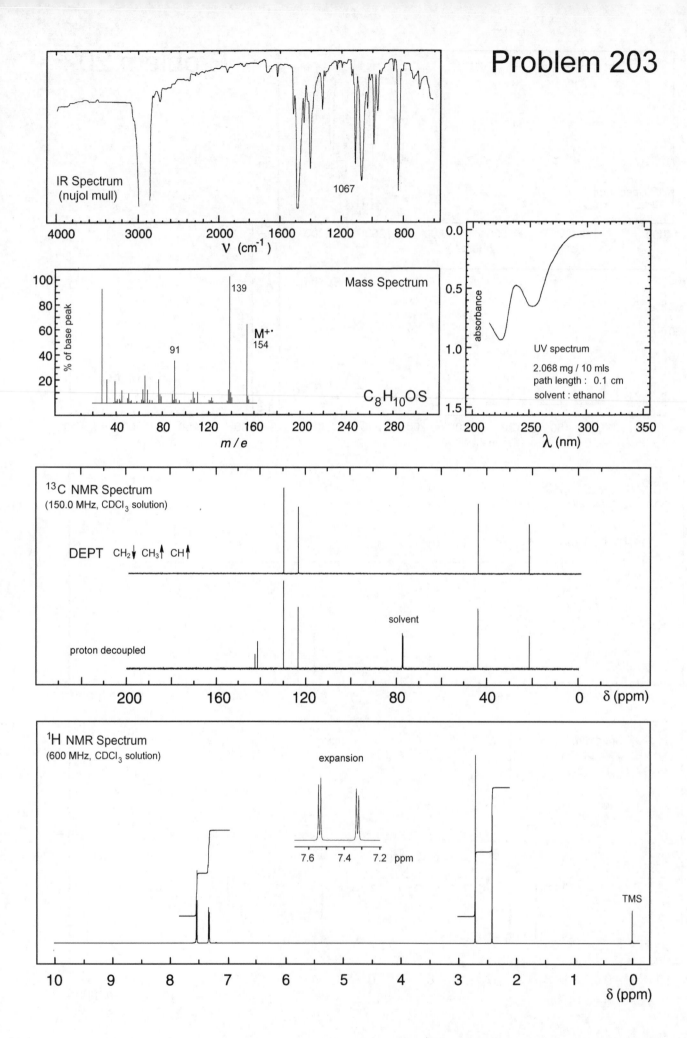

Problem 203

IR Spectrum
(nujol mull)

1067

ν (cm⁻¹)

Mass Spectrum

139

91

M⁺˙
154

C₈H₁₀OS

m/e

UV spectrum
2.068 mg / 10 mls
path length : 0.1 cm
solvent : ethanol

λ (nm)

¹³C NMR Spectrum
(150.0 MHz, CDCl₃ solution)

DEPT   CH₂↓ CH₃↑ CH↑

solvent

proton decoupled

δ (ppm)

¹H NMR Spectrum
(600 MHz, CDCl₃ solution)

expansion

7.6   7.4   7.2  ppm

TMS

δ (ppm)

# Problem 204

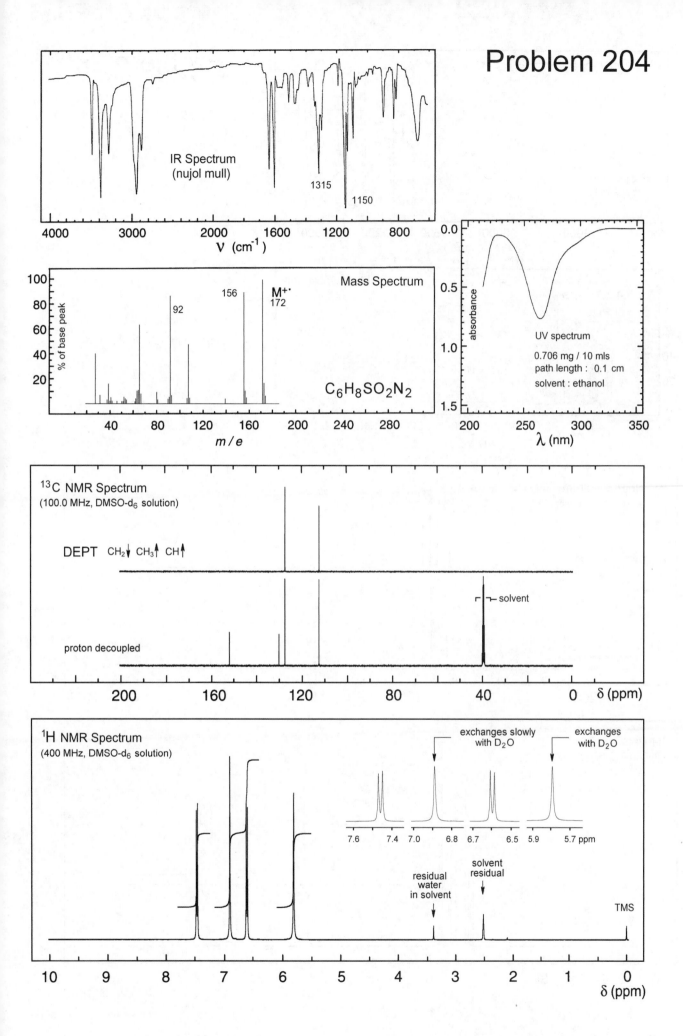

IR Spectrum
(nujol mull)

1315
1150

ν (cm⁻¹)

Mass Spectrum

92
156
M⁺·
172

% of base peak

C₆H₈SO₂N₂

m/e

UV spectrum

0.706 mg / 10 mls
path length : 0.1 cm
solvent : ethanol

absorbance

λ (nm)

¹³C NMR Spectrum
(100.0 MHz, DMSO-d₆ solution)

DEPT   CH₂↓ CH₃↑ CH↑

solvent

proton decoupled

δ (ppm)

¹H NMR Spectrum
(400 MHz, DMSO-d₆ solution)

exchanges slowly
with D₂O

exchanges
with D₂O

7.6   7.4   7.0   6.8   6.7   6.5   5.9   5.7 ppm

residual
water
in solvent

solvent
residual

TMS

δ (ppm)

# Problem 205

IR Spectrum
(liquid film)

1320  1135

$\nu$ (cm$^{-1}$)

Mass Spectrum

75

% of base peak

M$^{+\cdot}$ = 118 (< 1%)

$C_4H_6O_2S$

$m/e$

No significant UV
absorption above 220 nm

$^{13}C$ NMR Spectrum
(50.0 MHz, CDCl$_3$ solution)

proton coupled

proton decoupled

solvent

$\delta$ (ppm)

$^1H$ NMR Spectrum
(200 MHz, CDCl$_3$ solution)

expansion

7.0  6.6  6.2  ppm

TMS

$\delta$ (ppm)

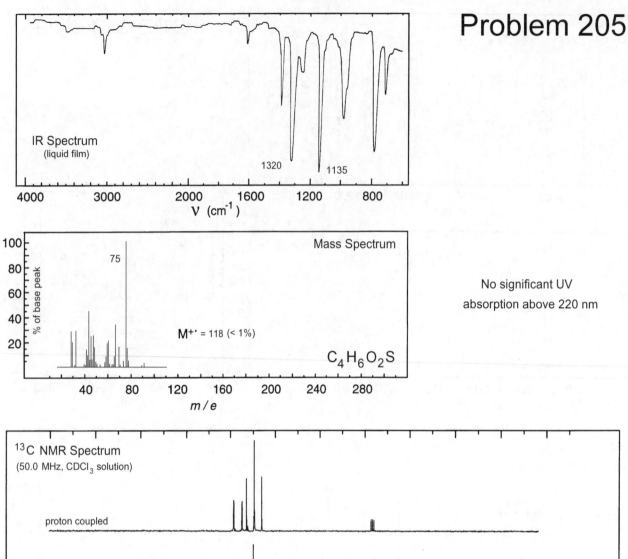
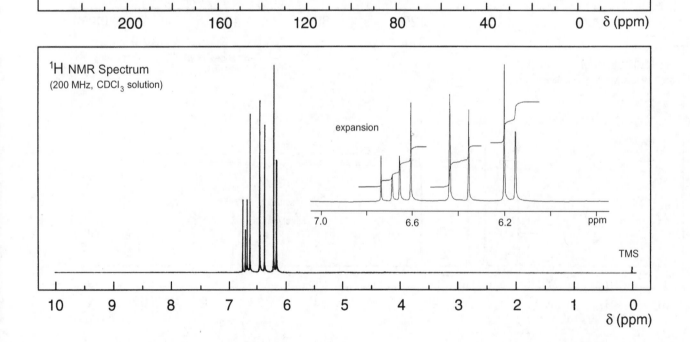

316

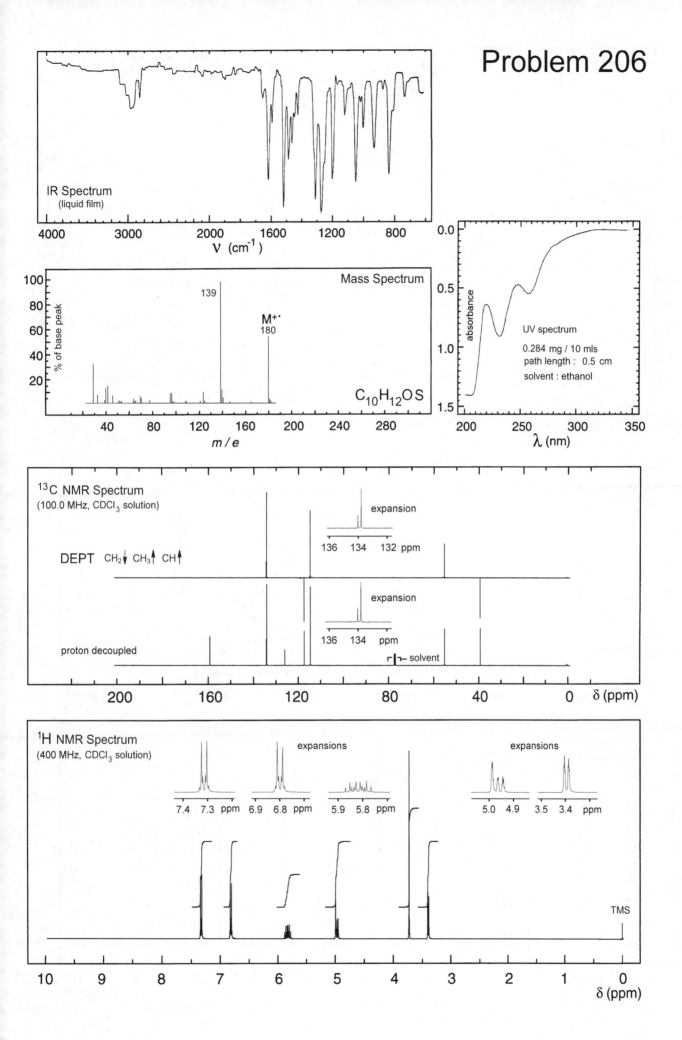

# Problem 206

IR Spectrum
(liquid film)

ν (cm⁻¹)

Mass Spectrum

139

M⁺·
180

% of base peak

m / e

$C_{10}H_{12}OS$

UV spectrum

0.284 mg / 10 mls
path length : 0.5 cm
solvent : ethanol

absorbance

λ (nm)

¹³C NMR Spectrum
(100.0 MHz, CDCl₃ solution)

DEPT  CH₂↓ CH₃↑ CH↑

expansion

136  134  132 ppm

expansion

136  134  ppm

proton decoupled

solvent

δ (ppm)

¹H NMR Spectrum
(400 MHz, CDCl₃ solution)

expansions

7.4  7.3 ppm   6.9  6.8 ppm   5.9  5.8 ppm

expansions

5.0  4.9 ppm   3.5  3.4 ppm

TMS

δ (ppm)

# Problem 207

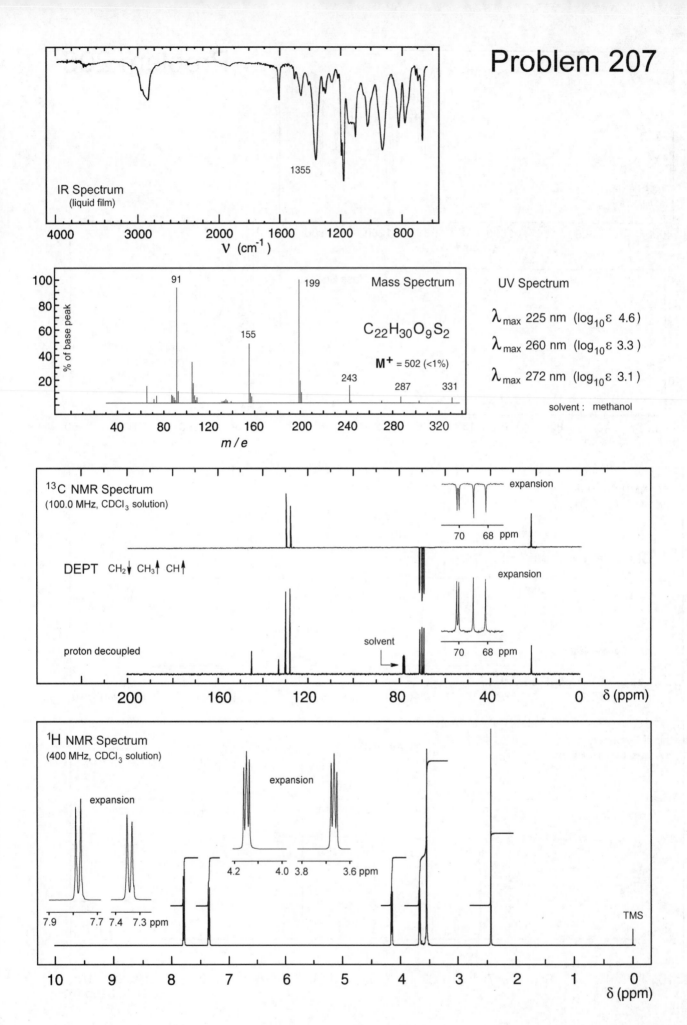

IR Spectrum
(liquid film)

4000  3000  2000  1600  1200  800

ν (cm⁻¹)

Mass Spectrum

C₂₂H₃₀O₉S₂

M⁺ = 502 (<1%)

91

199

155

243   287   331

m/e

UV Spectrum

λ_max 225 nm (log₁₀ε 4.6)

λ_max 260 nm (log₁₀ε 3.3)

λ_max 272 nm (log₁₀ε 3.1)

solvent : methanol

¹³C NMR Spectrum
(100.0 MHz, CDCl₃ solution)

expansion

DEPT   CH₂↓  CH₃↑  CH↑

expansion

proton decoupled

solvent

70   68 ppm

70   68 ppm

200   160   120   80   40   0   δ (ppm)

¹H NMR Spectrum
(400 MHz, CDCl₃ solution)

expansion

expansion

4.2   4.0  3.8   3.6 ppm

7.9   7.7  7.4  7.3 ppm

TMS

10   9   8   7   6   5   4   3   2   1   0
δ (ppm)

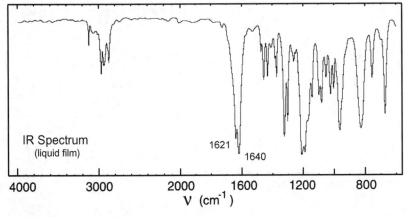

# Problem 208

IR Spectrum
(liquid film)

1621
1640

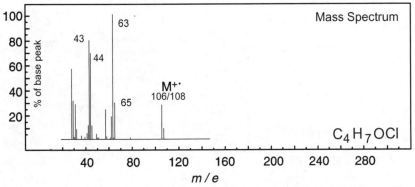

Mass Spectrum

63
43
44
65
M⁺·
106/108

% of base peak

m/e

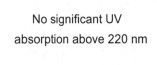

$C_4H_7OCl$

No significant UV
absorption above 220 nm

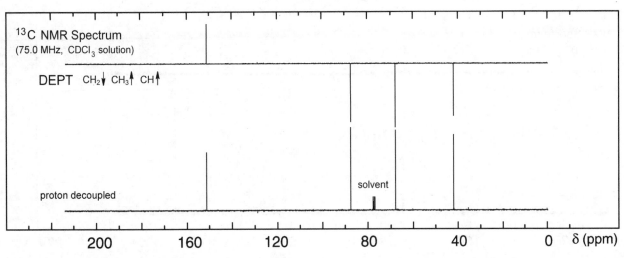

¹³C NMR Spectrum
(75.0 MHz, CDCl₃ solution)

DEPT  CH₂↓ CH₃↑ CH↑

solvent

proton decoupled

δ (ppm)

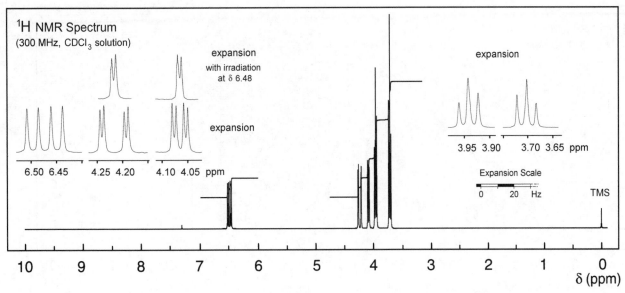

¹H NMR Spectrum
(300 MHz, CDCl₃ solution)

expansion
with irradiation
at δ 6.48

expansion

expansion

6.50 6.45    4.25 4.20    4.10 4.05 ppm

3.95 3.90    3.70 3.65 ppm

Expansion Scale
0    20   Hz

TMS

δ (ppm)

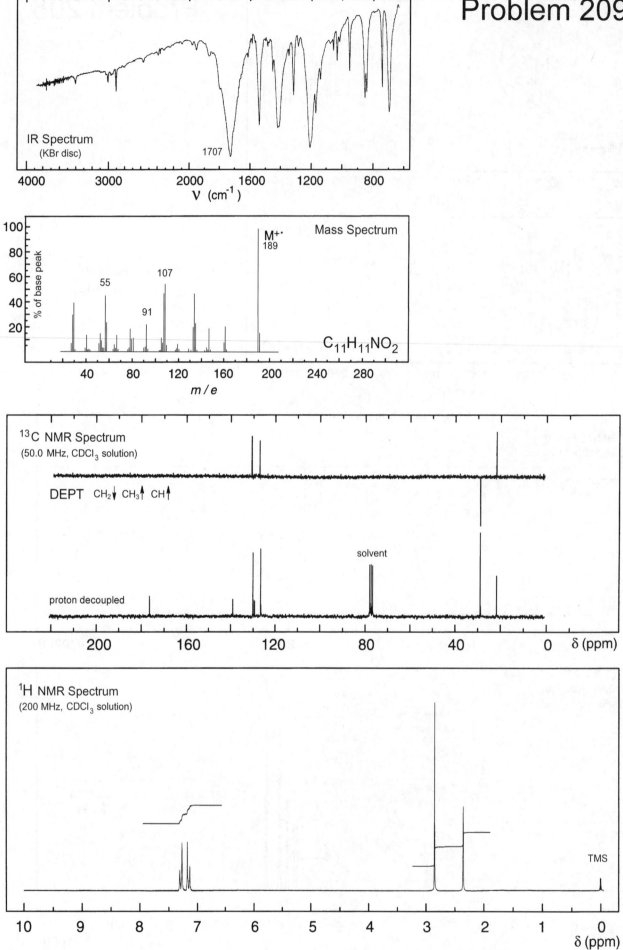

# Problem 209

**IR Spectrum**
(KBr disc)

1707

ν (cm⁻¹)

**Mass Spectrum**

M⁺·
189

107

55

91

$C_{11}H_{11}NO_2$

% of base peak

m/e

**¹³C NMR Spectrum**
(50.0 MHz, CDCl₃ solution)

**DEPT**  CH₂↓ CH₃↑ CH↑

solvent

proton decoupled

δ (ppm)

**¹H NMR Spectrum**
(200 MHz, CDCl₃ solution)

TMS

δ (ppm)

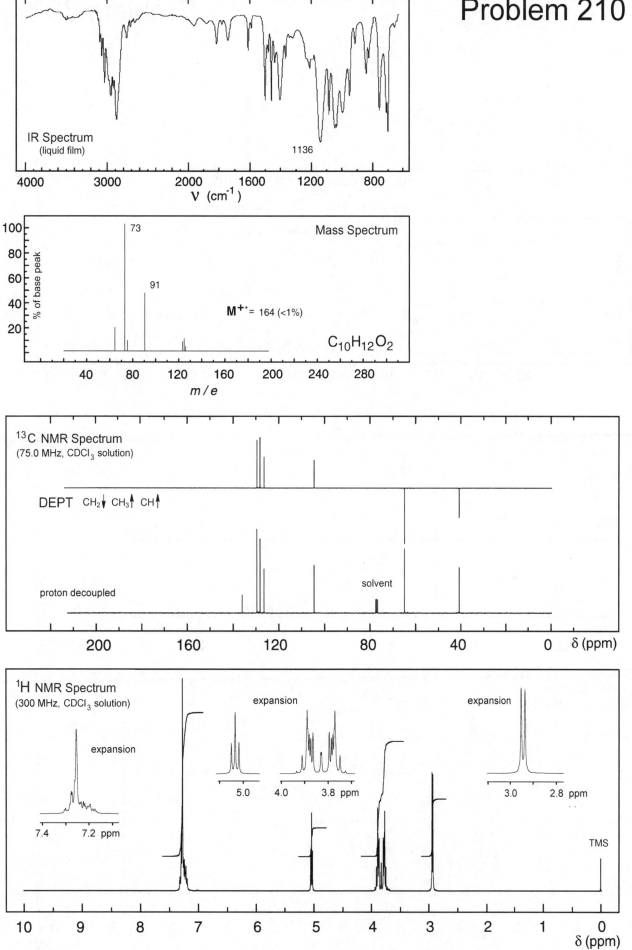

# Problem 210

**IR Spectrum**
(liquid film)

1136

ν (cm⁻¹)

4000   3000   2000   1600   1200   800

**Mass Spectrum**

73

91

**M⁺·** = 164 (<1%)

$C_{10}H_{12}O_2$

% of base peak

40   80   120   160   200   240   280

m / e

**¹³C NMR Spectrum**
(75.0 MHz, CDCl₃ solution)

DEPT   CH₂↓  CH₃↑  CH↑

proton decoupled

solvent

200   160   120   80   40   0   δ (ppm)

**¹H NMR Spectrum**
(300 MHz, CDCl₃ solution)

expansion

expansion

expansion

expansion

7.4   7.2 ppm

5.0   4.0   3.8 ppm

3.0   2.8 ppm

TMS

10   9   8   7   6   5   4   3   2   1   0   δ (ppm)

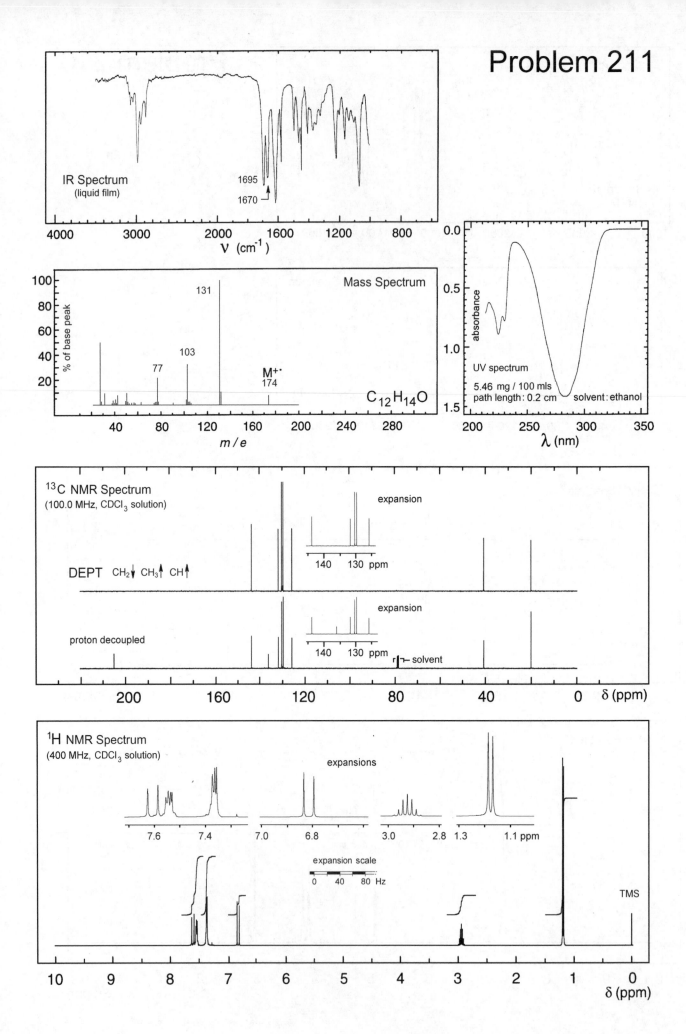

# Problem 211

**IR Spectrum** (liquid film)

1695
1670

$\nu$ (cm$^{-1}$)

**Mass Spectrum**

131
103
77
M$^{+\cdot}$
174

$C_{12}H_{14}O$

% of base peak

$m/e$

**UV spectrum**
5.46 mg / 100 mls
path length: 0.2 cm    solvent: ethanol

absorbance

$\lambda$ (nm)

**$^{13}$C NMR Spectrum** (100.0 MHz, CDCl$_3$ solution)

expansion
140   130 ppm

DEPT   CH$_2\downarrow$ CH$_3\uparrow$ CH$\uparrow$

expansion
140   130 ppm

proton decoupled

solvent

$\delta$ (ppm)

**$^1$H NMR Spectrum** (400 MHz, CDCl$_3$ solution)

expansions

7.6   7.4   7.0   6.8   3.0   2.8   1.3   1.1 ppm

expansion scale
0   40   80 Hz

TMS

$\delta$ (ppm)

# Problem 212

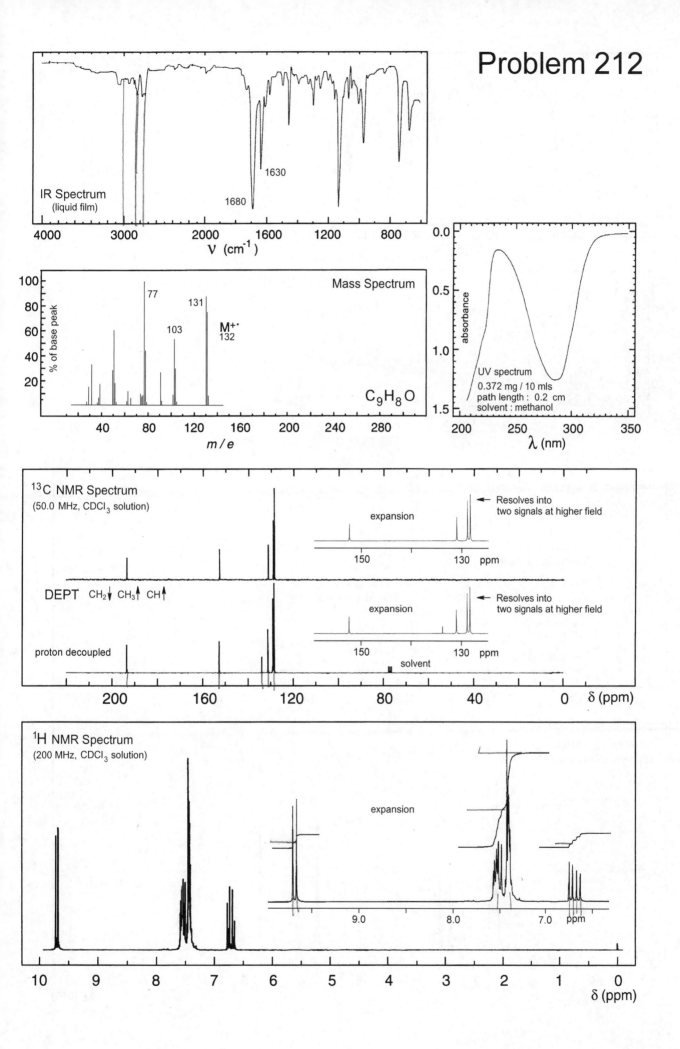

IR Spectrum
(liquid film)

1680
1630

ν (cm⁻¹)

Mass Spectrum

77
131
103
M⁺˙
132

C₉H₈O

% of base peak

m / e

UV spectrum
0.372 mg / 10 mls
path length : 0.2 cm
solvent : methanol

absorbance

λ (nm)

¹³C NMR Spectrum
(50.0 MHz, CDCl₃ solution)

Resolves into
two signals at higher field

expansion

150    130    ppm

DEPT  CH₂↓ CH₃↑ CH↑

Resolves into
two signals at higher field

expansion

150    130    ppm

proton decoupled

solvent

δ (ppm)

¹H NMR Spectrum
(200 MHz, CDCl₃ solution)

expansion

9.0    8.0    7.0    ppm

δ (ppm)

323

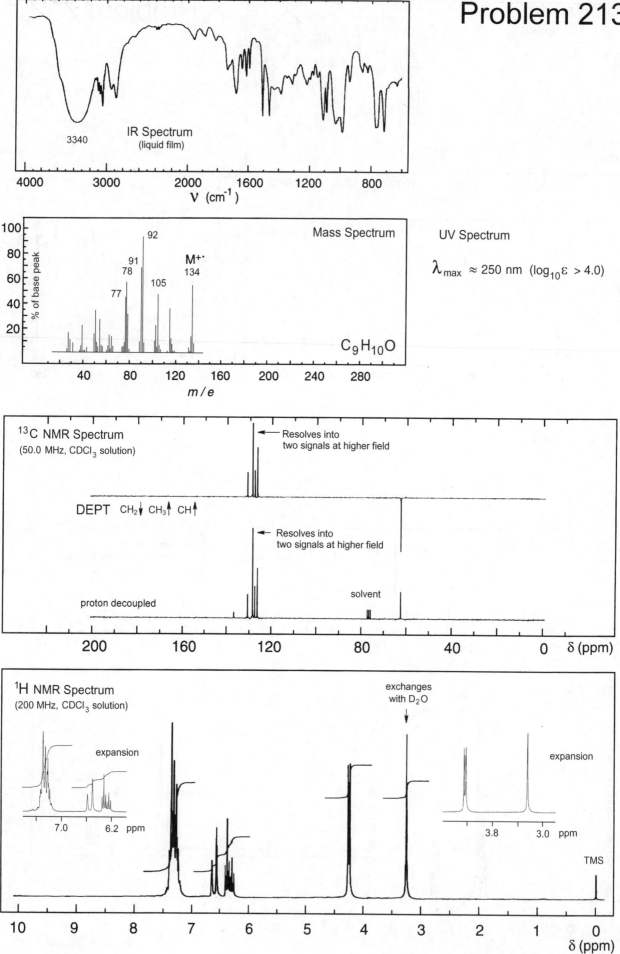

# Problem 213

**IR Spectrum**
(liquid film)

3340

ν (cm⁻¹)

**Mass Spectrum**

% of base peak

92
91
78
77
105
M⁺·
134

C₉H₁₀O

m/e

**UV Spectrum**

$\lambda_{max} \approx 250$ nm $(\log_{10}\varepsilon > 4.0)$

**¹³C NMR Spectrum**
(50.0 MHz, CDCl₃ solution)

Resolves into two signals at higher field

DEPT   CH₂↓ CH₃↑ CH↑

Resolves into two signals at higher field

proton decoupled

solvent

δ (ppm)

**¹H NMR Spectrum**
(200 MHz, CDCl₃ solution)

exchanges with D₂O

expansion

7.0    6.2 ppm

expansion

3.8    3.0 ppm

TMS

δ (ppm)

# Problem 214

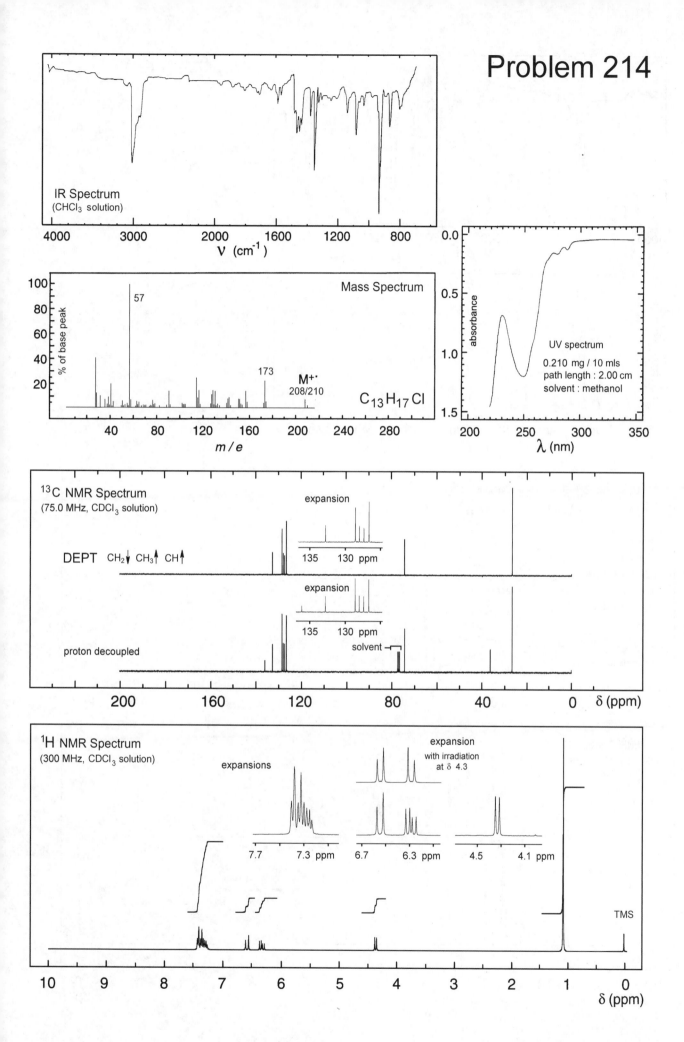

IR Spectrum
(CHCl$_3$ solution)

ν (cm$^{-1}$)

Mass Spectrum

% of base peak

57

173

M$^{+\cdot}$
208/210

C$_{13}$H$_{17}$Cl

m/e

UV spectrum
0.210 mg / 10 mls
path length : 2.00 cm
solvent : methanol

absorbance

λ (nm)

$^{13}$C NMR Spectrum
(75.0 MHz, CDCl$_3$ solution)

DEPT   CH$_2$↓ CH$_3$↑ CH↑

expansion

proton decoupled

solvent

expansion

δ (ppm)

$^1$H NMR Spectrum
(300 MHz, CDCl$_3$ solution)

expansions

expansion
with irradiation
at δ 4.3

7.7    7.3 ppm    6.7    6.3 ppm    4.5    4.1 ppm

TMS

δ (ppm)

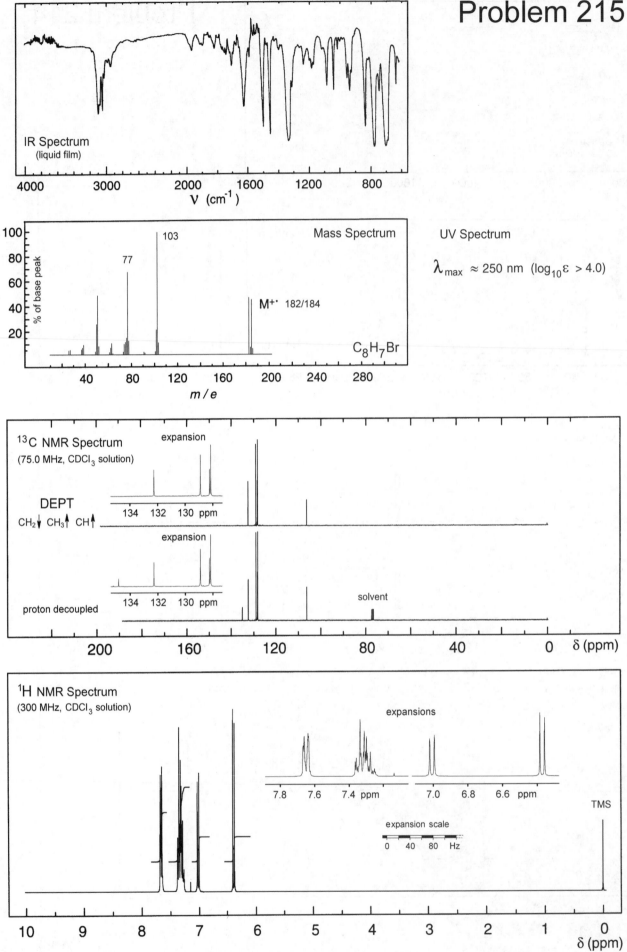

# Problem 215

**IR Spectrum** (liquid film)

$\nu$ (cm$^{-1}$)

**Mass Spectrum**

**UV Spectrum**

$\lambda_{max} \approx 250$ nm $(\log_{10}\varepsilon > 4.0)$

103

77

M$^{+\cdot}$ 182/184

% of base peak

$m/e$

$C_8H_7Br$

**$^{13}$C NMR Spectrum** (75.0 MHz, CDCl$_3$ solution)

expansion

**DEPT**

CH$_2\downarrow$ CH$_3\uparrow$ CH$\uparrow$

expansion

proton decoupled

solvent

$\delta$ (ppm)

**$^1$H NMR Spectrum** (300 MHz, CDCl$_3$ solution)

expansions

expansion scale

0   40   80  Hz

TMS

$\delta$ (ppm)

326

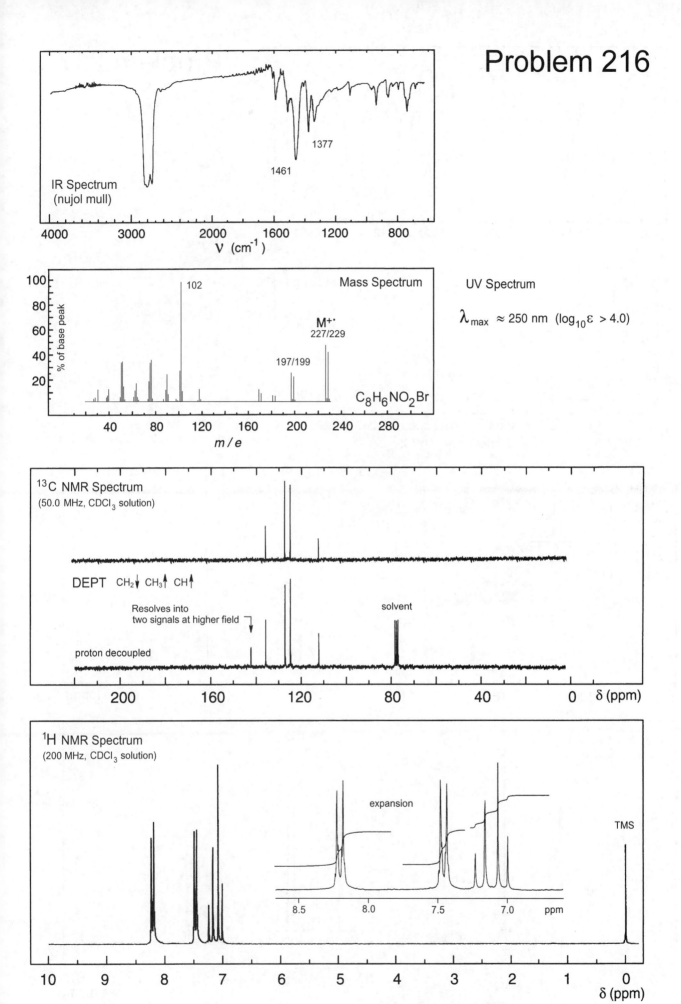

# Problem 216

**IR Spectrum (nujol mull)**

1461
1377

ν (cm⁻¹)

**Mass Spectrum**

102

M⁺·
227/229

197/199

C₈H₆NO₂Br

m/e

**UV Spectrum**

λ_max ≈ 250 nm (log₁₀ε > 4.0)

**¹³C NMR Spectrum** (50.0 MHz, CDCl₃ solution)

DEPT CH₂↓ CH₃↑ CH↑

Resolves into two signals at higher field

solvent

proton decoupled

δ (ppm)

**¹H NMR Spectrum** (200 MHz, CDCl₃ solution)

expansion

TMS

δ (ppm)

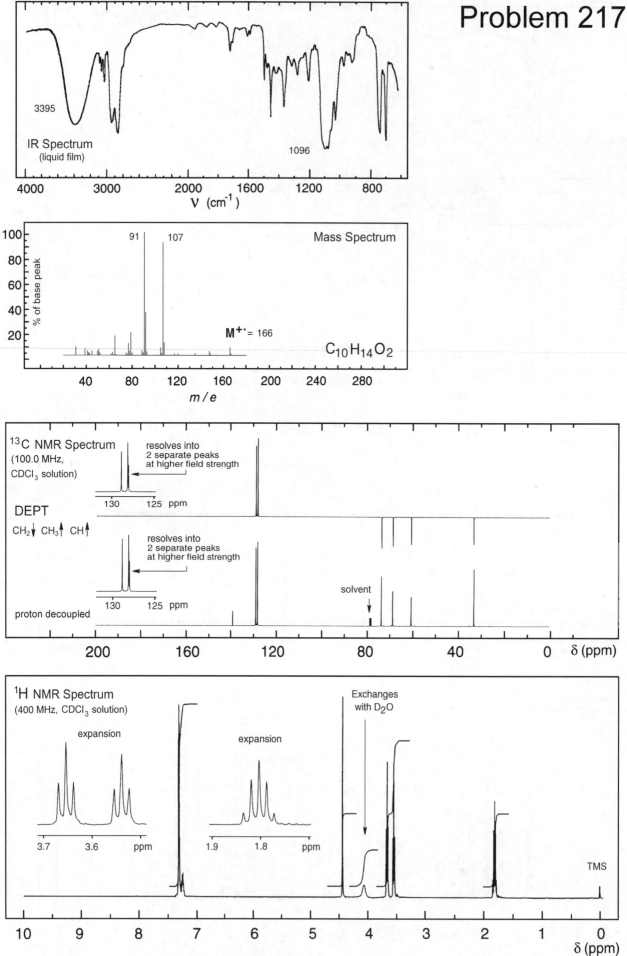

# Problem 217

**IR Spectrum** (liquid film)
3395
1096

ν (cm⁻¹)

**Mass Spectrum**
91
107
M⁺· = 166
$C_{10}H_{14}O_2$

% of base peak

m/e

**¹³C NMR Spectrum** (100.0 MHz, CDCl₃ solution)

resolves into 2 separate peaks at higher field strength

**DEPT**
CH₂↓ CH₃↑ CH↑

resolves into 2 separate peaks at higher field strength

solvent

proton decoupled

δ (ppm)

**¹H NMR Spectrum** (400 MHz, CDCl₃ solution)

expansion

expansion

Exchanges with D₂O

TMS

δ (ppm)

# Problem 218

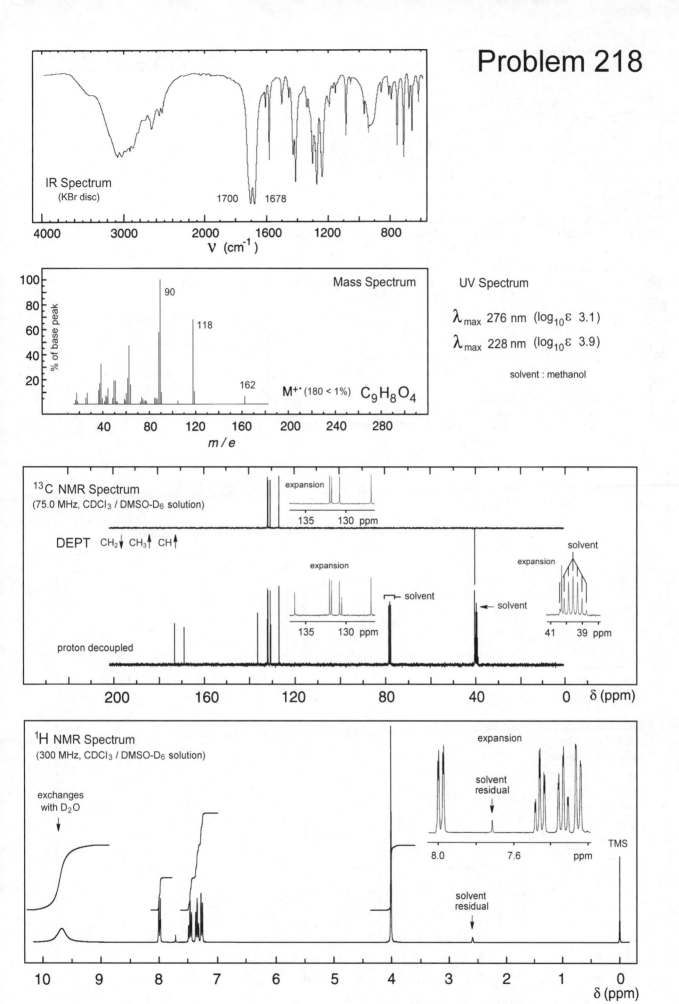

IR Spectrum
(KBr disc)

1700  1678

ν (cm⁻¹)

4000   3000   2000   1600   1200   800

Mass Spectrum

% of base peak

90

118

162

M⁺˙ (180 < 1%)   C₉H₈O₄

40   80   120   160   200   240   280

m/e

UV Spectrum

λ max 276 nm  (log₁₀ε  3.1)

λ max 228 nm  (log₁₀ε  3.9)

solvent : methanol

¹³C NMR Spectrum
(75.0 MHz, CDCl₃ / DMSO-D₆ solution)

expansion

135   130 ppm

DEPT  CH₂↓ CH₃↑ CH↑

expansion

solvent

solvent

expansion

solvent

135   130 ppm

41   39 ppm

proton decoupled

200   160   120   80   40   0   δ (ppm)

¹H NMR Spectrum
(300 MHz, CDCl₃ / DMSO-D₆ solution)

expansion

exchanges
with D₂O

solvent
residual

TMS

8.0   7.6   ppm

solvent
residual

10   9   8   7   6   5   4   3   2   1   0   δ (ppm)

329

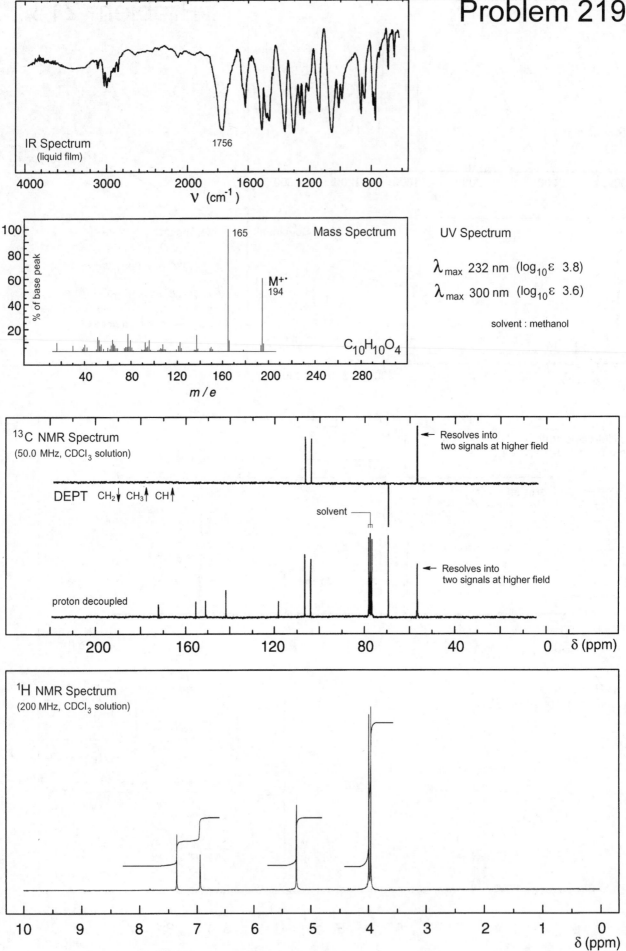

# Problem 219

**IR Spectrum** (liquid film)

1756

$\nu$ (cm$^{-1}$)

**Mass Spectrum**

165

M$^{+\cdot}$
194

$C_{10}H_{10}O_4$

% of base peak

$m/e$

**UV Spectrum**

$\lambda_{max}$ 232 nm (log$_{10}\varepsilon$ 3.8)

$\lambda_{max}$ 300 nm (log$_{10}\varepsilon$ 3.6)

solvent : methanol

$^{13}$C NMR Spectrum
(50.0 MHz, CDCl$_3$ solution)

Resolves into
two signals at higher field

DEPT   CH$_2\downarrow$ CH$_3\uparrow$ CH$\uparrow$

solvent

Resolves into
two signals at higher field

proton decoupled

$\delta$ (ppm)

$^1$H NMR Spectrum
(200 MHz, CDCl$_3$ solution)

$\delta$ (ppm)

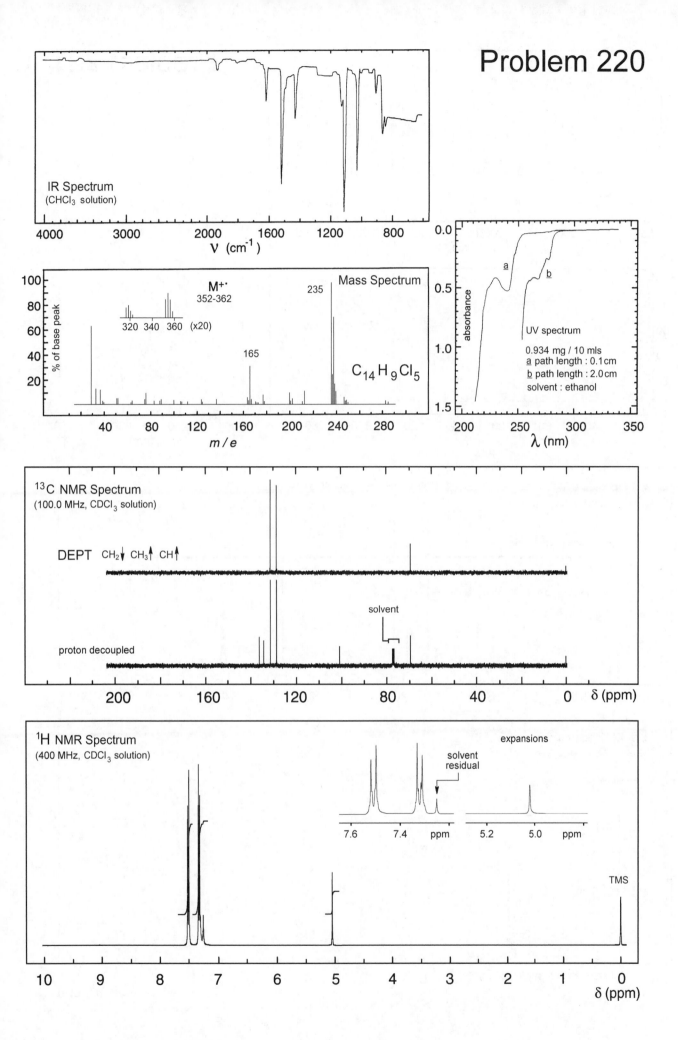

# Problem 220

**IR Spectrum**
(CHCl₃ solution)

ν (cm⁻¹)

**Mass Spectrum**

M⁺·
352-362

235

165

C₁₄H₉Cl₅

m/e

**UV spectrum**

0.934 mg / 10 mls
a path length : 0.1 cm
b path length : 2.0 cm
solvent : ethanol

λ (nm)

**¹³C NMR Spectrum**
(100.0 MHz, CDCl₃ solution)

DEPT   CH₂↓ CH₃↑ CH↑

solvent

proton decoupled

δ (ppm)

**¹H NMR Spectrum**
(400 MHz, CDCl₃ solution)

expansions

solvent
residual

TMS

δ (ppm)

# Problem 221

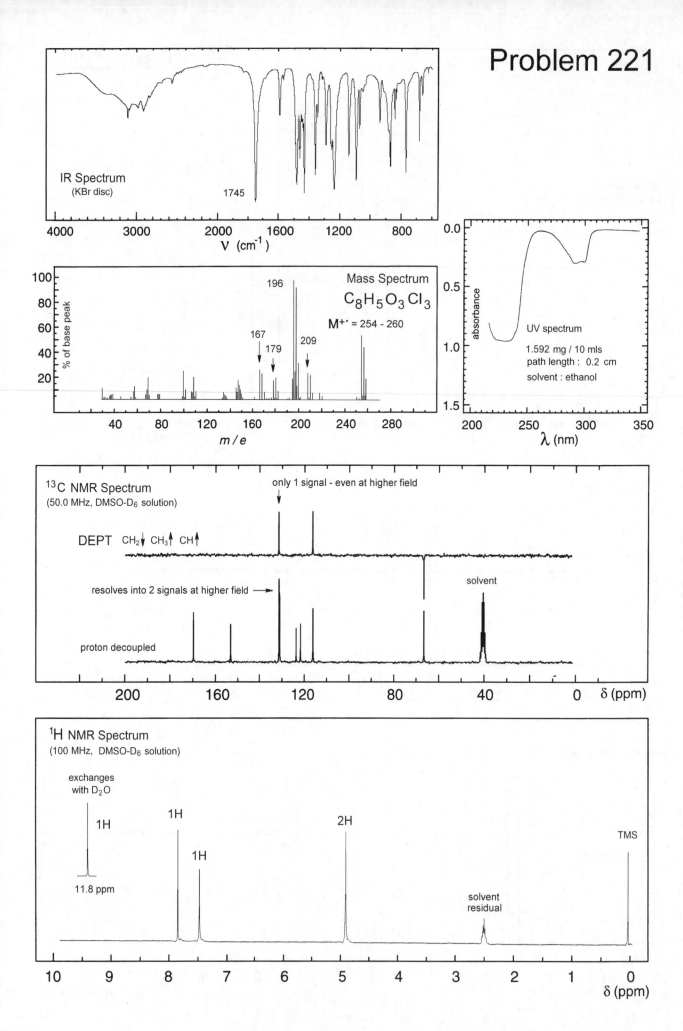

IR Spectrum
(KBr disc)

1745

ν (cm⁻¹)

Mass Spectrum

$C_8H_5O_3Cl_3$

$M^{+\cdot} = 254 - 260$

196

167

179

209

% of base peak

m/e

UV spectrum

1.592 mg / 10 mls
path length : 0.2 cm
solvent : ethanol

absorbance

λ (nm)

¹³C NMR Spectrum
(50.0 MHz, DMSO-D₆ solution)

only 1 signal - even at higher field

DEPT    CH₂↓ CH₃↑ CH↑

resolves into 2 signals at higher field →

solvent

proton decoupled

δ (ppm)

¹H NMR Spectrum
(100 MHz, DMSO-D₆ solution)

exchanges
with D₂O

1H

11.8 ppm

1H

1H

2H

solvent
residual

TMS

δ (ppm)

# Problem 222

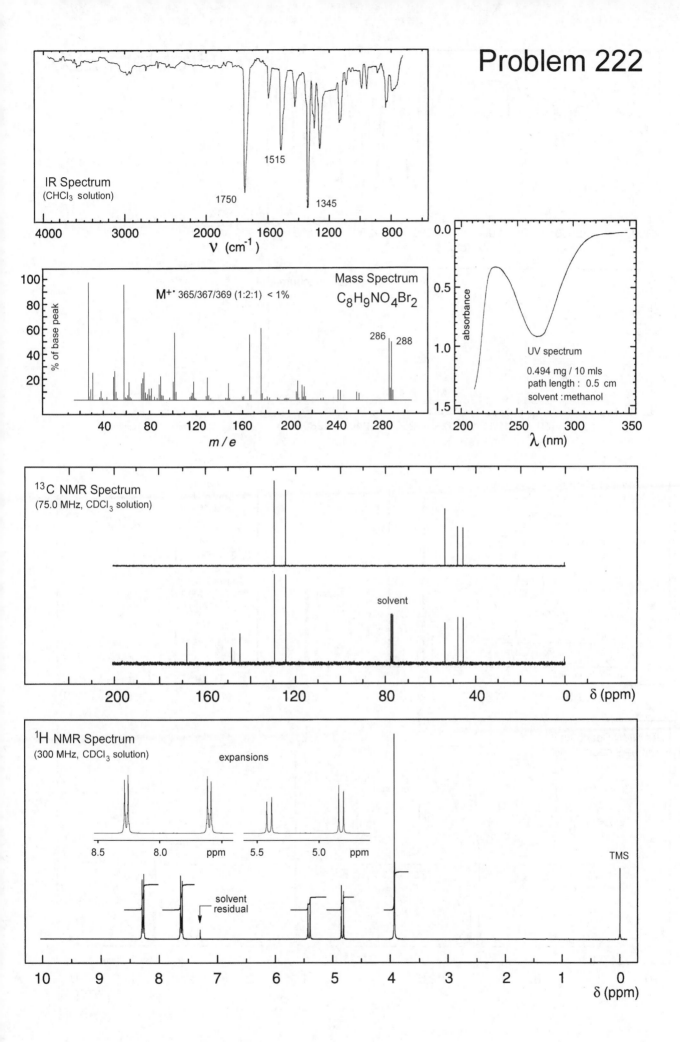

IR Spectrum
(CHCl₃ solution)

1750

1515

1345

ν (cm⁻¹)

Mass Spectrum
$C_8H_9NO_4Br_2$

M⁺˙ 365/367/369 (1:2:1) < 1%

% of base peak

286  288

m / e

UV spectrum

0.494 mg / 10 mls
path length :  0.5 cm
solvent : methanol

absorbance

λ (nm)

¹³C NMR Spectrum
(75.0 MHz, CDCl₃ solution)

solvent

δ (ppm)

¹H NMR Spectrum
(300 MHz, CDCl₃ solution)

expansions

8.5    8.0    ppm    5.5    5.0    ppm

solvent
residual

TMS

δ (ppm)

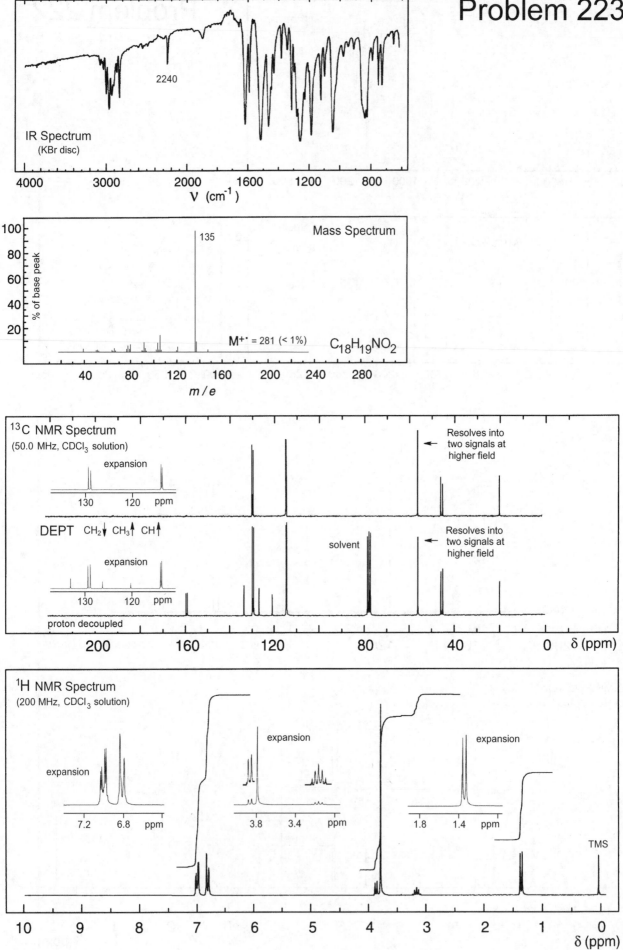

Problem 223

IR Spectrum
(KBr disc)

2240

$\nu$ (cm$^{-1}$)

Mass Spectrum

135

M$^{+\bullet}$ = 281 (< 1%)   C$_{18}$H$_{19}$NO$_2$

% of base peak

m / e

$^{13}$C NMR Spectrum
(50.0 MHz, CDCl$_3$ solution)

expansion

Resolves into
two signals at
higher field

DEPT   CH$_2\downarrow$ CH$_3\uparrow$ CH$\uparrow$

expansion

solvent

Resolves into
two signals at
higher field

proton decoupled

$\delta$ (ppm)

$^1$H NMR Spectrum
(200 MHz, CDCl$_3$ solution)

expansion

expansion

expansion

TMS

$\delta$ (ppm)

# Problem 224

IR Spectrum
(liquid film)

1725

ν (cm⁻¹)

Note: ozonolysis of this compound affords acetone

Mass Spectrum

99

155

M⁺·
200

$C_{10}H_{16}O_4$

% of base peak

m / e

UV spectrum

0.307 mg / 10 mls
path length : 0.5 cm
solvent : ethanol

absorbance

λ (nm)

<sup></sup>

$^{13}C$ NMR Spectrum
(100.0 MHz, CDCl₃ solution)

DEPT   CH₂↓ CH₃↑ CH↑

proton decoupled

solvent

δ (ppm)

$^1H$ NMR Spectrum
(400 MHz, CDCl₃ solution)

expansions

4.5          4.0  ppm        1.5          1.0  ppm

TMS

δ (ppm)

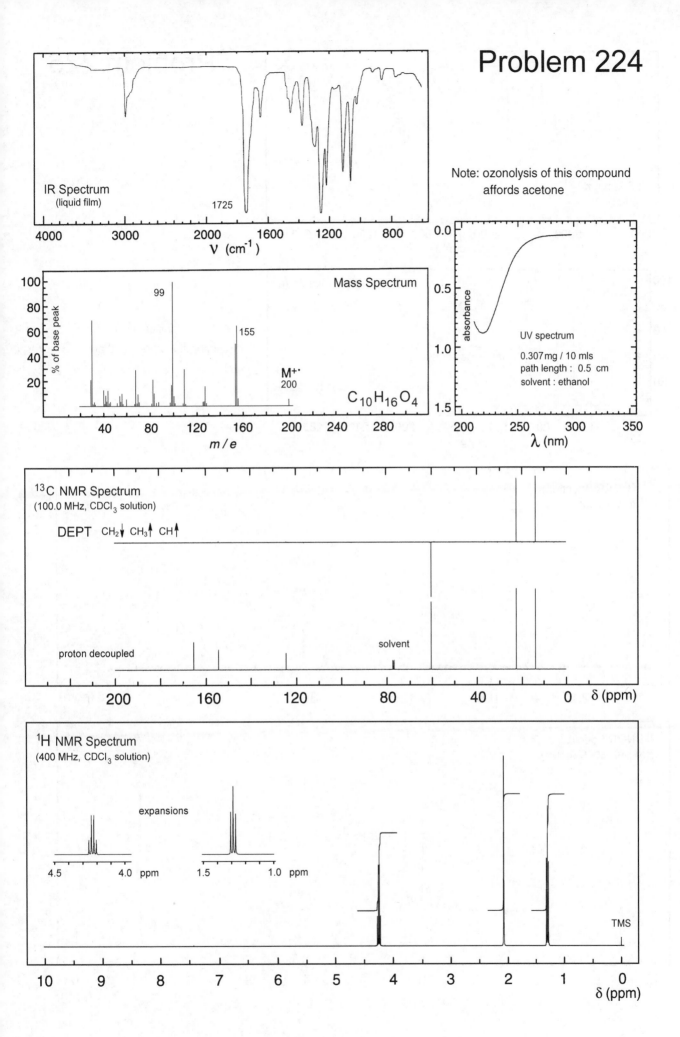

335

# Problem 225

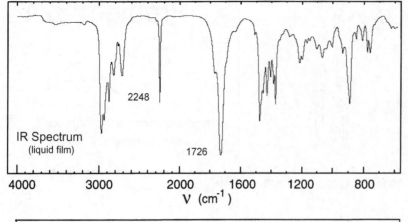

IR Spectrum
(liquid film)

2248

1726

ν (cm⁻¹)

4000  3000  2000  1600  1200  800

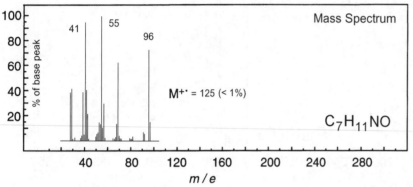

Mass Spectrum

% of base peak

41

55

96

M⁺· = 125 (< 1%)

$C_7H_{11}NO$

m/e

40  80  120  160  200  240  280

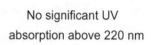

No significant UV
absorption above 220 nm

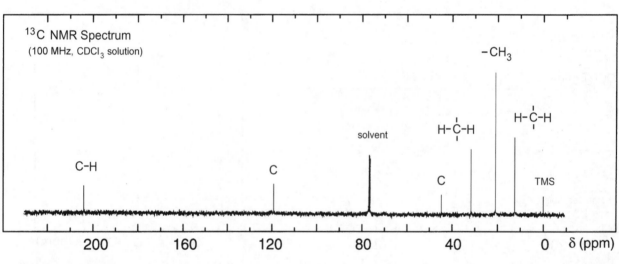

¹³C NMR Spectrum
(100 MHz, CDCl₃ solution)

−CH₃

H−C−H

C−H

C

H−C−H

C

solvent

TMS

δ (ppm)

200  160  120  80  40  0

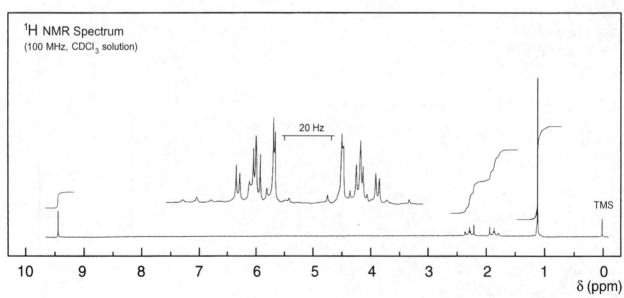

¹H NMR Spectrum
(100 MHz, CDCl₃ solution)

20 Hz

TMS

10  9  8  7  6  5  4  3  2  1  0

δ (ppm)

336

# Problem 226

IR Spectrum
(liquid film)

1725

ν (cm⁻¹)

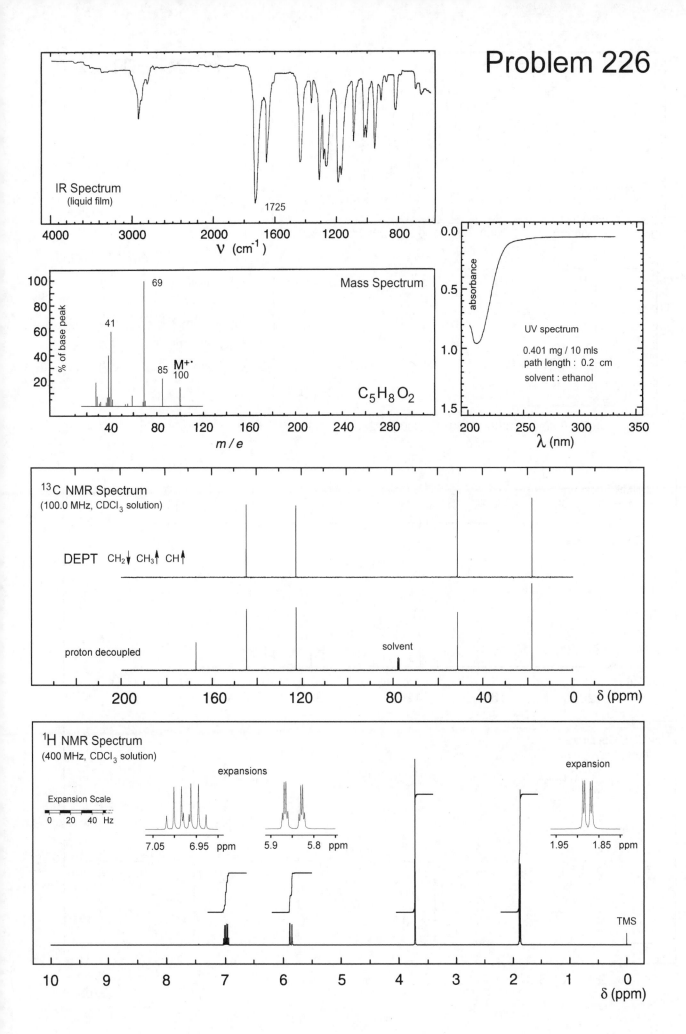

Mass Spectrum

69

41

85

M⁺•
100

$C_5H_8O_2$

% of base peak

m/e

UV spectrum

0.401 mg / 10 mls
path length : 0.2 cm
solvent : ethanol

absorbance

λ (nm)

¹³C NMR Spectrum
(100.0 MHz, CDCl₃ solution)

DEPT   CH₂↓ CH₃↑ CH↑

proton decoupled

solvent

δ (ppm)

¹H NMR Spectrum
(400 MHz, CDCl₃ solution)

expansions

Expansion Scale

0   20   40  Hz

7.05      6.95  ppm

5.9      5.8  ppm

expansion

1.95     1.85  ppm

TMS

δ (ppm)

# Problem 227

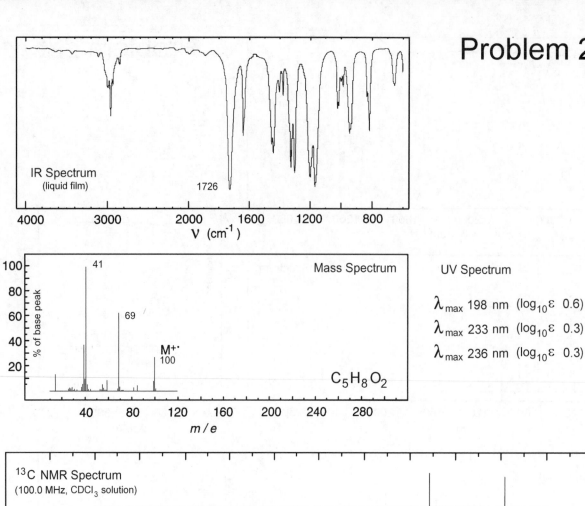

IR Spectrum
(liquid film)

1726

ν (cm⁻¹)

$\lambda_{max}$ 198 nm (log₁₀ε 0.6)

$\lambda_{max}$ 233 nm (log₁₀ε 0.3)

$\lambda_{max}$ 236 nm (log₁₀ε 0.3)

Mass Spectrum

UV Spectrum

41

69

M⁺·
100

$C_5H_8O_2$

% of base peak

m / e

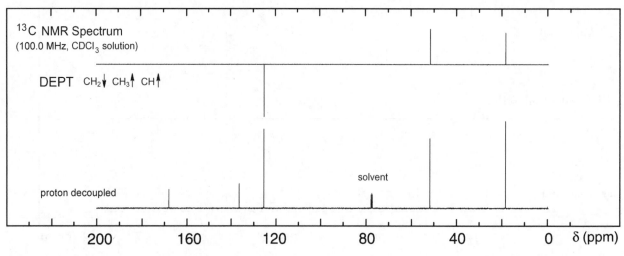

¹³C NMR Spectrum
(100.0 MHz, CDCl₃ solution)

DEPT   CH₂↓ CH₃↑ CH↑

solvent

proton decoupled

δ (ppm)

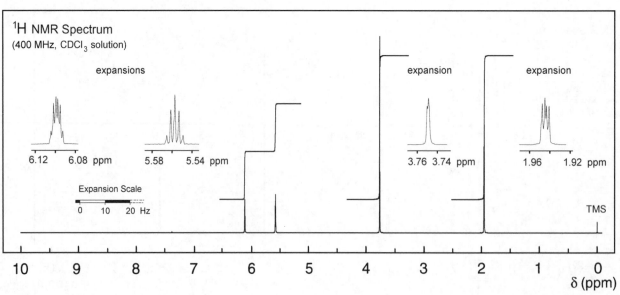

¹H NMR Spectrum
(400 MHz, CDCl₃ solution)

expansions

6.12   6.08 ppm

5.58   5.54 ppm

expansion

3.76 3.74 ppm

expansion

1.96   1.92 ppm

Expansion Scale

0   10   20 Hz

TMS

δ (ppm)

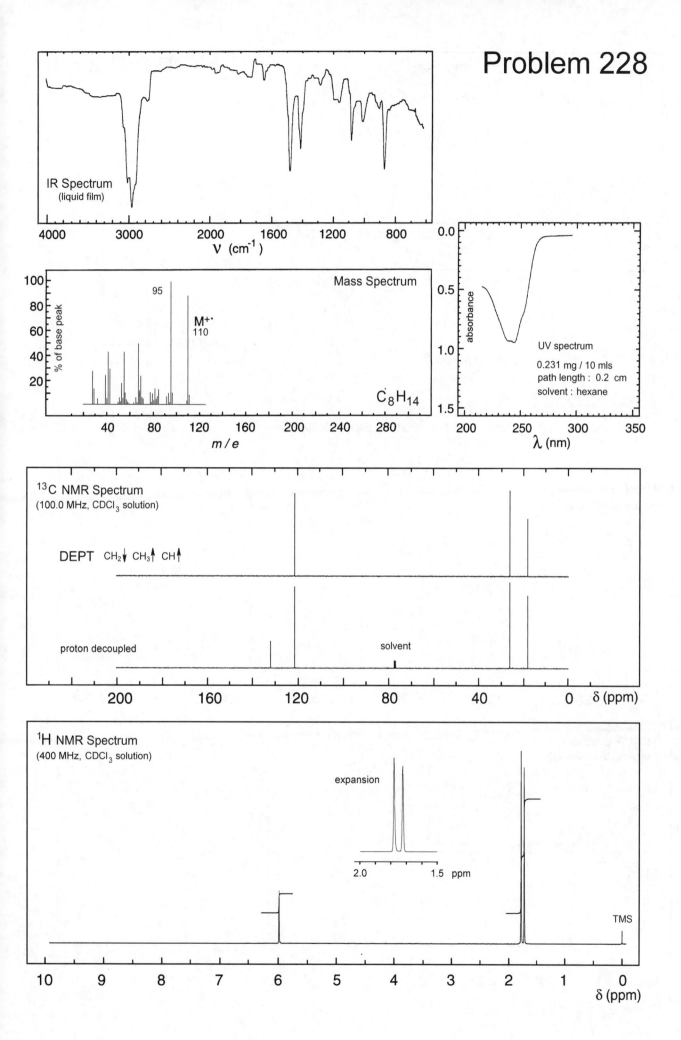

# Problem 228

IR Spectrum
(liquid film)

ν (cm⁻¹)

Mass Spectrum

95

M⁺·
110

% of base peak

m/e

$C_8H_{14}$

UV spectrum

0.231 mg / 10 mls
path length : 0.2 cm
solvent : hexane

λ (nm)

¹³C NMR Spectrum
(100.0 MHz, CDCl₃ solution)

DEPT CH₂↓ CH₃↑ CH↑

proton decoupled

solvent

δ (ppm)

¹H NMR Spectrum
(400 MHz, CDCl₃ solution)

expansion

2.0      1.5   ppm

TMS

δ (ppm)

## Problem 229

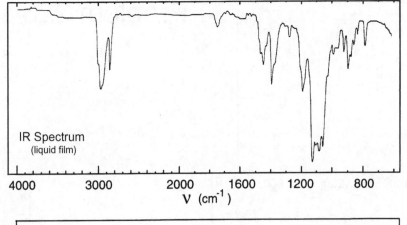

IR Spectrum
(liquid film)

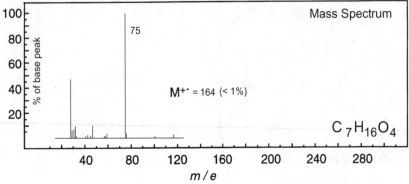

Mass Spectrum

75

$M^{+\cdot} = 164$ (< 1%)

$C_7H_{16}O_4$

No significant UV
absorption above 220 nm

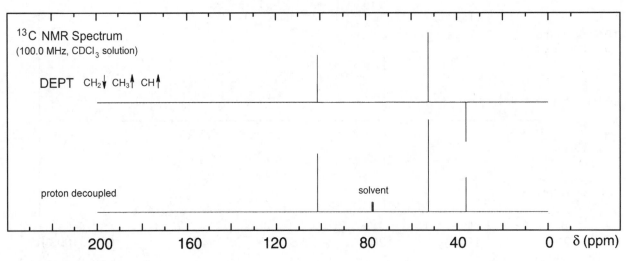

$^{13}C$ NMR Spectrum
(100.0 MHz, CDCl$_3$ solution)

DEPT   CH$_2\downarrow$  CH$_3\uparrow$  CH$\uparrow$

proton decoupled

solvent

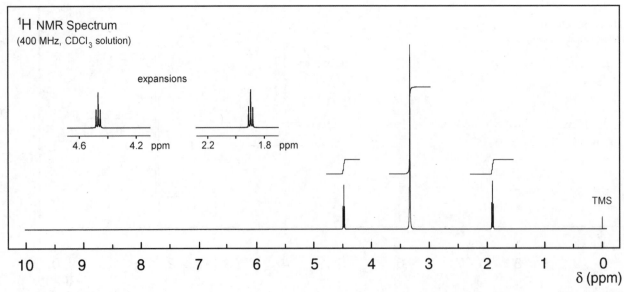

$^1H$ NMR Spectrum
(400 MHz, CDCl$_3$ solution)

expansions

TMS

# Problem 230

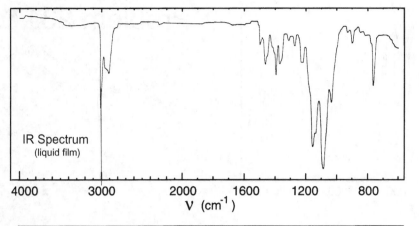

IR Spectrum
(liquid film)

ν (cm⁻¹)

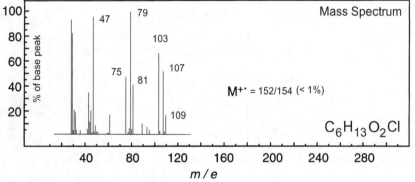

Mass Spectrum

% of base peak

M⁺˙ = 152/154 (< 1%)

$C_6H_{13}O_2Cl$

m/e

No significant UV
absorption above 220 nm

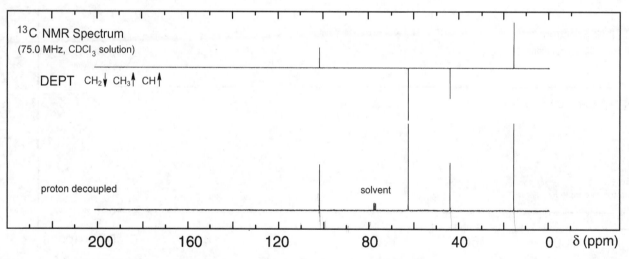

$^{13}C$ NMR Spectrum
(75.0 MHz, CDCl₃ solution)

DEPT   $CH_2\downarrow$  $CH_3\uparrow$  $CH\uparrow$

proton decoupled

solvent

δ (ppm)

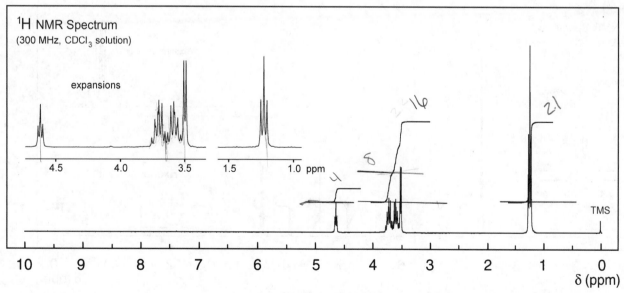

$^1H$ NMR Spectrum
(300 MHz, CDCl₃ solution)

expansions

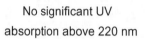

ppm

TMS

δ (ppm)

1:2:4:6

341

# Problem 231

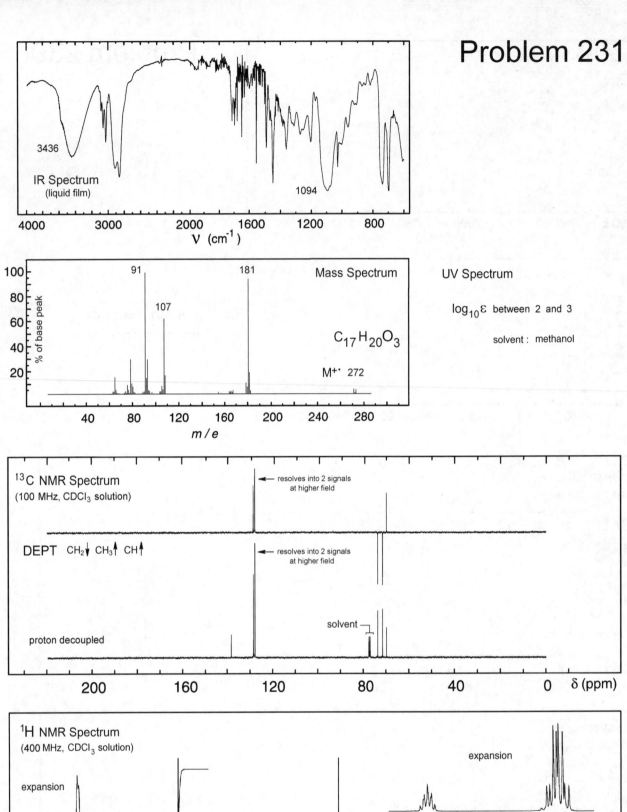

IR Spectrum
(liquid film)

3436

1094

ν (cm⁻¹)

Mass Spectrum

91

107

181

$C_{17}H_{20}O_3$

M⁺· 272

% of base peak

m/e

UV Spectrum

$\log_{10} \varepsilon$ between 2 and 3

solvent : methanol

¹³C NMR Spectrum
(100 MHz, CDCl₃ solution)

resolves into 2 signals
at higher field

DEPT  CH₂↓ CH₃↑ CH↑

resolves into 2 signals
at higher field

solvent

proton decoupled

δ (ppm)

¹H NMR Spectrum
(400 MHz, CDCl₃ solution)

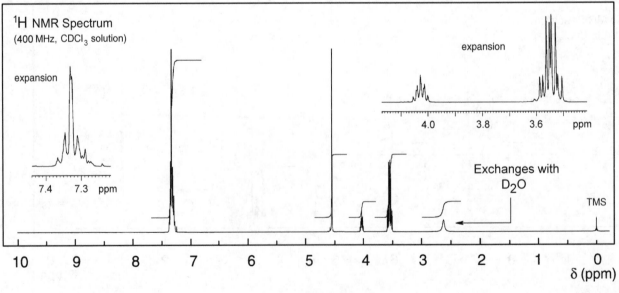

expansion

expansion

Exchanges with
D₂O

TMS

δ (ppm)

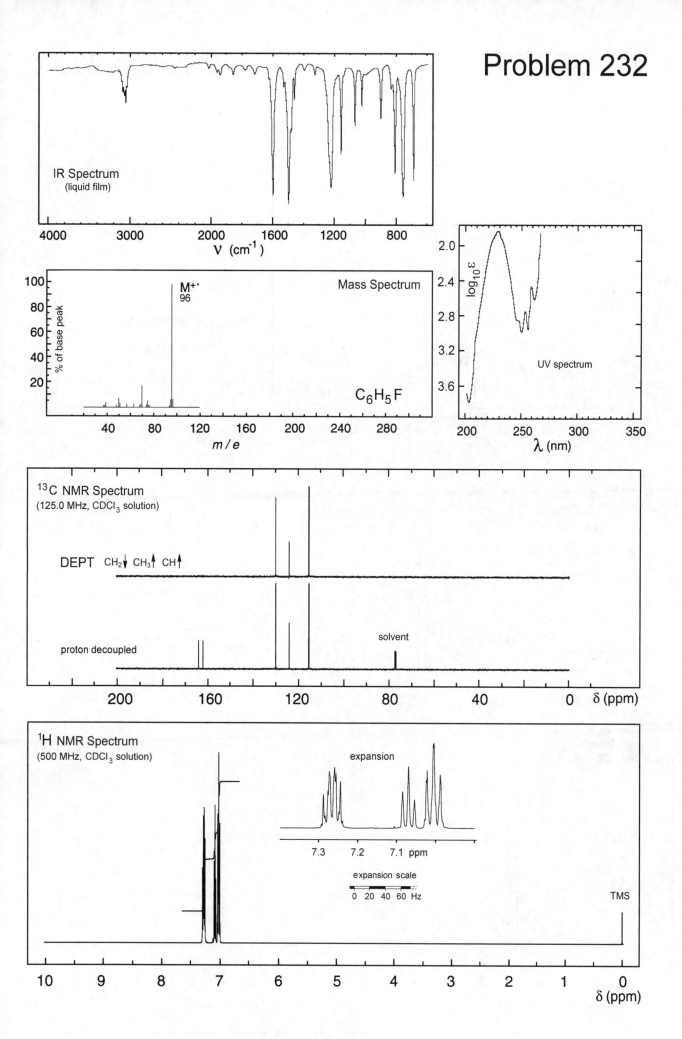

# Problem 232

**IR Spectrum** (liquid film)

ν (cm⁻¹)

**Mass Spectrum**

M⁺˙
96

% of base peak

C₆H₅F

m/e

**UV spectrum**

log₁₀ε

λ (nm)

**¹³C NMR Spectrum** (125.0 MHz, CDCl₃ solution)

DEPT  CH₂↓ CH₃↑ CH↑

proton decoupled

solvent

δ (ppm)

**¹H NMR Spectrum** (500 MHz, CDCl₃ solution)

expansion

7.3    7.2    7.1 ppm

expansion scale

0  20  40  60 Hz

TMS

δ (ppm)

# Problem 233

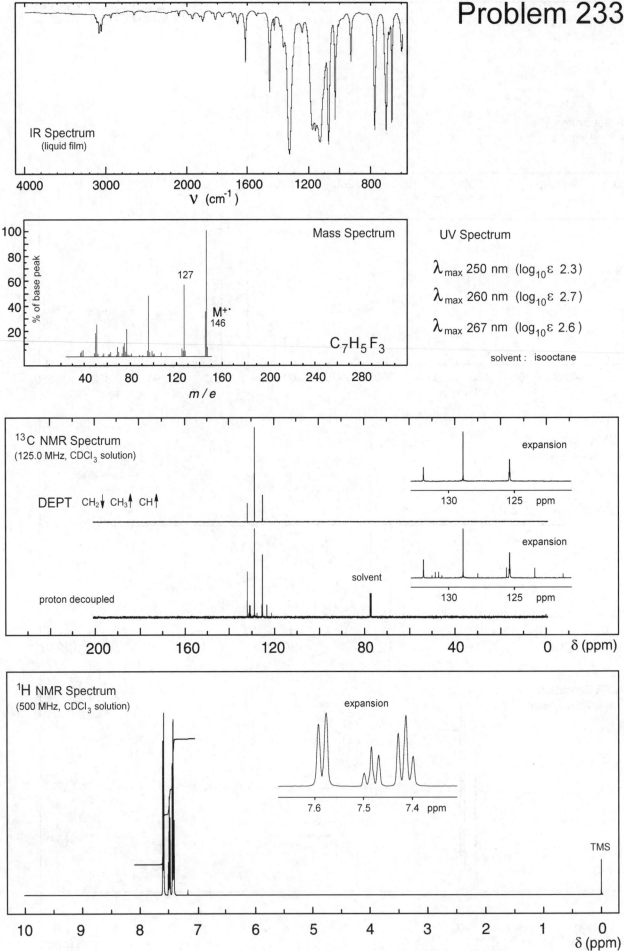

IR Spectrum
(liquid film)

$\nu$ (cm$^{-1}$)

Mass Spectrum

% of base peak

127

M$^{+\cdot}$
146

$C_7H_5F_3$

m/e

UV Spectrum

$\lambda_{max}$ 250 nm (log$_{10}\varepsilon$ 2.3)

$\lambda_{max}$ 260 nm (log$_{10}\varepsilon$ 2.7)

$\lambda_{max}$ 267 nm (log$_{10}\varepsilon$ 2.6)

solvent : isooctane

$^{13}$C NMR Spectrum
(125.0 MHz, CDCl$_3$ solution)

DEPT  CH$_2\downarrow$ CH$_3\uparrow$ CH$\uparrow$

expansion

proton decoupled

solvent

expansion

$\delta$ (ppm)

$^1$H NMR Spectrum
(500 MHz, CDCl$_3$ solution)

expansion

TMS

$\delta$ (ppm)

# Problem 234

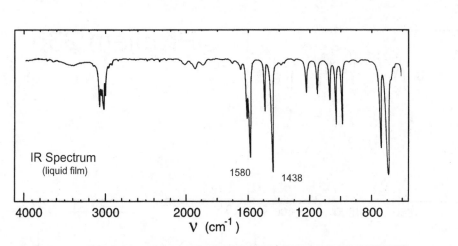

IR Spectrum
(liquid film)

1580   1438

ν (cm⁻¹)

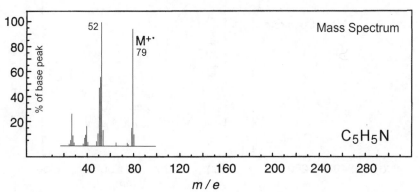

Mass Spectrum

52

M⁺·
79

C₅H₅N

% of base peak

m / e

UV Spectrum

$\lambda_{max}$ 257 nm  $(\log_{10}\varepsilon$ 3.4 )
$\lambda_{max}$ 270 nm  $(\log_{10}\varepsilon$ 2.6 )

solvent : methanol

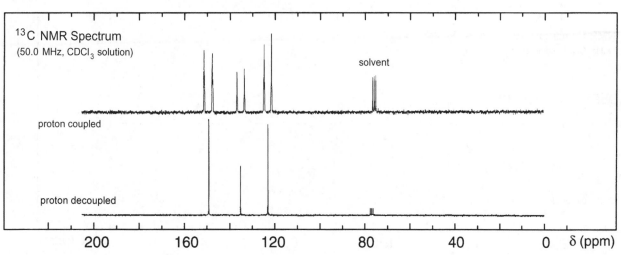

¹³C NMR Spectrum
(50.0 MHz, CDCl₃ solution)

solvent

proton coupled

proton decoupled

δ (ppm)

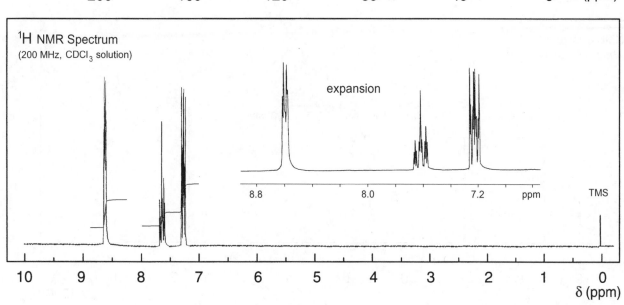

¹H NMR Spectrum
(200 MHz, CDCl₃ solution)

expansion

8.8   8.0   7.2   ppm

TMS

δ (ppm)

# Problem 235

IR Spectrum
(liquid film)

Mass Spectrum

M⁺·
93

66

92

C₆H₇N

UV spectrum

0.636 mg / 10 mls
path length : 0.5 cm

solvent : methanol

¹³C NMR Spectrum
(150.0 MHz, CDCl₃ solution)

DEPT  CH₂↓ CH₃↑ CH↑

proton decoupled

solvent

¹H NMR Spectrum
(600 MHz, CDCl₃ solution)

expansions

8.45    8.40 ppm

7.05    7.00 ppm

TMS

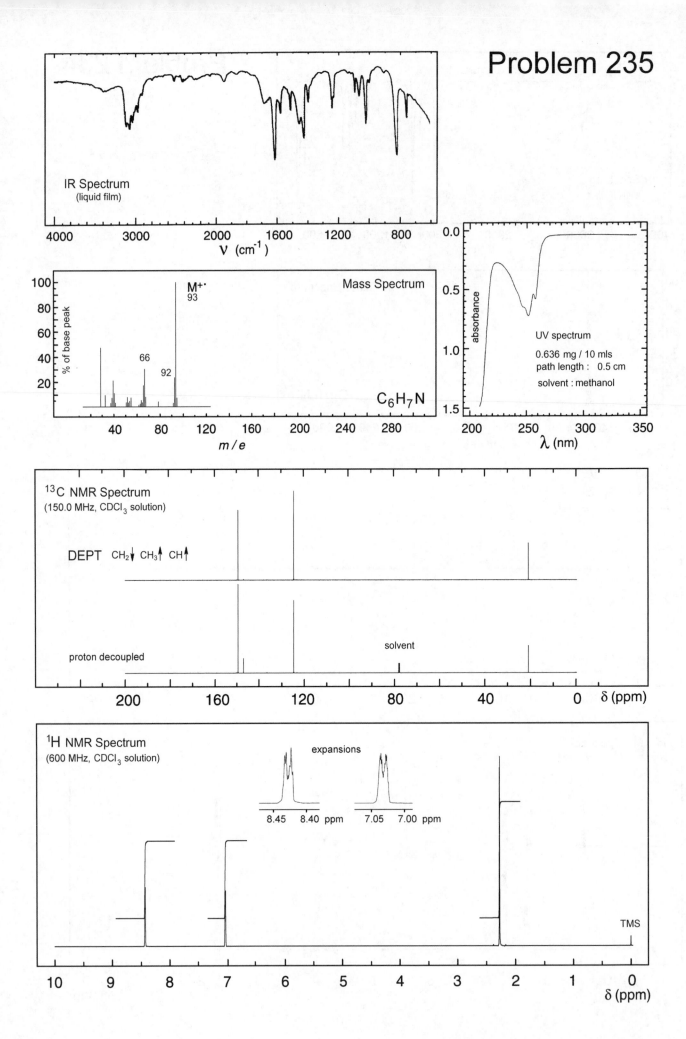

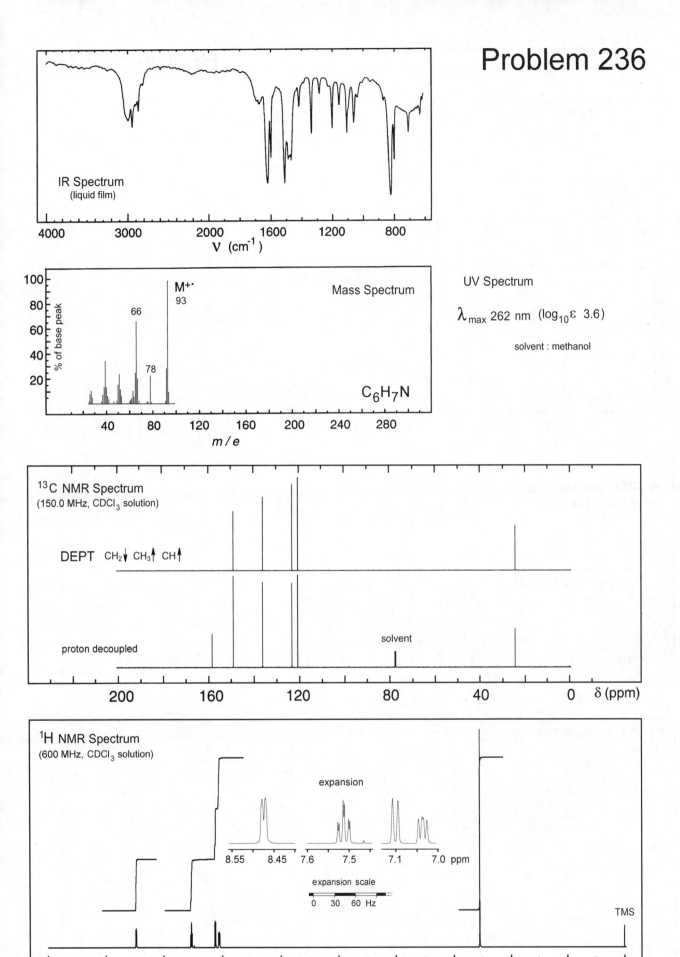

# Problem 236

IR Spectrum
(liquid film)

ν (cm⁻¹)

Mass Spectrum

M⁺·
93

66

78

C₆H₇N

m/e

UV Spectrum

λ_max 262 nm (log₁₀ε 3.6)

solvent : methanol

¹³C NMR Spectrum
(150.0 MHz, CDCl₃ solution)

DEPT   CH₂↓  CH₃↑  CH↑

proton decoupled

solvent

δ (ppm)

¹H NMR Spectrum
(600 MHz, CDCl₃ solution)

expansion

8.55  8.45     7.6   7.5     7.1   7.0  ppm

expansion scale

0   30   60 Hz

TMS

δ (ppm)

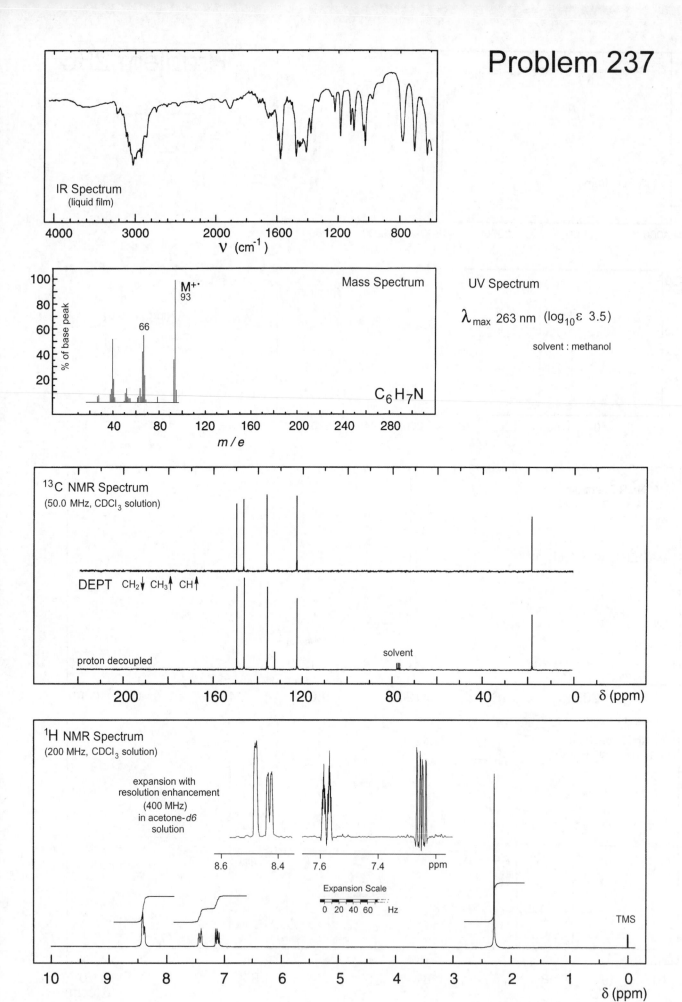

# Problem 237

**IR Spectrum**
(liquid film)

ν (cm⁻¹)

**Mass Spectrum**

M⁺˙
93

66

C₆H₇N

% of base peak

m / e

**UV Spectrum**

$\lambda_{max}$ 263 nm  $(\log_{10}\varepsilon$  3.5$)$

solvent : methanol

**¹³C NMR Spectrum**
(50.0 MHz, CDCl₃ solution)

DEPT  CH₂↓ CH₃↑ CH↑

proton decoupled

solvent

δ (ppm)

**¹H NMR Spectrum**
(200 MHz, CDCl₃ solution)

expansion with
resolution enhancement
(400 MHz)
in acetone-*d6*
solution

ppm

Expansion Scale

0  20  40  60    Hz

TMS

δ (ppm)

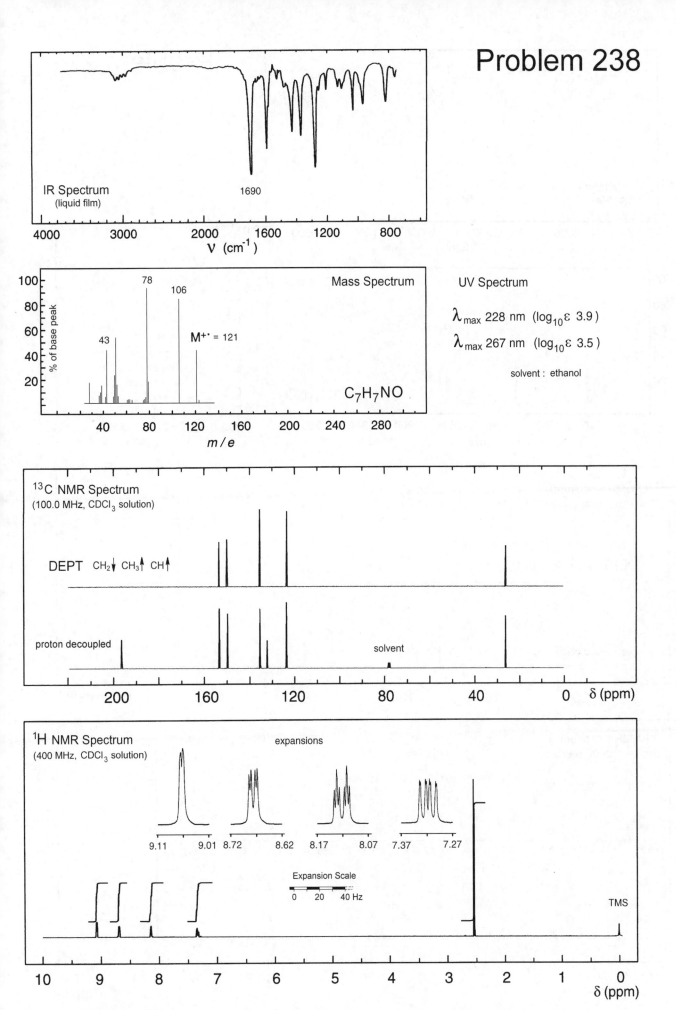

# Problem 238

IR Spectrum
(liquid film)

1690

ν (cm⁻¹)

4000  3000  2000  1600  1200  800

Mass Spectrum

UV Spectrum

λ$_{max}$ 228 nm (log$_{10}$ε 3.9)

λ$_{max}$ 267 nm (log$_{10}$ε 3.5)

solvent : ethanol

% of base peak

78

106

43

M⁺· = 121

$C_7H_7NO$

40  80  120  160  200  240  280

m / e

¹³C NMR Spectrum
(100.0 MHz, CDCl₃ solution)

DEPT  CH₂↓ CH₃↑ CH↑

proton decoupled

solvent

200  160  120  80  40  0  δ (ppm)

¹H NMR Spectrum
(400 MHz, CDCl₃ solution)

expansions

9.11  9.01  8.72  8.62  8.17  8.07  7.37  7.27

Expansion Scale

0  20  40 Hz

TMS

10  9  8  7  6  5  4  3  2  1  0  δ (ppm)

# Problem 239

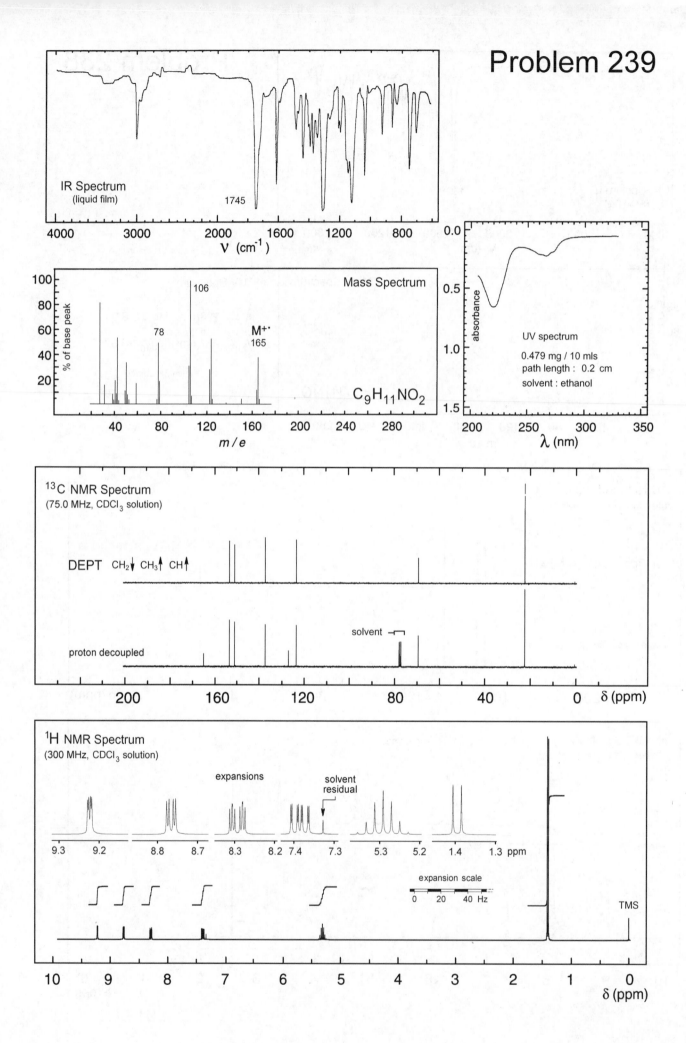

IR Spectrum
(liquid film)

1745

ν (cm⁻¹)

Mass Spectrum

106

78

M⁺˙
165

% of base peak

m / e

C₉H₁₁NO₂

UV spectrum

0.479 mg / 10 mls
path length : 0.2 cm
solvent : ethanol

absorbance

λ (nm)

¹³C NMR Spectrum
(75.0 MHz, CDCl₃ solution)

DEPT   CH₂↓ CH₃↑ CH↑

solvent

proton decoupled

δ (ppm)

¹H NMR Spectrum
(300 MHz, CDCl₃ solution)

expansions

solvent
residual

expansion scale

0   20   40 Hz

TMS

δ (ppm)

# Problem 240

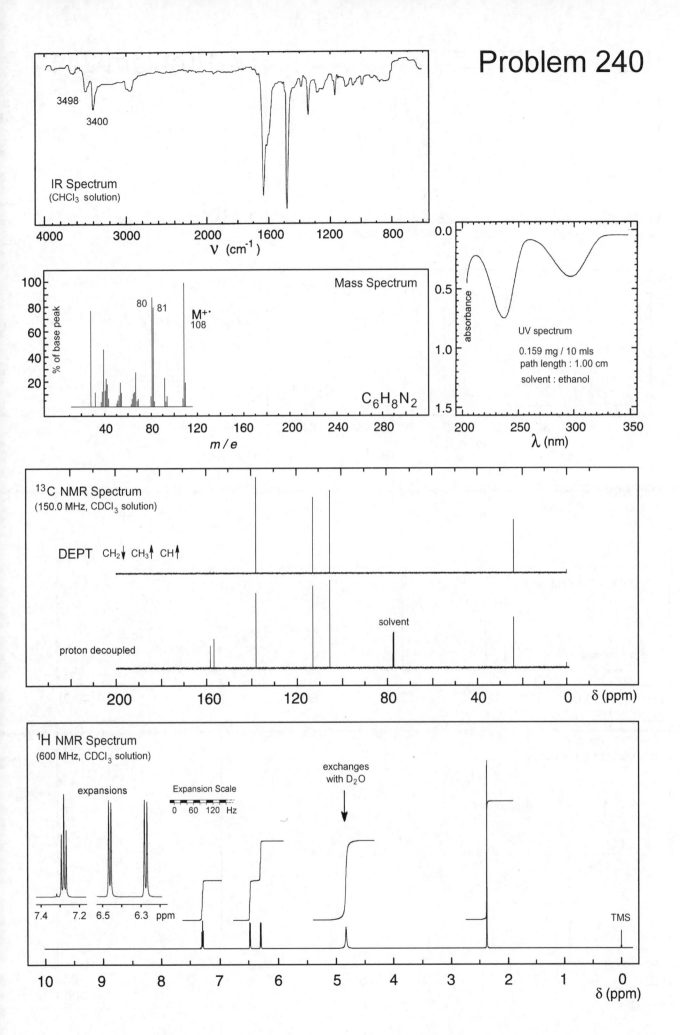

IR Spectrum
(CHCl₃ solution)
3498
3400

ν (cm⁻¹)

Mass Spectrum

% of base peak

80  81
M⁺·
108

C₆H₈N₂

m/e

UV spectrum

0.159 mg / 10 mls
path length : 1.00 cm
solvent : ethanol

absorbance

λ (nm)

¹³C NMR Spectrum
(150.0 MHz, CDCl₃ solution)

DEPT   CH₂↓ CH₃↑ CH↑

proton decoupled

solvent

δ (ppm)

¹H NMR Spectrum
(600 MHz, CDCl₃ solution)

expansions

Expansion Scale
0  60  120  Hz

exchanges
with D₂O

7.4   7.2   6.5   6.3  ppm

TMS

δ (ppm)

351

# Problem 241

IR Spectrum
(liquid film)

1587

$\nu$ (cm$^{-1}$)

Mass Spectrum

M$^{+\cdot}$
94

53
67
79

C$_5$H$_6$N$_2$

UV Spectrum

$\lambda_{max}$ 270 nm (log$_{10}\varepsilon$ 2.5)

$\lambda_{max}$ 240 nm (log$_{10}\varepsilon$ 3.4)

solvent : methanol

$^{13}$C NMR Spectrum
(50.0 MHz, CDCl$_3$ solution)

DEPT  CH$_2\downarrow$ CH$_3\uparrow$ CH$\uparrow$

proton decoupled

solvent

$\delta$ (ppm)

$^1$H NMR Spectrum
(200 MHz, CDCl$_3$ solution)

expansions

9.0   8.4 ppm      7.6   7.0 ppm

TMS

$\delta$ (ppm)

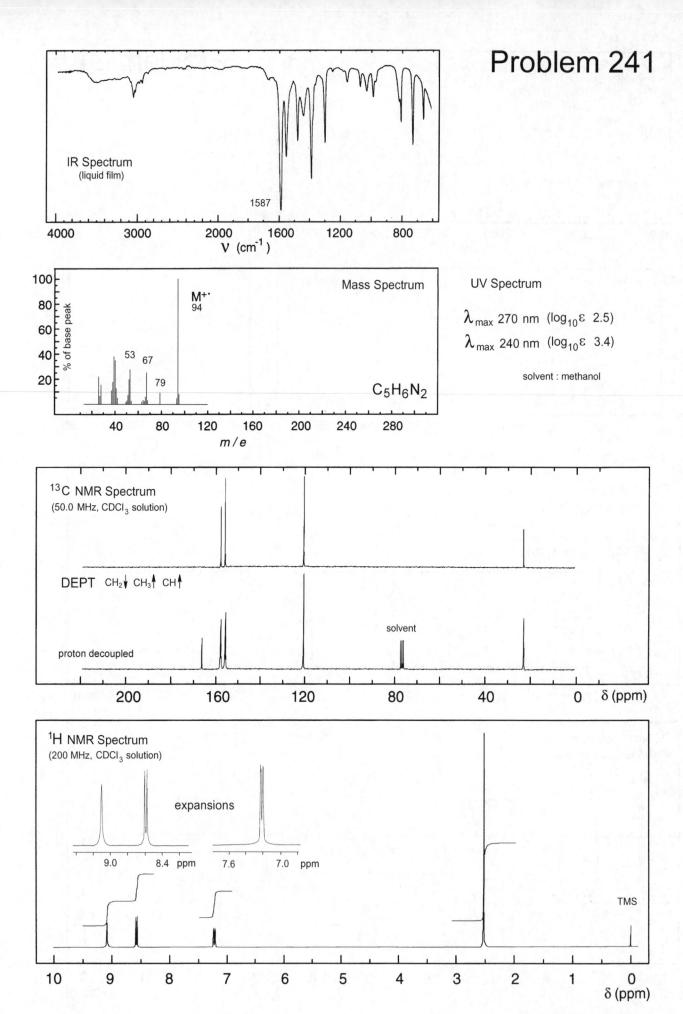

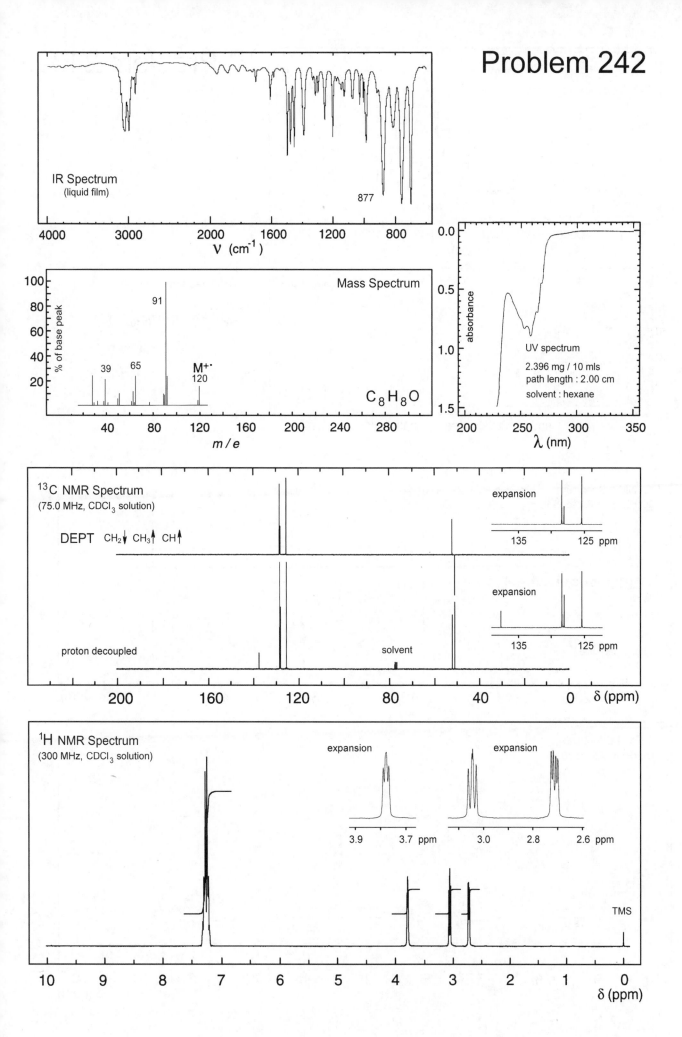

# Problem 242

**IR Spectrum** (liquid film)

877

$\nu$ (cm$^{-1}$)

**Mass Spectrum**

91

39

65

M$^{+\cdot}$
120

$C_8H_8O$

$m/e$

**UV spectrum**
2.396 mg / 10 mls
path length : 2.00 cm
solvent : hexane

absorbance

$\lambda$ (nm)

## $^{13}$C NMR Spectrum
(75.0 MHz, CDCl$_3$ solution)

DEPT  CH$_2\downarrow$ CH$_3\uparrow$ CH$\uparrow$

expansion

135        125 ppm

expansion

135        125 ppm

proton decoupled

solvent

$\delta$ (ppm)

## $^1$H NMR Spectrum
(300 MHz, CDCl$_3$ solution)

expansion

3.9        3.7 ppm

expansion

3.0        2.8        2.6 ppm

TMS

$\delta$ (ppm)

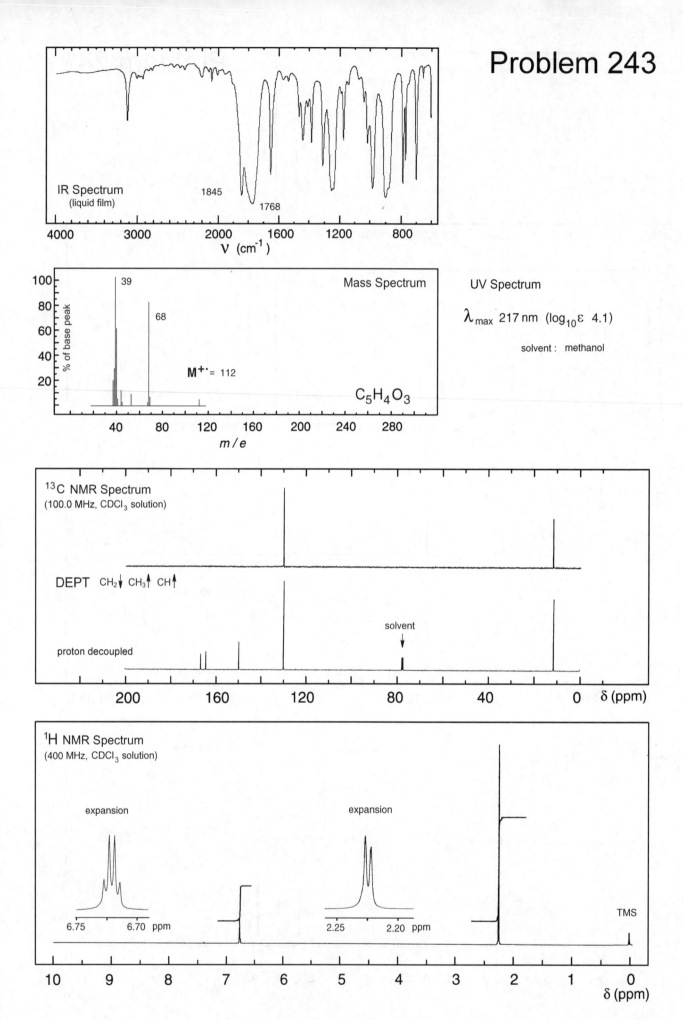

## Problem 243

**IR Spectrum** (liquid film)
1845
1768

ν (cm⁻¹)

**Mass Spectrum**
39
68
M⁺· = 112
$C_5H_4O_3$

**UV Spectrum**
$\lambda_{max}$ 217 nm (log₁₀ε 4.1)
solvent : methanol

**¹³C NMR Spectrum** (100.0 MHz, CDCl₃ solution)

DEPT CH₂↓ CH₃↑ CH↑

solvent

proton decoupled

δ (ppm)

**¹H NMR Spectrum** (400 MHz, CDCl₃ solution)

expansion
6.75    6.70  ppm

expansion
2.25    2.20  ppm

TMS

δ (ppm)

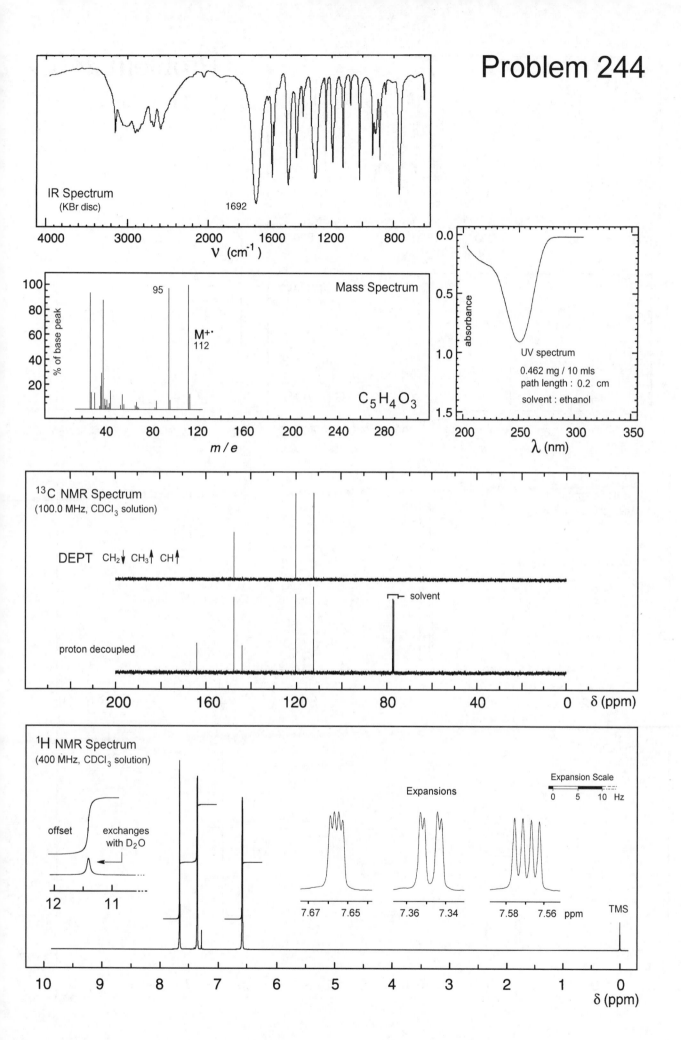

# Problem 244

IR Spectrum
(KBr disc)

1692

ν (cm⁻¹)

Mass Spectrum

95

M⁺·
112

$C_5H_4O_3$

% of base peak

m/e

UV spectrum

0.462 mg / 10 mls
path length : 0.2 cm

solvent : ethanol

absorbance

λ (nm)

¹³C NMR Spectrum
(100.0 MHz, CDCl₃ solution)

DEPT  CH₂↓ CH₃↑ CH↑

solvent

proton decoupled

δ (ppm)

¹H NMR Spectrum
(400 MHz, CDCl₃ solution)

Expansions

Expansion Scale

0   5   10  Hz

offset    exchanges
with D₂O

12    11

7.67   7.65      7.36   7.34      7.58   7.56  ppm

TMS

δ (ppm)

# Problem 245

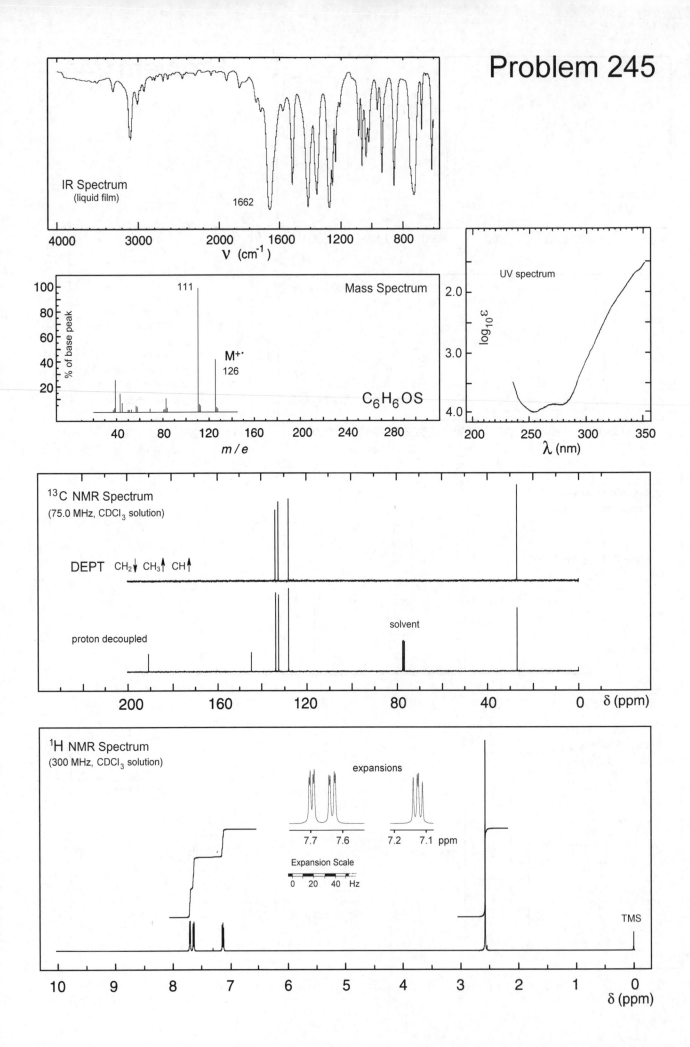

IR Spectrum
(liquid film)

1662

ν (cm⁻¹)

Mass Spectrum

111

M⁺·
126

C₆H₆OS

% of base peak

m/e

UV spectrum

log₁₀ ε

λ (nm)

¹³C NMR Spectrum
(75.0 MHz, CDCl₃ solution)

DEPT   CH₂↓ CH₃↑ CH↑

proton decoupled

solvent

δ (ppm)

¹H NMR Spectrum
(300 MHz, CDCl₃ solution)

expansions

7.7   7.6        7.2   7.1  ppm

Expansion Scale

0   20   40   Hz

TMS

δ (ppm)

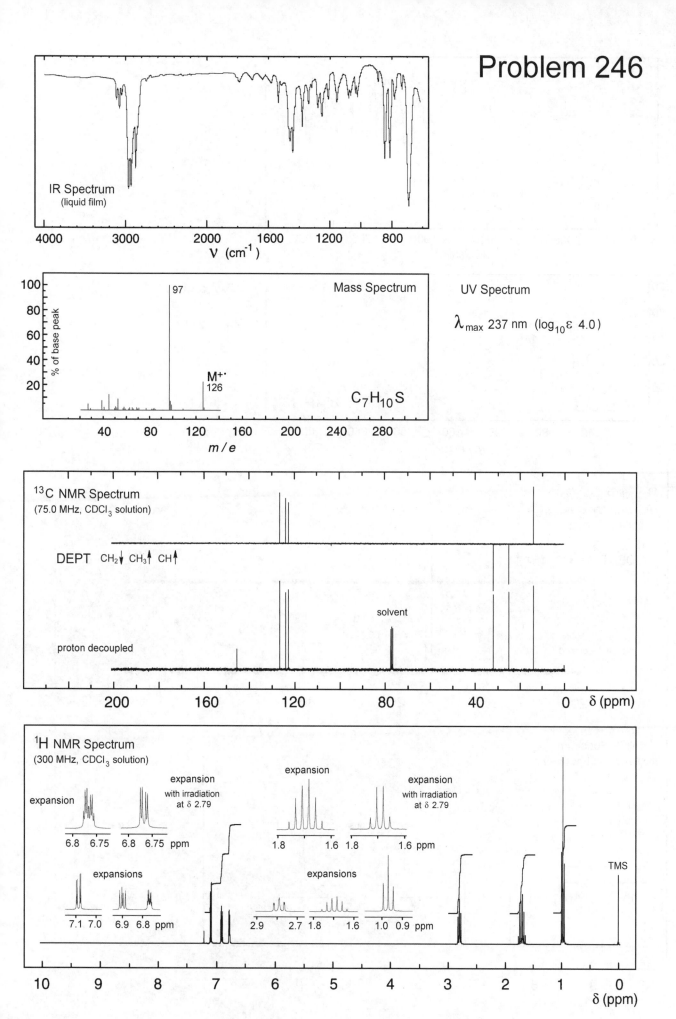

# Problem 246

**IR Spectrum**
(liquid film)

ν (cm⁻¹)

**Mass Spectrum**

% of base peak

97

M⁺˙
126

C₇H₁₀S

m/e

**UV Spectrum**

λ_max 237 nm (log₁₀ε 4.0)

**¹³C NMR Spectrum**
(75.0 MHz, CDCl₃ solution)

DEPT  CH₂↓ CH₃↑ CH↑

solvent

proton decoupled

δ (ppm)

**¹H NMR Spectrum**
(300 MHz, CDCl₃ solution)

expansion

expansion
with irradiation
at δ 2.79

expansion

expansion
with irradiation
at δ 2.79

6.8  6.75 ppm

6.8  6.75 ppm

1.8   1.6  1.8   1.6 ppm

expansion

expansions

expansions

TMS

7.1 7.0    6.9  6.8 ppm

2.9  2.7  1.8   1.6   1.0  0.9 ppm

δ (ppm)

# Problem 247

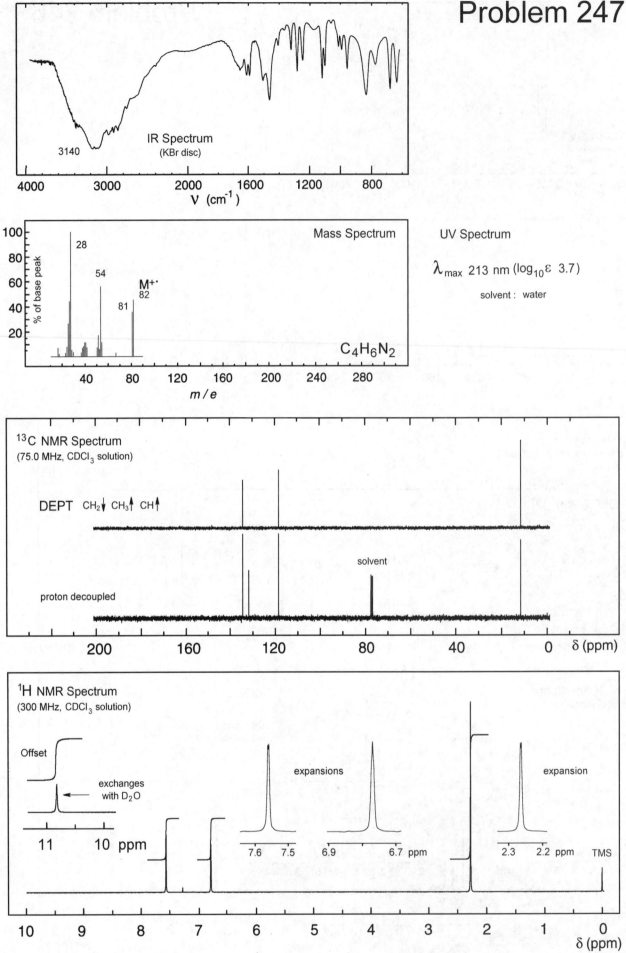

IR Spectrum
(KBr disc)

3140

$\nu$ (cm$^{-1}$)

Mass Spectrum

28

54

81

M$^{+\cdot}$
82

% of base peak

$m/e$

$C_4H_6N_2$

UV Spectrum

$\lambda_{max}$ 213 nm (log$_{10}\varepsilon$ 3.7)

solvent : water

$^{13}$C NMR Spectrum
(75.0 MHz, CDCl$_3$ solution)

DEPT CH$_2\downarrow$ CH$_3\uparrow$ CH$\uparrow$

solvent

proton decoupled

$\delta$ (ppm)

$^1$H NMR Spectrum
(300 MHz, CDCl$_3$ solution)

Offset

exchanges
with D$_2$O

expansions

expansion

11      10 ppm

7.6    7.5        6.9      6.7 ppm        2.3    2.2 ppm      TMS

$\delta$ (ppm)

# Problem 248

IR Spectrum
(KBr disc)

ν (cm⁻¹)

$\nu$ (cm$^{-1}$)

Mass Spectrum

M⁺· = 134

$M^{+\cdot}$ = 134

% of base peak

$C_8H_6S$

UV Spectrum

$\lambda_{max}$ 226 nm (log$_{10}$ε 4.4)

$\lambda_{max}$ 256 nm (log$_{10}$ε 3.7)

$\lambda_{max}$ 288 nm (log$_{10}$ε 3.3)

$\lambda_{max}$ 297 nm (log$_{10}$ε 3.5)

solvent : ethanol

$^{13}C$ NMR Spectrum
(150.0 MHz, C$_6$D$_6$ solution)

DEPT  CH$_2$↓ CH$_3$↑ CH↑

expansion

solvent

solvent

expansion

proton decoupled

δ (ppm)

$^1H$ NMR Spectrum
(600 MHz, C$_6$D$_6$ solution)

expansion

solvent
residual

expansion scale

0   60   120 Hz

TMS

δ (ppm)

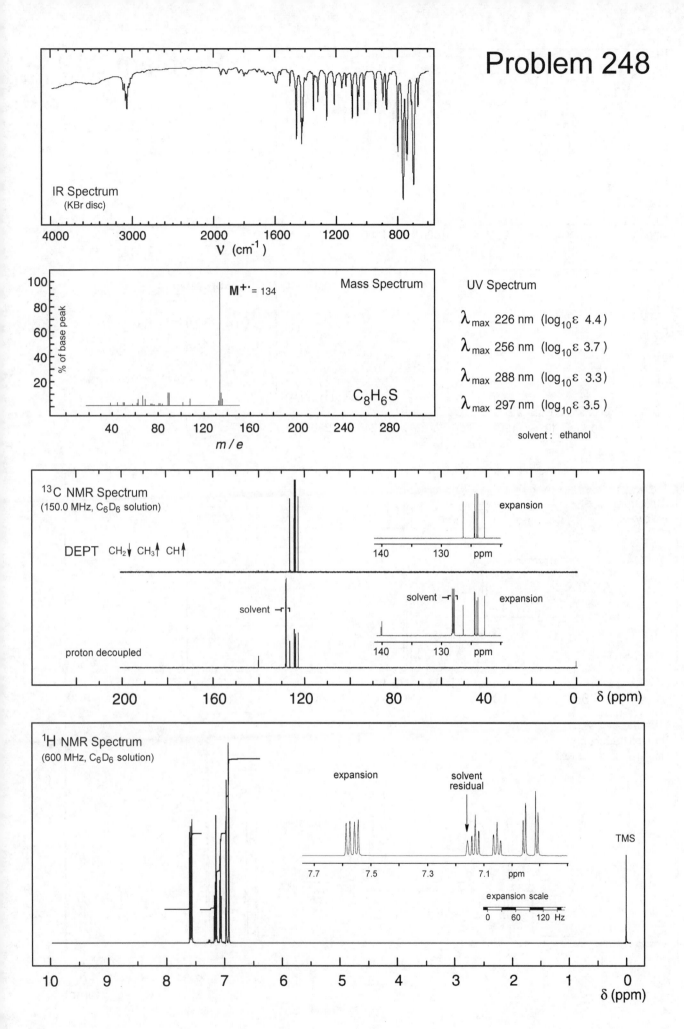

# Problem 249

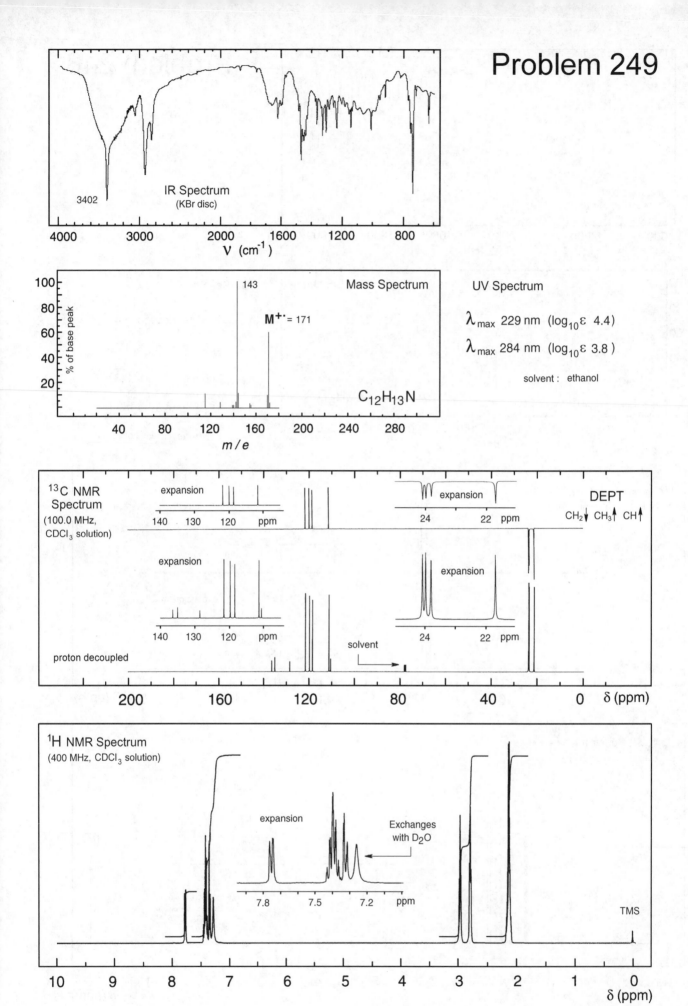

IR Spectrum
(KBr disc)
3402

ν (cm⁻¹)

Mass Spectrum
143
M⁺˙ = 171
C₁₂H₁₃N

% of base peak

m/e

UV Spectrum

$\lambda_{max}$ 229 nm (log₁₀ε 4.4)

$\lambda_{max}$ 284 nm (log₁₀ε 3.8)

solvent : ethanol

¹³C NMR Spectrum
(100.0 MHz, CDCl₃ solution)

expansion
DEPT
CH₂↓ CH₃↑ CH↑

expansion

proton decoupled
solvent

δ (ppm)

¹H NMR Spectrum
(400 MHz, CDCl₃ solution)

expansion
Exchanges with D₂O

TMS

δ (ppm)

360

# Problem 250

IR Spectrum
(liquid film)

1796

ν (cm⁻¹)

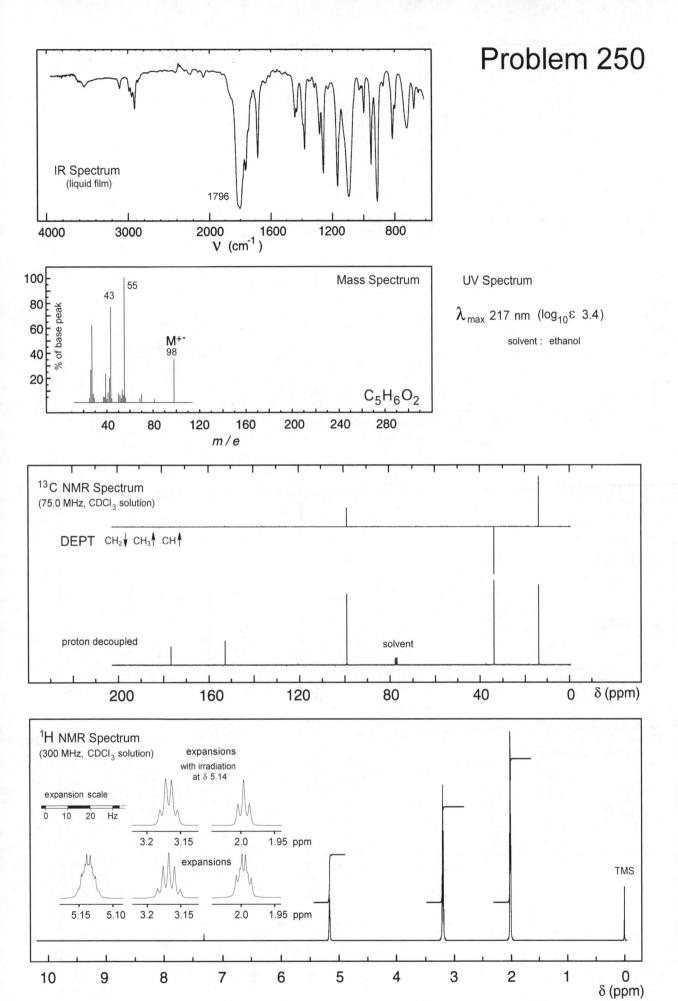

Mass Spectrum

100
80
60
40
20
% of base peak

43

55

M⁺·
98

C₅H₆O₂

m/e

UV Spectrum

$\lambda_{max}$ 217 nm (log₁₀ε 3.4)

solvent : ethanol

¹³C NMR Spectrum
(75.0 MHz, CDCl₃ solution)

DEPT  CH₂↓ CH₃↑ CH↑

proton decoupled

solvent

δ (ppm)

¹H NMR Spectrum
(300 MHz, CDCl₃ solution)

expansions

with irradiation
at δ 5.14

3.2    3.15      2.0    1.95 ppm

expansion scale

0   10   20  Hz

expansions

5.15   5.10    3.2    3.15      2.0    1.95 ppm

TMS

δ (ppm)

# Problem 251

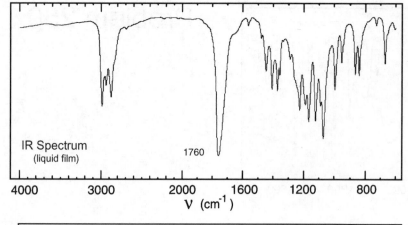

IR Spectrum
(liquid film)

1760

ν (cm⁻¹)

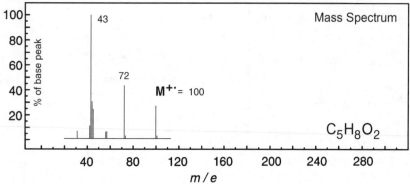

Mass Spectrum

43

72

**M⁺·** = 100

No significant UV
absorption above 220 nm

$C_5H_8O_2$

% of base peak

m/e

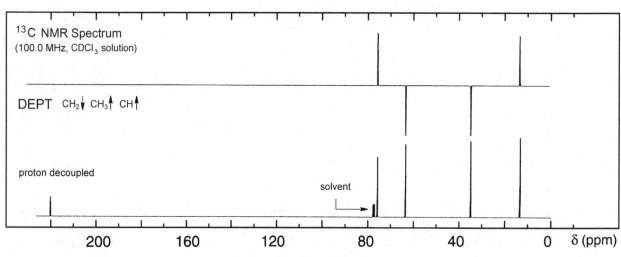

¹³C NMR Spectrum
(100.0 MHz, CDCl₃ solution)

DEPT   CH₂↓ CH₃↑ CH↑

proton decoupled

solvent

δ (ppm)

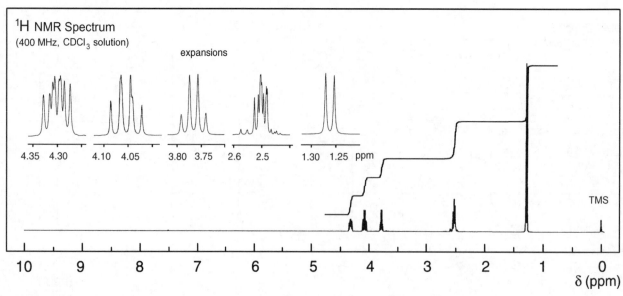

¹H NMR Spectrum
(400 MHz, CDCl₃ solution)

expansions

4.35   4.30     4.10   4.05     3.80   3.75     2.6   2.5     1.30   1.25   ppm

TMS

δ (ppm)

# Problem 252

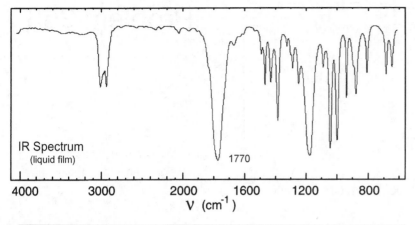

IR Spectrum
(liquid film)

1770

ν (cm⁻¹)

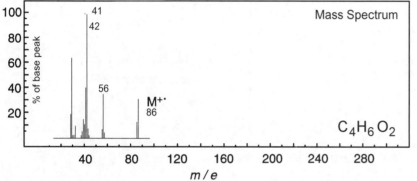

Mass Spectrum

100
80
60
40
20
% of base peak

41
42
56
M⁺·
86

m/e

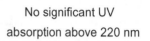

No significant UV
absorption above 220 nm

$C_4H_6O_2$

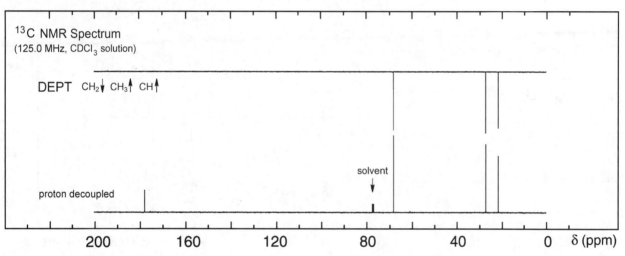

¹³C NMR Spectrum
(125.0 MHz, CDCl₃ solution)

DEPT  CH₂↓ CH₃↑ CH↑

solvent

proton decoupled

δ (ppm)

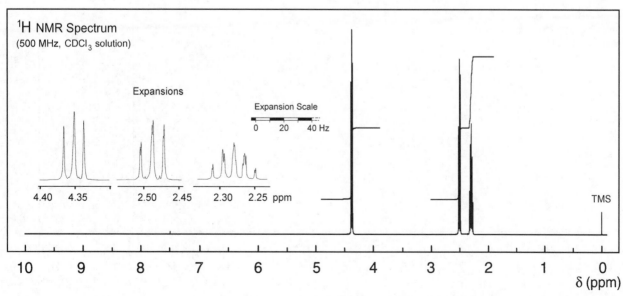

¹H NMR Spectrum
(500 MHz, CDCl₃ solution)

Expansions

Expansion Scale

0    20    40 Hz

4.40  4.35        2.50  2.45        2.30  2.25 ppm

TMS

δ (ppm)

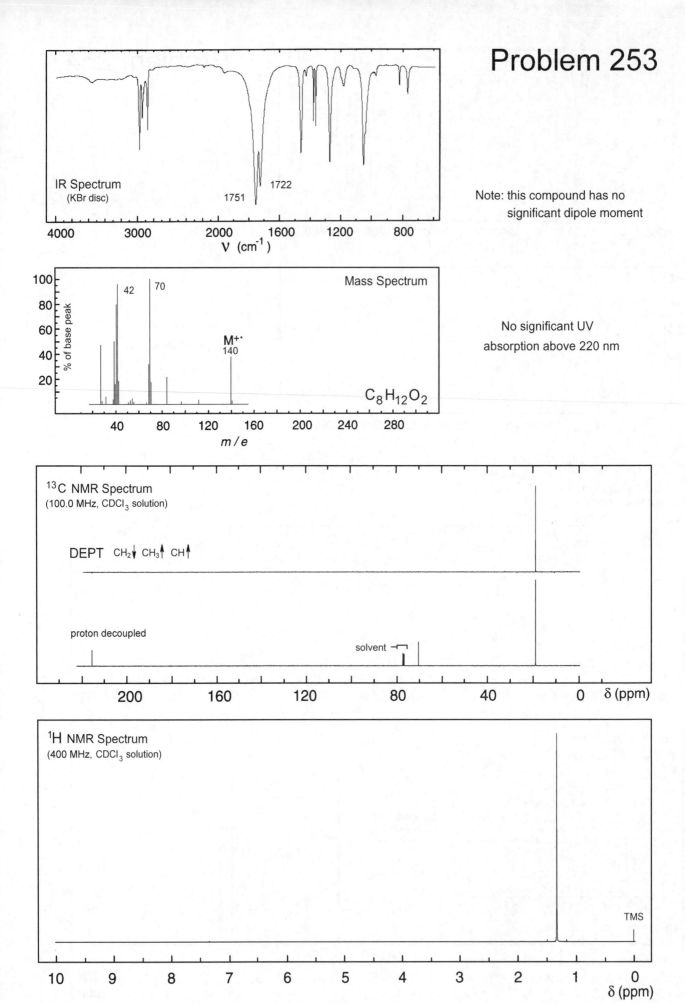

# Problem 253

## IR Spectrum
(KBr disc)

1751
1722

ν (cm⁻¹)
4000  3000  2000  1600  1200  800

Note: this compound has no
significant dipole moment

## Mass Spectrum

% of base peak

42  70

M⁺·
140

$C_8H_{12}O_2$

m/e
40  80  120  160  200  240  280

No significant UV
absorption above 220 nm

## ¹³C NMR Spectrum
(100.0 MHz, CDCl₃ solution)

DEPT  CH₂↓ CH₃↑ CH↑

proton decoupled

solvent

δ (ppm)
200  160  120  80  40  0

## ¹H NMR Spectrum
(400 MHz, CDCl₃ solution)

TMS

δ (ppm)
10  9  8  7  6  5  4  3  2  1  0

# Problem 254

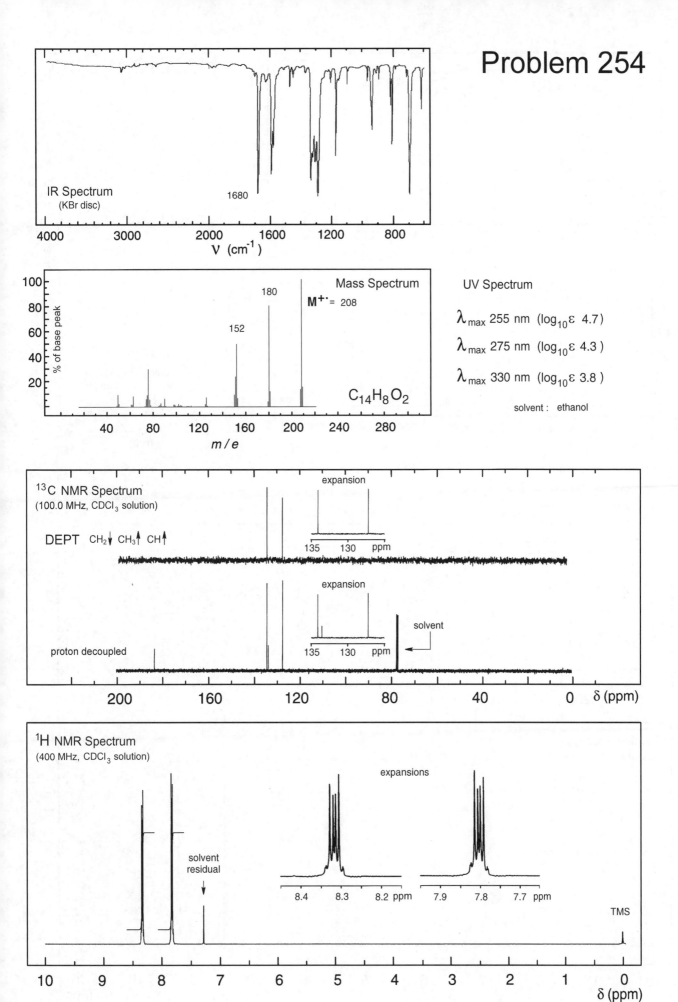

IR Spectrum
(KBr disc)
1680

ν (cm⁻¹)

Mass Spectrum
180
152
M⁺· = 208
100
80
60
40
20
% of base peak
C₁₄H₈O₂

m/e

UV Spectrum

λ max 255 nm (log₁₀ε 4.7)

λ max 275 nm (log₁₀ε 4.3)

λ max 330 nm (log₁₀ε 3.8)

solvent : ethanol

¹³C NMR Spectrum
(100.0 MHz, CDCl₃ solution)

DEPT CH₂↓ CH₃↑ CH↑

expansion
135 130 ppm

expansion
135 130 ppm

proton decoupled

solvent

δ (ppm)

¹H NMR Spectrum
(400 MHz, CDCl₃ solution)

expansions

solvent residual

8.4 8.3 8.2 ppm

7.9 7.8 7.7 ppm

TMS

δ (ppm)

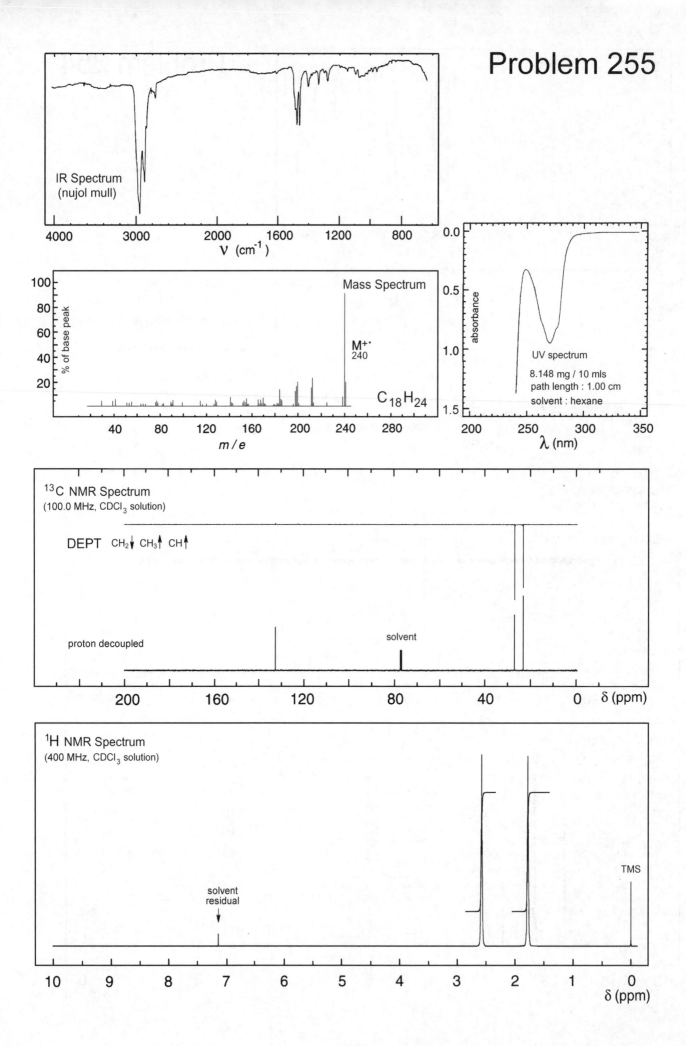

# Problem 255

IR Spectrum
(nujol mull)

ν (cm⁻¹)

Mass Spectrum

% of base peak

M⁺·
240

C₁₈H₂₄

m/e

UV spectrum

absorbance

8.148 mg / 10 mls
path length : 1.00 cm
solvent : hexane

λ (nm)

¹³C NMR Spectrum
(100.0 MHz, CDCl₃ solution)

DEPT   CH₂↓ CH₃↑ CH↑

proton decoupled

solvent

δ (ppm)

¹H NMR Spectrum
(400 MHz, CDCl₃ solution)

solvent
residual

TMS

δ (ppm)

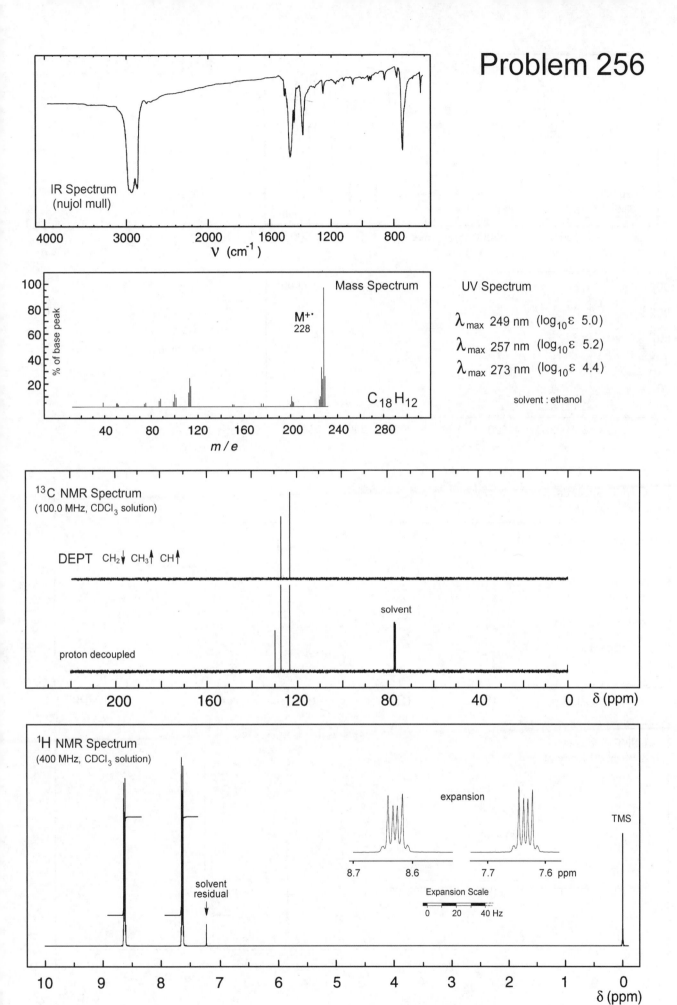

# Problem 256

IR Spectrum
(nujol mull)

ν (cm⁻¹)

Mass Spectrum

M⁺·
228

$C_{18}H_{12}$

% of base peak

m/e

UV Spectrum

$\lambda_{max}$ 249 nm  (log₁₀ε  5.0)

$\lambda_{max}$ 257 nm  (log₁₀ε  5.2)

$\lambda_{max}$ 273 nm  (log₁₀ε  4.4)

solvent : ethanol

¹³C NMR Spectrum
(100.0 MHz, CDCl₃ solution)

DEPT  CH₂↓ CH₃↑ CH↑

solvent

proton decoupled

δ (ppm)

¹H NMR Spectrum
(400 MHz, CDCl₃ solution)

expansion

8.7      8.6        7.7      7.6  ppm

Expansion Scale

0      20      40 Hz

TMS

solvent
residual

δ (ppm)

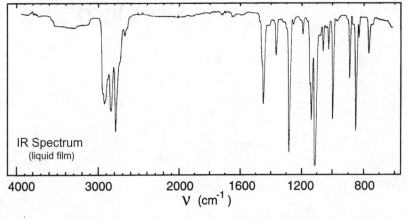

IR Spectrum
(liquid film)

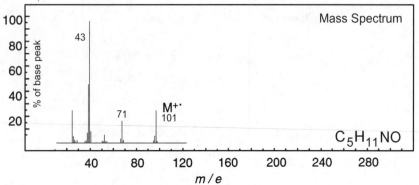

Mass Spectrum

No significant UV
absorption above 220 nm

M⁺·
101

$C_5H_{11}NO$

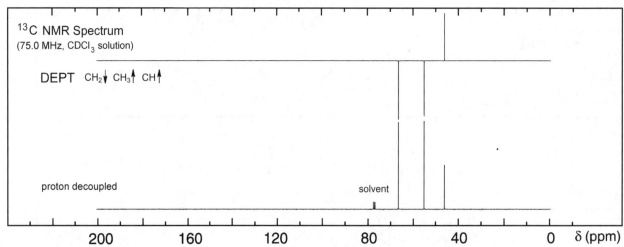

$^{13}C$ NMR Spectrum
(75.0 MHz, CDCl₃ solution)

DEPT   CH₂↓ CH₃↑ CH↑

proton decoupled

solvent

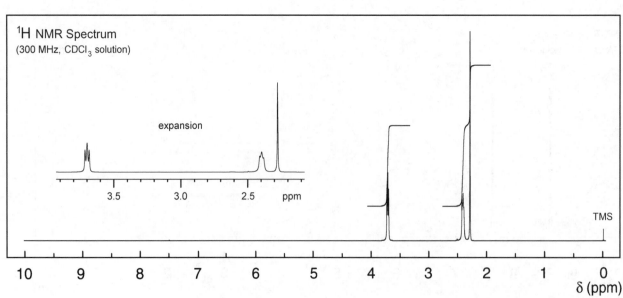

$^1H$ NMR Spectrum
(300 MHz, CDCl₃ solution)

expansion

TMS

# Problem 258

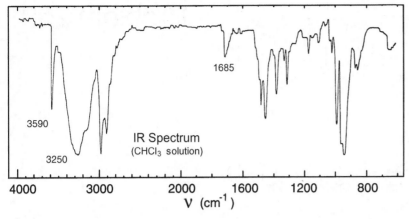

IR Spectrum
(CHCl₃ solution)

3590

3250

1685

$\nu$ (cm⁻¹)

4000 3000 2000 1600 1200 800

Mass Spectrum

No significant UV
absorption above 220 nm

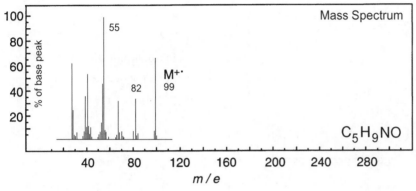

100
80
60
40
20

% of base peak

55

82

M⁺˙
99

$C_5H_9NO$

m/e

40 80 120 160 200 240 280

## ¹³C NMR Spectrum
(100.0 MHz, CDCl₃ solution)

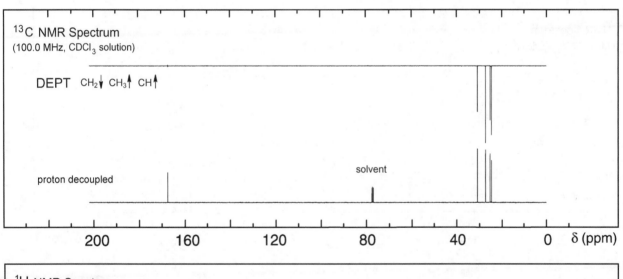

DEPT CH₂↓ CH₃↑ CH↑

proton decoupled

solvent

200 160 120 80 40 0 δ (ppm)

## ¹H NMR Spectrum
(400 MHz, CDCl₃ solution)

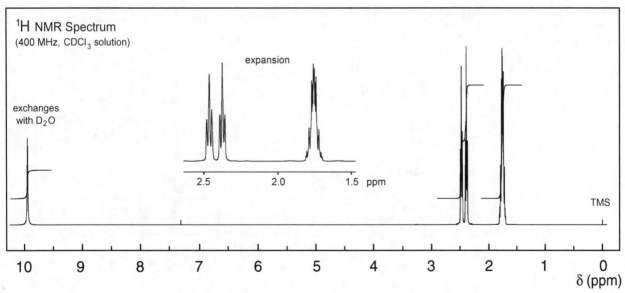

expansion

exchanges
with D₂O

2.5 2.0 1.5 ppm

TMS

10 9 8 7 6 5 4 3 2 1 0
δ (ppm)

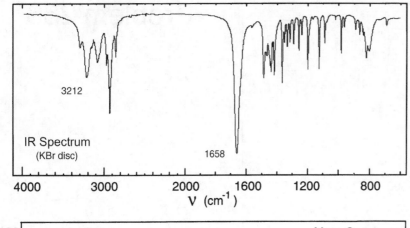

IR Spectrum
(KBr disc)

3212

1658

$V$ (cm$^{-1}$)

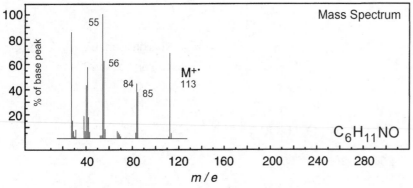

Mass Spectrum

% of base peak

55

56

84

85

M$^{+\cdot}$
113

$C_6H_{11}NO$

$m/e$

No significant UV
absorption above 220 nm

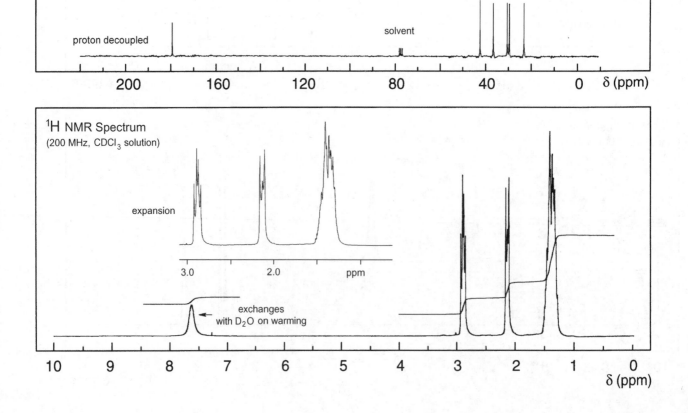

$^{13}$C NMR Spectrum
(50.0 MHz, CDCl$_3$ solution)

DEPT   CH$_2$↓ CH$_3$↑ CH↑

proton decoupled

solvent

δ (ppm)

$^1$H NMR Spectrum
(200 MHz, CDCl$_3$ solution)

expansion

3.0      2.0      ppm

exchanges
with D$_2$O on warming

δ (ppm)

370

# Problem 260

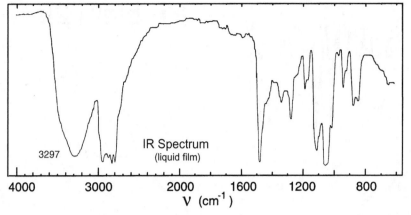

IR Spectrum
(liquid film)

3297

ν (cm⁻¹)

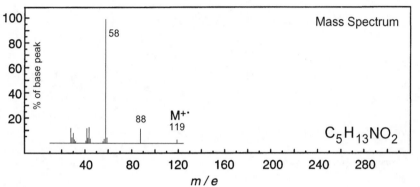

Mass Spectrum

% of base peak

58

88

M⁺·
119

$C_5H_{13}NO_2$

m/e

No significant UV
absorption above 220 nm

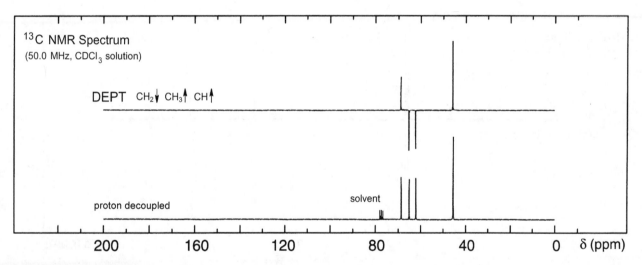

$^{13}C$ NMR Spectrum
(50.0 MHz, CDCl₃ solution)

DEPT   CH₂↓ CH₃↑ CH↑

proton decoupled

solvent

δ (ppm)

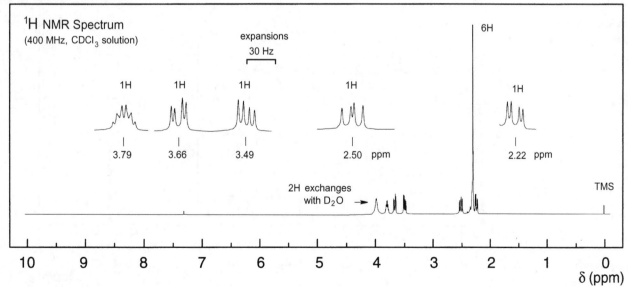

$^1H$ NMR Spectrum
(400 MHz, CDCl₃ solution)

expansions

30 Hz

6H

1H    1H    1H        1H              1H

3.79  3.66  3.49      2.50  ppm       2.22  ppm

2H exchanges
with D₂O

TMS

δ (ppm)

# Problem 261

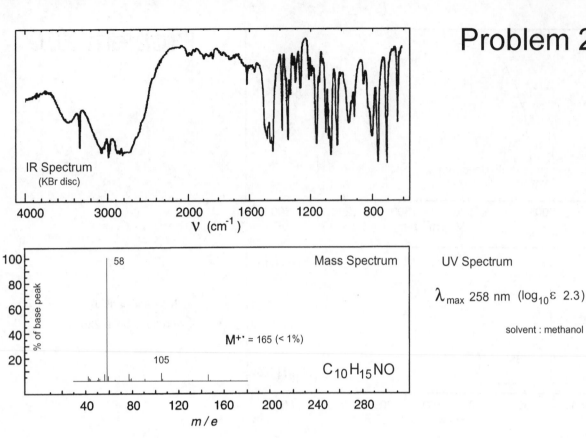

IR Spectrum
(KBr disc)

ν (cm⁻¹)

Mass Spectrum

UV Spectrum

$\lambda_{max}$ 258 nm (log₁₀ε 2.3)

solvent : methanol

M⁺˙ = 165 (< 1%)

C₁₀H₁₅NO

m/e

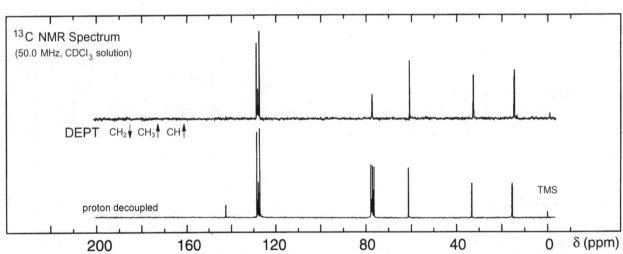

¹³C NMR Spectrum
(50.0 MHz, CDCl₃ solution)

DEPT   CH₂↓ CH₃↑ CH↑

proton decoupled

TMS

δ (ppm)

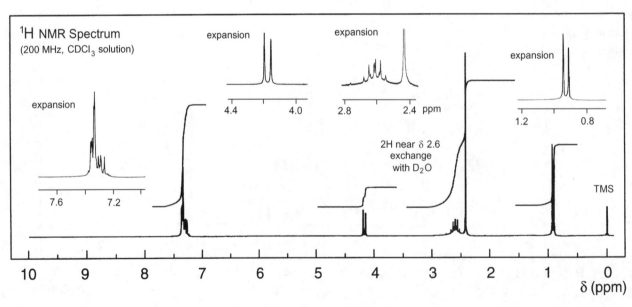

¹H NMR Spectrum
(200 MHz, CDCl₃ solution)

expansion

expansion

expansion

2H near δ 2.6
exchange
with D₂O

TMS

δ (ppm)

## Problem 262

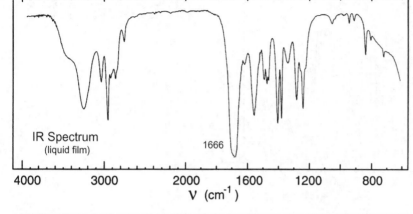

IR Spectrum
(liquid film)

1666

ν (cm⁻¹)

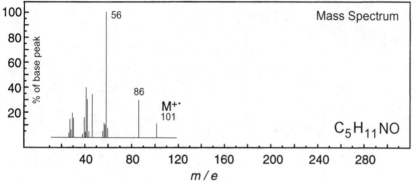

Mass Spectrum

56

86

M⁺˙
101

$C_5H_{11}NO$

% of base peak

m/e

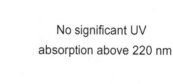

No significant UV
absorption above 220 nm

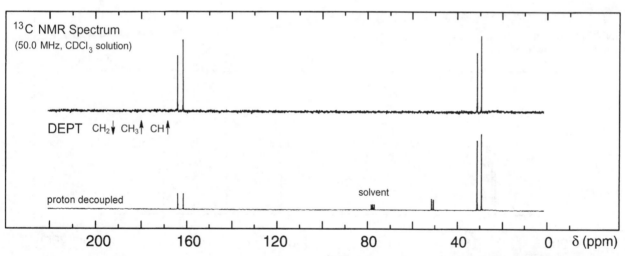

¹³C NMR Spectrum
(50.0 MHz, CDCl₃ solution)

DEPT  CH₂↓ CH₃↑ CH↑

solvent

proton decoupled

δ (ppm)

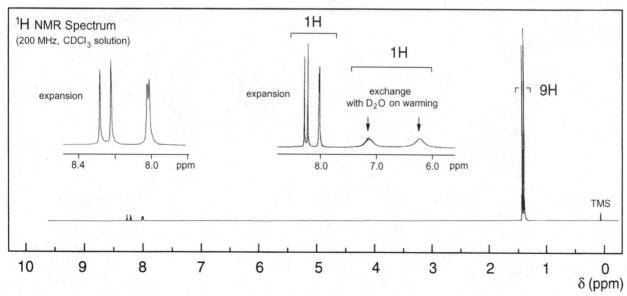

¹H NMR Spectrum
(200 MHz, CDCl₃ solution)

1H

1H

9H

expansion

expansion

exchange
with D₂O on warming

8.4      8.0   ppm

8.0   ppm

7.0      6.0   ppm

TMS

δ (ppm)

373

# Problem 263

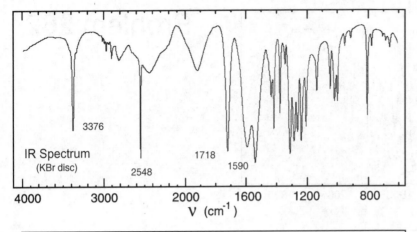

IR Spectrum
(KBr disc)

3376
2548
1718
1590

ν (cm⁻¹)

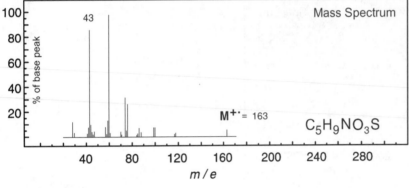

Mass Spectrum

43

% of base peak

$M^{+\cdot} = 163$

$C_5H_9NO_3S$

m/e

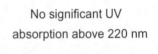

No significant UV
absorption above 220 nm

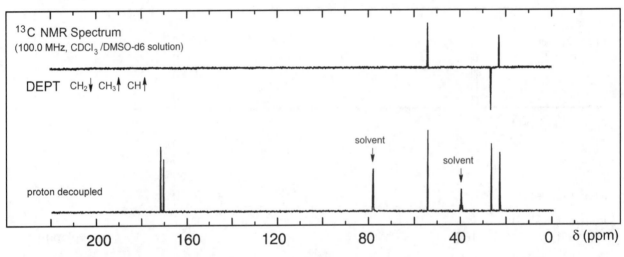

¹³C NMR Spectrum
(100.0 MHz, CDCl₃ /DMSO-d6 solution)

DEPT   CH₂↓ CH₃↑ CH↑

solvent

solvent

proton decoupled

δ (ppm)

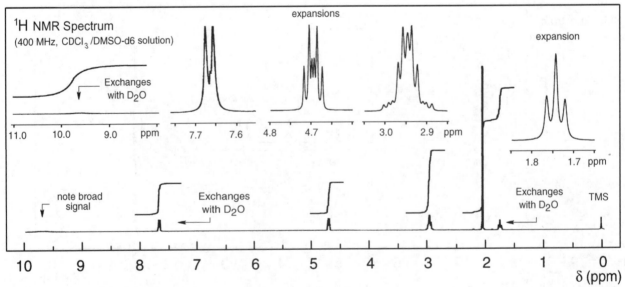

¹H NMR Spectrum
(400 MHz, CDCl₃ /DMSO-d6 solution)

expansions

expansion

Exchanges
with D₂O

Exchanges
with D₂O

note broad
signal

Exchanges
with D₂O

TMS

δ (ppm)

374

# Problem 264

Note: irradiation of the signal at δ 6.58 in the
$^1$H NMR produces an enhancement of the
signals at δ 6.66 and 4.45 ppm via the NOE

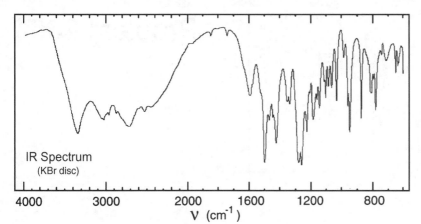

IR Spectrum
(KBr disc)

ν (cm$^{-1}$)

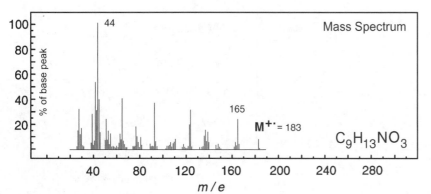

Mass Spectrum

% of base peak

M$^{+ \cdot}$ = 183

C$_9$H$_{13}$NO$_3$

m/e

UV Spectrum

$\lambda_{max}$ 220 nm (log$_{10}$ε 3.8)

$\lambda_{max}$ 280 nm (log$_{10}$ε 3.4)

solvent : H$_2$O pH 7

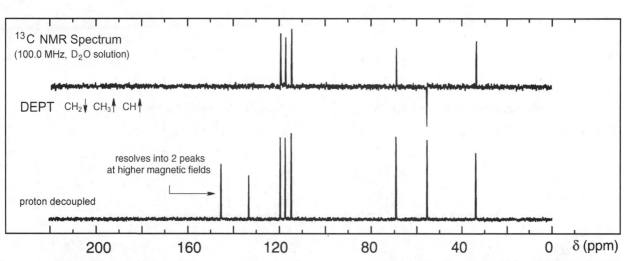

$^{13}$C NMR Spectrum
(100.0 MHz, D$_2$O solution)

DEPT  CH$_2$↓ CH$_3$↑ CH↑

resolves into 2 peaks
at higher magnetic fields

proton decoupled

δ (ppm)

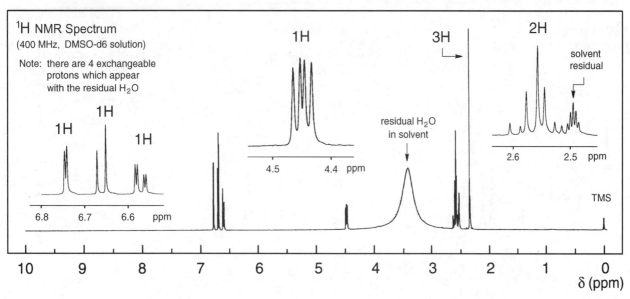

$^1$H NMR Spectrum
(400 MHz, DMSO-d6 solution)

Note: there are 4 exchangeable
protons which appear
with the residual H$_2$O

1H  1H  1H  1H

ppm

1H

4.5    4.4  ppm

residual H$_2$O
in solvent

3H

2H

solvent
residual

2.6    2.5  ppm

TMS

δ (ppm)

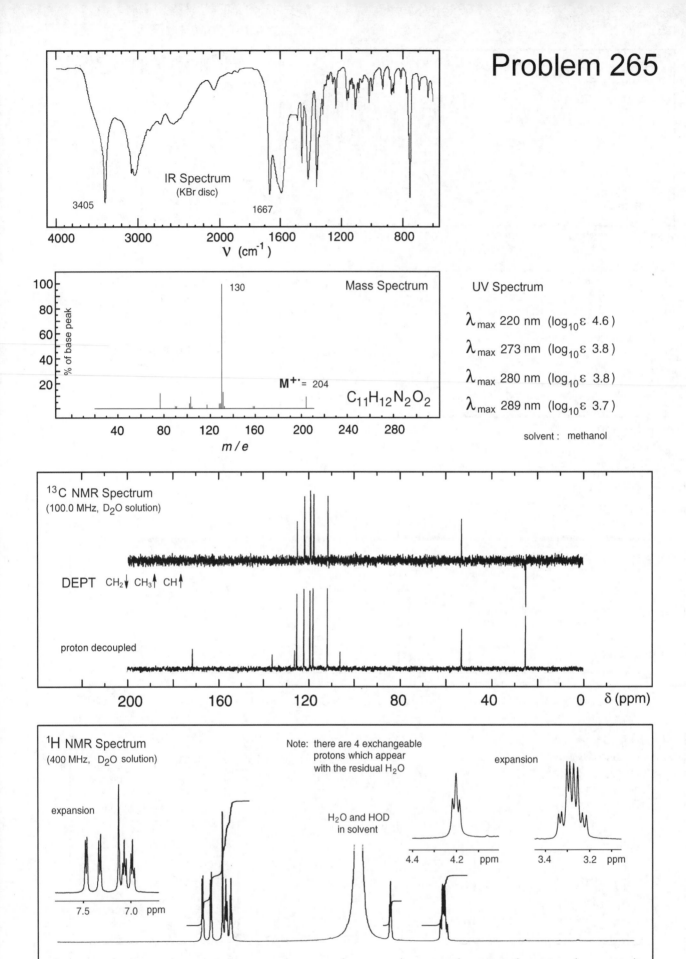

IR Spectrum
(KBr disc)

3405

1667

ν (cm⁻¹)

Mass Spectrum

130

M⁺· = 204

C₁₁H₁₂N₂O₂

% of base peak

m/e

UV Spectrum

λ max 220 nm (log₁₀ε 4.6)

λ max 273 nm (log₁₀ε 3.8)

λ max 280 nm (log₁₀ε 3.8)

λ max 289 nm (log₁₀ε 3.7)

solvent : methanol

¹³C NMR Spectrum
(100.0 MHz, D₂O solution)

DEPT  CH₂↓ CH₃↑ CH↑

proton decoupled

δ (ppm)

¹H NMR Spectrum
(400 MHz, D₂O solution)

Note: there are 4 exchangeable
protons which appear
with the residual H₂O

expansion

H₂O and HOD
in solvent

expansion

expansion

δ (ppm)

# Problem 266

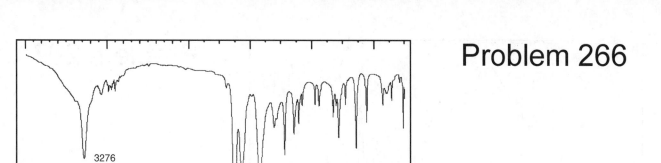

IR Spectrum
(KBr disc)

3276

1702  1666

ν (cm⁻¹)

4000   3000   2000   1600   1200   800

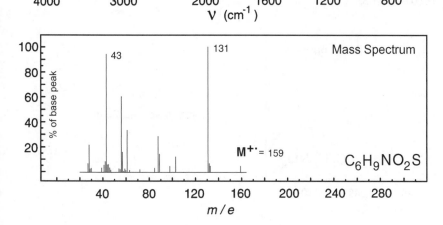

Mass Spectrum

43

131

M⁺· = 159

C₆H₉NO₂S

% of base peak

100
80
60
40
20

40   80   120   160   200   240   280

m/e

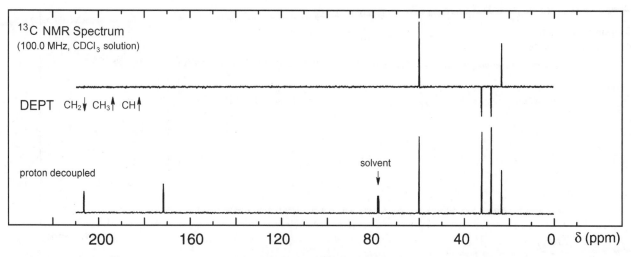

¹³C NMR Spectrum
(100.0 MHz, CDCl₃ solution)

DEPT   CH₂↓ CH₃↑ CH↑

proton decoupled

solvent

200   160   120   80   40   0   δ (ppm)

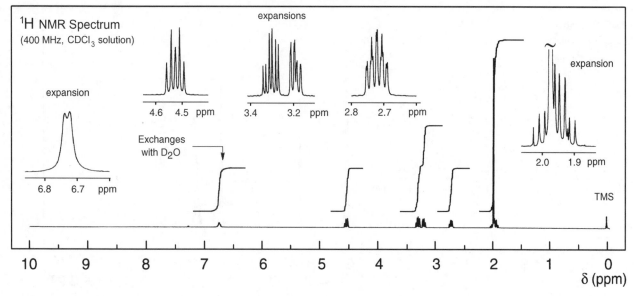

¹H NMR Spectrum
(400 MHz, CDCl₃ solution)

expansions

4.6   4.5 ppm      3.4   3.2 ppm      2.8   2.7 ppm

expansion

2.0   1.9 ppm

expansion

6.8   6.7 ppm

Exchanges
with D₂O

TMS

10   9   8   7   6   5   4   3   2   1   0   δ (ppm)

# Problem 267

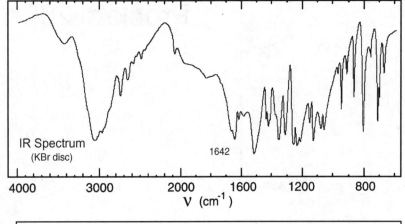

IR Spectrum
(KBr disc)

1642

ν (cm⁻¹)

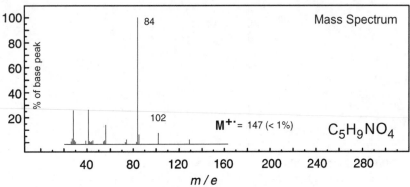

Mass Spectrum

84

102

M⁺˙ = 147 (< 1%)

$C_5H_9NO_4$

No significant UV
absorption above 220 nm

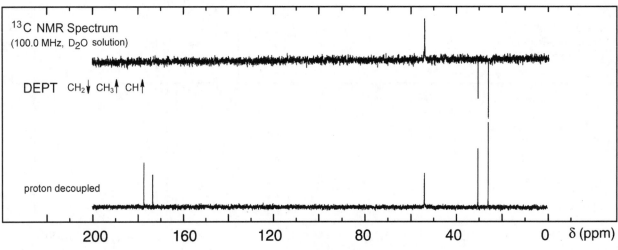

¹³C NMR Spectrum
(100.0 MHz, D₂O solution)

DEPT   CH₂↓ CH₃↑ CH↑

proton decoupled

δ (ppm)

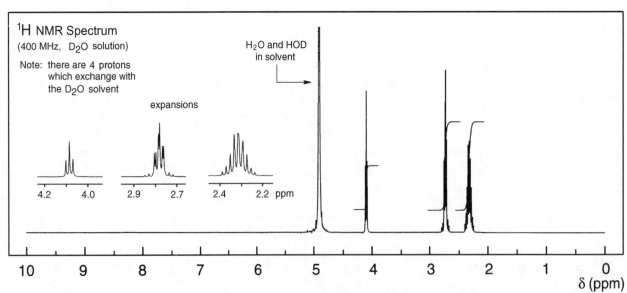

¹H NMR Spectrum
(400 MHz, D₂O solution)

Note: there are 4 protons
which exchange with
the D₂O solvent

H₂O and HOD
in solvent

expansions

δ (ppm)

378

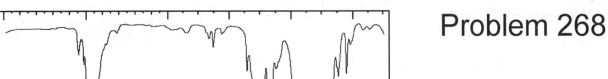

IR Spectrum
(liquid film)

ν (cm⁻¹)

57

85

% of base peak

Mass Spectrum

M⁺· = 130 (< 1%)

$C_7H_{14}O_2$

m/e

No significant UV
absorption above 220 nm

¹³C NMR Spectrum
(100.0 MHz, CDCl₃ solution)

DEPT  CH₂↓ CH₃↑ CH↑

solvent

proton decoupled

δ (ppm)

¹H NMR Spectrum
(400 MHz, CDCl₃ solution)

expansions

expansions

4.8  4.7 ppm

3.6  3.5  3.4 ppm

1.2  1.1 ppm

expansions

5.8  5.7 ppm

5.3  5.2 ppm

TMS

δ (ppm)

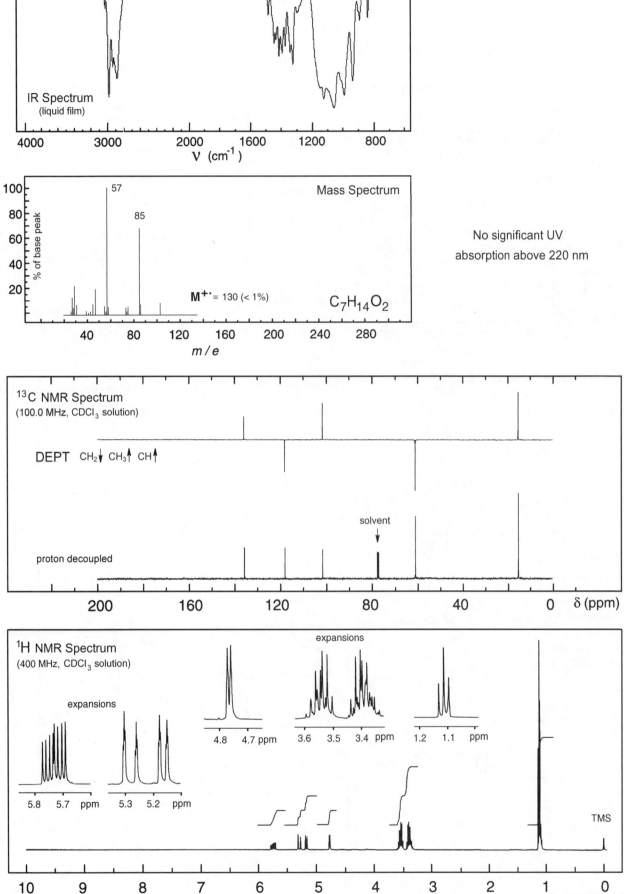

# Problem 269

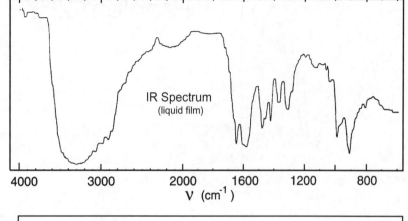

IR Spectrum
(liquid film)

ν (cm⁻¹)

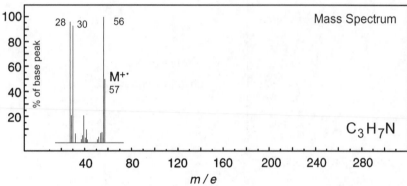

Mass Spectrum

M⁺˙
57

C₃H₇N

No significant UV
absorption above 220 nm

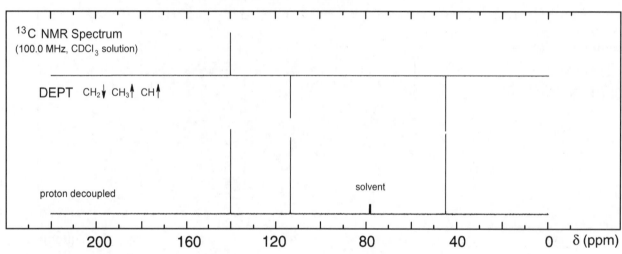

¹³C NMR Spectrum
(100.0 MHz, CDCl₃ solution)

DEPT   CH₂↓ CH₃↑ CH↑

proton decoupled

solvent

δ (ppm)

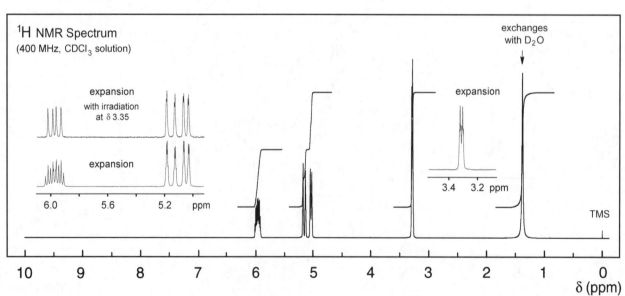

¹H NMR Spectrum
(400 MHz, CDCl₃ solution)

exchanges
with D₂O

expansion
with irradiation
at δ 3.35

expansion

expansion

6.0    5.6    5.2   ppm

3.4    3.2 ppm

TMS

δ (ppm)

380

# Problem 270

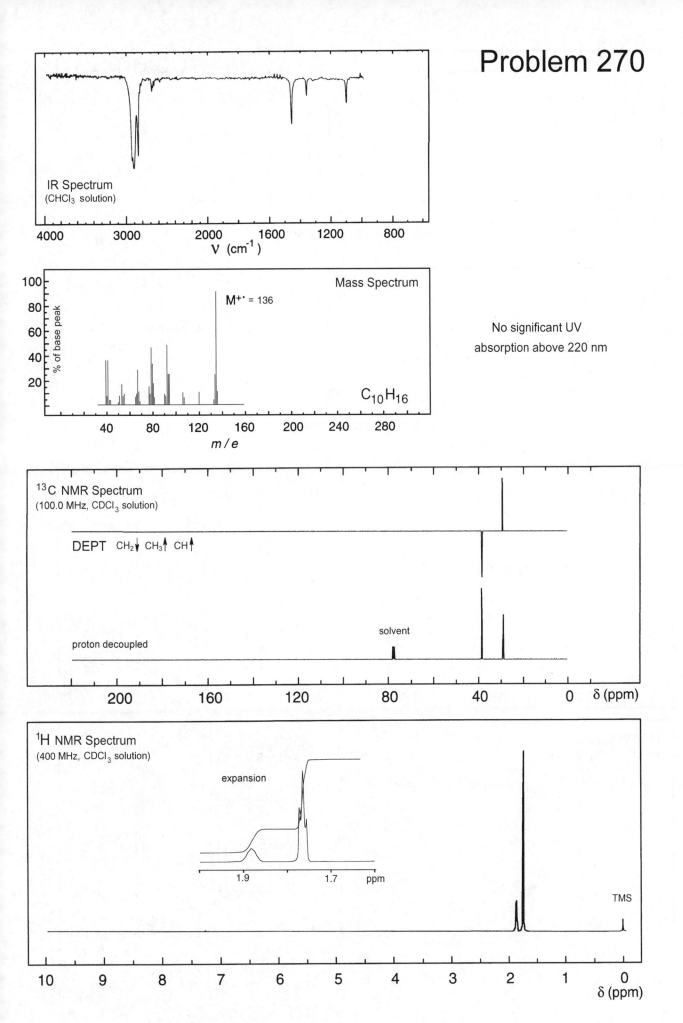

IR Spectrum
(CHCl₃ solution)

$\nu$ (cm⁻¹)

Mass Spectrum

M⁺˙ = 136

No significant UV
absorption above 220 nm

% of base peak

$C_{10}H_{16}$

m/e

¹³C NMR Spectrum
(100.0 MHz, CDCl₃ solution)

DEPT   CH₂↓ CH₃↑ CH↑

solvent

proton decoupled

$\delta$ (ppm)

¹H NMR Spectrum
(400 MHz, CDCl₃ solution)

expansion

1.9        1.7    ppm

TMS

$\delta$ (ppm)

# Problem 271

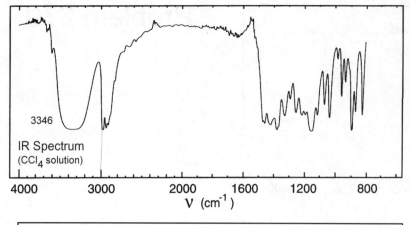

IR Spectrum
(CCl$_4$ solution)

3346

$\nu$ (cm$^{-1}$)

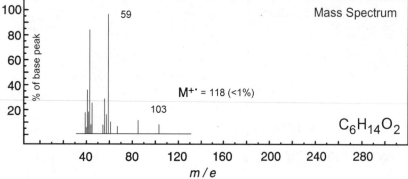

Mass Spectrum

59

M$^{+\cdot}$ = 118 (<1%)

103

C$_6$H$_{14}$O$_2$

% of base peak

$m/e$

No significant UV
absorption above 220 nm

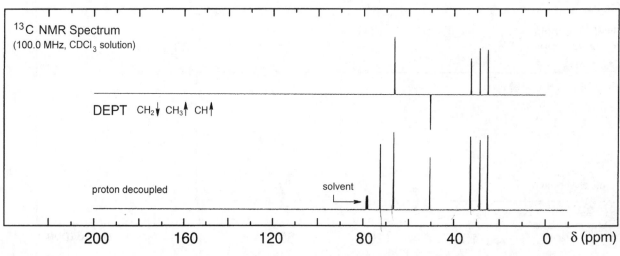

$^{13}$C NMR Spectrum
(100.0 MHz, CDCl$_3$ solution)

DEPT  CH$_2$↓ CH$_3$↑ CH↑

proton decoupled

solvent

$\delta$ (ppm)

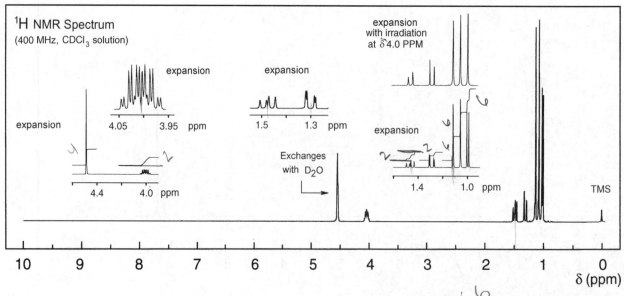

$^1$H NMR Spectrum
(400 MHz, CDCl$_3$ solution)

expansion
with irradiation
at $\delta$ 4.0 PPM

expansion

expansion

4.05    3.95  ppm

1.5    1.3  ppm

expansion

expansion

1.4    1.0  ppm

4.4    4.0  ppm

Exchanges
with D$_2$O

TMS

$\delta$ (ppm)

# Problem 272

383

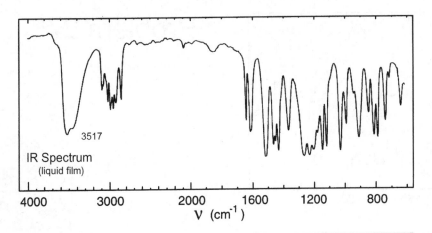

IR Spectrum
(liquid film)

3517

ν (cm⁻¹)

$\nu$ (cm$^{-1}$)

Note: irradiation of the signal at δ 6.45 in the
¹H NMR produces an enhancement of the
signals at δ 3.4 and 3.7 ppm via the NOE

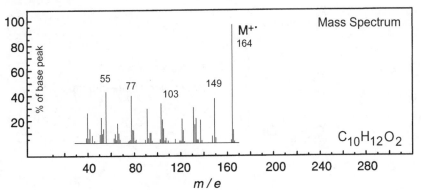

Mass Spectrum

M⁺·
164

% of base peak

55  77  103  149

$C_{10}H_{12}O_2$

m/e

UV Spectrum

$\lambda_{max}$ 281 nm ($\log_{10}\varepsilon$ 3.5)
$\lambda_{max}$ 230 nm ($\log_{10}\varepsilon$ 3.6)

solvent : methanol

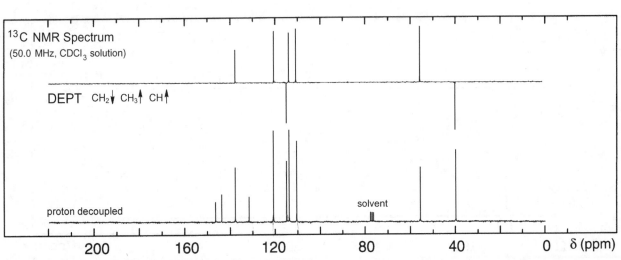

¹³C NMR Spectrum
(50.0 MHz, CDCl₃ solution)

DEPT  CH₂↓ CH₃↑ CH↑

proton decoupled

solvent

δ (ppm)

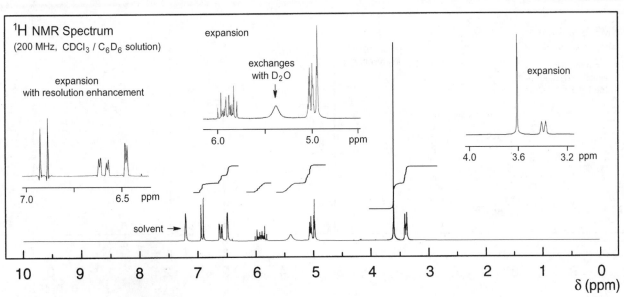

¹H NMR Spectrum
(200 MHz, CDCl₃ / C₆D₆ solution)

expansion

exchanges
with D₂O

expansion
with resolution enhancement

expansion

solvent

δ (ppm)

# Problem 273

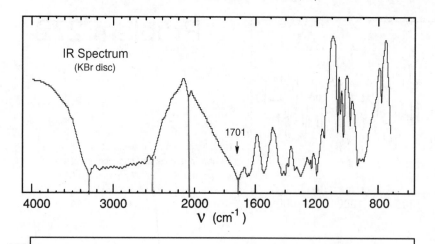

IR Spectrum
(KBr disc)

1701

V (cm⁻¹)

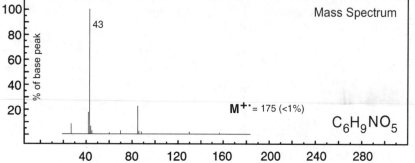

Mass Spectrum

43

No significant UV
absorption above 220 nm

M⁺˙ = 175 (<1%)

$C_6H_9NO_5$

m/e

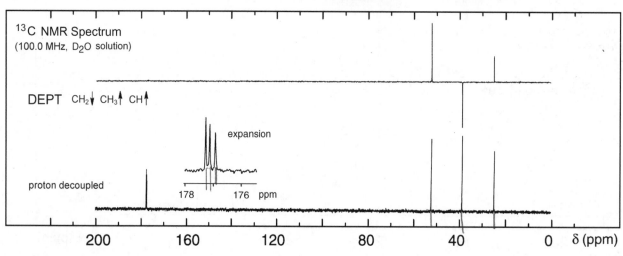

¹³C NMR Spectrum
(100.0 MHz, D₂O solution)

DEPT   CH₂↓ CH₃↑ CH↑

expansion

178    176   ppm

proton decoupled

δ (ppm)

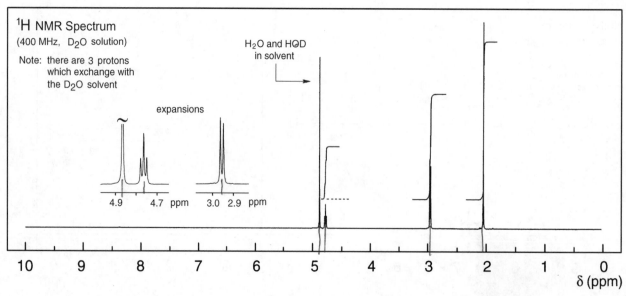

¹H NMR Spectrum
(400 MHz,  D₂O solution)

Note: there are 3 protons
which exchange with
the D₂O solvent

H₂O and HOD
in solvent

expansions

4.9    4.7  ppm      3.0  2.9  ppm

δ (ppm)

# Problem 274

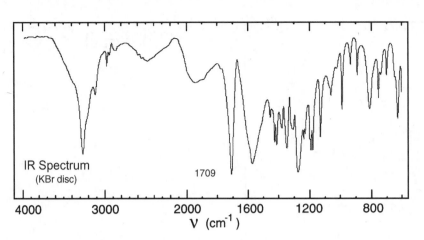

IR Spectrum
(KBr disc)

1709

ν (cm⁻¹)

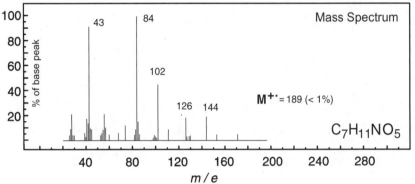

Mass Spectrum

% of base peak

43
84
102
126
144

M⁺· = 189 (< 1%)

$C_7H_{11}NO_5$

m / e

No significant UV
absorption above 220 nm

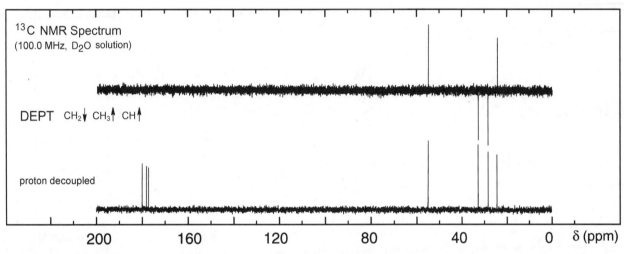

¹³C NMR Spectrum
(100.0 MHz, D₂O solution)

DEPT   CH₂↓ CH₃↑ CH↑

proton decoupled

δ (ppm)

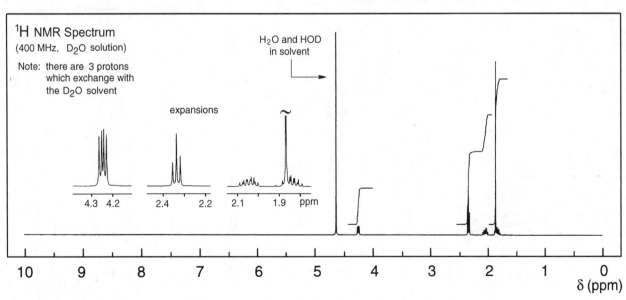

¹H NMR Spectrum
(400 MHz, D₂O solution)

Note: there are 3 protons
which exchange with
the D₂O solvent

H₂O and HOD
in solvent

expansions

4.3  4.2        2.4  2.2        2.1  1.9  ppm

δ (ppm)

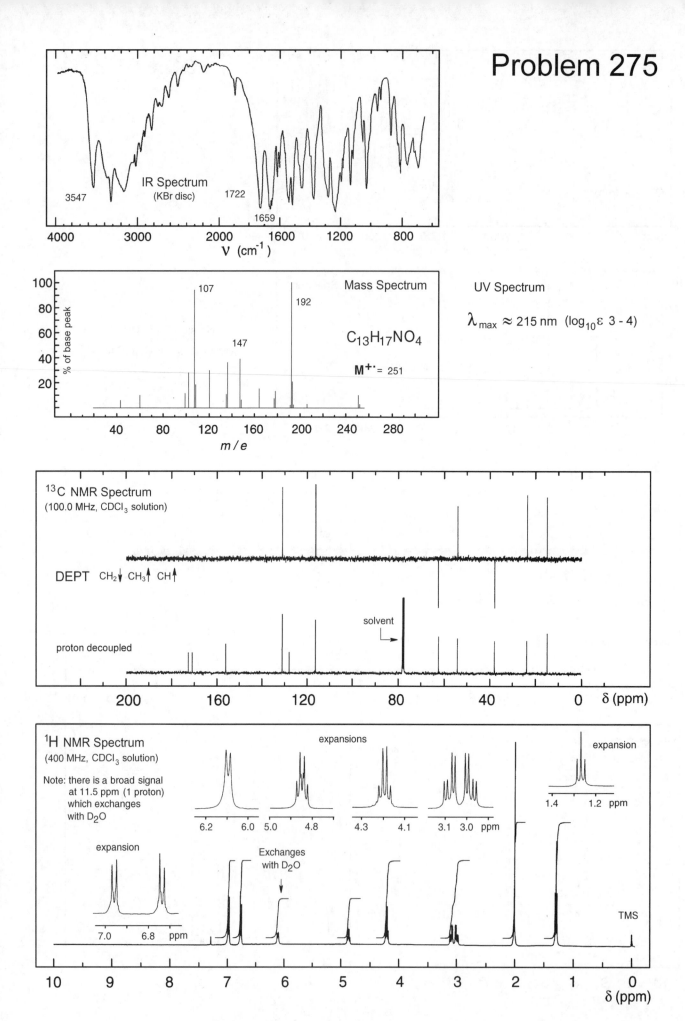

# Problem 275

**IR Spectrum**
(KBr disc)

3547   1722   1659

**Mass Spectrum**

107   192   147

$C_{13}H_{17}NO_4$

$M^{+\cdot} = 251$

**UV Spectrum**

$\lambda_{max} \approx 215$ nm  $(\log_{10}\varepsilon\ 3\text{ - }4)$

**$^{13}C$ NMR Spectrum**
(100.0 MHz, CDCl$_3$ solution)

DEPT   CH$_2\downarrow$ CH$_3\uparrow$ CH$\uparrow$

solvent

proton decoupled

**$^1$H NMR Spectrum**
(400 MHz, CDCl$_3$ solution)

Note: there is a broad signal
at 11.5 ppm (1 proton)
which exchanges
with D$_2$O

expansions

6.2  6.0   5.0  4.8   4.3  4.1   3.1  3.0 ppm

expansion

1.4   1.2 ppm

expansion

7.0   6.8 ppm

Exchanges
with D$_2$O

TMS

# Problem 276

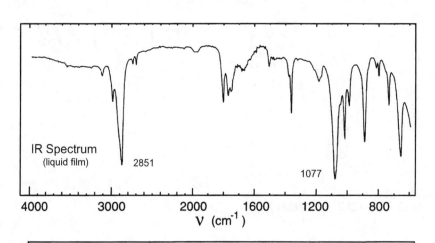

IR Spectrum
(liquid film)

2851

1077

ν (cm⁻¹)

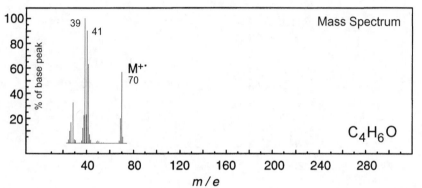

Mass Spectrum

39
41

M⁺·
70

% of base peak

C₄H₆O

m/e

No significant UV
absorption above 220 nm

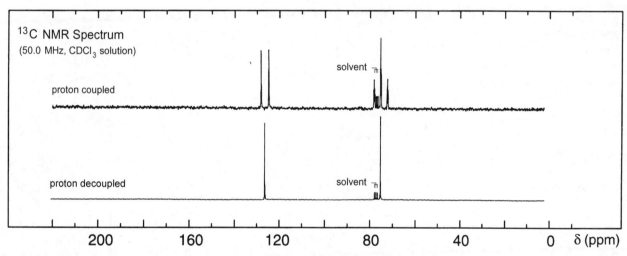

¹³C NMR Spectrum
(50.0 MHz, CDCl₃ solution)

solvent

proton coupled

solvent

proton decoupled

δ (ppm)

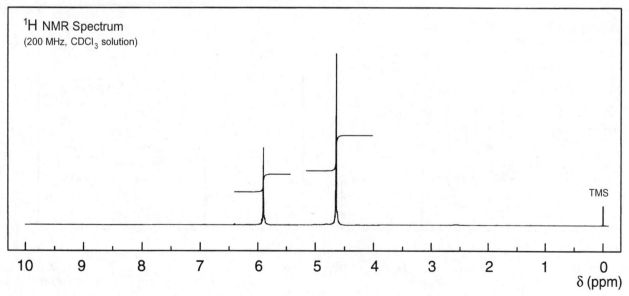

¹H NMR Spectrum
(200 MHz, CDCl₃ solution)

TMS

δ (ppm)

387

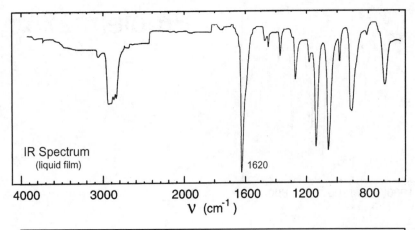

# Problem 277

IR Spectrum
(liquid film)

$\nu$ (cm$^{-1}$)

1620

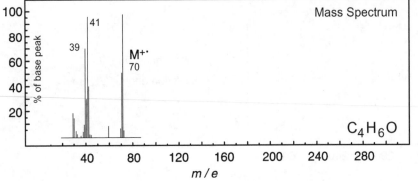

Mass Spectrum

No significant UV
absorption above 220 nm

100
80
60
40
20

% of base peak

39
41
M$^{+\cdot}$
70

$C_4H_6O$

$m/e$

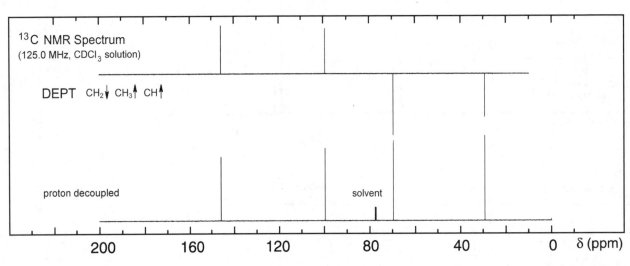

$^{13}$C NMR Spectrum
(125.0 MHz, CDCl$_3$ solution)

DEPT   CH$_2\downarrow$ CH$_3\uparrow$ CH$\uparrow$

proton decoupled

solvent

$\delta$ (ppm)

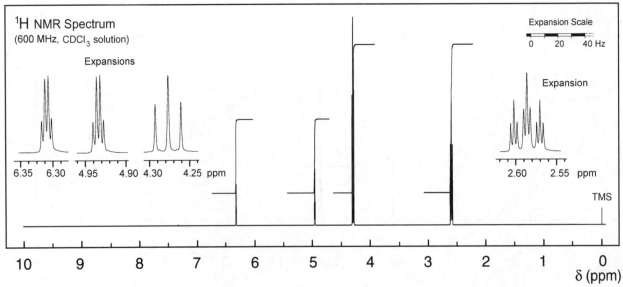

$^1$H NMR Spectrum
(600 MHz, CDCl$_3$ solution)

Expansions

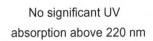

Expansion Scale

0   20   40 Hz

Expansion

6.35   6.30      4.95   4.90      4.30   4.25 ppm

2.60   2.55 ppm

TMS

$\delta$ (ppm)

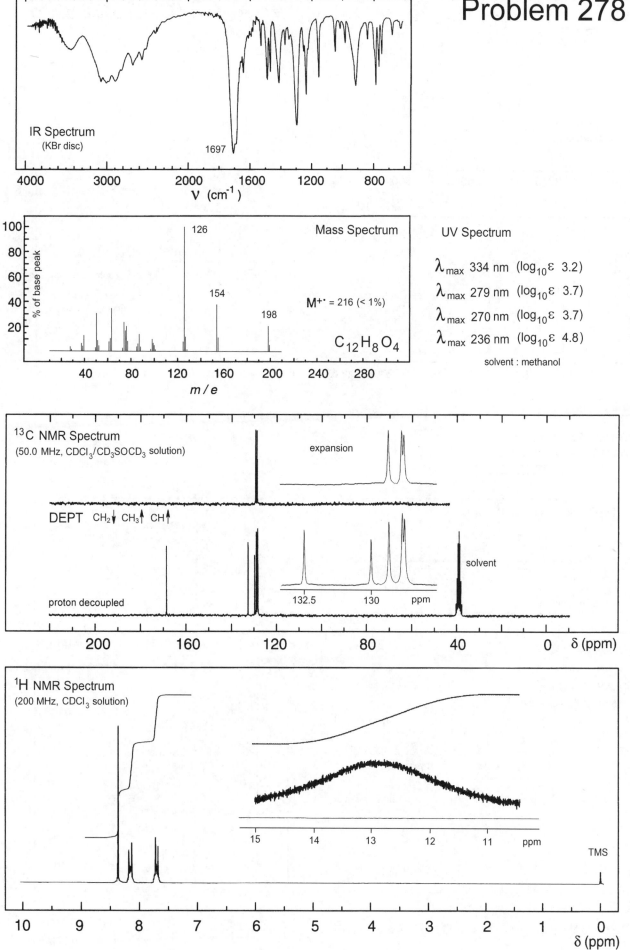

**IR Spectrum** (KBr disc)

1697

ν (cm⁻¹)

**Mass Spectrum**

126

154

198

M⁺· = 216 (< 1%)

$C_{12}H_8O_4$

% of base peak

m/e

**UV Spectrum**

$\lambda_{max}$ 334 nm (log₁₀ε 3.2)

$\lambda_{max}$ 279 nm (log₁₀ε 3.7)

$\lambda_{max}$ 270 nm (log₁₀ε 3.7)

$\lambda_{max}$ 236 nm (log₁₀ε 4.8)

solvent : methanol

**¹³C NMR Spectrum**
(50.0 MHz, CDCl₃/CD₃SOCD₃ solution)

expansion

DEPT   CH₂↓ CH₃↑ CH↑

proton decoupled

solvent

132.5   130   ppm

δ (ppm)

**¹H NMR Spectrum**
(200 MHz, CDCl₃ solution)

15   14   13   12   11   ppm

TMS

δ (ppm)

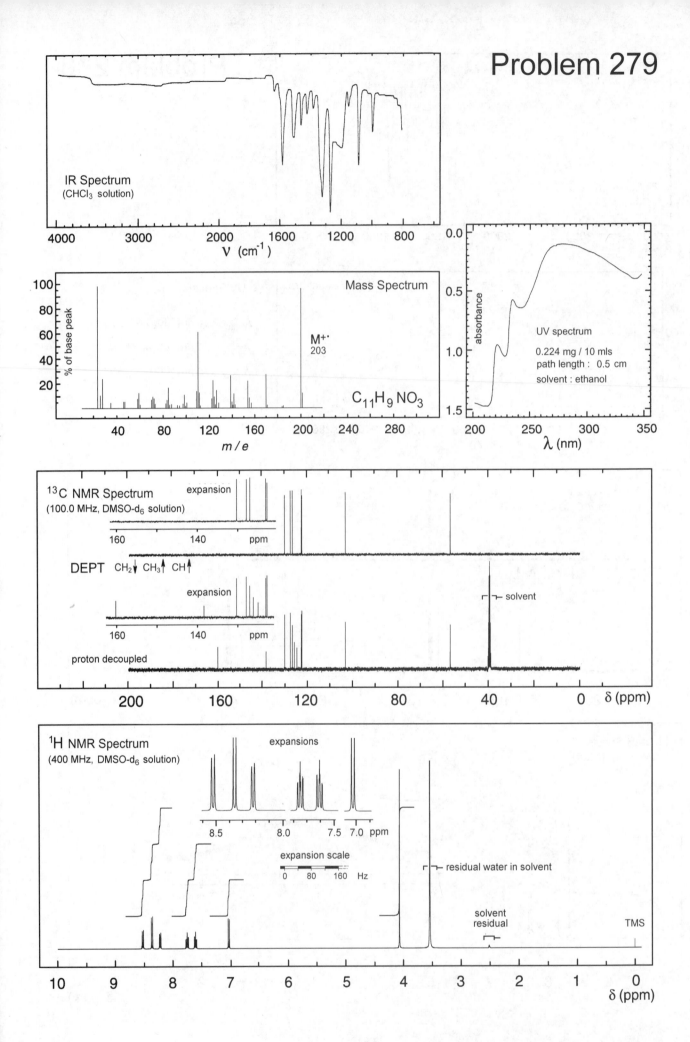

# Problem 279

IR Spectrum
(CHCl₃ solution)

Mass Spectrum

M⁺·
203

$C_{11}H_9NO_3$

UV spectrum

0.224 mg / 10 mls
path length : 0.5 cm
solvent : ethanol

¹³C NMR Spectrum
(100.0 MHz, DMSO-d₆ solution)

expansion

DEPT   CH₂↓ CH₃↑ CH↑

expansion

solvent

proton decoupled

¹H NMR Spectrum
(400 MHz, DMSO-d₆ solution)

expansions

expansion scale

residual water in solvent

solvent
residual

TMS

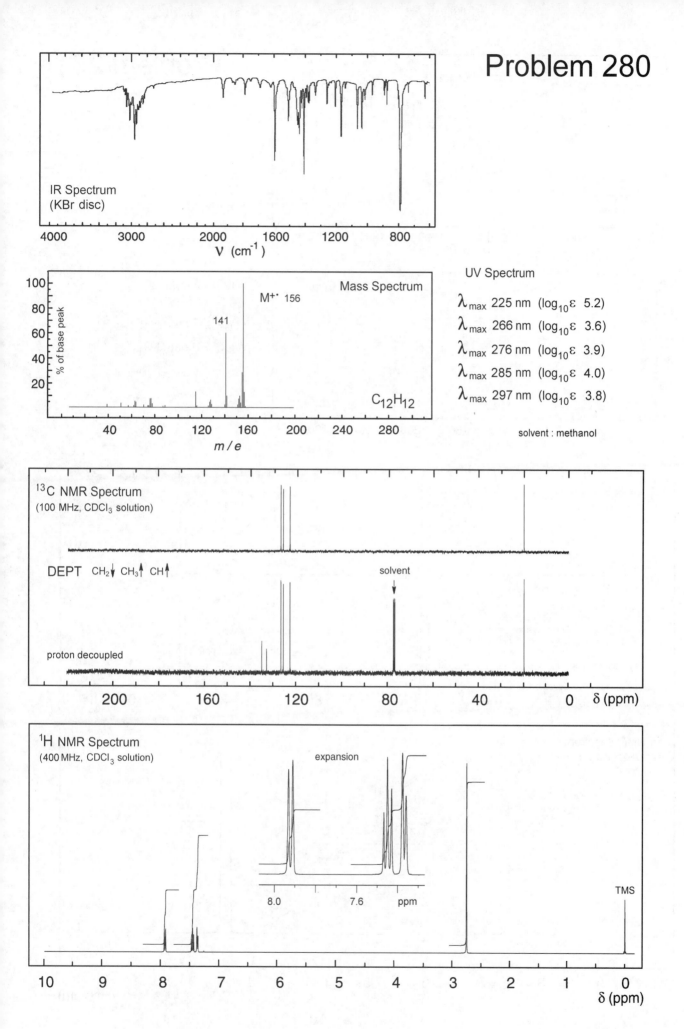

IR Spectrum
(KBr disc)

ν (cm⁻¹)

Mass Spectrum

M⁺· 156

141

$C_{12}H_{12}$

m/e

UV Spectrum

$\lambda_{max}$ 225 nm (log₁₀ε 5.2)

$\lambda_{max}$ 266 nm (log₁₀ε 3.6)

$\lambda_{max}$ 276 nm (log₁₀ε 3.9)

$\lambda_{max}$ 285 nm (log₁₀ε 4.0)

$\lambda_{max}$ 297 nm (log₁₀ε 3.8)

solvent : methanol

¹³C NMR Spectrum
(100 MHz, CDCl₃ solution)

DEPT CH₂↓ CH₃↑ CH↑

solvent

proton decoupled

δ (ppm)

¹H NMR Spectrum
(400 MHz, CDCl₃ solution)

expansion

8.0    7.6    ppm

TMS

δ (ppm)

# Problem 281

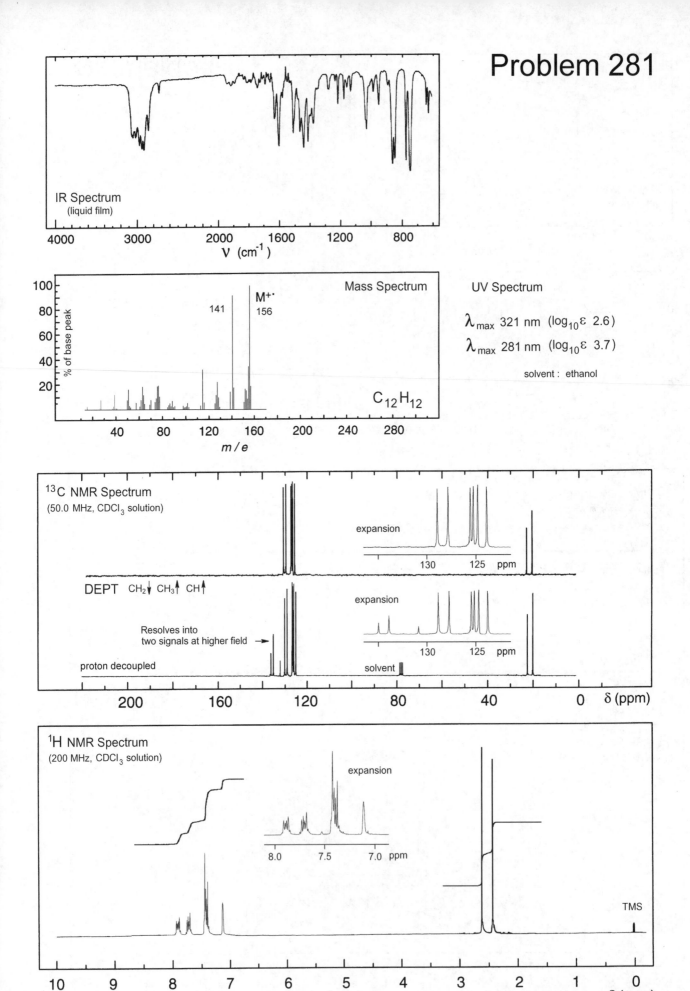

IR Spectrum
(liquid film)

ν (cm⁻¹)

Mass Spectrum

141

M⁺˙
156

% of base peak

m / e

$C_{12}H_{12}$

UV Spectrum

$\lambda_{max}$ 321 nm (log₁₀ε 2.6)

$\lambda_{max}$ 281 nm (log₁₀ε 3.7)

solvent : ethanol

¹³C NMR Spectrum
(50.0 MHz, CDCl₃ solution)

expansion

130    125  ppm

DEPT   CH₂↓ CH₃↑ CH↑

expansion

130    125  ppm

Resolves into
two signals at higher field →

proton decoupled

solvent

δ (ppm)

¹H NMR Spectrum
(200 MHz, CDCl₃ solution)

expansion

8.0    7.5    7.0  ppm

TMS

δ (ppm)

# Problem 282

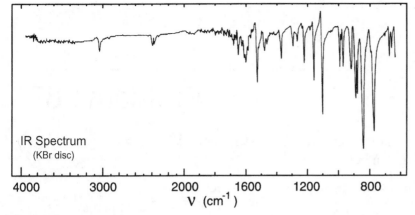

IR Spectrum
(KBr disc)

ν (cm$^{-1}$)

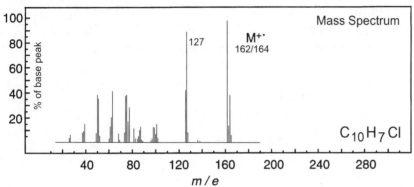

Mass Spectrum

127

M$^{+\cdot}$
162/164

% of base peak

$C_{10}H_7Cl$

m/e

UV Spectrum

$\lambda_{max}$ 311 nm (log$_{10}\varepsilon$ 2.6)

$\lambda_{max}$ 289 nm (log$_{10}\varepsilon$ 3.7)

$\lambda_{max}$ 225 nm (log$_{10}\varepsilon$ 5.0)

solvent : methanol

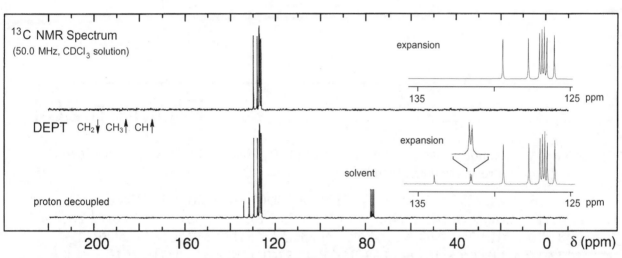

$^{13}$C NMR Spectrum
(50.0 MHz, CDCl$_3$ solution)

expansion

DEPT  CH$_2$↓ CH$_3$↑ CH↑

expansion

solvent

proton decoupled

δ (ppm)

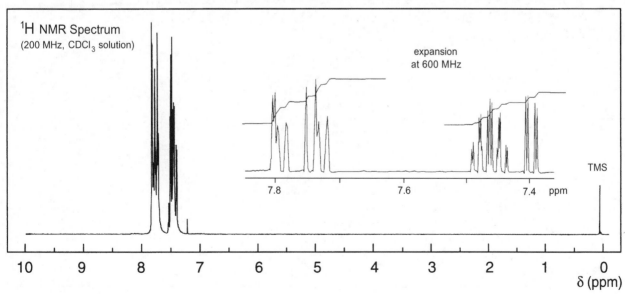

$^1$H NMR Spectrum
(200 MHz, CDCl$_3$ solution)

expansion
at 600 MHz

TMS

δ (ppm)

# Problem 283

An organic compound has the molecular formula $C_{10}H_{14}$. Identify the compound using the spectroscopic data given below.

$\nu_{max}$ (liquid film): no significant features in the infrared spectrum.
$\lambda_{max}$: 265 (log $\varepsilon$: 2.3) nm. $^1$H NMR (CDCl$_3$ solution): $\delta$ 7.1, m, 5H; 2.5, apparent sextet, $J$ 7 Hz, 1H; 1.6, apparent quintet, $J$ 7 Hz, 2H; 1.22, d, $J$ 7 Hz, 3H; 0.81, t, $J$ 7 Hz, 3H ppm. $^{13}$C{$^1$H} NMR (CDCl$_3$ solution): $\delta$ 148.4 (C), 129.3, 127.9, 126.1, 42.3 (CH), 31.7 (CH$_2$), 22.2, 12.2 (CH$_3$) ppm. Mass spectrum: $m/e$ 134 (M$^{+\cdot}$, 20), 119(8), 105(100), 77(10).

# Problem 284

An organic compound has the molecular formula $C_{12}H_{17}NO$. Identify the compound using the spectroscopic data given below.

$\nu_{max}$ (KBr disc): 3296m, 1642s cm$^{-1}$. $^1$H NMR (CDCl$_3$ solution): $\delta$ 7.23-7.42, m, 5H; 5.74, br s, exch. D$_2$O, 1H; 5.14, q, $J$ 6.7 Hz, 1H; 2.15, t, $J$ 7.1 Hz, 2H; 1.66, m, 2H; 1.48, d, $J$ 6.7 Hz, 3H; 0.93, t, $J$ 7.3 Hz, 3H ppm. $^{13}$C{$^1$H} NMR (CDCl$_3$ solution): $\delta$ 172.0 (C), 143.3 (C), 128.6, 127.3, 126.1, 48.5 (CH), 38.8 (CH$_2$), 21.7 (CH$_3$), 19.1 (CH$_2$), 13.7 (CH$_3$) ppm. Mass spectrum: $m/e$ 191(M$^{+\cdot}$, 40), 120(33), 105(58), 104(100), 77(18), 43(46).

# Problem 285

An organic compound has the molecular formula $C_{16}H_{30}O_4$. Identify the compound using the spectroscopic data given below.

$\nu_{max}$ (CHCl$_3$ solution): 1733 cm$^{-1}$. $^1$H NMR (CDCl$_3$ solution): $\delta$ 4.19, q, $J$ 7.2 Hz, 4H; 3.35, s, 1H; 1.20, t, $J$ 7.2 Hz, 6H; 1.25-1.29, m, 10H; 1.10, s, 6H; 0.88, t, $J$ 6.8 Hz, 3H ppm. $^{13}$C{$^1$H} NMR (CDCl$_3$ solution): $\delta$ 168.5 (C), 60.8 (CH$_2$), 59.6 (CH), 41.1 (CH$_2$), 36.3 (C), 31.8 (CH$_2$), 29.9 (CH$_2$), 25.1 (CH$_3$), 23.6 (CH$_2$), 22.6 (CH$_2$), 14.1 (CH$_3$), 14.0 (CH$_3$), ppm. Mass spectrum: $m/e$ 286 (M$^{+\cdot}$, 70), 241(25), 201(38), 160(100), 115(53).

# Problem 286

An organic compound has the molecular formula $C_8H_{13}NO_3$. Identify the compound using the spectroscopic data given below.

$\nu_{max}$ (nujol mull): 1690-1725s cm$^{-1}$. $\lambda_{max}$: no significant features in the ultraviolet spectrum. $^1$H NMR (CDCl$_3$ solution): $\delta$ 4.25, q, $J$ 6.7 Hz, 2H; 3.8, t, $J$ 7 Hz, 4H; 2.45, t, $J$ 7 Hz, 4H; 1.3, t, $J$ 6.7 Hz , 3H ppm. $^{13}$C{$^1$H} NMR (CDCl$_3$ solution): $\delta$ 207 (C); 155 (C); 62 (CH$_2$), 43 (CH$_2$), 41 (CH$_2$), 15 (CH$_3$) ppm. Mass spectrum: $m/e$ 171 (M$^{+\cdot}$, 15), 142(25), 56(68), 42(100).

# Problem 287

An organic compound has the molecular formula $C_{12}H_{13}NO_3$. Identify the compound using the spectroscopic data given below.

$\nu_{max}$ (nujol mull): 3338, 1715, 1592 cm$^{-1}$. $\lambda_{max}$: 254 (log ε: 4.3) nm. $^1$H NMR (DMSO-$d_6$ solution): δ 12.7, broad s, exch. $D_2O$, 1H; 8.42, d, $J$ 6.1 Hz, 1H; 7.45-7.25, m, 5H; 6.63 dd, $J$ 15.9, 1.2 Hz, 1H; 6.30, dd, $J$ 15.9, 6.7 Hz, 1H; 4.93 ddd, $J$ 6.7, 6.1, 1.2 Hz, 1H; 1.90, s, 3H ppm. $^{13}$C{$^1$H} NMR (DMSO-$d_6$ solution): δ 171.9 (C), 169.0 (C), 135.9 (C), 131.7 (CH), 128.7 (CH), 127.9 (CH), 126.3 (CH), 124.6 (CH), 54.4 (CH), 22.3 (CH$_3$) ppm. Mass spectrum: $m/e$ 219 (M$^{+\cdot}$, 25), 175(10), 132(100), 131(94), 103(35), 77(46), 43(83).

# Problem 288

An organic compound has the molecular formula $C_{13}H_{16}O_4$. Identify the compound using the spectroscopic data given below.

$\nu_{max}$ (KBr disc): 3479s, 1670s, cm$^{-1}$. $\lambda_{max}$: 250 (log ε: 4) nm. $^1$H NMR (CDCl$_3$ solution): δ 1.40, s, 3H; 2.00, bs exch., 1H; 2.71, d, $J$ 15.7 Hz, 1H; 2.77, d, $J$ 15.7 Hz, 1H; 2.91, d, $J$ 17.8 Hz , 1H; 3.14, d, $J$ 17.8 Hz, 1H; 3.79, s, 3H; 3.84, s, 3H; 6.80, d, $J$ 9.0 Hz, 1H; 6.98, d, $J$ 9.0 Hz, 1H ppm. $^{13}$C{$^1$H} NMR (CDCl$_3$ solution) : δ 196.0 (C); 154.0 (C); 157.7 (C); 131.5 (C); 122.0 (C); 116.0 (CH); 110.5 (CH); 70.8 (C); 56.3 (CH$_3$); 55.9 (CH$_3$); 54.1 (CH$_2$); 37.7 (CH$_2$); 29.2 (CH$_3$) ppm. Mass spectrum: $m/e$ 236 (M$^{+\cdot}$, 87), 218(33), 178(100), 163(65).

# 10.2

# THE ANALYSIS OF MIXTURES

# Problem 289

A 400 MHz $^1$H NMR spectrum of a mixture of ethanol ($C_2H_6O$) δ 1.24, δ 1.78, δ 3.72 and bromoethane ($C_2H_5Br$) δ 1.68 and δ 3.44 is given below.  Estimate the relative proportions (mole %) of the 2 components from the integrals in the spectrum.

$$CH_3-CH_2-OH \qquad CH_3-CH_2-Br$$

ethanol                     bromoethane

$^1$H NMR Spectrum
(400 MHz, CDCl$_3$ solution)

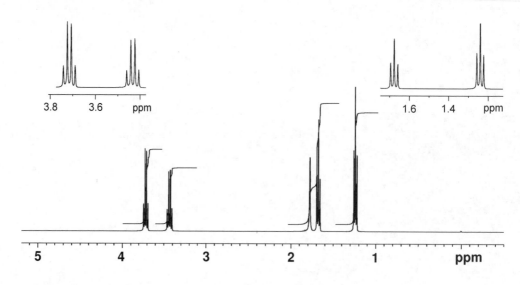

| Compound | Mole % |
|---|---|
| ethanol | |
| bromoethane | |

# Problem 290

A 400 MHz $^1$H NMR spectrum of a mixture of common organic solvents consisting of benzene ($C_6H_6$) $\delta$ 7.37; diethyl ether ($C_4H_{10}O$) $\delta$ 3.49 and $\delta$ 1.22; and dichloromethane ($CH_2Cl_2$) $\delta$ 5.30 is given below. Estimate the relative proportions (mole %) of the 3 components from the integrals in the spectrum.

$$CH_3-CH_2-O-CH_2-CH_3 \qquad Cl-CH_2-Cl$$

benzene             diethyl ether                    dichloromethane

$^1$H NMR Spectrum
(400 MHz, CDCl$_3$ solution)

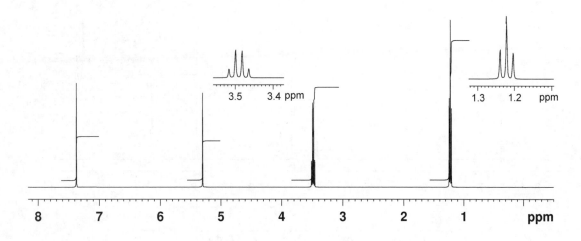

| Compound | Mole % |
|---|---|
| benzene | |
| diethyl ether | |
| dichloromethane | |

# Problem 291

A 400 MHz $^1$H NMR spectrum of a mixture of benzene ($C_6H_6$) δ 7.37, ethyl acetate
($C_4H_8O_2$) δ 4.13, δ 2.05, δ 1.26 and dioxane ($C_4H_8O_2$) δ 3.70 is given below.
Estimate the relative proportions (mole %) of the 3 components from the integrals in
the spectrum.

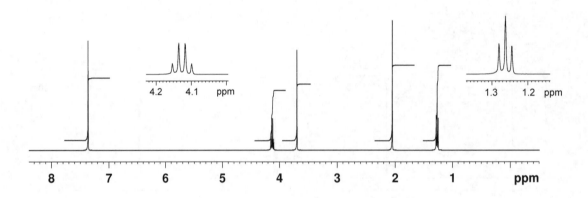

benzene          ethyl acetate                    dioxane

$^1$H NMR Spectrum
(400 MHz, CDCl$_3$ solution)

| Compound | Mole % |
|---|---|
| benzene | |
| ethyl acetate | |
| dioxane | |

# Problem 292

A 100 MHz $^{13}$C NMR spectrum of a mixture of ethanol ($C_2H_6O$) δ 18.3 ($CH_3$), δ 57.8 ($CH_2$) and bromoethane ($C_2H_5Br$) δ 19.5 ($CH_3$) and δ 27.9 ($CH_2$) in $CDCl_3$ solution is given below.  The spectrum was recorded with a long relaxation delay (300 seconds) between acquisitions and with the NOE suppressed.  Estimate the relative proportions (mole %) of the 2 components from the peak intensities in the spectrum.

$$CH_3 - CH_2 - OH \qquad CH_3 - CH_2 - Br$$

ethanol                    bromoethane

$^{13}$C NMR Spectrum
(100 MHz, $CDCl_3$ solution)

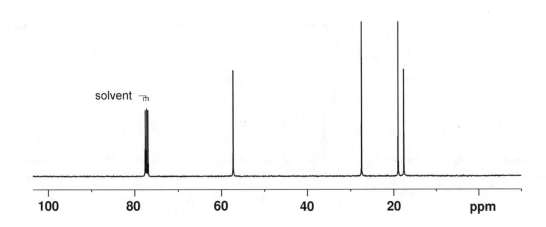

| Compound | Mole % |
|----------|--------|
| ethanol |  |
| bromoethane |  |

401

# Problem 293

A 100 MHz $^{13}$C NMR spectrum of a mixture of benzene ($C_6H_6$) δ 128.7 (CH), diethyl ether ($C_4H_{10}O$) δ 67.4 ($CH_2$) and δ 17.1 ($CH_3$) and dichloromethane ($CH_2Cl_2$) δ 53.7 in $CDCl_3$ solution is given below. The spectrum was recorded with a long relaxation delay (300 seconds) between acquisitions and with the NOE suppressed. Estimate the relative proportions (mole %) of the 3 components from the peak intensities in the spectrum.

$CH_3-CH_2-O-CH_2-CH_3$         $Cl-CH_2-Cl$

benzene                    diethyl ether                    dichloromethane

$^{13}$C NMR Spectrum
(100 MHz, $CDCl_3$ solution)

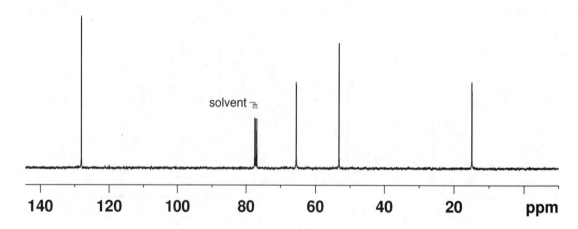

| Compound | Mole % |
|---|---|
| benzene | |
| diethyl ether | |
| dichloromethane | |

# Problem 294

A 100 MHz $^{13}$C NMR spectrum of a mixture of benzene ($C_6H_6$) δ 128.7 (CH), ethyl acetate ($CH_3CH_2OCOCH_3$) δ 170.4 (C=O), δ 60.1 ($CH_2$), δ 20.1 ($CH_3$), δ 14.3 ($CH_3$) and dioxane ($C_4H_8O_2$) δ 66.3 ($CH_2$) in $CDCl_3$ solution is given below.  The spectrum was recorded with a long relaxation delay (300 seconds) between acquisitions and with the NOE suppressed.  Estimate the relative proportions (mole %) of the 3 components from the peak intensities in the spectrum.

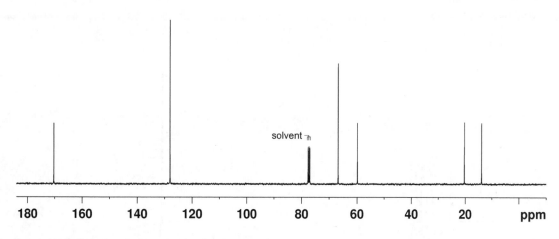

benzene            ethyl acetate                dioxane

$^{13}$C NMR Spectrum
(100 MHz, CDCl$_3$ solution)

| Compound | Mole % |
|---|---|
| benzene | |
| ethyl acetate | |
| dioxane | |

403

# Problem 295

Oxidation of fluorene ($C_{13}H_{10}$) with chromic acid gives fluorenone ($C_{13}H_8O$).  If the reaction does not go to completion, then a mixture of the starting material and the product is usually obtained.  The $^1$H NMR spectrum below is from a partially oxidized sample of fluorene so it contains a mixture of fluorene and fluorenone.  Determine the relative amounts (mole %) of fluorene and fluorenone in the mixture.

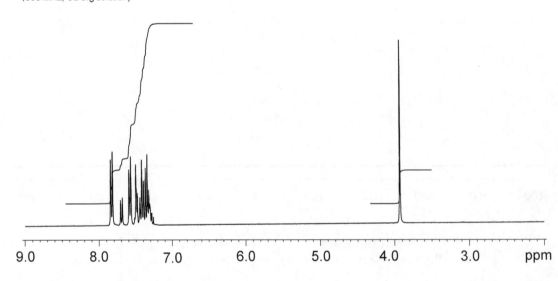

fluorene                               fluorenone

$CrO_3 / H^+$

$^1$H NMR Spectrum
(300 MHz, CDCl$_3$ solution)

| Compound | Mole % |
|---|---|
| fluorene | |
| fluorenone | |

# Problem 296

Careful nitration of anisole ($CH_3OC_6H_5$) with a new nitrating reagent gives a mixture of 4-nitroanisole and 2-nitroanisole. The section of the $^1$H NMR spectrum below is from the aromatic region of the crude reaction mixture which is a mixture of the 4- and 2-nitroanisoles. Determine the relative amounts of the two products in the reaction mixture from the integrals in the spectrum.

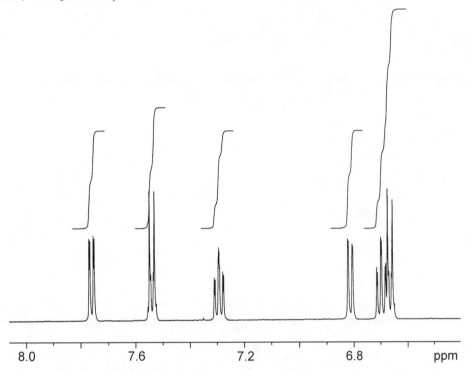

$^1$H NMR Spectrum
(500 MHz, CDCl$_3$ solution)

| Compound | Mole % |
|---|---|
| 4-nitroanisole | |
| 2-nitroanisole | |

# 10.3

# PROBLEMS IN 2-DIMENSIONAL NMR

# Problem 297

The $^1$H and $^{13}$C NMR spectra of 1-propanol ($C_3H_8O$) recorded in CDCl$_3$ solution at 298K, are given below.  The 2-dimensional $^1$H-$^1$H COSY spectrum and the C-H correlation spectrum are given on the facing page.  From the COSY spectrum, assign the proton spectrum and then use the C-H correlation spectrum to assign the $^{13}$C spectrum *i.e.* determine the chemical shift corresponding to each of the protons and each of the carbons in the molecule.

1-propanol

$$\overset{3}{C}H_3-\overset{2}{C}H_2-\overset{1}{C}H_2-\overset{4}{O}H$$

| Proton | Chemical Shift ($\delta$) in ppm | Carbon | Chemical Shift ($\delta$) in ppm |
|--------|--------|--------|--------|
| H1 | | C1 | |
| H2 | | C2 | |
| H3 | | C3 | |
| H4 | | | |

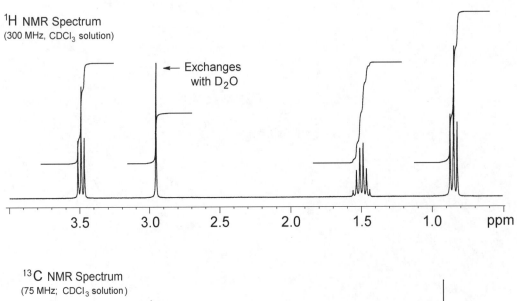

$^1$H NMR Spectrum
(300 MHz, CDCl$_3$ solution)

← Exchanges with D$_2$O

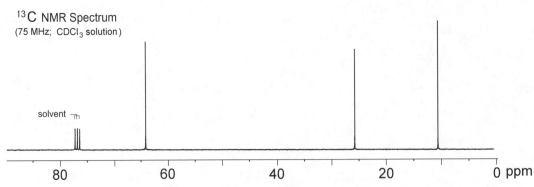

$^{13}$C NMR Spectrum
(75 MHz; CDCl$_3$ solution)

solvent

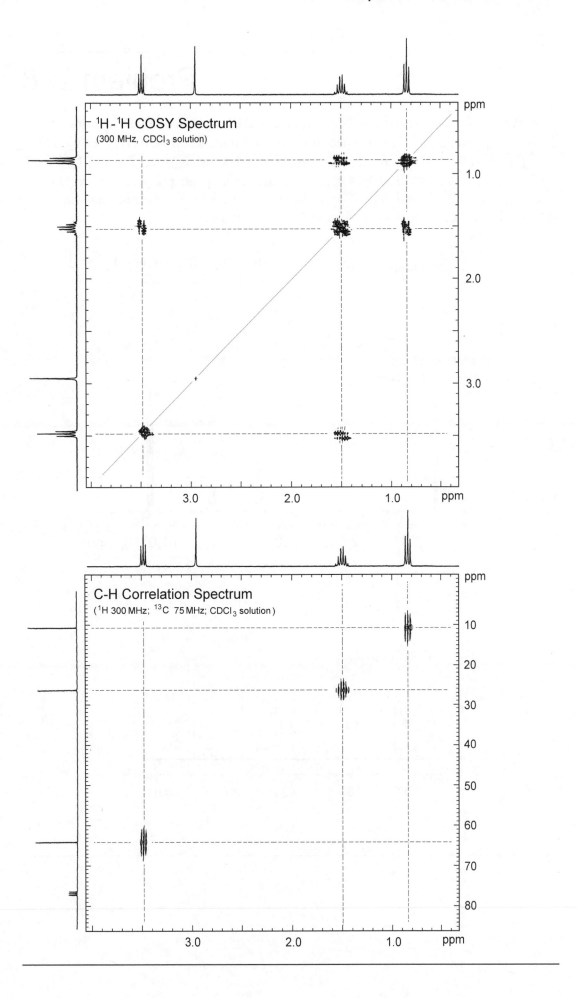

**$^1$H-$^1$H COSY Spectrum**
(300 MHz,  CDCl$_3$ solution)

**C-H Correlation Spectrum**
($^1$H 300 MHz;  $^{13}$C  75 MHz; CDCl$_3$ solution)

# Problem 298

The $^1$H and $^{13}$C NMR spectra of 1-iodobutane ($C_4H_9I$) recorded in CDCl$_3$ solution at 298K, are given below.  The $^1$H spectrum contains signals at δ 3.20 (H1), 1.80 (H2), 1.42 (H3) and 0.93 (H4) ppm.  The $^{13}$C spectrum contains signals at δ 6.7 (C1), 35.5 (C2), 23.6 (C3) and 13.0 (C4) ppm.  On the facing page, produce a schematic diagram of the COSY and the C-H correlation spectra for this molecule showing where all of the cross peaks and diagonal peaks would be.

1-iodobutane

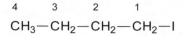

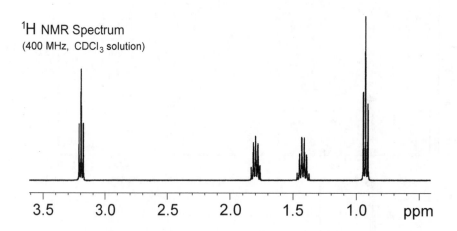

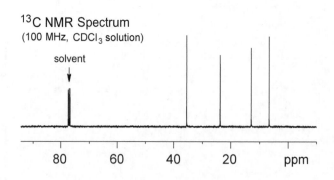

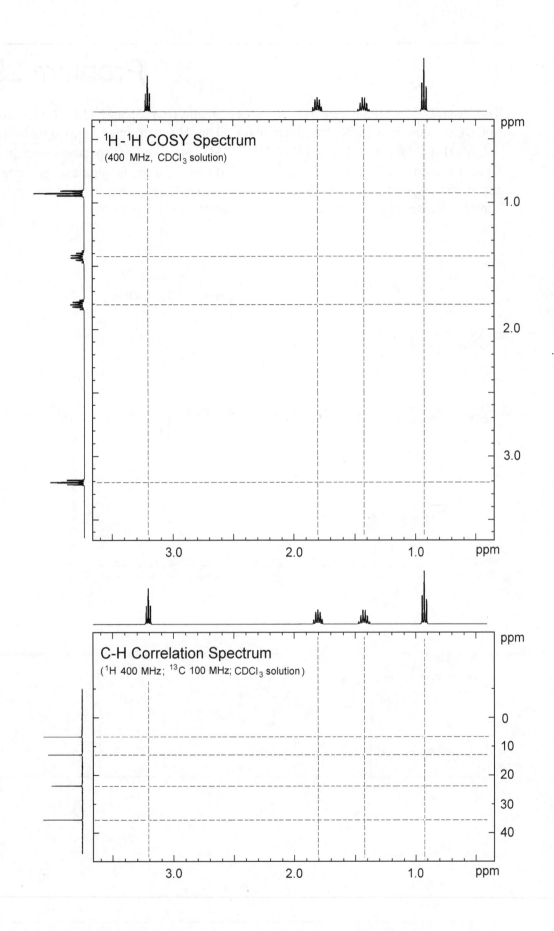

¹H-¹H COSY Spectrum
(400 MHz, CDCl₃ solution)

C-H Correlation Spectrum
(¹H 400 MHz; ¹³C 100 MHz; CDCl₃ solution)

# Problem 299

The $^1$H and $^{13}$C NMR spectra of isobutanol (2-methyl-1-propanol, $C_4H_{10}O$) recorded in CDCl$_3$ solution at 298K, are given below.  The $^1$H spectrum contains signals at δ 3.28 (H1), 2.98 (OH), 1.68 (H2) and 0.83 (H3) ppm.  The $^{13}$C spectrum contains signals at δ 69.3 (C1), 30.7 (C2) and 18.7 (C3) ppm.  On the facing page, produce a schematic diagram of the COSY and the C-H correlation spectra for this molecule showing where all of the cross peaks and diagonal peaks would be.

isobutanol

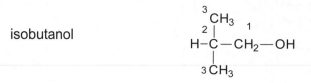

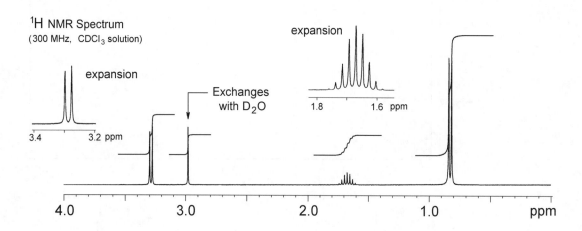

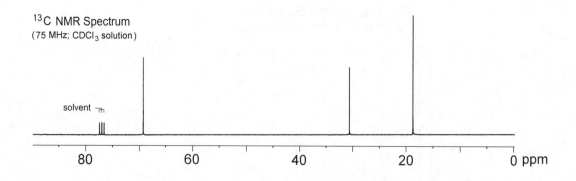

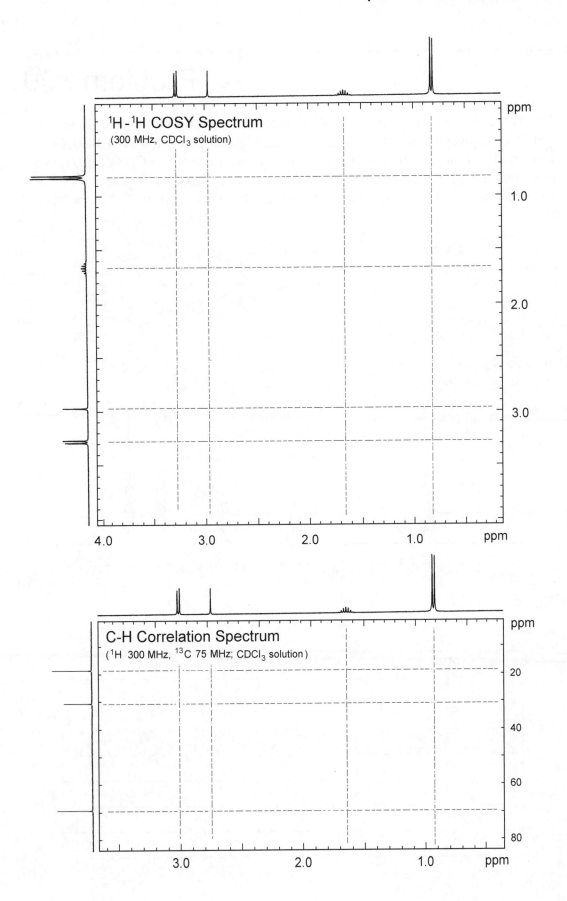

# Problem 300

The $^1$H spectrum of 3-heptanone ($C_7H_{14}O$) recorded in $C_6D_6$ solution at 298K at 500 MHz, is given below.  The $^1$H spectrum has signals at δ 0.79, 0.91, 1.14, 1.44, 1.94 and 1.97 (partly overlapped) ppm.  The 2-dimensional $^1$H-$^1$H COSY spectrum is given on the facing page.  From the COSY spectrum, assign the proton spectrum *i.e.* determine the chemical shift corresponding to each of the protons in the molecule.

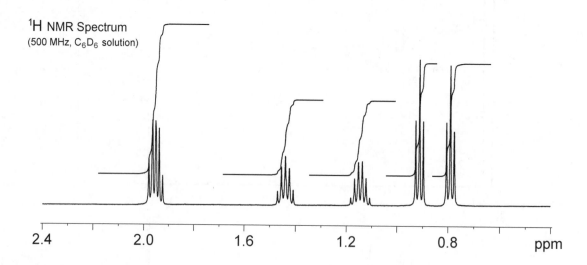

3-heptanone

$$\overset{1}{C}H_3-\overset{2}{C}H_2-\overset{3}{C}-\overset{4}{C}H_2-\overset{5}{C}H_2-\overset{6}{C}H_2-\overset{7}{C}H_3$$

$^1$H NMR Spectrum
(500 MHz, $C_6D_6$ solution)

| Proton | Chemical Shift (δ) in ppm |
|--------|---------------------------|
| H1     |                           |
| H2     |                           |
|        |                           |
| H4     |                           |
| H5     |                           |
| H6     |                           |
| H7     |                           |

$^1$H COSY spectrum of 3-heptanone (recorded in $C_6D_6$ solution at 298K, at 500 MHz).

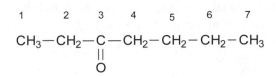

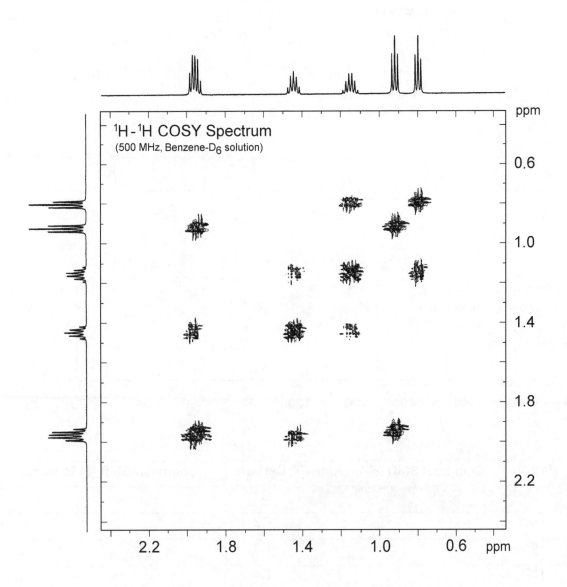

# Problem 301

The $^1$H and $^{13}$C NMR spectra of δ-valerolactone ($C_5H_8O_2$) recorded at 600 MHz in $C_6D_6$ solution at 298K, are given below.  The $^1$H spectrum has signals at δ 1.08, 1.16, 2.08, and 3.71 ppm.  The $^{13}$C spectrum has signals at δ 19.0, 22.2, 29.9, 68.8 and 170.0 ppm.  The 2-dimensional $^1$H-$^1$H COSY spectrum and the C–H correlation spectrum are given on the facing page.  From the COSY spectrum, assign the proton spectrum and use this information to assign the $^{13}$C spectrum.

δ-valerolactone

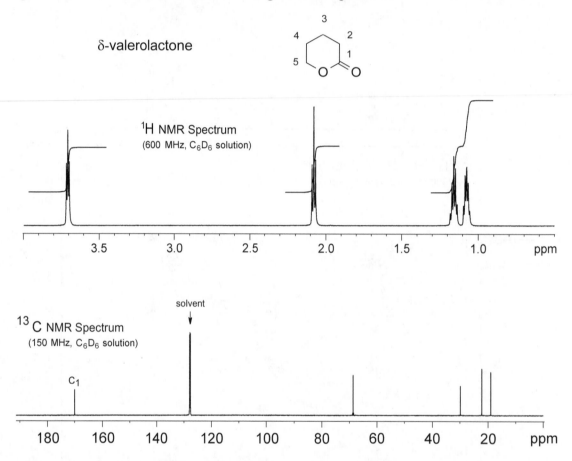

| Proton | Chemical Shift (δ) in ppm | Carbon | Chemical Shift (δ) in ppm |
|--------|---------------------------|--------|---------------------------|
|        |                           | C1     |                           |
| H2     |                           | C2     |                           |
| H3     |                           | C3     |                           |
| H4     |                           | C4     |                           |
| H5     |                           | C5     |                           |

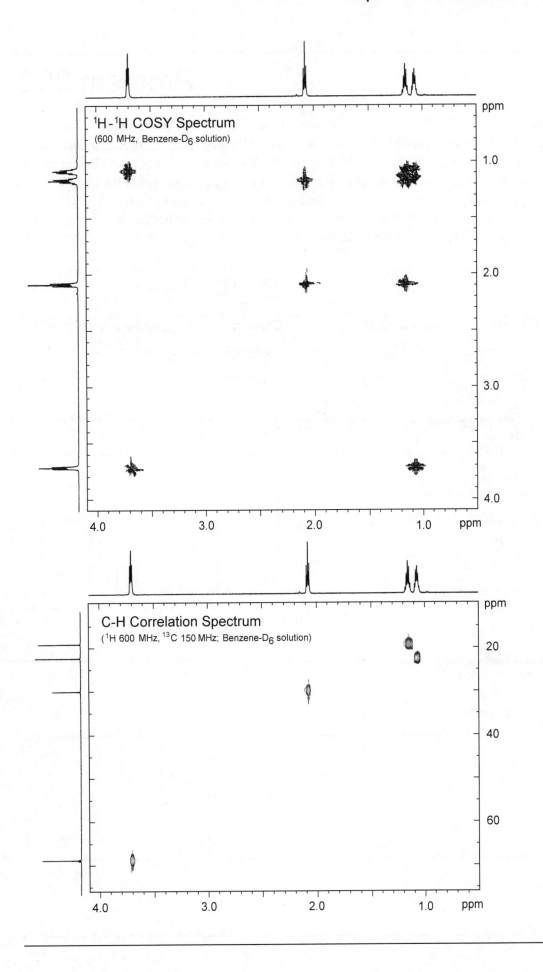

¹H-¹H COSY Spectrum
(600 MHz, Benzene-D₆ solution)

C-H Correlation Spectrum
(¹H 600 MHz, ¹³C 150 MHz; Benzene-D₆ solution)

417

# Problem 302

The $^1$H and $^{13}$C NMR spectra of 1-bromobutane ($C_4H_9Br$) are given below. The $^1$H spectrum has signals at δ 0.91, 1.45, 1.82, and 3.39 ppm. The $^{13}$C spectrum has signals at δ 13.2, 21.4, 33.4 and 34.7 ppm. The 2-dimensional $^1$H-$^1$H COSY spectrum and the C-H correlation spectrum are given on the facing page. From the COSY spectrum, assign the proton spectrum and use this information to assign the $^{13}$C spectrum and then draw in the strong peaks that you would expect to see in the schematic HMBC on the following page.

1-bromobutane

$$\overset{4}{C}H_3 - \overset{3}{C}H_2 - \overset{2}{C}H_2 - \overset{1}{C}H_2 - Br$$

| Proton | Chemical Shift (δ) in ppm | Carbon | Chemical Shift (δ) in ppm |
|--------|---------------------------|--------|---------------------------|
| H1 | | C1 | |
| H2 | | C2 | |
| H3 | | C3 | |
| H4 | | C4 | |

$^1$H NMR Spectrum
(400 MHz, CDCl$_3$ solution)

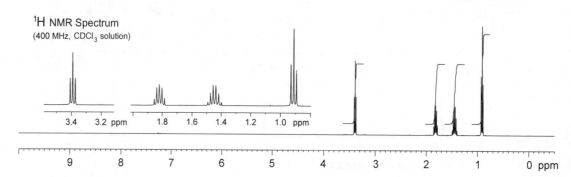

$^{13}$C NMR Spectrum
(100 MHz, CDCl$_3$ solution)

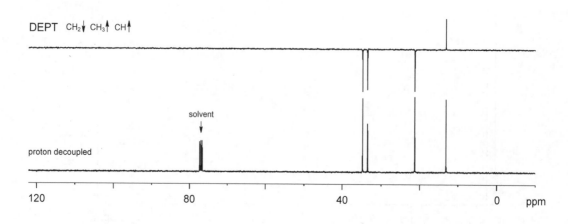

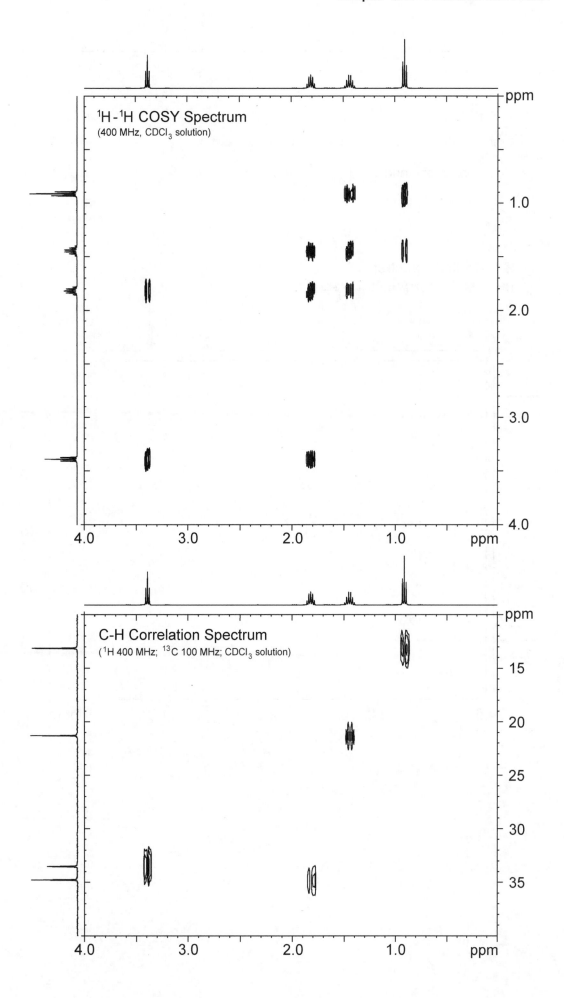

**¹H-¹H COSY Spectrum**
(400 MHz, CDCl₃ solution)

C-H Correlation Spectrum
(¹H 400 MHz; ¹³C 100 MHz; CDCl₃ solution)

1-bromobutane

$$\overset{4}{CH_3}-\overset{3}{CH_2}-\overset{2}{CH_2}-\overset{1}{CH_2}-Br$$

## C-H  HMBC Spectrum
($^1$H 400 MHz; $^{13}$C 100 MHz; CDCl$_3$ solution)

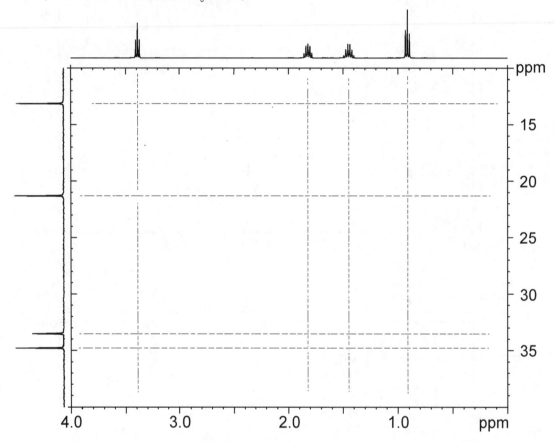

# Problem 303

The $^1$H and $^{13}$C NMR spectra of butyl ethyl ether ($C_6H_{14}O$) recorded at 298K in CDCl$_3$ solution, are given below. The $^1$H spectrum has signals at δ 0.87, 1.11, 1.36, 1.52, 3.27 and 3.29 (partly overlapped) ppm. The $^{13}$C spectrum has signals at δ 13.5, 15.0, 19.4, 32.1, 66.0 and 70.1 ppm. The 2-dimensional $^1$H-$^1$H COSY spectrum and the C-H correlation spectrum are given on the facing page. From the COSY spectrum, assign the proton spectrum and use this information to assign the $^{13}$C spectrum and then draw in the strong peaks that you would expect to see in the schematic HMBC on the following page.

butyl ethyl ether

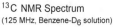

$$\overset{1}{CH_3}-\overset{2}{CH_2}-\overset{3}{CH_2}-\overset{4}{CH_2}-O-\overset{5}{CH_2}-\overset{6}{CH_3}$$

| Proton | Chemical Shift (δ) in ppm | Carbon | Chemical Shift (δ) in ppm |
|--------|---------------------------|--------|---------------------------|
| H1 |  | C1 |  |
| H2 |  | C2 |  |
| H3 |  | C3 |  |
| H4 |  | C4 |  |
| H5 |  | C5 |  |
| H6 |  | C6 |  |

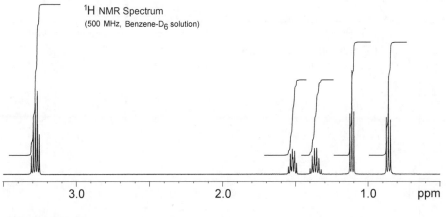

$^1$H NMR Spectrum
(500 MHz, Benzene-D$_6$ solution)

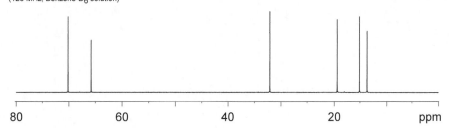

$^{13}$C NMR Spectrum
(125 MHz, Benzene-D$_6$ solution)

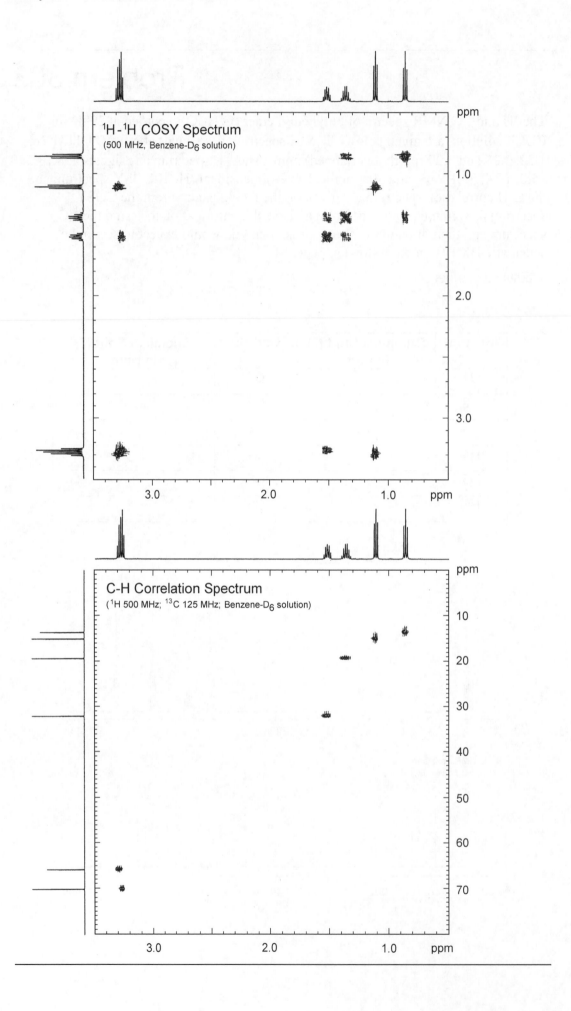

¹H-¹H COSY Spectrum
(500 MHz, Benzene-D₆ solution)

C-H Correlation Spectrum
(¹H 500 MHz; ¹³C 125 MHz; Benzene-D₆ solution)

butyl ethyl ether

## C-H  HMBC Spectrum
($^1$H 500 MHz; $^{13}$C 125 MHz; Benzene-D$_6$ solution)

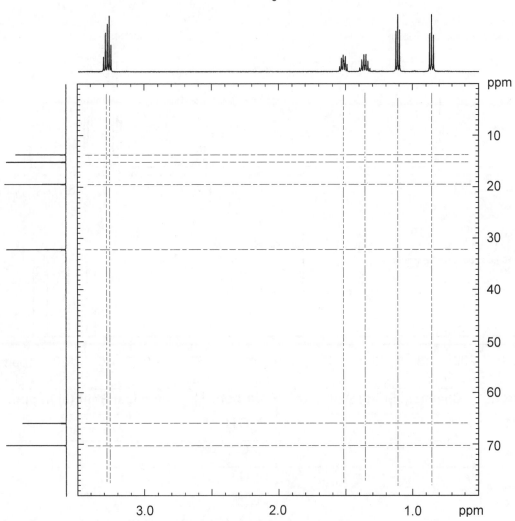

# Problem 304

The $^{1}$H and $^{13}$C spectra of 3-octanone ($C_8H_{16}O$) recorded in $C_6D_6$ solution at 298K at 600 MHz, are given below.  The $^{1}$H spectrum has signals at δ 0.82, 0.92, 1.11, 1.19, 1.47, 1.92 and 1.94 (partly overlapped) ppm.  The $^{13}$C spectrum has signals at δ 7.8, 14.0, 22.7, 23.7, 31.7, 35.4, 42.1 and 209.0 ppm.  The 2-D $^{1}$H-$^{1}$H COSY spectrum and the C–H correlation spectrum are given on the facing page.  From the COSY spectrum, assign the proton spectrum and use this information to assign the $^{13}$C spectrum.

3-octanone

$$\underset{O}{\overset{1}{CH_3}-\overset{2}{CH_2}-\overset{3}{\underset{\parallel}{C}}-\overset{4}{CH_2}-\overset{5}{CH_2}-\overset{6}{CH_2}-\overset{7}{CH_2}-\overset{8}{CH_3}}$$

$^{1}$H NMR Spectrum
(600 MHz, Benzene-D$_6$ solution)

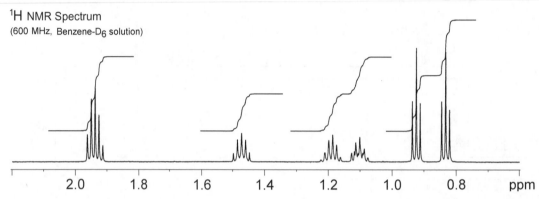

$^{13}$C NMR Spectrum
(150 MHz, Benzene-D$_6$ solution)

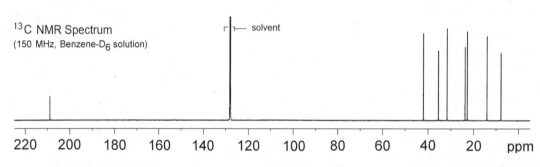

| Proton | Chemical Shift (δ) in ppm | Carbon | Chemical Shift (δ) in ppm |
|---|---|---|---|
| H1 | | C1 | |
| H2 | | C2 | |
|  |  | C3 | |
| H4 | | C4 | |
| H5 | | C5 | |
| H6 | | C6 | |
| H7 | | C7 | |
| H8 | | C8 | |

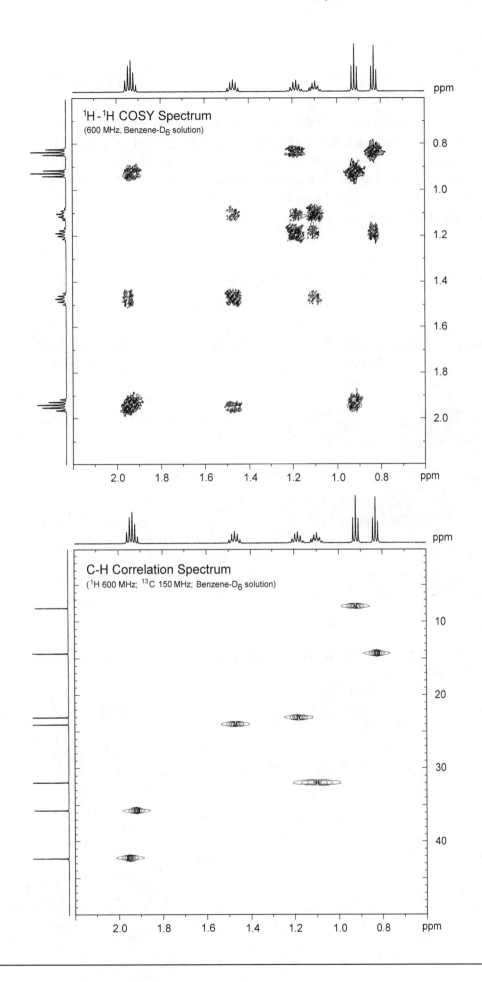

¹H - ¹H COSY Spectrum
(600 MHz, Benzene-D₆ solution)

C-H Correlation Spectrum
(¹H 600 MHz; ¹³C 150 MHz; Benzene-D₆ solution)

# Problem 305

The $^1$H and $^{13}$C NMR spectra of diethyl diethylmalonate ($C_{11}H_{20}O_4$) recorded 298K in CDCl$_3$ solution, are given below.  The $^1$H spectrum has signals at $\delta$ 0.76, 1.19, 1.88 and 4.13 ppm.  The $^{13}$C spectrum has signals at $\delta$ 8.1, 14.0, 24.5, 58.0, 60.8 and 171.9 ppm.  The 2-dimensional $^1$H-$^1$H COSY spectrum and the C-H correlation spectrum are given on the facing page.  From the COSY spectrum, assign the proton spectrum and use this information to assign the $^{13}$C spectrum.

diethyl diethylmalonate

$$
\begin{array}{c}
\overset{3\quad 2\quad 1}{COOCH_2CH_3} \\[2pt]
\underset{9\quad 10\quad 11}{\overset{4\quad 5\quad 6\mid\; 7\quad 8}{CH_3CH_2-C-CH_2CH_3}} \\[2pt]
COOCH_2CH_3
\end{array}
$$

| Proton | Chemical Shift ($\delta$) in ppm | Carbon | Chemical Shift ($\delta$) in ppm |
|--------|--------------------------------|--------|--------------------------------|
| H1 | | C1 | |
| H2 | | C2 | |
| | | C3 | |
| H4 | | C4 | |
| H5 | | C5 | |
| | | C6 | |
| H7 | | C7 | |
| H8 | | C8 | |
| | | C9 | |
| H10 | | C10 | |
| H11 | | C11 | |

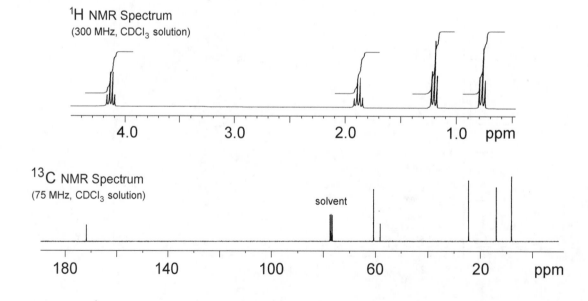

$^1$H NMR Spectrum
(300 MHz, CDCl$_3$ solution)

$^{13}$C NMR Spectrum
(75 MHz, CDCl$_3$ solution)

solvent

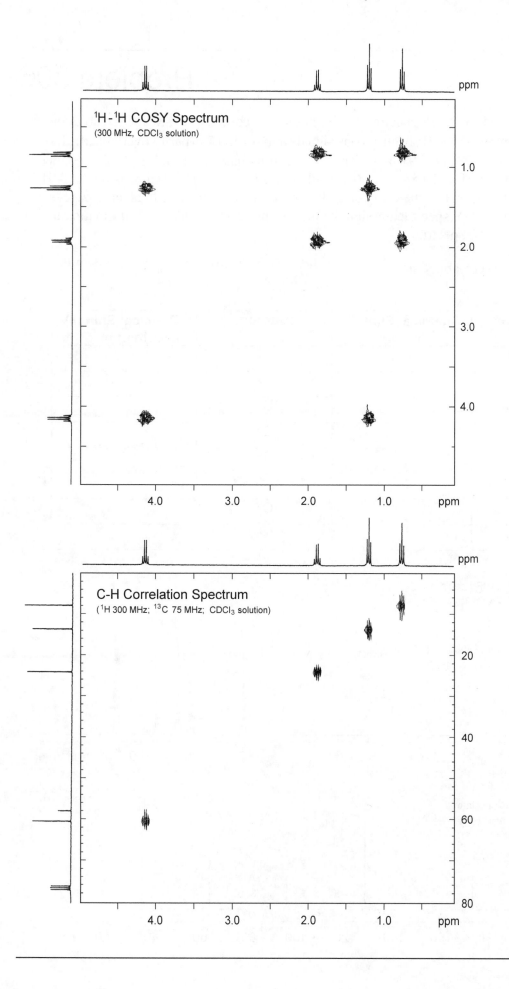

¹H -¹H COSY Spectrum
(300 MHz, CDCl₃ solution)

C-H Correlation Spectrum
(¹H 300 MHz; ¹³C 75 MHz; CDCl₃ solution)

427

# Problem 306

The $^1$H and $^{13}$C NMR spectra of butyl butyrate recorded at 298K in $C_6D_6$ solution are given below.  The $^1$H spectrum has signals at δ 0.75, 0.79 (partly overlapped), 1.19, 1.40, 1.52, 2.08 and 3.97 ppm.  The $^{13}$C spectrum has signals at δ 13.9 (2 overlapped resonances), 19.0, 19.5, 31.2, 36.2, 64.0 and 172.8 ppm.  The 2-dimensional $^1$H-$^1$H COSY spectrum and the C-H correlation spectrum are given on the facing page. From the COSY spectrum, assign the proton spectrum and use this information to assign the $^{13}$C spectrum.

butyl butyrate

$$\overset{1}{CH_3}-\overset{2}{CH_2}-\overset{3}{CH_2}-\overset{4}{CH_2}-O-\overset{5}{\underset{\underset{O}{\|}}{C}}-\overset{6}{CH_2}-\overset{7}{CH_2}-\overset{8}{CH_3}$$

| Proton | Chemical Shift (δ) in ppm | Carbon | Chemical Shift (δ) in ppm |
|--------|---------------------------|--------|---------------------------|
| H1 |  | C1 |  |
| H2 |  | C2 |  |
| H3 |  | C3 |  |
| H4 |  | C4 |  |
|  |  | C5 |  |
| H6 |  | C6 |  |
| H7 |  | C7 |  |
| H8 |  | C8 |  |

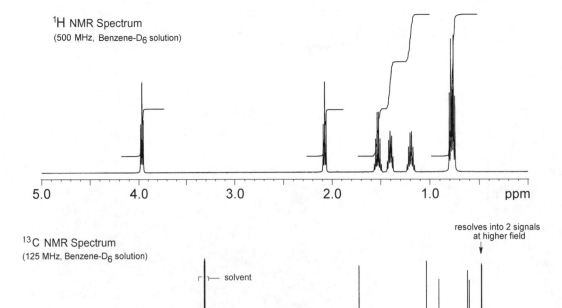

$^1$H NMR Spectrum
(500 MHz, Benzene-D$_6$ solution)

$^{13}$C NMR Spectrum
(125 MHz, Benzene-D$_6$ solution)

resolves into 2 signals at higher field

solvent

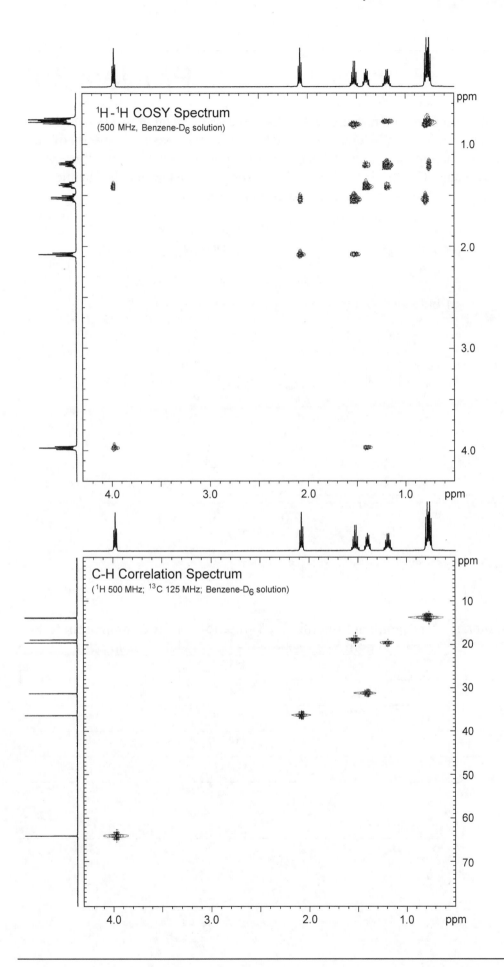

**¹H - ¹H COSY Spectrum**
(500 MHz, Benzene-D₆ solution)

**C-H Correlation Spectrum**
(¹H 500 MHz; ¹³C 125 MHz; Benzene-D₆ solution)

# Problem 307

The $^1$H NMR spectrum of a mixture of 1-iodobutane and 1-butanol recorded at 298K in CDCl$_3$ solution, is given below.  There is some overlap between the spectra of the components of the mixture.  The TOCSY spectrum and the COSY spectrum are given on the facing page.  Use the TOCSY and COSY spectra to determine the chemical shifts of all of the protons in 1-butanol and 1-iodobutane.

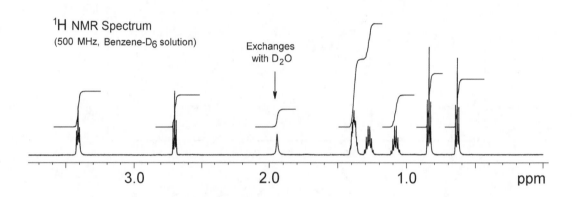

1-iodobutane

$$\overset{4}{CH_3}-\overset{3}{CH_2}-\overset{2}{CH_2}-\overset{1}{CH_2}-I$$

1-butanol

$$\overset{4^*}{CH_3}-\overset{3^*}{CH_2}-\overset{2^*}{CH_2}-\overset{1^*}{CH_2}-OH$$

$^1$H NMR Spectrum
(500 MHz, Benzene-D$_6$ solution)

Exchanges with D$_2$O

| 1-iodobutane | $^1$H Chemical Shift ($\delta$) in ppm | 1-butanol | $^1$H Chemical Shift ($\delta$) in ppm |
|:---:|:---:|:---:|:---:|
| H1 | | H1* | |
| H2 | | H2* | |
| H3 | | H3* | |
| H4 | | H4* | |
| | | -OH | |

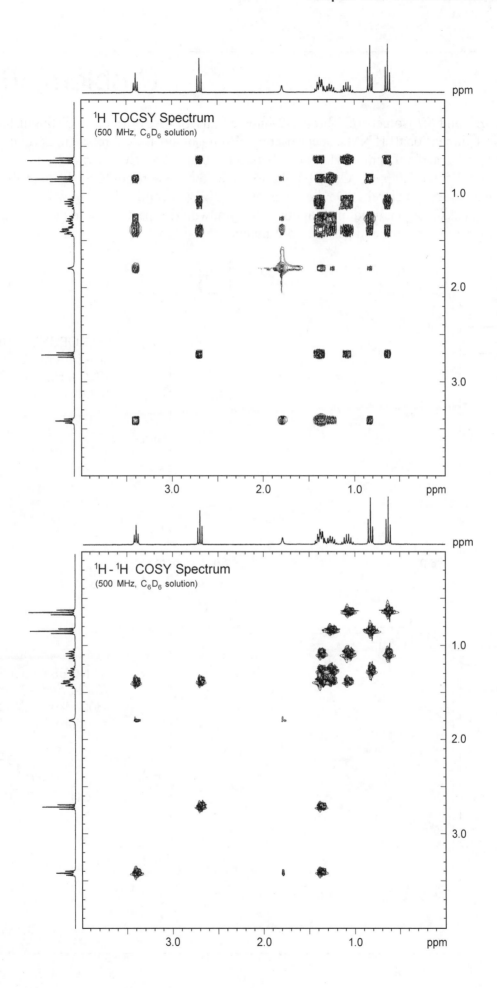

¹H TOCSY Spectrum
(500 MHz, C₆D₆ solution)

¹H - ¹H COSY Spectrum
(500 MHz, C₆D₆ solution)

# Problem 308

The *(E)*- and *(Z)*-isomers of 2-bromo-2-butene ($C_4H_7Br$) are difficult to distinguish by basic 1-dimensional $^1H$ NMR spectroscopy.  Both isomers have 3 resonances (one between 5.5 and 6.0 ppm, -C<u>H</u>=C; one between 2.0 and 2.5, -CC<u>H</u>$_3$Br; and one between 1.5 and 2.0 ppm, -CHC<u>H</u>$_3$).  In principle, the isomers could be distinguished using a NOESY spectrum.  On the schematic NOESY spectra below, draw in the strong peaks (diagonal and off-diagonal) that you would expect to see in the spectra of *(E)*-2-bromo-2-butene and *(Z)*-bromo-2-butene.

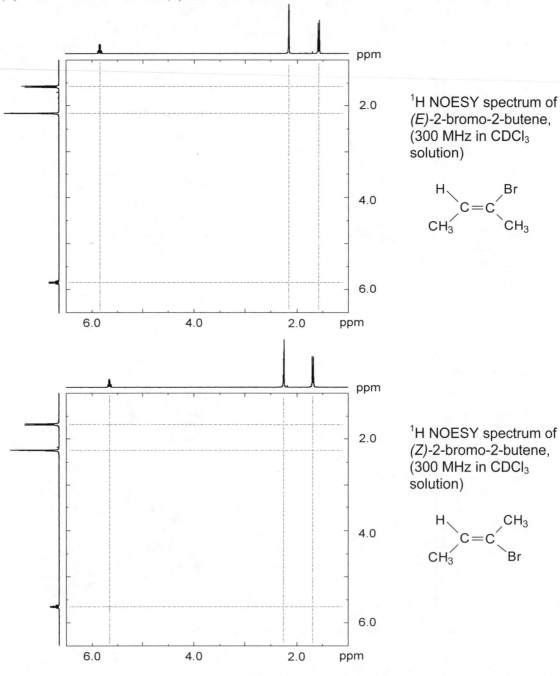

$^1H$ NOESY spectrum of *(E)*-2-bromo-2-butene, (300 MHz in CDCl$_3$ solution)

$^1H$ NOESY spectrum of *(Z)*-2-bromo-2-butene, (300 MHz in CDCl$_3$ solution)

# Problem 309

The $^1$H NMR spectrum of one stereoisomer of 3-methylpent-2-en-4-yn-1-ol [HC≡C(CH$_3$)C=CHCH$_2$OH], (C$_6$H$_8$O) is given below. The 2-dimensional $^1$H NOESY spectrum is also given. Determine the stereochemistry of the compound and draw a structural formula for the compound indicating the stereochemistry.

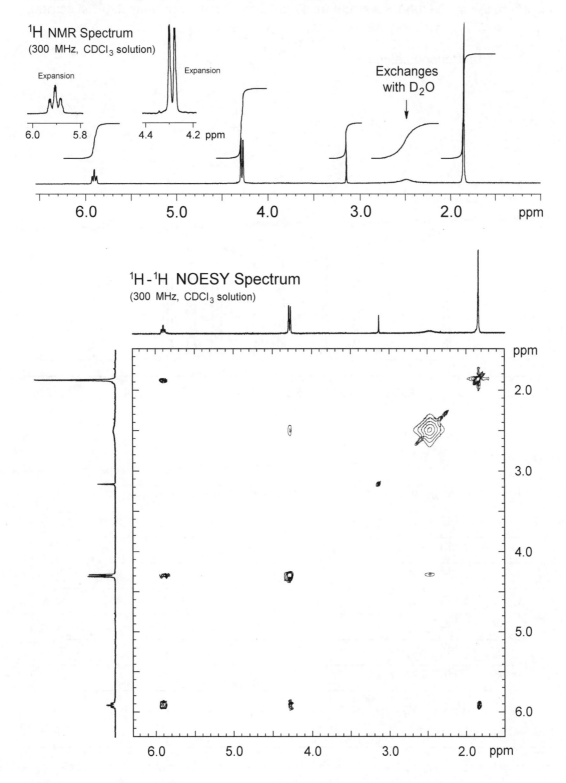

$^1$H NMR Spectrum
(300 MHz, CDCl$_3$ solution)

Expansion

Expansion

6.0   5.8

4.4   4.2 ppm

Exchanges
with D$_2$O

6.0        5.0        4.0        3.0        2.0        ppm

$^1$H-$^1$H NOESY Spectrum
(300 MHz, CDCl$_3$ solution)

ppm

2.0

3.0

4.0

5.0

6.0

6.0        5.0        4.0        3.0        2.0  ppm

# Problem 310

The $^1$H NMR spectrum of 1-nitronaphthalene ($C_{10}H_7NO_2$), recorded at 298K in CDCl$_3$ solution, is given below.  The 2-dimensional $^1$H NOESY spectrum is given on the facing page.  Given that the nitro group at position 1 will always extensively deshield the proton at position 8 such that it will appear as the resonance at lowest field in the spectrum, use the NOESY spectrum to assign all of the other protons in the spectrum.

1-nitronaphthalene

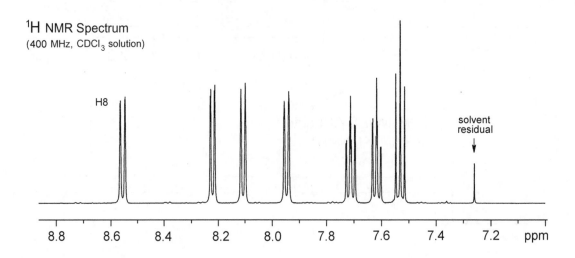

$^1$H NMR Spectrum
(400 MHz, CDCl$_3$ solution)

| Proton | Chemical Shift ($\delta$) in ppm |
|--------|----------------------------------|
| H2 | |
| H3 | |
| H4 | |
| H5 | |
| H6 | |
| H7 | |
| H8 | 8.56 |

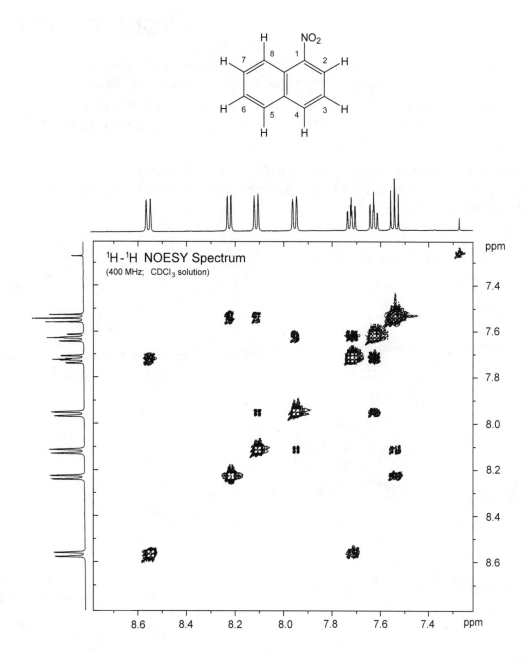

# Problem 311

The $^1$H (400 MHz) and $^{13}$C (100 MHz) NMR spectra of 2-bromo-5-nitrotoluene (C$_7$H$_6$BrNO$_2$), recorded at 298K in CDCl$_3$ solution, are given below. The $^1$H spectrum has resonances at δ 2.5 (H7), 7.7 (H3), 7.9 (H4) and 8.1 (H6) ppm. The $^{13}$C spectrum has resonances at δ 23.1 (C7), 122.1 (C4), 125.3 (C6), 132.2 (C2), 133.2 (C3), 139.8 (C1) and 147.1 (C5) ppm.

On the schematic 2D spectra on the facing page, draw in the strong peaks that you would expect to see in the C-H Correlation spectrum and in the HMBC spectrum of 2-bromo-5-nitrotoluene.

2-bromo-5-nitrotoluene

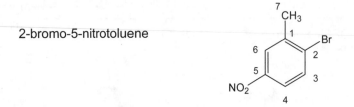

$^1$H NMR Spectrum
(300 MHz, CDCl$_3$ solution)

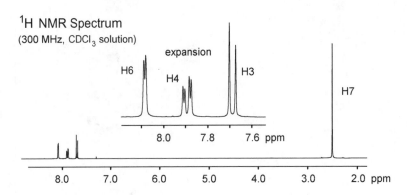

$^{13}$C NMR Spectrum
(75.0 MHz, CDCl$_3$ solution)

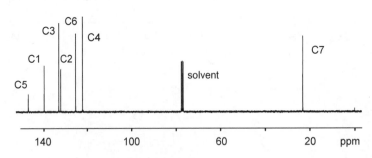

## C-H Correlation Spectrum
($^1$H 300 MHz; $^{13}$C 75 MHz; CDCl$_3$ solution)

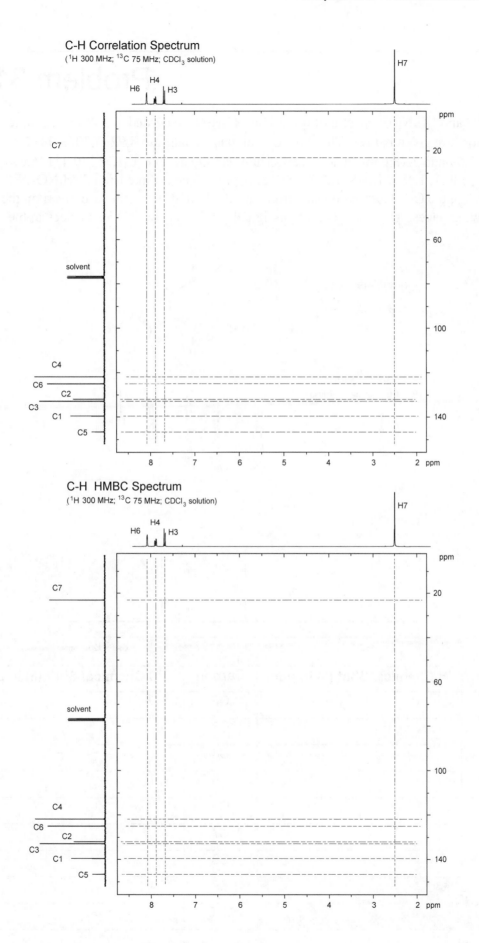

## C-H HMBC Spectrum
($^1$H 300 MHz; $^{13}$C 75 MHz; CDCl$_3$ solution)

# Problem 312

The $^1$H and $^{13}$C NMR spectra of quinoline, (C$_9$H$_7$N), recorded at 298K in acetone-d$_6$ solution, are given below. The $^1$H spectrum has signals at $\delta$ 8.92, 8.30, 8.08, 7.94, 7.75, 7.59 and 7.48 ppm. The $^{13}$C spectrum has signals at $\delta$ 151.3, 149.4, 136.6, 130.3, 130.1, 129.2, 128.8, 127.3 and 122.1 ppm. The 2-dimensional $^1$H NOESY spectrum, the C-H Correlation spectrum and the HMBC spectrum are given on the following pages. Use the spectra to assign all of the $^1$H and $^{13}$C resonances in the spectrum.

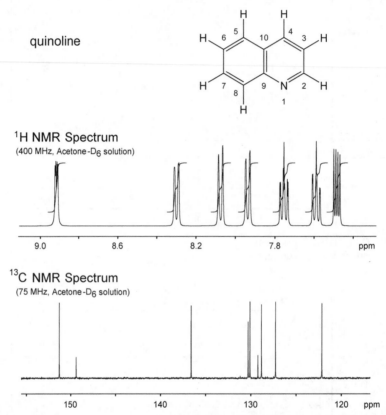

| Proton | $^1$H Chemical Shift ($\delta$) in ppm | Carbon | $^{13}$C Chemical Shift ($\delta$) in ppm |
|--------|----------------------------------------|--------|-------------------------------------------|
| H2     |                                        | C2     |                                           |
| H3     |                                        | C3     |                                           |
| H4     |                                        | C4     |                                           |
| H5     |                                        | C5     |                                           |
| H6     |                                        | C6     |                                           |
| H7     |                                        | C7     |                                           |
| H8     |                                        | C8     |                                           |
|        |                                        | C9     |                                           |
|        |                                        | C10    |                                           |

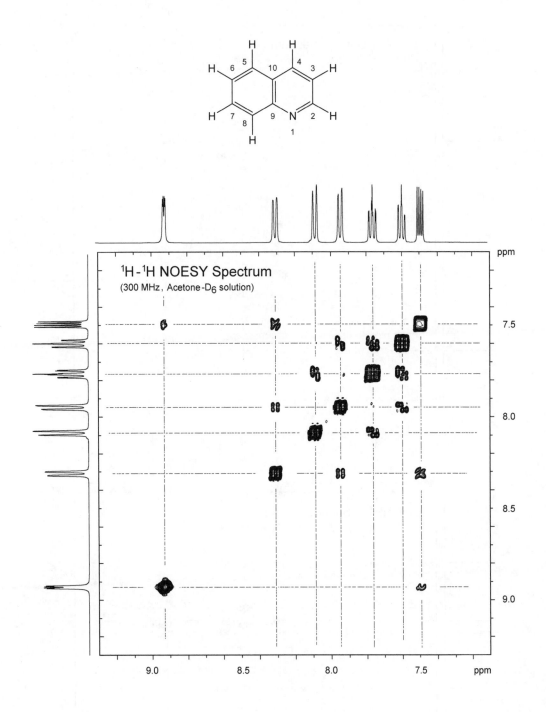

<sup>1</sup>H -<sup>1</sup>H NOESY Spectrum
(300 MHz, Acetone-D<sub>6</sub> solution)

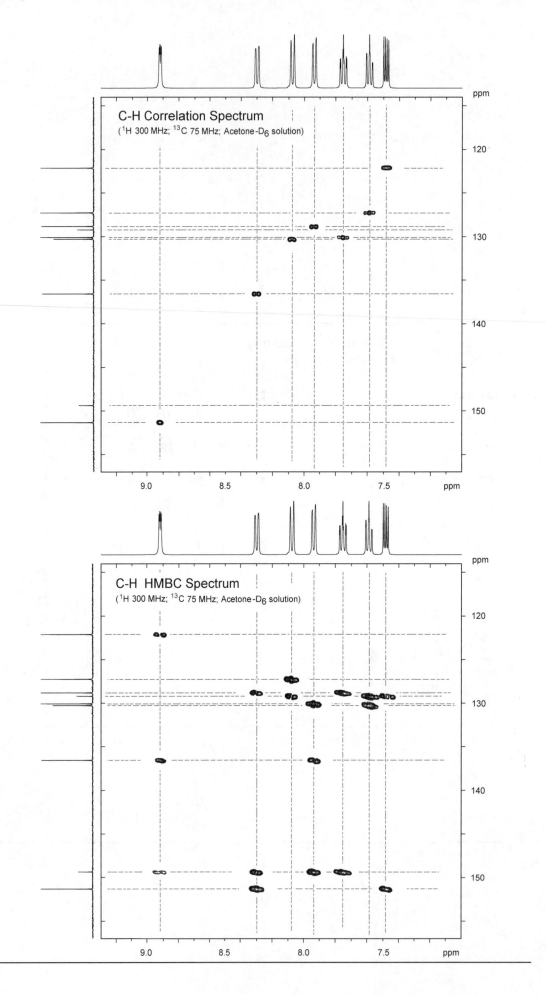

## C-H Correlation Spectrum
($^1$H 300 MHz; $^{13}$C 75 MHz; Acetone-D$_6$ solution)

## C-H HMBC Spectrum
($^1$H 300 MHz; $^{13}$C 75 MHz; Acetone-D$_6$ solution)

# Problem 313

The $^1$H and $^{13}$C NMR spectra of diethyleneglycol ethyl ether acetate ($C_8H_{16}O_4$), recorded at 298K in CDCl$_3$ solution, are given below. The $^1$H spectrum has signals at δ 4.23, 3.71, 3.66, 3.60, 3.54, 2.08 and 1.22 ppm. The $^{13}$C spectrum has signals at δ 171.0, 70.8, 69.8, 69.2, 66.7, 63.6, 20.9 and 15.2 ppm.

The 2-dimensional $^1$H COSY spectrum, the C-H Correlation spectrum and the HMBC spectrum are given on the following pages. Use the spectra to assign all of the $^1$H and $^{13}$C resonances in the spectrum.

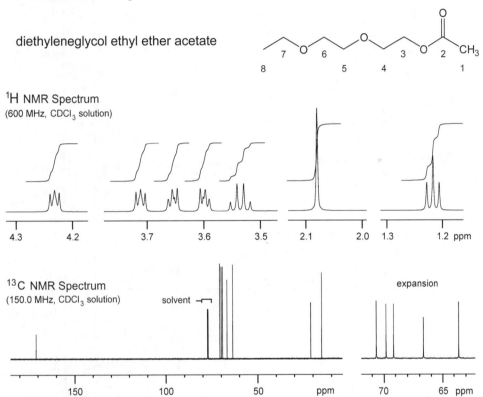

diethyleneglycol ethyl ether acetate

$^1$H NMR Spectrum
(600 MHz, CDCl$_3$ solution)

$^{13}$C NMR Spectrum
(150.0 MHz, CDCl$_3$ solution)     solvent

expansion

| Proton | $^1$H Chemical Shift (δ) in ppm | Carbon | $^{13}$C Chemical Shift (δ) in ppm |
|--------|-------------------------------|--------|-----------------------------------|
| H1 |  | C1 |  |
|  |  | C2 |  |
| H3 |  | C3 |  |
| H4 |  | C4 |  |
| H5 |  | C5 |  |
| H6 |  | C6 |  |
| H7 |  | C7 |  |
| H8 |  | C8 |  |

441

## $^1$H-$^1$H COSY Spectrum
(600 MHz, CDCl$_3$ solution)

## C-H Correlation Spectrum
($^1$H 600 MHz, $^{13}$C 150 MHz, CDCl$_3$ solution)

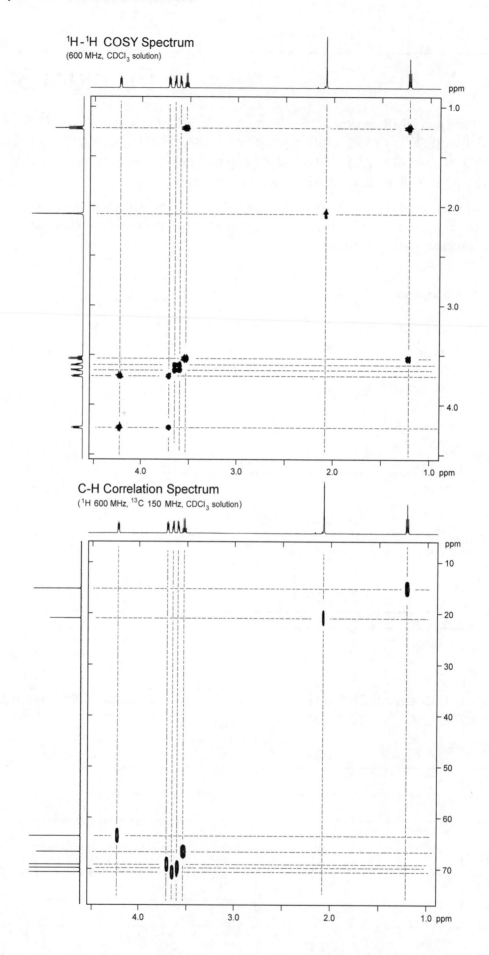

C-H  HMBC Spectrum
($^1$H 600 MHz, $^{13}$C 150 MHz, CDCl$_3$ solution)

C-H  HMBC Spectrum (expansion)
($^1$H 600 MHz, $^{13}$C 150 MHz, CDCl$_3$ solution)

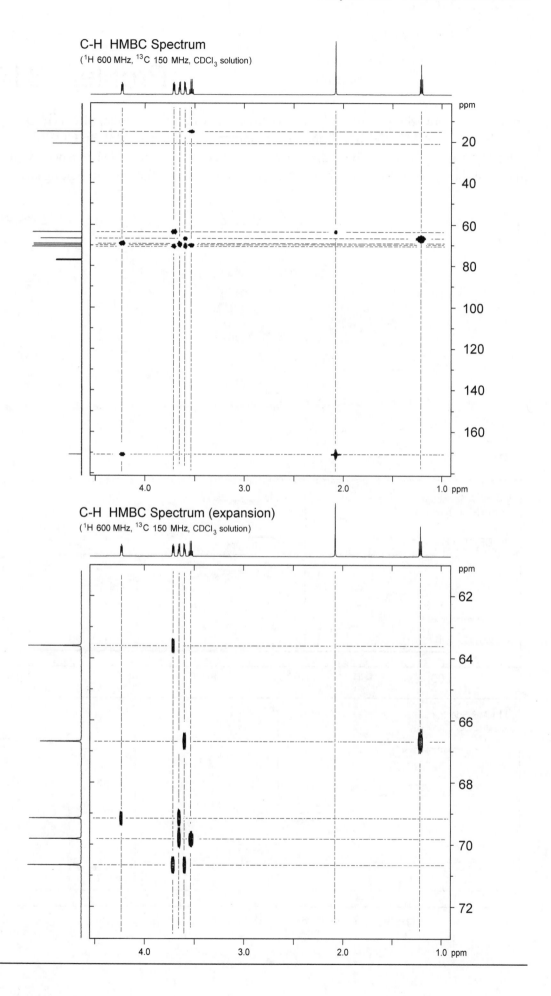

# Problem 314

Given below are 18 aromatic carbonyl compounds which are isomers of $C_{10}H_{12}O$. The $^1H$ and $^{13}C$ NMR spectra of one of the isomers, recorded at 298K in $CDCl_3$ solution, is given below. The 2-dimensional C-H correlation and HMBC spectra are given on the facing page. To which of these compounds do the spectra belong?

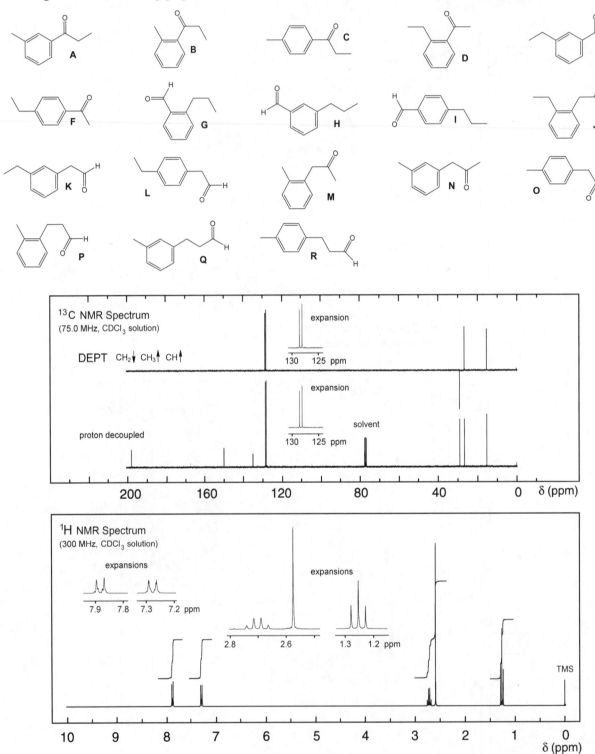

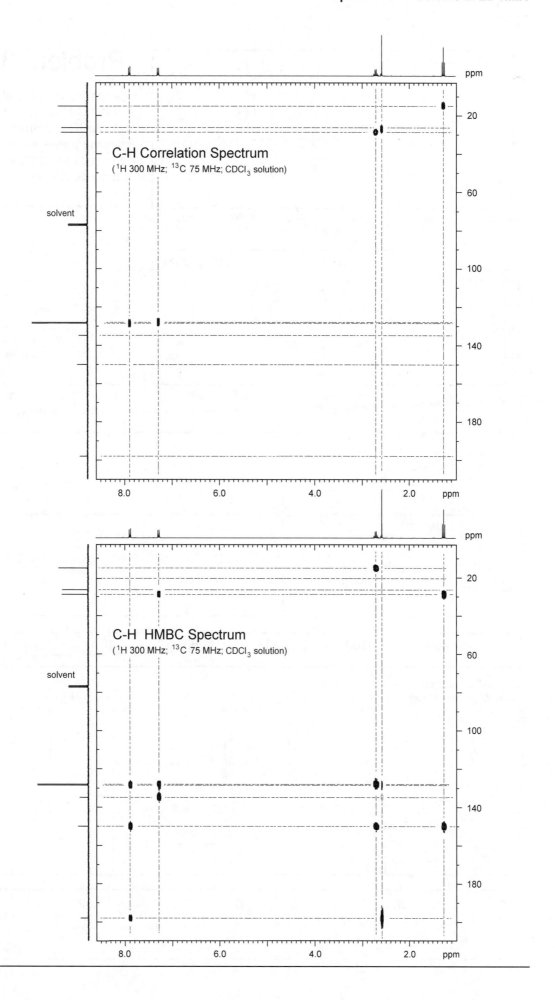

C-H Correlation Spectrum
($^1$H 300 MHz; $^{13}$C 75 MHz; CDCl$_3$ solution)

solvent

C-H  HMBC Spectrum
($^1$H 300 MHz; $^{13}$C 75 MHz; CDCl$_3$ solution)

solvent

# Problem 315

Use the basic spectral data and the C-H Correlation and HMBC spectra on the following pages to determine the structure of this compound.

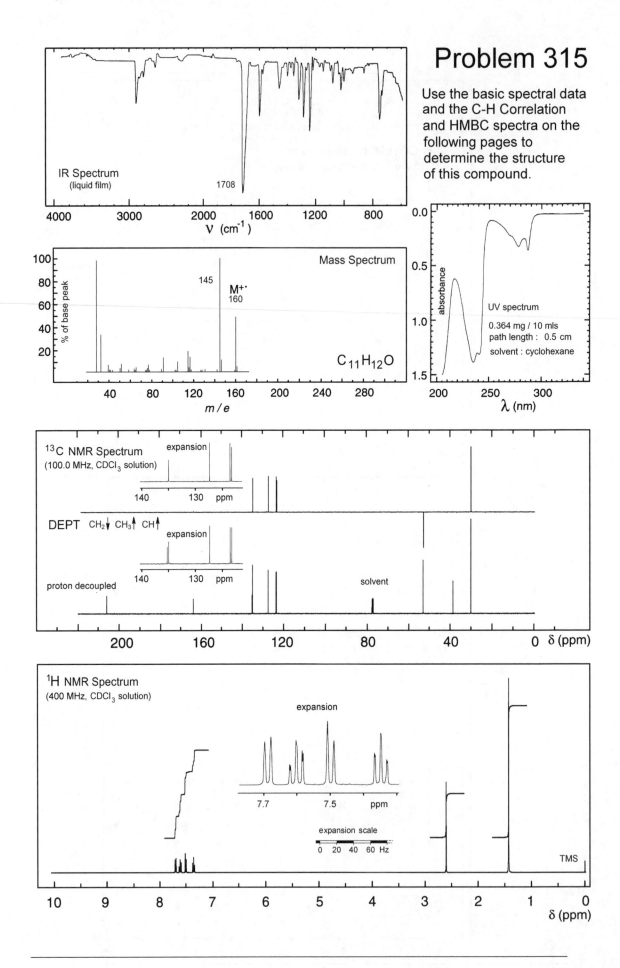

IR Spectrum
(liquid film)

1708

ν (cm⁻¹)

Mass Spectrum

145

M⁺·
160

C₁₁H₁₂O

m/e

UV spectrum

0.364 mg / 10 mls
path length : 0.5 cm

solvent : cyclohexane

λ (nm)

¹³C NMR Spectrum
(100.0 MHz, CDCl₃ solution)

expansion

DEPT   CH₂↓ CH₃↑ CH↑

expansion

proton decoupled

solvent

δ (ppm)

¹H NMR Spectrum
(400 MHz, CDCl₃ solution)

expansion

expansion scale

0  20  40  60 Hz

TMS

δ (ppm)

# C-H Correlation Spectrum
($^1$H 400 MHz; $^{13}$C 100 MHz; CDCl$_3$ solution)

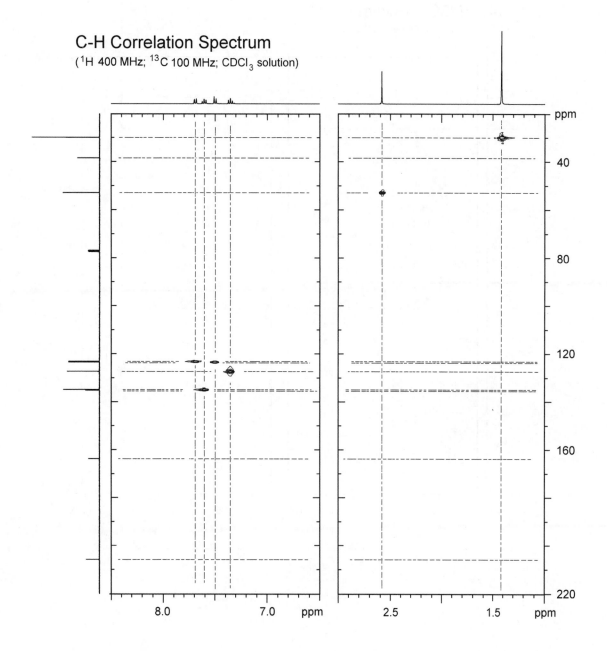

## C-H  HMBC Spectrum
($^1$H 400 MHz; $^{13}$C 100 MHz; CDCl$_3$ solution)

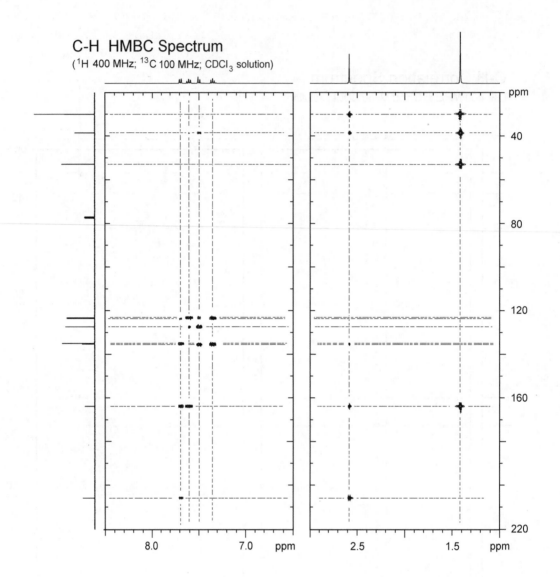

# Problem 316

Given below are 4 aromatic compounds which are isomers of $C_{15}H_{14}O$.  The $^1H$ and $^{13}C$ NMR spectra of one of the isomers, recorded at 298K in acetone $-D_6$ solution, are given below.  The 2-dimensional C-H correlation and HMBC spectra are given on the following pages.  To which of these compounds do the spectra belong?

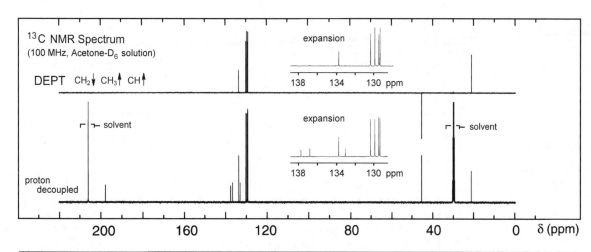

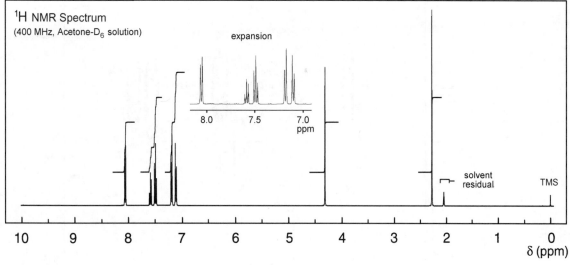

### C-H Correlation Spectrum
($^1$H 400 MHz, $^{13}$C 100 MHz, Acetone-D$_6$ solution)

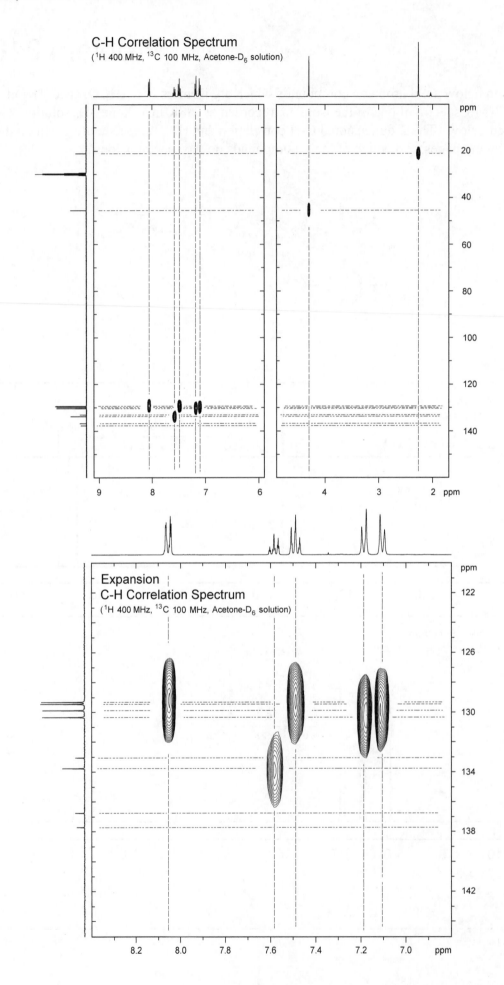

Expansion
C-H Correlation Spectrum
($^1$H 400 MHz, $^{13}$C 100 MHz, Acetone-D$_6$ solution)

## C-H HMBC Spectrum
($^1$H 400 MHz, $^{13}$C 100 MHz, Acetone-D$_6$ solution)

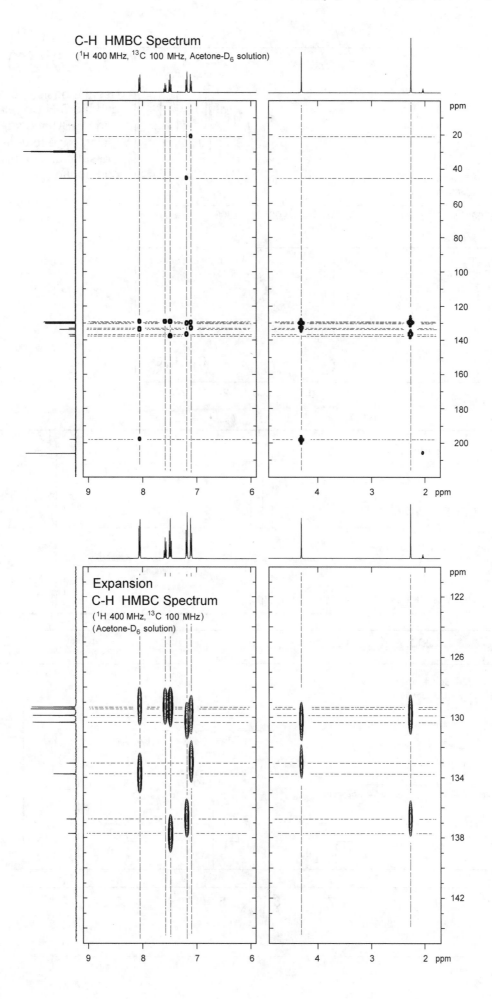

### Expansion
### C-H HMBC Spectrum
($^1$H 400 MHz, $^{13}$C 100 MHz)
(Acetone-D$_6$ solution)

# Problem 317

Use the basic spectral data and the C-H Correlation and HMBC spectra on the facing page to determine the structure of this compound.

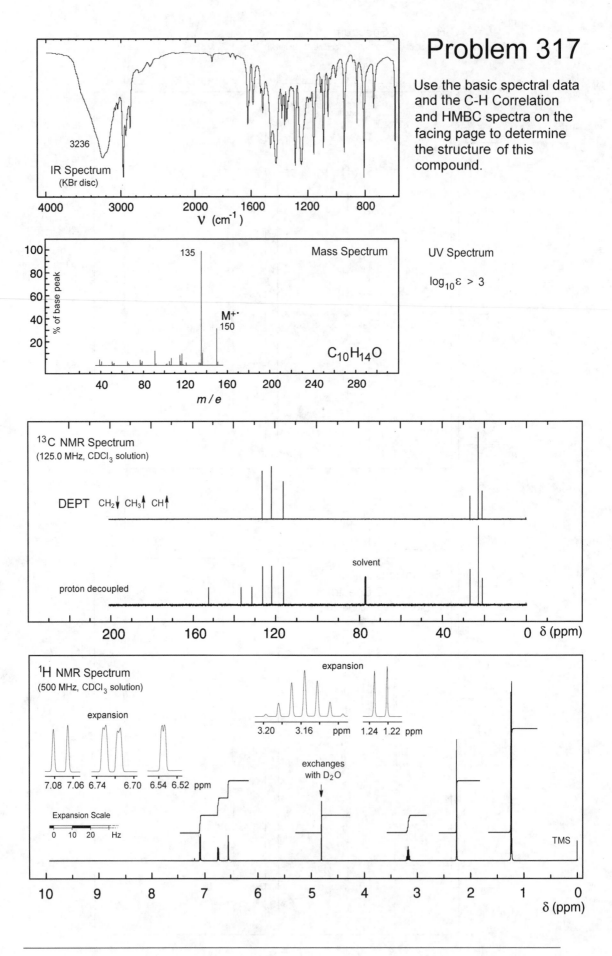

IR Spectrum
(KBr disc)

3236

4000   3000   2000   1600   1200   800

ν (cm⁻¹)

Mass Spectrum

% of base peak

135

M⁺·
150

$C_{10}H_{14}O$

40   80   120   160   200   240   280

m/e

UV Spectrum

$\log_{10}\varepsilon > 3$

$^{13}C$ NMR Spectrum
(125.0 MHz, CDCl₃ solution)

DEPT   CH₂↓ CH₃↑ CH↑

proton decoupled

solvent

200   160   120   80   40   0   δ (ppm)

$^1H$ NMR Spectrum
(500 MHz, CDCl₃ solution)

expansion

3.20   3.16   ppm   1.24   1.22 ppm

expansion

7.08 7.06   6.74   6.70   6.54 6.52 ppm

exchanges
with D₂O

Expansion Scale

0   10   20   Hz

TMS

10   9   8   7   6   5   4   3   2   1   0

δ (ppm)

## C-H Correlation Spectrum
($^1$H 300 MHz; $^{13}$C 75 MHz; CDCl$_3$ solution)

## C-H  HMBC Spectrum
($^1$H 300 MHz; $^{13}$C 75 MHz; CDCl$_3$ solution)

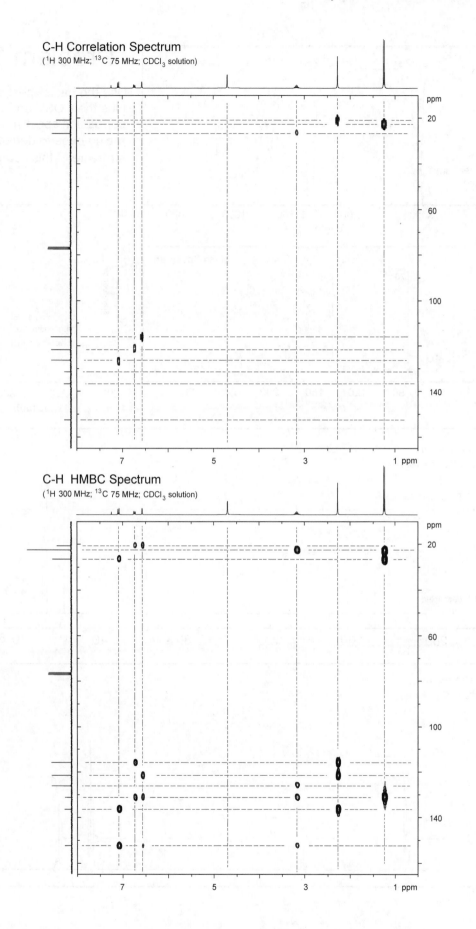

453

# Problem 318

Use the basic spectral data plus the COSY and C-H correlation spectra on the facing page to deduce the structure of this compound.

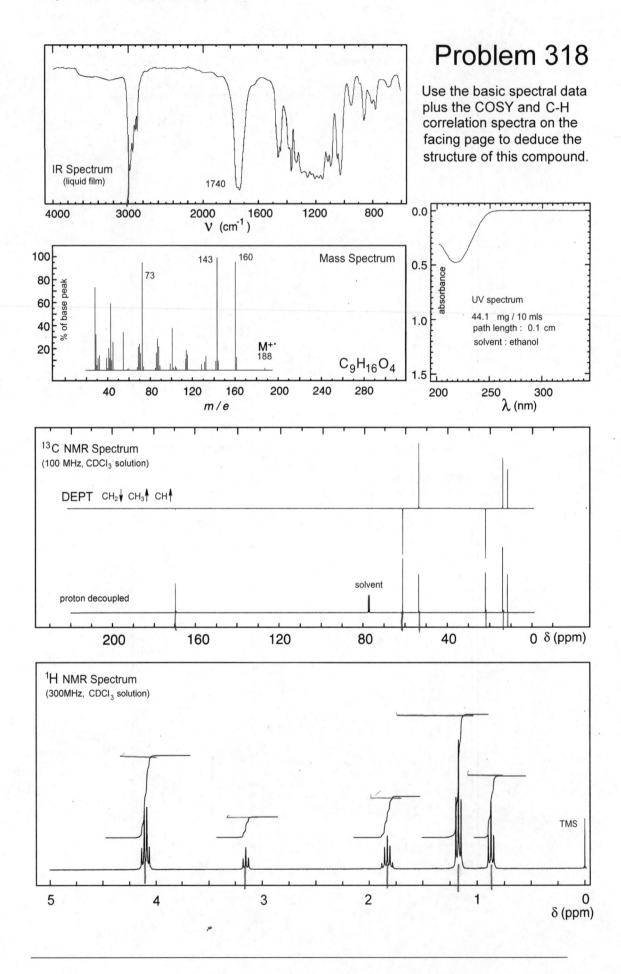

IR Spectrum (liquid film)

1740

$\nu$ (cm$^{-1}$)

Mass Spectrum

73

143

160

% of base peak

M$^{+\cdot}$ 188

$C_9H_{16}O_4$

$m/e$

UV spectrum
44.1   mg / 10 mls
path length :  0.1 cm
solvent : ethanol

absorbance

$\lambda$ (nm)

$^{13}C$ NMR Spectrum
(100 MHz, CDCl$_3$ solution)

DEPT   CH$_2$↓ CH$_3$↑ CH↑

solvent

proton decoupled

$\delta$ (ppm)

$^1H$ NMR Spectrum
(300MHz,  CDCl$_3$ solution)

TMS

$\delta$ (ppm)

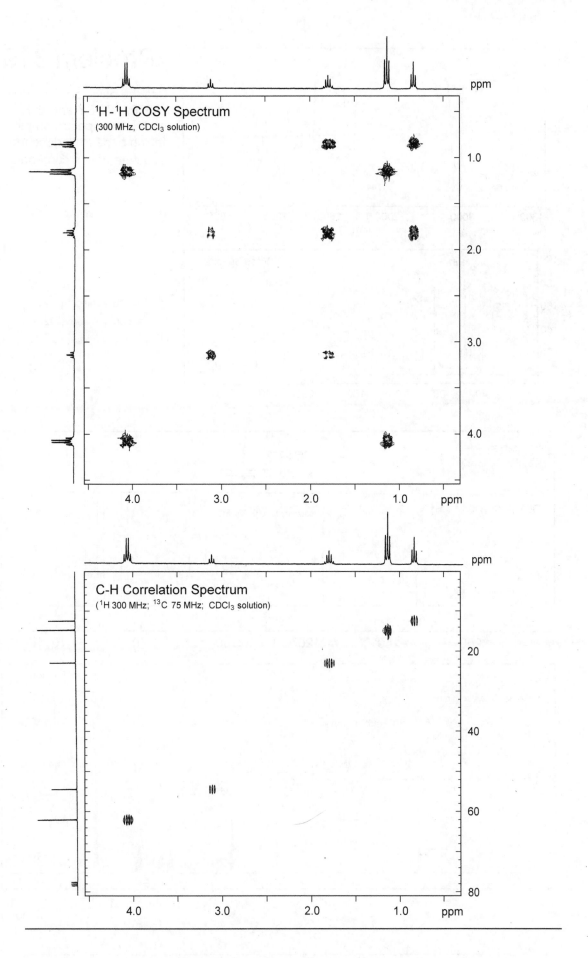

¹H - ¹H COSY Spectrum
(300 MHz, CDCl₃ solution)

C-H Correlation Spectrum
(¹H 300 MHz; ¹³C 75 MHz; CDCl₃ solution)

# Problem 319

Use the basic spectral data plus the COSY and C-H correlation spectra on the facing page to deduce the structure of this compound.

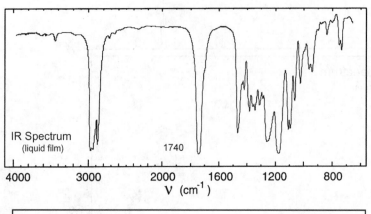

IR Spectrum
(liquid film)

1740

ν (cm$^{-1}$)

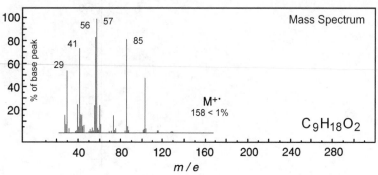

Mass Spectrum

100
80
60
40
20

% of base peak

29
41
56
57
85

M$^{+\cdot}$
158 < 1%

C$_9$H$_{18}$O$_2$

m/e

No significant UV absorption above 220 nm

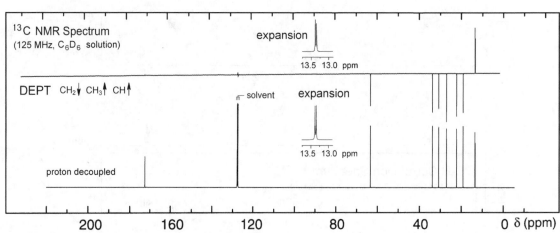

$^{13}$C NMR Spectrum
(125 MHz, C$_6$D$_6$ solution)

expansion

13.5  13.0 ppm

DEPT  CH$_2$↓ CH$_3$↑ CH↑

solvent

expansion

13.5  13.0 ppm

proton decoupled

200    160    120    80    40    0 δ (ppm)

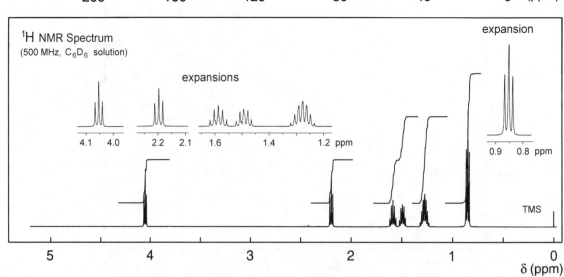

$^1$H NMR Spectrum
(500 MHz, C$_6$D$_6$ solution)

expansions

expansion

4.1  4.0

2.2  2.1

1.6    1.4    1.2 ppm

0.9  0.8 ppm

TMS

5    4    3    2    1    0

δ (ppm)

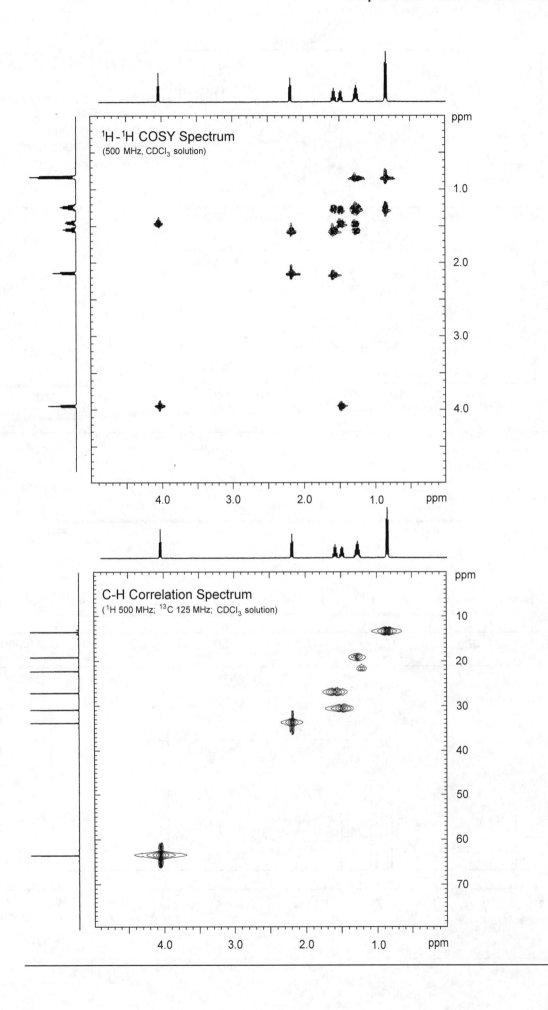

¹H-¹H COSY Spectrum
(500 MHz, CDCl₃ solution)

C-H Correlation Spectrum
(¹H 500 MHz; ¹³C 125 MHz; CDCl₃ solution)

# Problem 320

Use the basic spectral data and the NOESY spectrum on the facing page to deduce the structure of this compound.

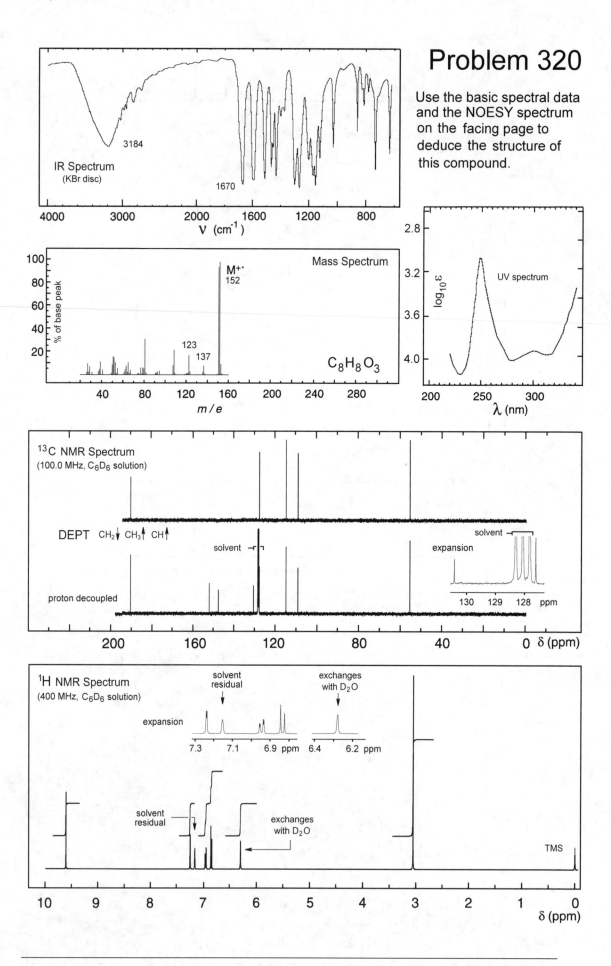

IR Spectrum
(KBr disc)

3184

1670

ν (cm⁻¹)

Mass Spectrum

M⁺˙
152

123

137

$C_8H_8O_3$

% of base peak

m/e

UV spectrum

$\log_{10}\varepsilon$

λ (nm)

¹³C NMR Spectrum
(100.0 MHz, $C_6D_6$ solution)

DEPT  CH₂↓ CH₃↑ CH↑

solvent

expansion

solvent

proton decoupled

δ (ppm)

¹H NMR Spectrum
(400 MHz, $C_6D_6$ solution)

solvent residual

exchanges with $D_2O$

expansion

solvent residual

exchanges with $D_2O$

TMS

δ (ppm)

## $^1$H-$^1$H NOESY Spectrum

(300 MHz, Benzene-D$_6$ solution)

Diagonal peaks plotted with reduced intensity

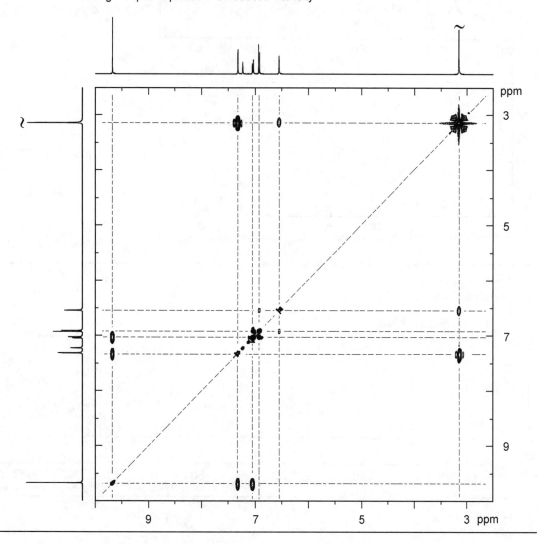

# Problem 321

Use the basic spectral data plus the COSY and NOESY spectra on the facing page to deduce the structure, including stereochemistry of this compound.

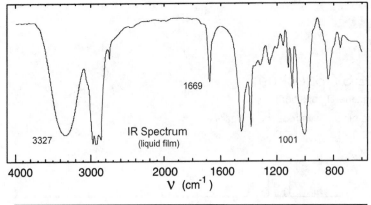

IR Spectrum
(liquid film)

3327

1669

1001

ν (cm⁻¹)

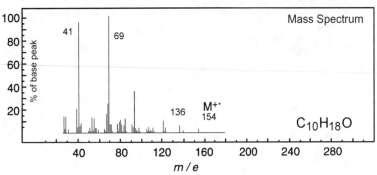

Mass Spectrum

41

69

136

M⁺·
154

C₁₀H₁₈O

No significant UV absorption above 220 nm

% of base peak

m/e

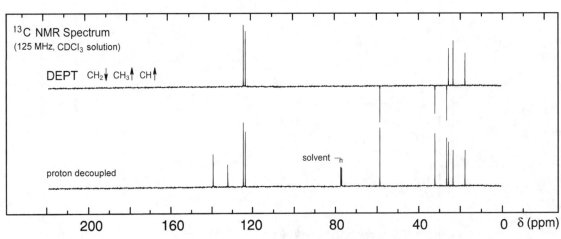

¹³C NMR Spectrum
(125 MHz, CDCl₃ solution)

DEPT   CH₂↓ CH₃↑ CH↑

proton decoupled

solvent

δ (ppm)

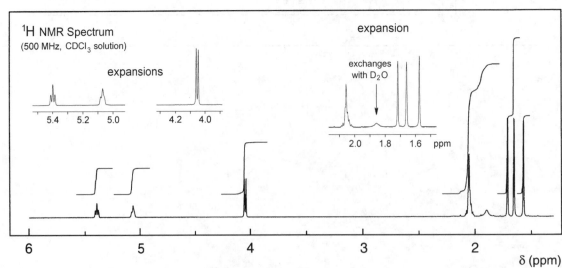

¹H NMR Spectrum
(500 MHz, CDCl₃ solution)

expansions

expansion

exchanges
with D₂O

δ (ppm)

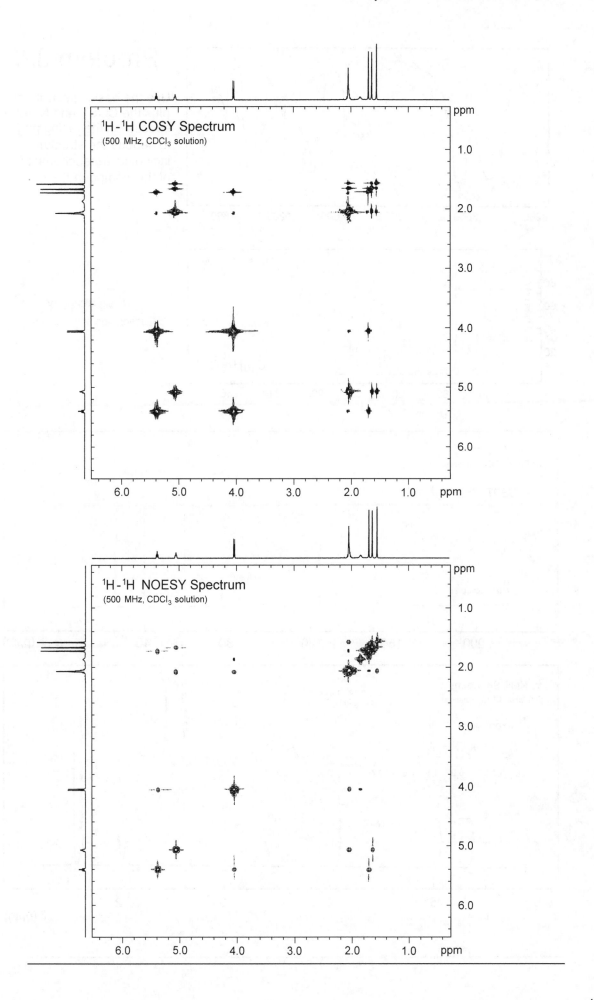

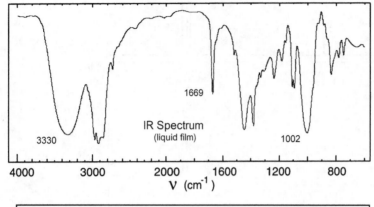

# Problem 322

Use the basic spectral data plus the COSY and NOESY spectra on the facing page to deduce the structure, including stereochemistry of this compound.

IR Spectrum
(liquid film)

3330
1669
1002

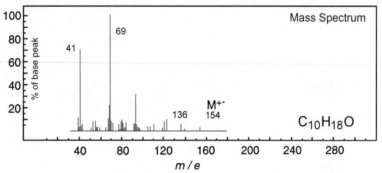

Mass Spectrum

No significant UV
absorption above 220 nm

41
69
136
M+·
154

$C_{10}H_{18}O$

% of base peak

m/e

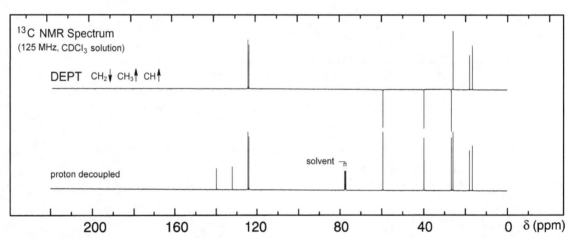

$^{13}C$ NMR Spectrum
(125 MHz, $CDCl_3$ solution)

DEPT   $CH_2$↓ $CH_3$↑ CH↑

proton decoupled

solvent

δ (ppm)

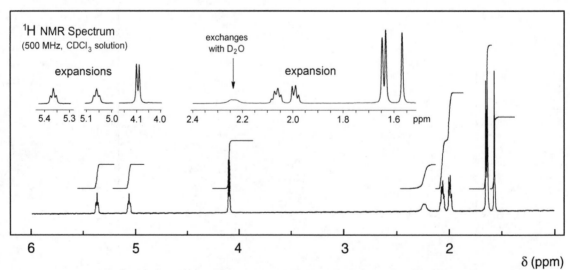

$^1H$ NMR Spectrum
(500 MHz, $CDCl_3$ solution)

exchanges
with $D_2O$

expansions

expansion

ppm

δ (ppm)

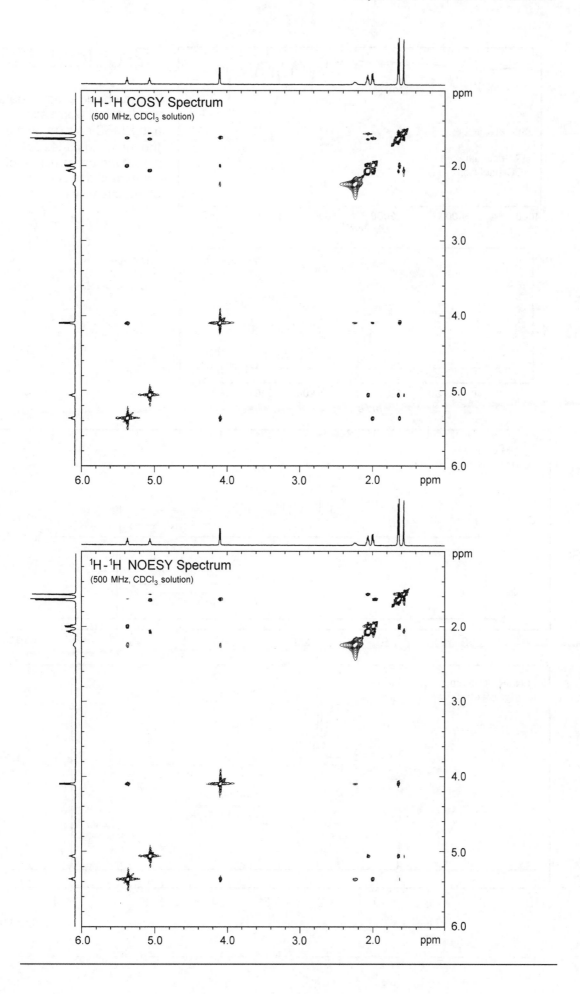

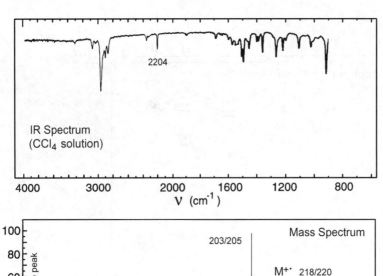

2204

IR Spectrum
(CCl₄ solution)

$\nu$ (cm⁻¹)

# Problem 323

Use the basic spectral data and the C-H Correlation and HMBC spectra on the the following pages to determine the structure of the compound, including its stereochemistry.

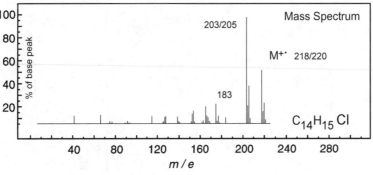

Mass Spectrum

203/205

M⁺˙  218/220

183

$C_{14}H_{15}Cl$

% of base peak

m/e

UV Spectrum

$\log_{10}\varepsilon$ > 4.5

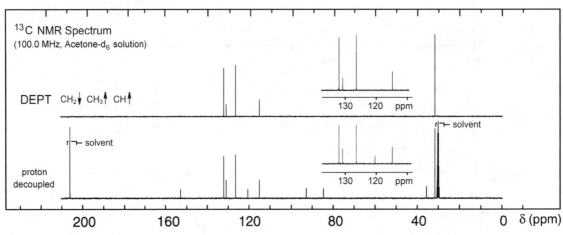

¹³C NMR Spectrum
(100.0 MHz, Acetone-d₆ solution)

DEPT   CH₂↓ CH₃↑ CH↑

solvent

proton decoupled

solvent

$\delta$ (ppm)

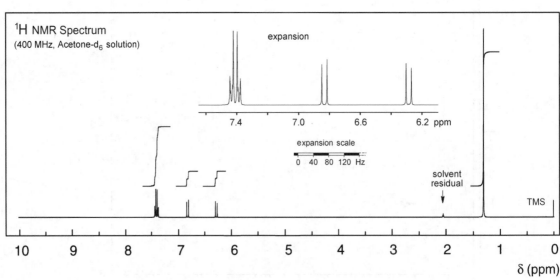

¹H NMR Spectrum
(400 MHz, Acetone-d₆ solution)

expansion

expansion scale

0  40  80  120  Hz

solvent residual

TMS

$\delta$ (ppm)

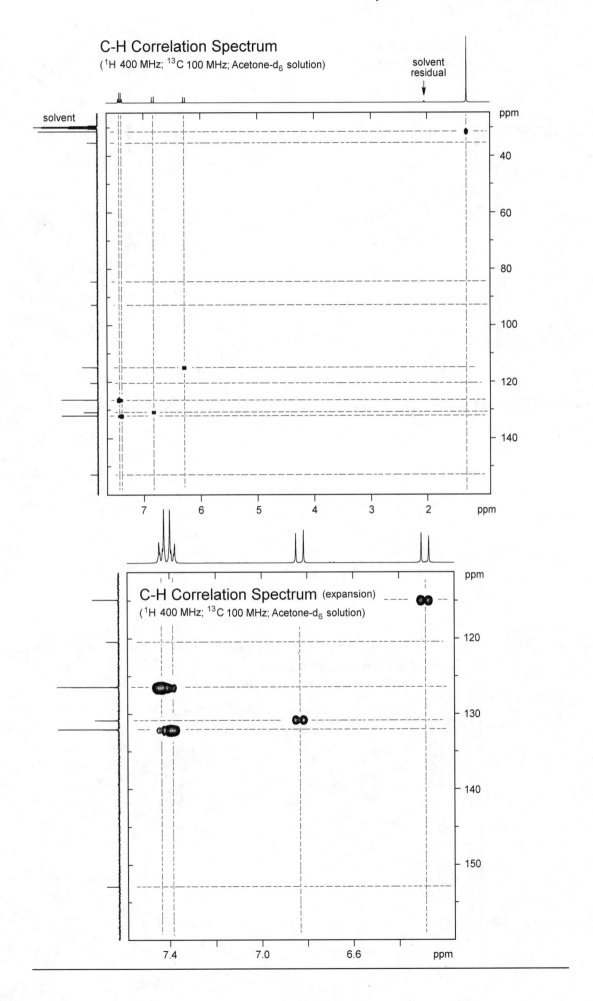

## C-H Correlation Spectrum
($^1$H 400 MHz; $^{13}$C 100 MHz; Acetone-d$_6$ solution)

## C-H Correlation Spectrum (expansion)
($^1$H 400 MHz; $^{13}$C 100 MHz; Acetone-d$_6$ solution)

## C-H  HMBC Spectrum
($^1$H 400 MHz; $^{13}$C 100 MHz; Acetone-d$_6$ solution)

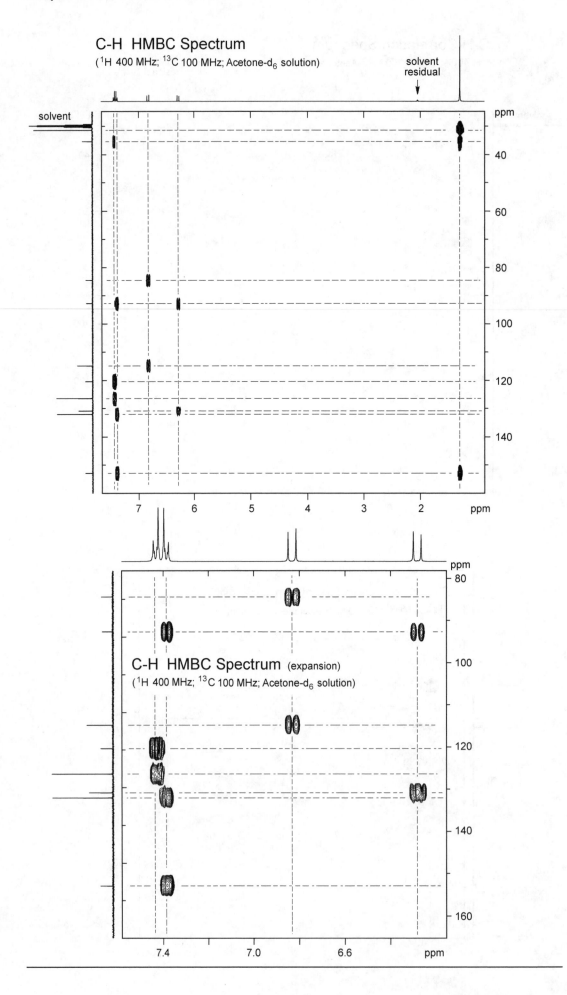

## C-H  HMBC Spectrum (expansion)
($^1$H 400 MHz; $^{13}$C 100 MHz; Acetone-d$_6$ solution)

# 10.4

# NMR SPECTRAL ANALYSIS

# Problem 324

Give the <u>number of different chemical environments</u> for the magnetic nuclei $^1H$ and $^{13}C$ in the following compounds.  Assume that any conformational processes are fast on the NMR timescale unless otherwise indicated.

| Structure | Number of $^1H$ environments | Number of $^{13}C$ environments |
|---|---|---|
| $CH_3-CO-CH_2CH_2CH_3$ | | |
| $CH_3CH_2-CO-CH_2CH_3$ | | |
| $CH_2=CHCH_2CH_3$ | | |
| cis- $CH_3CH=CHCH_3$ | | |
| trans-$CH_3CH=CHCH_3$ | | |
| (benzene) | | |
| (chlorobenzene) —Cl | | |
| (1,2-dibromobenzene) Br, Br | | |
| Cl (1,3-dichlorobenzene) Cl | | |
| Br— —Br | | |
| Br— —Cl | | |
| Cl, OCH$_3$ | | |
| (cyclohexane) Assuming **slow** chair-chair interconversion | | |
| (cyclohexane) Assuming **fast** chair-chair interconversion | | |
| H, Cl Assuming the molecule to be *conformationally rigid* | | |

# Problem 325

Draw a schematic (line) representation of the pure first-order spectrum (AMX) corresponding to the following parameters:

**Frequencies** (Hz from TMS):      $\nu_A = 300$;  $\nu_M = 240$;  $\nu_X = 120$.

**Coupling constants** (Hz):      $J_{AM} = 10$;   $J_{AX} = 2$;  $J_{MX} = 8$.

Assume that the spectrum is a pure first-order spectrum and ignore small distortions in relative intensities of lines that would be apparent in a "real" spectrum.

*(a)*     Sketch in "splitting diagrams" above the schematic spectrum to indicate which splittings correspond to which coupling constants.

*(b)*     Give the chemical shifts on the $\delta$ scale corresponding to the above spectrum obtained with an instrument operating at 60 MHz for protons.

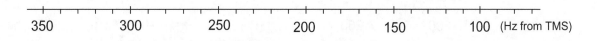

# Problem 326

Draw a schematic (line) representation of the pure first-order spectrum (AMX) corresponding to the following parameters:

**Frequencies** (Hz from TMS):     $v_A = 180;\ v_M = 220;\ v_X = 300.$

**Coupling constants** (Hz):     $J_{AM} = 10;\ J_{AX} = 12;\ J_{MX} = 5.$

Assume that the spectrum is a pure first-order spectrum and ignore small distortions in relative intensities of lines that would be apparent in a "real" spectrum.

*(a)*     Sketch in "splitting diagrams" above the schematic spectrum to indicate which splittings correspond to which coupling constants.

*(b)*     Give the chemical shifts on the $\delta$ scale corresponding to the above spectrum obtained with an instrument operating at 200 MHz for protons.

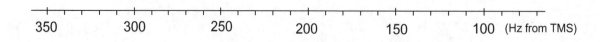

# Problem 327

Draw a schematic (line) representation of the pure first-order spectrum ($AX_2$) corresponding to the following parameters:

**Frequencies** (Hz from TMS):       $\nu_A = 150$;  $\nu_X = 300$.

**Coupling constants** (Hz):       $J_{AX} = 20$.

Assume that the spectrum is a pure first-order spectrum and ignore small distortions in relative intensities of lines that would be apparent in a "real" spectrum.

*(a)*     Sketch in "splitting diagrams" above the schematic spectrum to indicate which splittings correspond to which coupling constants.

*(b)*     Give the chemical shifts on the $\delta$ scale corresponding to the above spectrum obtained with an instrument operating at 400 MHz for protons.

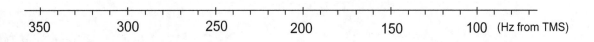

| | | | | | |
|---|---|---|---|---|---|
| 350 | 300 | 250 | 200 | 150 | 100   (Hz from TMS) |

# Problem 328

A 60 MHz $^1$H NMR spectrum of diethyl ether is given below.

Note that the spectrum is calibrated only in parts per million (ppm) from tetramethylsilane (TMS), *i.e.* in δ units.

*(a)*   Assign the signals due to the $-CH_2-$ and $-CH_3$ groups respectively using three independent criteria (the relative areas of the signals, the multiplicity of each signal and the chemical shift of each signal).

*(b)*   Obtain the chemical shift of each group in ppm, then convert to Hz at 60 MHz from TMS (see Section 5.5).

*(c)*   Obtain the value of the first-order coupling constants $^3J_{H-H}$ (in Hz).

*(d)*   Demonstrate that first-order analysis was justified (see Section 5.9).

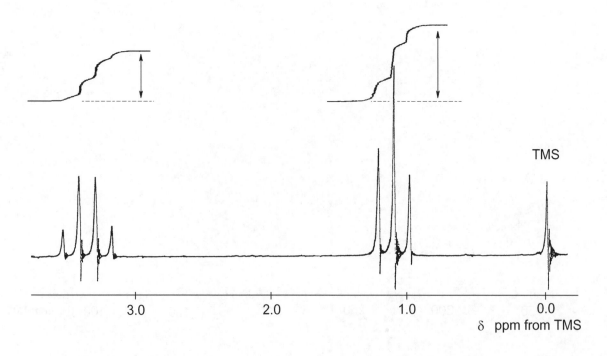

TMS

3.0          2.0          1.0          0.0

δ  ppm from TMS

# Problem 329

A 100 MHz $^1$H NMR spectrum of a 3-proton system is given below.

*(a)*   Draw a splitting diagram and analyse this spectrum by first-order methods, *i.e.* extract all relevant coupling constants (*J* in Hz) and chemical shifts (δ in ppm) by direct measurement.

*(b)*   Justify the use of a first-order analysis (see Section 5.9).

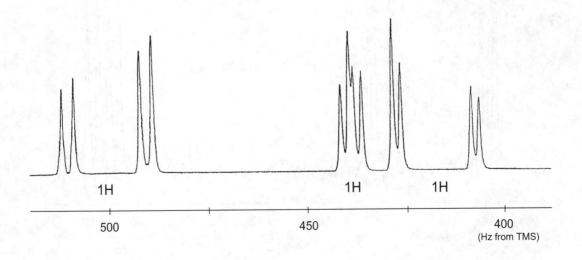

1H                    1H        1H

500              450              400
                           (Hz from TMS)

# Problem 330

Portion of the 60 MHz NMR spectrum 2-furoic
acid in CDCl₃ is shown below. Only the
resonances due to the three aromatic protons
(H$_A$, H$_M$ and H$_X$) are shown.

2-Furoic Acid

*(a)*    Draw a splitting diagram and analyse this spectrum by first-order methods,
        *i.e.* extract all relevant coupling constants (*J* in Hz) and chemical shifts (δ in
        ppm) by direct measurement.

*(b)*    Justify the use of a first-order analysis (see Section 5.9).

**Note**:   This is a <u>60 MHz spectrum</u>.

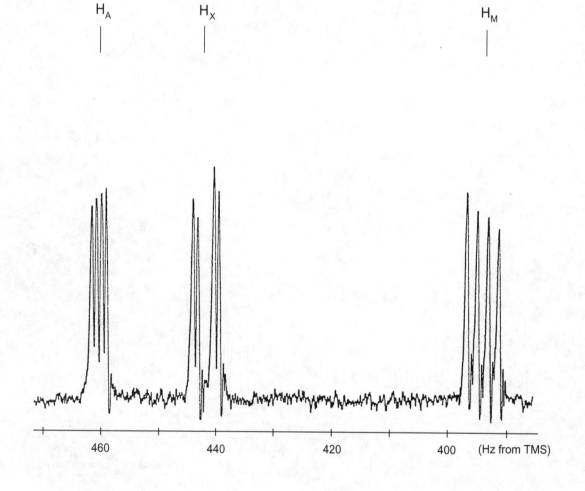

H$_A$              H$_X$                                              H$_M$

460              440              420              400      (Hz from TMS)

# Problem 331

A portion of the 100 MHz $^1$H NMR spectrum of
2-amino-5-chlorobenzoic acid in $CD_3OD$ is
given below.  Only the resonances due to the
three aromatic protons are shown.

*(a)*     Draw a splitting diagram and analyse this spectrum by first-order methods,
          *i.e.* extract all relevant coupling constants ($J$ in Hz) and chemical shifts ($\delta$ in
          ppm) by direct measurement.

*(b)*     Justify the use of a first-order analysis (see Section 5.9).

*(c)*     Assign the three multiplets to $H_3$, $H_4$ and $H_6$ given:

  - the characteristic ranges for coupling constants between aromatic
    protons (see Section 5.9);

  - the fact that $H_3$ will give rise to the resonance at the highest field due to
    the strong influence of the amino group (see Table 5.6).

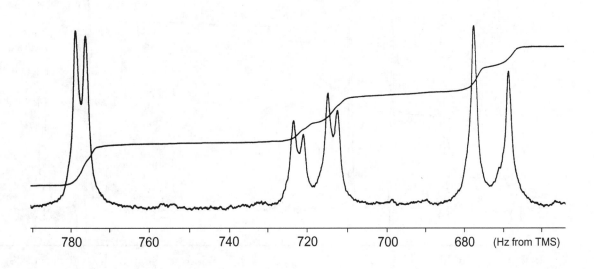

# Problem 332a

Portion of 100 MHz $^1$H NMR spectrum of methyl acrylate
(5% in $C_6D_6$) is given below.  Only the part of the
spectrum containing the resonances of the olefinic protons
$H_A$, $H_B$ and $H_C$ is shown.

methyl acrylate

*(a)*    Draw a splitting diagram.

*(b)*    Analyse this spectrum by first-order methods, *i.e.* extract all relevant coupling
constants (*J* in Hz) and chemical shifts (δ in ppm) by direct measurement.

*(c)*    Justify the statement that "this spectrum is really a borderline second-order
(strongly coupled) case".  Point out the most conspicuous deviation from
first-order character in this spectrum (see Section 5.9).

*(d)*    Assign the three multiplets to $H_A$, $H_B$ and $H_C$ on the basis of coupling
constants only (see Section 5.9).

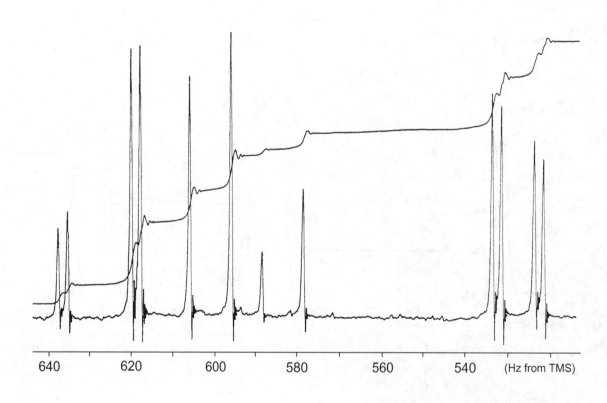

# Problem 332b

This is the **computer-simulated spectrum** corresponding to the complete analysis of the spectrum shown in Problem 318a, *i.e.* an exact analysis in which first-order assumptions were not made. The simulated spectrum fits the experimental spectrum, verifying that the analysis was correct. Compare your (first-order) results from Problem 318a with the actual solution given here.

| | |
|---|---|
| Number of SPINS | = 3 |
| F(1) | = + 528.500 Hz |
| F(2) | = + 594.531 Hz |
| F(3) | = + 626.093 Hz |
| | |
| J(1,2) | = + 10.539 Hz |
| J(1,3) | = + 1.589 Hz |
| J(2,3) | = + 17.278 Hz |
| | |
| START of simulation | = + 750.000 Hz |
| FINISH of simulation | = + 500.000 Hz |
| LINE WIDTH | = + 0.427 Hz |

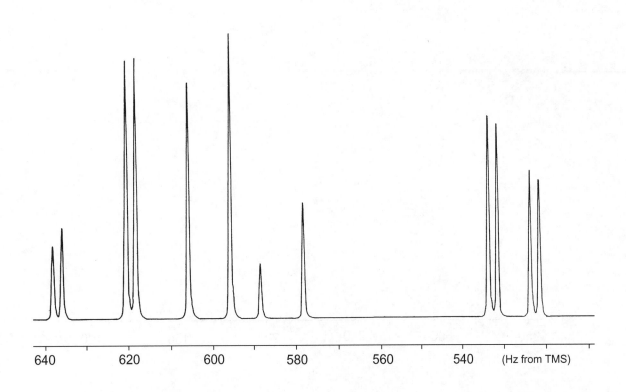

|   |   |   |   |   |   |   |
|---|---|---|---|---|---|---|
| 640 | 620 | 600 | 580 | 560 | 540 | (Hz from TMS) |

# Problem 333

Draw a schematic (line) representation of the pure first-order spectrum ($AX_3$) corresponding to the following parameters:

**Frequencies** (Hz from TMS): $\quad\quad v_A = 160;\ v_X = 280.$

**Coupling constants** (Hz): $\quad\quad J_{AX} = 15.$

Assume that the spectrum is a pure first-order spectrum and ignore small distortions in relative intensities of lines that would be apparent in a "real" spectrum.

*(a)* Sketch in "splitting diagrams" above the schematic spectrum to indicate which splittings correspond to which coupling constants.

*(b)* Give the chemical shifts on the $\delta$ scale corresponding to the above spectrum obtained with an instrument operating at 60 MHz for protons.

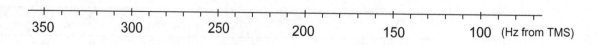

# Problem 334

A 100 MHz $^1$H NMR spectrum of a 4-proton system is given below.

*(a)*     Draw a splitting diagram.

*(b)*     Analyse this spectrum by first-order methods, *i.e.* extract all relevant coupling constants ($J$ in Hz) and chemical shifts ($\delta$ in ppm) by direct measurement.

*(c)*     Justify the use of first-order analysis (see Section 5.9).

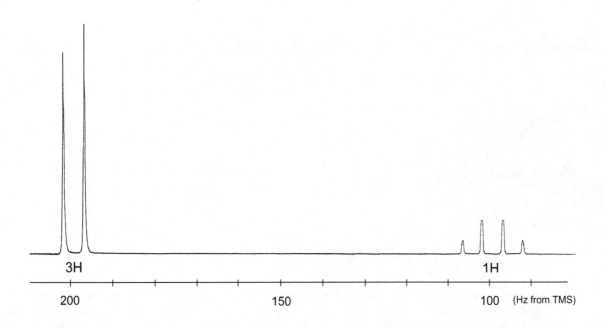

# Problem 335

Draw a schematic (line) representation of the pure first-order spectrum ($AMX_2$) corresponding to the following parameters:

**Frequencies** (Hz from TMS):  $\nu_A = 340$;  $\nu_M = 240$;  $\nu_X = 100$.

**Coupling constants** (Hz):  $J_{AM} = 10$;  $J_{AX} = 2$;  $J_{MX} = 6$.

Assume that the spectrum is a pure first-order spectrum and ignore small distortions in relative intensities of lines that would be apparent in a "real" spectrum.

*(a)*    Sketch in "splitting diagrams" above the schematic spectrum to indicate which splittings correspond to which coupling constants.

*(b)*    Give the chemical shifts on the δ scale corresponding to the above spectrum obtained with an instrument operating at 60 MHz for protons.

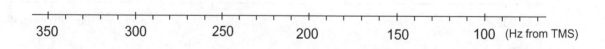

# Problem 336

Draw a schematic (line) representation of the pure first-order spectrum (AM$_2$X) corresponding to the following parameters:

**Frequencies** (Hz from TMS):        $v_A = 110$;  $v_M = 200$;  $v_X = 290$.

**Coupling constants** (Hz):        $J_{AM} = 10$;  $J_{AX} = 12$;  $J_{MX} = 3$.

Assume that the spectrum is a pure first-order spectrum and ignore small distortions in relative intensities of lines that would be apparent in a "real" spectrum.

*(a)*    Sketch in "splitting diagrams" above the schematic spectrum to indicate which splittings correspond to which coupling constants.

*(b)*    Give the chemical shifts on the $\delta$ scale corresponding to the above spectrum obtained with an instrument operating at 60 MHz for protons.

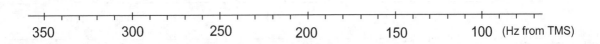

350        300        250        200        150        100  (Hz from TMS)

# Problem 337

A 100 MHz $^1$H NMR spectrum of a 4-proton system is given below.

*(a)*  Draw a splitting diagram.

*(b)*  Analyse this spectrum by first-order methods, *i.e.* extract all relevant coupling constants ($J$ in Hz) and chemical shifts ($\delta$ in ppm) by direct measurement.

*(c)*  Justify the use of first-order analysis (see Section 5.9).

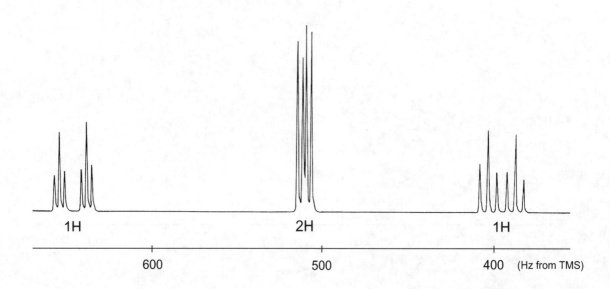

1H          2H          1H

600          500          400   (Hz from TMS)

# Problem 338

A 100 MHz $^1$H NMR spectrum of a 4-proton system is given below.

*(a)*   Draw a splitting diagram.

*(b)*   Analyse this spectrum by first-order methods, *i.e.* extract all relevant coupling constants (*J* in Hz) and chemical shifts ($\delta$ in ppm) by direct measurement.

*(c)*   Justify the use of first-order analysis (see Section 5.9).

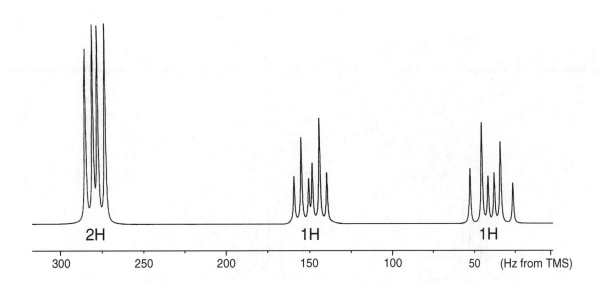

| 2H | 1H | 1H |

| 300 | 250 | 200 | 150 | 100 | 50 | (Hz from TMS) |

# Problem 339

A 100 MHz $^1$H NMR spectrum of a 4-proton system is given below.

*(a)*     Draw a splitting diagram.

*(b)*     Analyse this spectrum by first-order methods, *i.e.* extract all relevant coupling constants ($J$ in Hz) and chemical shifts ($\delta$ in ppm) by direct measurement.

*(c)*     Justify the use of first-order analysis (see Section 5.9).

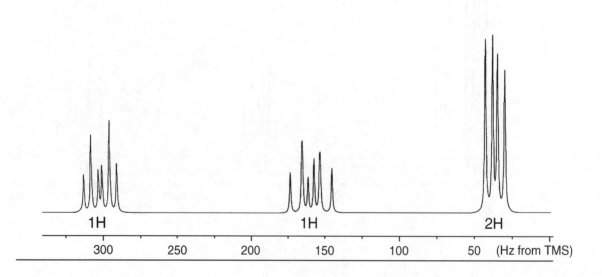

# Problem 340

A 200 MHz $^1$H NMR spectrum of a 5-proton system is given below.

*(a)* Draw a splitting diagram.

*(b)* Analyse this spectrum by first-order methods, *i.e.* extract all relevant coupling constants ($J$ in Hz) and chemical shifts ($\delta$ in ppm) by direct measurement.

*(c)* Justify the use of first-order analysis (see Section 5.9).

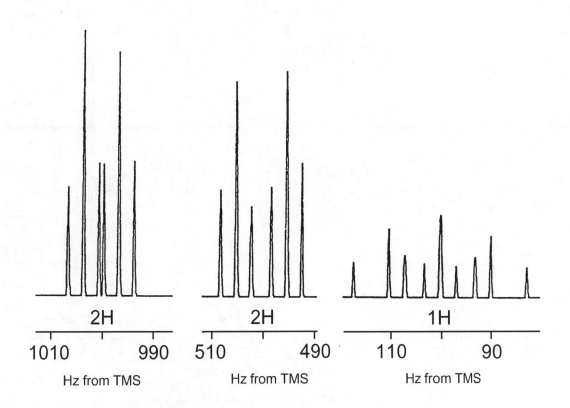

# Problem 341

Portion of 100 MHz NMR spectrum of crotonic acid in CDCl$_3$ is given below.  The upfield part of the spectrum, which is due to the methyl group, is less amplified to fit the page.

crotonic acid

*(a)*   Draw a splitting diagram and analyse this spectrum by first-order methods, *i.e.* extract all relevant coupling constants (*J* in Hz) and chemical shifts ($\delta$ in ppm) by direct measurement.  Justify the use of first-order analysis.

*(b)*   There are certain conventions used for naming spin-systems (*e.g.* AMX, AMX$_2$, AM$_2$X$_3$).  Note that this is a *5-spin system* and name the spin system responsible for this spectrum (see Section 5.9).

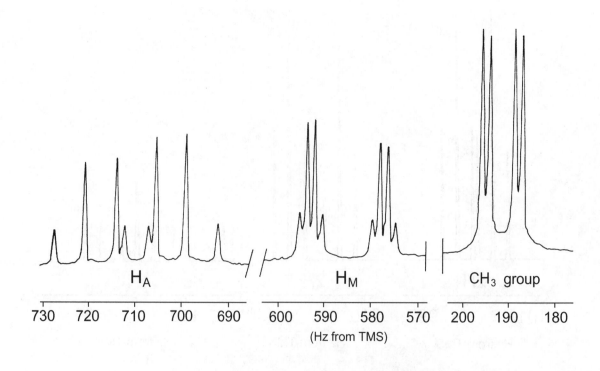

730   720   710   700   690      600   590   580   570      200   190   180

H$_A$                              H$_M$                    CH$_3$ group

(Hz from TMS)

# Problem 342

The 100 MHz $^1$H NMR spectrum (5% in CDCl$_3$) of an α,β-unsaturated aldehyde C$_4$H$_6$O is given below.

*(a)* Draw a splitting diagram and analyse this spectrum by first-order methods, *i.e.* extract all relevant coupling constants ($J$ in Hz) and chemical shifts (δ in ppm) by direct measurement.

*(b)* Justify the use of a first-order analysis (see Section 5.9).

*(c)* Use the coupling constants to obtain the structure of the compound, including the stereochemistry about the double bond (see Section 5.9).

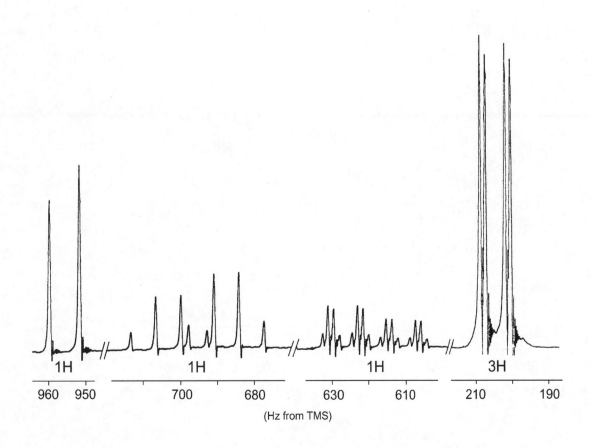

1H        1H        1H        3H

960  950        700      680      630      610      210      190

(Hz from TMS)

487

# Problem 343

Draw a schematic (line) representation of the pure first-order spectrum $(AMX_3)$ corresponding to the following parameters:

**Frequencies** (Hz from TMS):          $v_A = 80$;  $v_M = 220$;  $v_X = 320$.

**Coupling constants** (Hz):          $J_{AM} = 10$;  $J_{AX} = 12$;  $J_{MX} = 0$.

Assume that the spectrum is a pure first-order spectrum and ignore small distortions in relative intensities of lines that would be apparent in a "real" spectrum.

*(a)*     Sketch in "splitting diagrams" above the schematic spectrum to indicate which splittings correspond to which coupling constants.

*(b)*     Give the chemical shifts on the $\delta$ scale corresponding to the above spectrum obtained with an instrument operating at 60 MHz for protons.

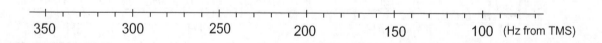

|   |   |   |   |   |   |
|---|---|---|---|---|---|
| 350 | 300 | 250 | 200 | 150 | 100  (Hz from TMS) |

# Problem 344

A portion of the 90 MHz $^1$H NMR spectrum (5% in CDCl$_3$) of one of the six possible isomeric dibromoanilines is given below.  Only the resonances of the aromatic protons are shown.

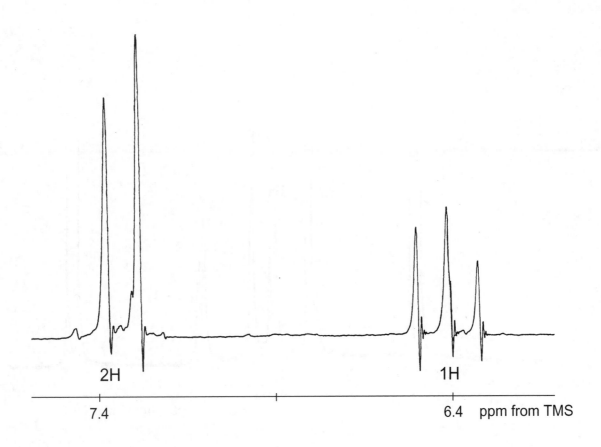

Determine which is the correct structure for this compound using arguments based on symmetry and the magnitudes of spin-spin coupling constants (see Section 5.9).

# Problem 345

The 400 MHz $^1$H NMR spectrum (5% in CDCl$_3$ after D$_2$O exchange) of one of the six possible isomeric hydroxycinnamic acids is given below.

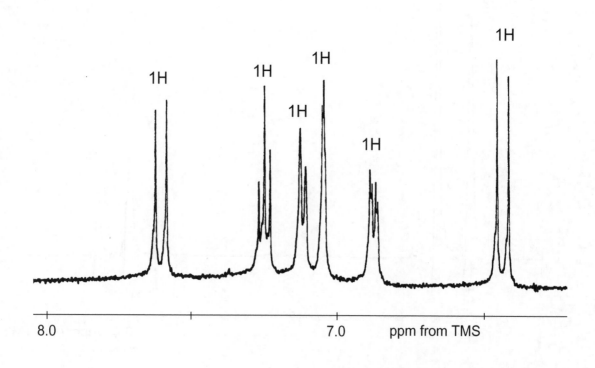

Determine which is the correct structure for this compound using arguments based on symmetry and the magnitudes of spin-spin coupling constants (see Section 5.9).

# Problem 346

In a published paper, the 90 MHz $^1$H NMR spectrum given below was assigned to 1,5-dichloronaphthalene, $C_{10}H_6Cl_2$.

1,5-dichloronaphthalene

*(a)*    Why can't this spectrum belong to 1,5-dichloronaphthalene?

*(b)*    Suggest two alternative dichloronaphthalenes that would have structures consistent with the spectrum given.

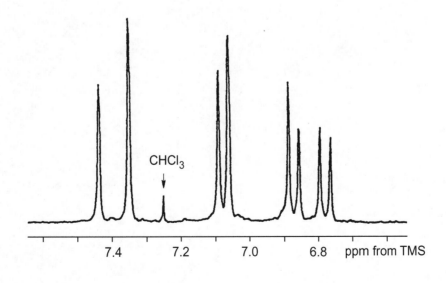

# Subject Index

**Key:**    $^{13}$C NMR = Carbon 13 nuclear magnetic resonance spectroscopy
$^{1}$H NMR = Proton nuclear magnetic resonance spectroscopy
2D NMR = 2-dimensional NMR
IR = Infrared spectroscopy
MS = Mass spectrometry
UV = Ultraviolet spectroscopy

## Subject Index